PATHWAYS TOWARDS HABITABLE PLANETS

COVER ILLUSTRATION:

Artwork for conference poster created by Malabars (www.malabars.com).

ASTRONOMICAL SOCIETY OF THE PACIFIC
CONFERENCE SERIES

Volume 430

EDITORIAL STAFF

Managing Editor: Joseph Jensen
Associate Managing Editor: Jonathan Barnes
Publication Manager: Pepita Ridgeway
Editorial Assistant: Cindy Moody
Web Developer/Technical Consultant: Jared M. Bellows
LaTeX Consultant: T. J. Mahoney

MS 179, Utah Valley University, 800 W. University Parkway, Orem, Utah 84058-5999
Phone: 801-863-8804 E-mail: aspcs@aspbooks.org
E-book site: http://www.aspbooks.org

PUBLICATION COMMITTEE

Lynne Hillenbrand (2009), Chair
California Institute of Technology

Marsha J. Bishop (2011)
National Radio Astronomy Observatory

Daniela Calzetti (2010)
University of Massachusetts

Gary J. Ferland (2008)
University of Kentucky

Ed Guinan (2010)
Villanova University

Luis Ho (2010)
The Observatories of the Carnegie
Institution of Washington

Scott J. Kenyon (2009)
Smithsonian Astrophysical Observatory

Joe Patterson (2009)
Columbia University

Catherine A. Pilachowski (2009)
Indiana University

René Racine (2008)
Université de Montréal

ASPCS volumes may be found online with color images at http://www.aspbooks.org.
ASP monographs may be found online at http://www.aspmonographs.org.

For a complete list of ASPCS volumes, ASP monographs, and
other ASP publications see http://www.astrosociety.org/pubs.html.

All book order and subscription inquiries should be directed to the ASP at
800-335-2626 (toll-free within the USA) or 415-337-2126,
or email service@astrosociety.org

ASTRONOMICAL SOCIETY OF THE PACIFIC
CONFERENCE SERIES

Volume 430

PATHWAYS TOWARDS HABITABLE PLANETS

Proceedings of a workshop held by the Spanish CSIC, the Catalan IEEC, the NASA
Exoplanet Science Institute, and the Blue Dots Initiative at CosmoCaixa, the
Barcelona Museum of Science in Barcelona, Spain
14-18 September 2009

Edited by

Vincent Coudé du Foresto
Observatoire de Paris - LESIA, Meudon, France

Dawn M. Gelino
NASA Exoplanet Science Institute,
California Institute of Technology, Pasadena, CA, USA

Ignasi Ribas
Institut de Ciències de l'Espai (CSIC-IEEC), Barcelona, Spain

SAN FRANCISCO

ASTRONOMICAL SOCIETY OF THE PACIFIC
390 Ashton Avenue
San Francisco, California, 94112-1722, USA

Phone: 415-337-1100
Fax: 415-337-5205
E-mail: service@astrosociety.org
Web site: www.astrosociety.org
E-books: www.aspbooks.org

First Edition
© 2010 by Astronomical Society of the Pacific
ASP Conference Series
All rights reserved.

No part of the material protected by this copyright notice may be reproduced or utilized in any form or by any means—graphic, electronic, or mechanical, including photocopying, taping, recording, or by any information storage and retrieval system—without written permission from the Astronomical Society of the Pacific.

ISBN: 978-1-58381-740-7
e-book ISBN: 978-1-58381-741-4

Library of Congress (LOC) Cataloging in Publication (CIP) Data:
Main entry under title
Library of Congress Control Number (LCCN): 2010927736

Printed in the United States of America by Sheridan Books, Ann Arbor, Michigan.
This book is printed on acid-free paper.

Contents

Foreword xvi

Preface xviii

Organizing Committee Members xx

Participants xxi

Conference Photograph xxix

Meeting Sponsors xxx

Part I. Welcome

How to Find a Habitable Planet 3
    J. F. Kasting

Part II. Reports from Prospective Exercises

The Blue Dots Initiative and Roadmapping Exercise 15
    V. Coudé du Foresto for the Blue Dots participants

Exoplanets Forum 2008 21
    W. A. Traub, P. R. Lawson, S. C. Unwin, M. W. Muterspaugh, R. Soummer, W. C. Danchi, P. Hinz, B. S. Gaudi, G. Torres, D. Deming, J. Lazio, and A. Dressler

Part III. Working Group Reviews

Multi-Aperture Imaging of Extrasolar Planetary Systems 29
    O. Absil

Scientific Output of Single Aperture Imaging of Exoplanets 37
    R. Gratton, M. Bonavita, S. Desidera, A. Boccaletti, M. Kasper, and F. Kerber

Blue Dots Team Transits Working Group Review 45
    A. Sozzetti, C. Afonso, R. Alonso, D. L. Blank, C. Catala, H. Deeg,
    J. L. Grenfell, C. Hellier, D. W. Latham, D. Minniti, F. Pont, and
    H. Rauer
Habitability of Planets 55
    F. Forget and R. D. Wordsworth
Characterizing Exoplanet Atmospheres and Surfaces 65
    T. Encrenaz
Habitable Planets: Targets and their Environments 76
    M. Güdel
Review from the Blue Dots Astrometry Working Group 84
    F. Malbet, A. Sozzetti, P. Lazorenko, R. Launhardt, D. Ségransan,
    F. Delplancke, N. Elias, M. Muterspaugh, A. Quirrenbach, S. Reffert,
    and G. van Belle
The (Un)Lonely Planet Guide: Formation and Evolution
of Planetary Systems from a "Blue Dots" Perspective 93
    M. R. Meyer

Part IV. Astrophysical Themes

General Milankovitch Cycles 109
    D. S. Spiegel, S. Raymond, C. D. Dressing, C. A. Scharf, J. L. Mitchell,
    and K. Menou
Exploring Extrasolar Worlds Today and Tomorrow 115
    G. Tinetti and C. A. Griffith
Uniformly Derived Orbital Parameters of Exo-planets using EXOFIT . . 122
    S. T. Balan, G. Lever, and O. Lahav
Pathways towards Neptune-mass Planets around Very Low-mass Stars . . 127
    S. Dreizler, J. Bean, A. Seifahrt, H. Hartman, H. Nilsson, G. Wiedemann,
    A. Reiners, and T. J. Henry
Tidal Constraints on Planetary Habitability 133
    R. Barnes, B. Jackson, R. Greenberg, S. N. Raymond, and R. Heller
Pathways towards Habitable Moons 139
    D. M. Kipping, S. J. Fossey, G. Campanella, J. Schneider, and G. Tinetti
Pressure-temperature Phase Diagram of the Earth 145
    E. G. Jones and C. H. Lineweaver
Abiotic Buildup of Ozone 152
    S. D. Domagal-Goldman and V. S. Meadows
CoRoT-7b: The First Transiting Super-Earth Fully Characterized in
Radius and Mass 158
    D. Rouan, A. Léger, J. Schneider, P. Barge, M. Fridlund, B. Samuel,
    D. Queloz, C. Moutou, F. Bouchy, A. Hatzes, and all the CoRoT
    Exoplanet Science Team

Part V. Methods and Projects

Pathways towards Habitable Planets: JWST's Capabilities for
Exoplanet Science 167
    M. Clampin

Radioastronomy and the Study of Exoplanets 175
    P. Zarka

NAHUAL: A Next-Generation Near Infrared Spectrograph for the GTC . 181
    E. L. Martín, E. Guenther, C. del Burgo, F. Rodler, C. Álvarez, C. Baffa,
    V. J. S. Béjar, J. A. Caballero, R. Deshpande, P. Esparza, M. López
    Morales, A. Moitinho, D. Montes, M. M. Montgomery, E. Pallé, R. Tata,
    L. Valdivielso, and M. R. Zapatero Osorio

Infrared Detection and Characterization of Debris Disks, Exozodiacal Dust,
and Exoplanets: The FKSI Mission Concept 188
    W. C. Danchi, R. K. Barry, B. Lopez, S. A. Rinehart, O. Absil,
    J.-C. Augereau, H. Beust, X. Bonfils, P. Bordé, D. Defrère, P. Kern,
    P. Lawson, A. Léger, J.-L. Monin, D. Mourard, M. Ollivier, R. Petrov,
    W. Traub, S. Unwin, and F. Vakili

Japanese Space Activity on Exoplanets and SPICA 195
    T. Nakagawa

SOFIA: On the Pathway toward Habitable Worlds 201
    R. D. Gehrz, D. Angerhausen, E. E. Becklin, M. A. Greenhouse,
    S. Horner, A. Krabbe, M. R. Swain, and E. T. Young

Probing the Extrasolar Planet Diversity with SEE-COAST 207
    P. Baudoz, A. Boccaletti, J. Schneider, R. Galicher, and G. Tinetti

The SIM Lite Mission 213
    M. Shao

Darwin in the Context of Cosmic Vision 2015-2025 219
    R. Liseau

MOA-II Microlensing Exoplanet Survey 225
    T. Sumi

Direct Detection of Giant Extrasolar Planets with SPHERE on the VLT . 231
    J.-L. Beuzit, A. Boccaletti, M. Feldt, K. Dohlen, D. Mouillet, P. Puget,
    F. Wildi, L. Abe, J. Antichi, A. Baruffolo, P. Baudoz, M. Carbillet,
    J. Charton, R. Claudi, S. Desidera, M. Downing, C. Fabron, P. Feautrier,
    E. Fedrigo, T. Fusco, J.-L. Gach, E. Giro, R. Gratton, T. Henning,
    N. Hubin, F. Joos, M. Kasper, A.-M. Lagrange, M. Langlois, R. Lenzen,
    C. Moutou, A. Pavlov, C. Petit, J. Pragt, P. Rabou, F. Rigal, S. Rochat,
    R. Roelfsema, G. Rousset, M. Saisse, H.-M. Schmid, E. Stadler,
    C. Thalmann, M. Turatto, S. Udry, F. Vakili, A. Vigan, and R. Waters

Resolved Imaging of Extra-Solar Photosynthesis Patches with a
"Laser Driven Hypertelescope Flotilla" 239
    A. Labeyrie, H. Le Coroller, S. Residori, U. Bortolozzo, J.-P. Huignard,
    and P. Riaud

Astrometric-Radial-Velocity and Coronagraph-Imaging Double-Blind Studies 249
W. A. Traub

Gaia and the Astrometry of Giant Planets 253
M. G. Lattanzi and A. Sozzetti

PLATO : PLAnetary Transits and Oscillations of Stars - The Exoplanetary System Explorer 260
C. Catala, T. Arentoft, M. Fridlund, R. Lindberg, J. M. Mas-Hesse, G. Micela, D. Pollacco, E. Poretti, H. Rauer, I. Roxburgh, A. Stankov, and S. Udry

EUCLID: Dark Universe Probe and Microlensing Planet Hunter 266
J. P. Beaulieu, D. P. Bennett, V. Batista, A. Cassan, D. Kubas, P. Fouqué, E. Kerrins, S. Mao, J. Miralda-Escudé, J. Wambsganss, B. S. Gaudi, A. Gould, and S. Dong

The Habitable Zone Planet Finder Project: A Proposed High Resolution NIR Spectrograph for the Hobby Eberly Telescope (HET) to Discover Low Mass Exoplanets around M Stars 272
S. Mahadevan, L. Ramsey, S. Redman, S. Zonak, J. Wright, A. Wolszczan, M. Endl, and B. Zhao

The Fresnel Interferometric Imager 278
L. Koechlin, J.-P. Rivet, R. Gili, P. Deba, T. Raksasataya, and D. Serre

SPICA Coronagraph 284
K. Enya, T. Kotani, T. Nakagawa, H. Kataza, K. Komatsu, H. Uchida, K. Haze, S. Higuchi, T. Miyata, S. Sako, T. Nakamura, T. Yamashita, N. Narita, M. Tamura, J. Nishikawa, H. Hayano, S. Oya, E. Kokubo, Y. Itoh, M. Fukagawa, H. Shibai, M. Honda, N. Baba, N. Murakami, M. Takami, T. Matsuo, S. Ida, L. Abe, O. Guyon, M. Venet, T. Yamamuro, and P. Bierden

Part VI. Synthesis of Panel Discusssions

Do we Need to Solve the Exozodi Question? If Yes, How to Best Solve It? 293
O. Absil, C. Eiroa, J.-C. Augereau, C. A. Beichman, W. C. Danchi, D. Defrère, M. Fridlund, and A. Roberge

How to Consolidate Efforts within the Community and the Related Agencies? 300
H. Zinnecker

Designating Habitable Planets for Follow-up Study - What are the Relative Parameter Spaces of Radial Velocities and Astrometry? 302
I. Ribas and F. Malbet

Part VII. A Broader Picture

Exoplanets and Earth Sciences 311
    E. Pallé
Life in the Solar System 317
    A. Giménez
The Extraterrestrial Life Debate in Different Cultures 324
    J. Schneider
Looking for Life Elsewhere: A Biologist's Perspective 331
    M.-C. Maurel
The ITER Pathway: Useful Lessons for Exoplanet Projects? 336
    J. Jacquinot

Part VIII. Conclusion

A Personal Insight on the Conference 345
    P. J. Léna

Part IX. Special Session: Advanced Strategic Mission Concept Studies

The New Worlds Observer: The Astrophysics Strategic Mission Concept
    Study 353
    W. Cash and the New Worlds Study Team
Advanced Technology Large-Aperture Space Telescope (ATLAST):
    Characterizing Habitable Worlds 361
    M. Postman, W. Traub, J. Krist, K. Stapelfeldt, R. Brown, W. Oegerle,
    A. Lo, M. Clampin, R. Soummer, J. Wiseman, and M. Mountain
Dilute Aperture Visible Nulling Coronagraphic Imager (DAViNCI) 368
    M. Shao and B. M. Levine
ACCESS: A Concept Study for the Direct Imaging and Spectroscopy
    of Exoplanetary Systems 375
    J. Trauger, K. Stapelfeldt, W. Traub, J. Krist, D. Moody, E. Serabyn,
    D. Mawet, L. Pueyo, S. Shaklan, C. Henry, P. Park, R. Gappinger,
    P. Brugarolas, J. Alexander, V. Mireles, O. Dawson, O. Guyon, J. Kasdin,
    B. Vanderbei, D. Spergel, R. Belikov, G. Marcy, R. Brown, J. Schneider,
    B. Woodgate, G. Matthews, R. Egerman, P. Voyer, P. Vallone, J. Elias,
    Y. Conturie, R. Polidan, C. Lillie, C. Spittler, D. Lee, R. Hejal,
    A. Bronowick, N. Saldivar, M. Ealey, and T. Price
The ExtraSolar Planetary Imaging Coronagraph 383
    M. Clampin and R. Lyon

The Pupil Mapping Exoplanet Coronagraphic Observer (PECO) 390
    O. Guyon, T. Greene, K. Cahoy, S. Shaklan, D. Tenerelli, and the PECO Team

Part X. Poster Presentations

Interactions between a Massive Planet and a Disc 399
    P. Amaro-Seoane, I. Ribas, U. Löckmann, and H. Baumgardt

Ground-based Astrometric Planet Searches using Medium-sized Telescopes 401
    G. Anglada-Escudé, A. P. Boss, and A. J. Weinberger

The Fourier-Kelvin Stellar Interferometer: Exploring Exoplanetary Systems with an Infrared Probe-class Mission 403
    R. K. Barry, W. C. Danchi, B. Lopez, S. A. Rinehart, O. Absil, J.-C. Augereau, H. Beust, X. Bonfils, P. Bordé, D. Defrére, P. Kern, A. Léger, J.-L. Monin, D. Mourard, M. Ollivier, R. Petrov, F. Vakili, and the FKSI Consortium

Towards Astrometric Detection of Neptune- to Earth-Mass Planets around M-Stars 405
    C. Bergfors, W. Brandner, M. Janson, N. Kudryavtseva, S. Daemgen, S. Hippler, F. Hormuth, and T. Henning

How Common are Extrasolar, Late Heavy Bombardments? 407
    M. Booth, M. C. Wyatt, A. Morbidelli, A. Moro-Martín, and H. F. Levison

Occurrence, Physical Conditions, and Observations of Super-Ios and Hyper-Ios 409
    D. Briot and J. Schneider

First Steps on the Design of the Optical Differentiation Coronagraph ... 411
    M. P. Cagigal, V. F. Canales, P. J. Valle, E. Sanchez-Blanco, M. Maldonado, and M. L. Garcia-Vargas

The Search for Exoplanets in India 413
    A. Chakraborty, B. G. Anadarao, and S. Mahadevan

Characterization of Extrasolar Planets with High Contrast Imaging ... 416
    R. Claudi, M. Bonavita, R. Gratton, S. Desidera, G. Tinetti, J.-L. Beuzit, and M. Kasper

ALADDIN: An Antarctic-based Nulling Interferometer for Exozodi Characterization 418
    V. Coudé du Foresto

A Photometric Transit Search for Planets around Cool Stars from the Italian Alps: Results from a Feasibility Study 420
    M. Damasso, P. Calcidese, A. Bernagozzi, E. Bertolini, P. Giacobbe, M. G. Lattanzi, R. Smart, and A. Sozzetti

Influence of Exozodiacal Dust Clouds on Mid-IR Earth-like Planet Detection 422
    D. Defrère, O. Absil, R. den Hartog, C. Hanot, and C. Stark

Future Technical Developments Concerning Nulling Interferometry at the
    Institut d'Astrophysique Spatiale (Orsay) 424
    O. Demangeon
Ground-based Search of Earth-mass Exoplanets using Transit-Timing
    Variations 426
    J. M. Fernandez
The Fabra-ROA Baker-Nunn Camera at Observatori Astronòmic
    del Montsec: A Wide-field Imaging Facility for Exoplanet
    Transit Detection 428
    *O. Fors, J. Núñez, J. L. Muiños, F. J. Montojo, R. Baena, M. Merino,
    R. Morcillo, and V. Blanco*
Dynamical Stability in the Habitable Zones of Nearby Extrasolar
    Planetary Systems 430
    B. Funk, Á. Süli, E. Pilat-Lohinger, R. Schwarz and S. Eggl
The NULLTIMATE Testbed: A Progress Report 432
    P. Gabor
Low-mass Objects in Moving Groups 434
    *M. C. Gálvez-Ortiz, J. R. A. Clarke, D. J. Pinfield, S. L. Folkes,
    J. S. Jenkins, A. E. García Pérez, B. Burningham, A. C. Day-Jones, and
    H. R. A. Jones*
Characterization of High-Energy Emissions of GKM Stars and the Effects
    on Planet Habitability 437
    A. Garcés, I. Ribas, and S. Catalán
Looking for Transits of Jupiter-Size Planets Orbiting Stars in Habitable
    Zones 439
    E. García-Melendo and I. Ribas
Modeling the Atmospheres of Earth-like Planets 441
    A. García Muñoz, E. Pallé, and E. L. Martín
On the Detectability of Biomarkers in Extrasolar Super-Earth
    Atmospheres 443
    *S. Gebauer, M. Godolt, J. L. Grenfell, P. Hedelt, P. von Paris, H. Rauer,
    F. Selsis, and A. Belu*
Influence of the Spectral Stellar Flux Distribution on Atmospheric
    Dynamics of Extrasolar Earth-like Planets 445
    *M. Godolt, J. L. Grenfell, A. Hamann-Reinus, M. Kunze, U. Langematz,
    and H. Rauer*
Exoplanetary Systems with SAFARI: A Far Infrared Imaging
    Spectrometer for SPICA 448
    *J. R. Goicoechea and B. Swinyard on behalf of the SPICA/SAFARI
    science teams*
SIM Lite Instrument Description, Operation, and Performance 450
    R. Goullioud
Testing PIAA Coronagraphs at NASA Ames 453
    *T. Greene, R. Belikov, M. Connelley, O. Guyon, D. Lynch, M. McKelvey,
    E. Pluzhnik, and F. Witteborn*

Transfer of Meteorites from Earth to the Interesting Objects within the
Solar System and the Extrasolar Planets 455
    T. Hara, T. Takagi, and D. Kajiura

A Coronagraph Experiment in a High Thermal Stability Environment with
a Binary Shaped Pupil Mask 457
    K. Haze, K. Enya, T. Kotani, L. Abe, T. Nakagawa, S. Higuchi, T. Sato,
    T. Wakayama, and T. Yamamuro

Performance of the Cophasing System of Persee, a Dynamic Nulling
Demonstrator 460
    K. Houairi, F. Cassaing, J. M. Le Duigou, J. P. Amans, M. Barillot,
    V. Coudé du Foresto, F. Hénault, S. Jacquinod, J. Lozi, J. Montri,
    M. Ollivier, J. M. Reess, and B. Sorrente

Search for Exoplanets using TTVs in the Southern Hemisphere 462
    S. Hoyer, P. Rojo, and M. López-Morales

First Results from the Transit Ephemeris Refinement and
Monitoring Survey (TERMS) 464
    S. R. Kane, S. Mahadevan, K. von Braun, G. Laughlin, A. Howard, and
    D. R. Ciardi

Flux and Polarization Signals of Water Clouds on Earth–Like Exoplanets 466
    T. Karalidi, D. M. Stam, and J. W. Hovenier

Using Polarization to Detect and Characterize Exoplanets 469
    L. Kedziora-Chudczer and J. Bailey

Signatures of Resonant Terrestrial Planets in Long-Period Systems 472
    G. F. Kennedy and R. A. Mardling

Influence of Clouds on the Emission Spectra of Earth-like
Extrasolar Planets 475
    D. Kitzmann, A. B. C. Patzer, P. von Paris, M. Godolt, J. L. Grenfell,
    and H. Rauer

A Wavefront Correction System for the SPICA Coronagraph Instrument . 477
    T. Kotani, K. Enya, T. Nakagawa, L. Abe, T. Miyata, S. Sako,
    T. Nakamura, K. Haze, S. Higuchi, and Y. Tange

A Laboratory Demonstration of High Dynamic Range Imaging using the
Single-mode Fiber Pupil Remapping System (FIRST) 480
    T. Kotani, E. Choquet, S. Lacour, G. Perrin, F. Marchis, and G. Duchêne

Stellar Activity Characteristics at FUV and Radio Wavelengths 483
    M. Leitzinger, P. Odert, A. Hanslmeier, I. Ribas, A. A. Konovalenko,
    M. Vanko, M. L. Khodachenko, H. Lammer, and H. O. Rucker

Visible Nulling Coronagraph Progress Report 485
    R. G. Lyon, M. Clampin, R. A. Woodruff, G. Vasudevan, P. Thompson,
    P. Petrone, T. Madison, M. Rizzo, G. Melnick, and V. Tolls

Checking Stability of Planet Orbits in Multiple-planet Systems 488
    F. Malbet, J. Catanzarite, M. Shao, and C. Zhai

Spectroscopic Observations of Nearby Cool Stars: The DUNES Sample . . 491
    J. Maldonado, C. Eiroa, R.M. Martínez-Arnáiz, and D. Montes

A Mid-infrared Space Observatory for Characterizing Exoplanets 493
    S. R. Martin, D. P. Scharf, R. Wirz, G. Purcell, and J. Rodriguez

The Subaru Coronagraphic Extreme AO Project 497
    F. Martinache and O. Guyon

Spectral Imaging with Nulling Interferometer 499
    T. Matsuo, W. A. Traub, M. Tamura, and H. Makoto

Multiple Aperture Imaging and its Application to Exo-Earth Imager . . . 501
    T. Matsuo, W. A. Traub, M. Tamura, and H. Makoto

The Systemic Console Package 503
    S. Meschiari, A. S. Wolf, E. Rivera, G. Laughlin, S. Vogt, and P. Butler

The Earthshine Project: Applications to the Search of Exoearths 505
    P. Montañés-Rodríguez and E. Pallé Bagó

Estimating the Age of Exoplanet's Host Stars by their Membership
in Moving Groups and Young Associations 507
    *D. Montes, J. A. Caballero, C. J. Fernández-Rodríguez, L. J. Alloza,
S. Bertrán de Lis, A. Garrido-Rubio, R. Greciano, J. E. Herranz-Luque,
I. Juárez-Martínez, E. Manjavacas, F. Ocaña, B. Pila-Díez,
A. Sánchez de Miguel, and C. E. Tapia-Ayuga*

Eight-octant Phase-mask Coronagraph for Detecting Earth-like
Exoplanets around Partially Resolved Stars 510
    *N. Murakami, J. Nishikawa, K. Yokochi, M. Tamura, N. Baba,
N. Hashimoto, and L. Abe*

A Coronagraph with an Unbalanced Nulling Interferometer and Adaptive
Optics 513
    *J. Nishikawa, K. Yokochi, N. Murakami, L. Abe, T. Kotani, M. Tamura,
T. Kurokawa, A. V. Tavrov, and M. Takeda*

M-Type Stars as Hosts for Habitable Planets 515
    *P. Odert, M. Leitzinger, A. Hanslmeier, H. Lammer, M. L. Khodachenko,
and I. Ribas*

Wide Angle Telescope Transit Search (WATTS): A Low-Elevation
Component of the TrES Network 517
    *B. Oetiker, M. Kowalczyk, B. Nietfeld, G. Mandushev,
and E. W. Dunham*

Where Can We Find Super-Earths? 519
    E. Podlewska-Gaca and E. Szuszkiewicz

CARMENES: Calar Alto High-Resolution Search for M dwarfs with
Exo-earths with a Near-infrared Echelle Spectrograph 521
    *A. Quirrenbach, P. J. Amado, H. Mandel, J. A. Caballero, I. Ribas,
A. Reiners, R. Mundt, and the CARMENES Consortium*

The Balloon Experimental Twin Telescope for Infrared Interferometry . . 524
    S. A. Rinehart and the BETTII Team

Detecting Planets around Very Cool Stars at Near Infrared Wavelengths with the Radial Velocity Technique 526
F. Rodler, C. del Burgo, E. L. Martín, and C. Álvarez

Metallicity of M-dwarfs from NIR Spectra 528
B. Rojas-Ayala and J. P. Lloyd

X-Exoplanets: An X-ray and EUV Database for Exoplanets 530
J. Sanz-Forcada, D. García-Álvarez, A. Velasco, E. Solano, I. Ribas, G. Micela, and A. Pollock

Pathway Toward an Infrared Interferometer 532
J. Schneider

Characterization of Exoplanet Atmospheres in the Solar Neighbourhood with E-ELT/METIS 534
C. Schnupp, W. Brandner, C. Bergfors, K. Geißler, S. Daemgen, S. Hippler, F. Hormuth, R. Lenzen, T. Henning, M. Janson, and E. Pantin

US and European Technology Roadmap for a Mid-infrared Space Interferometer 536
P. A. Schuller, P. R. Lawson, O. P. Lay, A. Léger, and S. R. Martin

Habitable Planets in Compact Close-in Planetary Systems 539
R. Schwarz, E. Pilat-Lohinger, B. Funk, and G. Wuchterl

Far-Infrared Interferometric Experiment (FITE): Toward the First Flight 541
H. Shibai, M. Fukagawa, E. Kato, T. Kanoh, T. Kohyama, Y. Itoh, K. Yamamoto, M. Kawada, T. Watabe, A. Nakashima, M. Tanabe, R. Kanoh, and M. Narita

Remote Detection of Biological Activity using Circular Polarization of Light 543
W. B. Sparks

The Inner Boundary of the Habitable Zone for Earth-like Planets 545
B. Stracke, J. L. Grenfell, P. von Paris, A. B. C. Patzer, and H. Rauer

The Role of the Solar XUV Flux and Impacts in the Earliest Atmosphere of the Earth 547
J. M. Trigo-Rodríguez, I. Ribas, and J. Llorca

Venus Near-Infrared Spectra: SCIAMACHY-Observations and Modeling . 549
M. Vasquez, F. Schreier, M. Gottwald, S. Slijkhuis, S. Gimeno-García, E. Krieg, and G. Lichtenberg

Constraints on Secondary Eclipse Probabilities of Long-Period Exoplanets from Orbital Elements 551
K. von Braun and S. R. Kane

Extrasolar Planets in the Gliese 581 System - Model Atmospheres and Implications for Habitability 554
P. von Paris, B. Stracke, A. B. C. Patzer, H. Rauer, J. L. Grenfell, P. Hedelt, M. Godolt, S. Gebauer, and D. Kitzmann

Detection of Transiting Super-Earths around Active Stars 556
    J. Weingrill, H. Lammer, M. Khodachenko, and A. Hanslmeier

Habitability on the Outer Edge: Three-dimensional Climate Modeling of
Early Mars and Gliese 581d 558
    R. D. Wordsworth, F. Forget, E. Millour, J.-B. Madeleine, V. Eymet, and
    F. Selsis

The SEDs of Circumsubstellar Protoplanetary Disks 560
    O. V. Zakhozhay and V. A. Zakhozhay

Author Index 565

Foreword
to the Contributions presented at
the Barcelona Conference (September 2009)
Pathways Towards Habitable Planets

The participants met to discuss and compare
the different prospective exercises
which have been carried out since 2007
in the United States, Europe, Japan, and other countries.

They consider:

- (c1) The question of whether life exists on other worlds is among humankind's deepest questions, whose answer has profound implications;

- (c2) That we live at a turning point in the history of knowledge when this answer is within our reach;

- (c3) That the search for other worlds has already provided astounding successes, with the discovery of (to date) more than 370 planets in over 300 diverse systems;

- (c4) That the field of exoplanets has strong inspirational power to draw new generations into science.

They agree:

- (a1) That our long term scientific goal is to obtain a comprehensive understanding of Earth-comparable planets, which contributes to understanding our own Earth in a broader context;

- (a2) That exoplanet science is an interdisciplinary field, requiring a great diversity of theories and observations, as well as intermediate steps in support of the ultimate goal;

- (a3) That the field has matured to a point where we can develop an integrated, flexible, and international roadmap towards achieving our goal;

...and therefore recommend:

- (r1) As exoplanet space missions play a decisive role, choosing the best ones for the next decade is a critical aspect of the roadmap; complementarity of the choices to be made by national agencies is essential, with crossed participations whenever possible;

- (r2) Ground programs require existing 10 m-class telescopes and future extremely large telescopes: specific instrumentation and observing time is therefore essential to complement space missions;
- (r3) At least one flagship space mission, which should be defined through broad international collaboration, will ultimately be required for a comprehensive understanding of other habitable worlds in our galactic neighbourhood.

Preface

In September 2009, 208 exoplanet scientists from around the world gathered in Barcelona to debate and reach a consensus on defining guidelines for a new roadmap with the ultimate objective of finding habitable, and potentially inhabited, planets outside of our Solar System. The Pathways Towards Habitable Planets Conference integrated prospective work and community organizing efforts on both sides of the Atlantic, such as the Exoplanets Forum in the United States and the Blue Dots initiative in Europe, and was set to provide timely input to the prospective excersises being carried out by the major space agencies.

The conference spanned a full week and was organized around a main session in the morning and early afternoon, while the late afternoon was devoted to parallel satellite meetings, including discussion panels on some of the key issues. The main sessions included reports from current prospective exercises (Part 2 of these proceedings), as well as reviews from the Blue Dots working groups (Part 3). More specific topics were subsequently covered in the Astrophysical Themes (Part 4) and Methods and Projects (Part 5) sessions.

The Friday sessions were devoted to synthetic summaries of the discussion panels (Part 6), an exploration of the topic of the conference from a broader (non-astronomical) perspective (Part 7) and the concluding summary talks (Part 8), followed by a discussion in which the participants unanimously adopted the resolution which is presented as the Foreword to this conference proceedings. A summary of the special satellite meeting dedicated to the NASA's Advanced Study Mission Concepts was also presented on Friday (Part 9).

The slides of the oral presentations can be downloaded from the conference web site at http://www.pathways2009.net (in the Scientific Program section) and they provide a rich visual complement (in color) to the written material contained in these proceedings. In addition to oral presentations, 116 posters were displayed at the conference (the corresponding contributions are in Part 10 of this volume): this number is a tribute to the extraordinary diversity and dynamism of the fields of research covered by this conference.

The Editors wish to thank the Scientific Organizing Committee for their hard work and dedication in preparing the science contents of the conference and helping to make difficult choices given the large number of talk requests. We are extremely grateful to the Ajuntament de Barcelona and Observatori Fabra for the organization of the event. All local organization matters were efficiently dealt with by an excellent team of people (Josep Colomé, Ane Garcés, Xavier Francisco, Lluís Gómez, Alina Hirschmann, Jordi Isern, Josep Maria Codina,

Juan Carlos Morales) who ensured that the conference ran smoothly with the help of Armengol i Associats. Ane Garcés was the meeting photographer and she is credited with the pictures appearing in this book. Financial sponsorship was kindly provided by the following institutions: Consejo Superior de Investigaciones Científicas (CSIC); Institut d'Estudis Espacials de Catalunya (IEEC); Departament d'Innovació, Universitats i Empresa of the Generalitat de Catalunya; Ministerio de Ciencia e Innovación of the Spanish Government; Instituto Nacional de Técnica Aeroespacial (INTA); Centro de Astrobiología (CAB); Centre National d'Études Spatiales (CNES); Centre National de la Recherche Scientifique (CNRS); Observatoire de Paris; Europlanet Research Infrastructure; and NASA Exoplanet Science Institute (NExScI). The auditorium of the Museu de la Ciència of the Fundació "La Caixa" with its first-class facilities provided a wonderful venue for the conference.

Publication of this book would not have been possible without Ellen O'Leary at NExScI. Her diligence and extraordinary hard work, along with that of Karen Willacy at NExScI, allowed as many Pathways contributors as possible to have their articles published while getting the publication into the hands of the readers in a timely manner. We would also like to thank the session chairs (Helmut Lammer, Alessandro Sozzetti, Wes Traub, Alan Boss, Hiroshi Shibai, Chas Beichman, Michael Devirian, Gerard van Belle, Franck Selsis), and others (Virginie Batista, Alain Léger, Marc Ollivier, Denis Defrere, Raphael Galicher, Stephen Ridgway) who served as science readers for the proceedings contributions. We are also grateful to NExScI for sponsoring the publication of the proceedings for this important conference.

The Pathways Conference marks a starting point for new great expectations within the exoplanet scientific community. For the first time, we have been able to gather a multidisciplinary team of scientists, including biologists, geologists, astronomers, and planetary scientists, as well as representatives of the different space agencies and ground observatories, in order to agree upon a future vision and strategy. Maybe, in the coming decades, we will able to address the question we have been seeking to answer for so long: Are we alone in the Universe?

The Pathways Editors
Vincent Coude du Foresto (Observatoire de Paris - LESIA)
Dawn M. Gelino (NASA Exoplanet Science Institute - NExScI)
Ignasi Ribas (Institut de Ciències de l'Espai - CSIC/IEEC)

Organizing Committee Members

Scientific Organizing Committee

Jean-Philippe Beaulieu (Institut d'Astrophysique de Paris, France), Chas Beichman (Jet Propulsion Laboratory, USA), Anthony Boccaletti (Paris Observatory, France), Abhijit Chakraborty (Physical Research Laboratory, India), Charles Cockell (The Open University, UK), Vincent Coud du Foresto (Paris Observatory, France; Co-chair), Bill Danchi (NASA Goddard Space Flight Center, USA), Zhao Gang (National Astronomical Observatory, China), Dawn Gelino (NASA Exoplanet Science Institute, Caltech, USA), Lisa Kaltenegger (Center for Astrophysics, Harvard, USA), Christoph Keller (Sterrekundig Institut Utrecht, Netherlands), Maxim Khodachenko (Space Research Institute, Austria), Helmut Lammer (Space Research Institute, Austria), Marc Ollivier (Institut d'Astrophysique Spatiale, France), Ignasi Ribas (Institut de Ciències de l'Espai, CSIC-IEEC, Spain; Co-chair), Nuno Santos (Porto University, Portugal), Dimitar Sasselov (Harvard University, USA), Sara Seager (Massachusetts Institute of Technology, USA), Franck Selsis (Bordeaux Observatory, France), Damien Ségransan (Geneva Observatory, Switzerland), Hiroshi Shibai (Osaka University & JAXA, Japan), Alessandro Sozzetti (INAF - Torino Astronomical Observatory, Italy), Ewa Szuszkiewicz (University of Szczecin, Poland), Giovanna Tinetti (University College London, UK), Wes Traub (Jet Propulsion Laboratory, USA), Gerard Van Belle (European Southern Observatory), Hans Zinnecker (Astrophysikalisches Institut Potsdam, Germany)

Local Organizing Committee

Josep Colomé (Institut de Ciències de l'Espai, CSIC-IEEC), Ane Garcés (Institut de Ciències de l'Espai, CSIC-IEEC), Lluís Gómez (Ajuntament de Barcelona), Alina Hirschmann (Institut de Ciències de l'Espai, CSIC-IEEC), Jordi Isern (Institut de Ciències de l'Espai, CSIC-IEEC), Josep Maria Codina (Observatori Fabra, RACAB), Juan Carlos Morales (Institut d'Estudis Espacials de Catalunya), Ignasi Ribas (Institut de Ciències de l'Espai, CSIC-IEEC; Chair)

Participants

O. ABSIL, University of Liège, ⟨absil@astro.ulg.ac.be⟩

P. AMARO-SEOANE, Max Planck Institute for Gravitational Physics (Albert Einstein Institute), ⟨pau@aei.mpg.de⟩

G. ANGLADA-ESCUDÉ, Department of Terrestrial Magnetism/Carnegie Institution of Washington, ⟨guillem.anglada@gmail.com⟩

P. ARRIAGADA, Pontificia Universidad Católica de Chile, ⟨parriaga@astro.puc.cl⟩

A. ARTIGAS, Institut de Ciències de l'Espai (IEEC/CSIC), ⟨artigas@ieec.cat⟩

N. BABKOVSKAIA, University of Helsinki, ⟨Natalia.Babkovskaia@helsinki.fi⟩

S. BALAN, Cavendish Laboratory, ⟨st452@cam.ac.uk⟩

R. BARNES, University of Washington, ⟨rory@astro.washington.edu⟩

D. BARRADO NAVASCUÉS, LAEX-CAB, ⟨barrado@laeff.inta.es⟩

R. BARRY, NASA – GSFC, ⟨Richard.K.Barry@nasa.gov⟩

P. BAUDOZ, Observatoire de Paris, ⟨pierre.baudoz@obspm.fr⟩

J.-P. BEAULIEU, University College London / Institut d'Astrophysique de Paris, ⟨beaulieu@iap.fr⟩

C. BEICHMAN, NASA Exoplanet Science Institute (Caltech/JPL), ⟨chas@pop.jpl.nasa.gov⟩

A. BELU, Université de Bordeaux, ⟨adrian.belu@u-bordeaux1.fr⟩

D. BENNETT, University of Notre Dame, ⟨bennett@nd.edu⟩

C. BERGFORS, Max Planck Institute for Astronomy, ⟨bergfors@mpia.de⟩

A. BERNAGOZZI, Astronomical Observatory of the Autonomous Region of the Aosta Valley, ⟨andrea.bernagozzi@unimi.it⟩

E. BERTOLINI, Astronomical Observatory of the Autonomous Region of Aosta Valley, ⟨direttore@oavda.it⟩

J.-L. BEUZIT, Observatoire de Grenoble, ⟨Jean-Luc.Beuzit@obs.ujf-grenoble.fr⟩

A. BOCCALETTI, Observatoire de Paris, ⟨anthony.boccaletti@obspm.fr⟩

M. BOOTH, Institute of Astronomy, Cambridge University, ⟨mbooth@ast.cam.ac.uk⟩

A. BOSS, Department of Terrestrial Magnetism/Carnegie Institution of Washington, ⟨boss@dtm.ciw.edu⟩

B. BRAAM, TNO Science and Industry, ⟨ben.braam@tno.nl⟩

M. CAGIGAL, Universidad de Cantabria, ⟨perezcm@unican.es⟩
W. CASH, University of Colorado, ⟨webster.cash@colorado.edu⟩
K. CASTEELS, IEEC, ⟨kcasteels@am.ub.es⟩
C. CATALA, LEISA – Observatoire de Paris, ⟨Claude.Catala@obspm.fr⟩
S. CATALAN, University of Hertfordshire, ⟨s.catalan@herts.ac.uk⟩
J. CATANZARITE, NASA – JPL/Caltech,
    ⟨Joseph.H.Catanzarite@jpl.nasa.gov⟩
M. CLAIRE, University of Washington, ⟨mclaire@astro.washington.edu⟩
M. CLAMPIN, NASA – GSFC, ⟨mark.clampin@nasa.gov⟩
R. CLAUDI, INAF Osservatorio Astronomico di Padova,
    ⟨riccardo.claudi@oapd.inaf.it⟩
J. COLOMÉ, Institut de Ciències de l'Espai (IEEC/CSIC), ⟨colome@ieec.cat⟩
V. COUDÉ DU FORESTO, LESIA / Observatoire de Paris,
    ⟨vincent.foresto@obspm.fr⟩
M. DAMASSO, Astronomical Observatory of the Autonomous Region of the
    Aosta Valley, ⟨m.damasso@gmail.com⟩
W. DANCHI, NASA – GSFC, ⟨william.c.danchi@nasa.gov⟩
D. DEFRÈRE, University of Liège, ⟨defrere@astro.ulg.ac.be⟩
O. DEMANGEON, Institut d'Astrophysique Spatiale,
    ⟨olivier.demangeon@ias.u-psud.fr⟩
D. DESPOIS, Observatoire de Bordeaux OASU/LAB,
    ⟨despois@obs.u-bordeaux1.fr⟩
M. DEVIRIAN, NASA – JPL, ⟨devirian@jpl.nasa.gov⟩
S. DOMAGAL-GOLDMAN, Virtual Planetary Laboratory/University of
    Washington, ⟨sgoldman@astro.washington.edu⟩
S. DREIZLER, University Goettingen,
    ⟨dreizler@astro.physik.uni-goettingen.de⟩
C. EIROA, Universidad Autonoma de Madrid, ⟨carlos.eiroa@uam.es⟩
T. ENCRENAZ, LESIA / Observatoire de Paris, ⟨therese.encrenaz@obspm.fr⟩
R. ENOCH, St. Andrews University, ⟨be12@st-andrews.ac.uk⟩
K. ENYA, Institute of Space and Astronautical Science / Japan Aerospace
    Exploration Agency, ⟨enya@ir.isas.jaxa.jp⟩
K. ERGENZINGER, Astrium, ⟨klaus.ergenzinger@astrium.eads.net⟩
B. FEMENÍA CASTELLÁ, Instituto de Astrofsica de Canarias,
    ⟨bfemenia@iac.es⟩
J. FERNANDEZ, Catolica University, Chile / Institute for Astrophysics
    Gttingen, ⟨jfernand@astro.puc.cl⟩
A. FONT-RIBERA, Insitut de Ciències de l'Espai (IEEC/CSIC),
    ⟨font@ieec.uab.es⟩
F. FORGET, LMD/IPSL, ⟨forget@lmd.jussieu.fr⟩

O. Fors, Observatori Fabra and Departament d'Astronomia i Meteorologia UB, ⟨ofors@am.ub.es⟩
C. V. M. Fridlund, ESA, ⟨malcolm.fridlund@esa.int⟩
M. Fukagawa, Osaka University, ⟨misato@iral.ess.osaka-u.ac.jp⟩
B. Funk, Eötvös Loránd University, ⟨funk@astro.univie.ac.at⟩
P. Gabor, Universite Paris XI, ⟨pavel.gabor@ias.u-psud.fr⟩
R. Galicher, Observatoire de Paris, ⟨raphael.galicher@obspm.fr⟩
M. C. Gálvez Ortiz, University of Hertfordshire, ⟨M.Galvez-Ortiz@herts.ac.uk⟩
A. Garcés, Institut de Ciències de l'Espai (IEEC/CSIC), ⟨garces@ieec.uab.es⟩
A. García Muñoz, Instituto de Astrofsica de Canarias, ⟨agm@iac.es⟩
E. García-Melendo, Esteve Duran Observatory Foundation, ⟨egarcia@foed.org⟩
T. Gautier, NASA – JPL / Caltech, ⟨thomas.n.gautier@jpl.nasa.gov⟩
R. Gehrz, University of Minnesota, ⟨gehrz@astro.umn.edu⟩
D. Gelino, NASA Exoplanet Science Institute, Caltech, ⟨dawn@ipac.caltech.edu⟩
A. Giménez, Centro de Astrobiologa, ⟨gimeneza@inta.es⟩
J. Goicoechea, Centro de Astrobiologia / CSIC-INTA, ⟨jr.goicoechea@gmail.com⟩
A. Golden, National University of Ireland, Galway, ⟨aaron.golden@nuigalway.ie⟩
N. Gomez-Perez, Department of Terrestrial Metabolism/Carnegie Institution of Washington, ⟨nataliag.co@gmail.com⟩
R. Goullioud, NASA – JPL, ⟨renaud.goullioud@jpl.nasa.gov⟩
R. Gratton, INAF / Osservatorio Astronomico di Padova, ⟨raffaele.gratton@oapd.inaf.it⟩
T. Greene, NASA/Ames Research Center, ⟨tom.greene@nasa.gov⟩
J. Grenfell, Technische Universität Berlin, ⟨lee.grenfell@dlr.de⟩
C. Grillmair, Caltech, ⟨carl@ipac.caltech.edu⟩
M. Guedel, ETH Zurich, ⟨guedel@astro.phys.ethz.ch⟩
O. Guyon, University of Arizona / Subaru Telescope, ⟨guyon@naoj.org⟩
T. Hara, Kyoto Sangyo University, ⟨hara@cc.kyoto-su.ac.jp⟩
A. Hatzes, Thueringer Landessternwarte Tautenburg, ⟨artie@tls-tautenburg.de⟩
K. Haze, Research Fellow of the Japan Society for the Promotion of Science, ⟨haze@ir.isas.jaxa.jp⟩
E. Herrero, IEEC, ⟨eherrero@ieec.cat⟩
S. Higuchi, University of Tokyo, ⟨higuchi@ir.tsas.jaxa.jp⟩

Participants

A. HIRSCHMANN, Institut de Ciències de l'Espai (IEEC/CSIC), ⟨alina@ieec.uab.es⟩

K. HOUAIRI, ONERA DOTA, ⟨kamel.houairi@onera.fr⟩

S. HOYER MIRANDA, Universidad de Chile, ⟨shoyer@das.uchile.cl⟩

D. HUDGINS, NASA, ⟨Douglas.M.Hudgins@nasa.gov⟩

J. ISERN, Institut de Ciències de lÉspai (IEEC/CSIC), ⟨isern@ieec.cat⟩

S. JACQUINOD, Institut d'Astrophysique Spatiale, ⟨sophie.jacquinod@ias.u-psud.fr⟩

J. JACQUINOT, CEA, ⟨jean.jacquinot@cea.fr⟩

M. JANSON, University of Toronto, ⟨janson@astro.utoronto.ca⟩

E. JONES, Australian National University, ⟨eriita@mso.anu.edu.au⟩

H. JONES, University of Hertfordshire, ⟨h.r.a.jones@herts.ac.uk⟩

S. KAFKA, Department of Terrestrial Magnetism/ Carnegie Institution of Washington / Caltech, ⟨stella@caltech.edu⟩

S. KANE, NASA Exoplanet Science Institute, Caltech, ⟨skane@ipac.caltech.edu⟩

T. KARALIDI, SRON-Netherlands Institute for Space Research, ⟨T.Karalidi@sron.nl⟩

J. KASTING, The Pennsylvania State University, ⟨kasting@essc.psu.edu⟩

L. KEDZIORA-CHUDCZER, University of New South Wales, ⟨lkedzior@Physics.usyd.edu.au⟩

G. KENNEDY, ICCUB, IEEC, ⟨gareth.f.kennedy@gmail.com⟩

N. KIANG, Goddard Institute for Space Studies, ⟨nkiang@giss.nasa.gov⟩

D. KIPPING, University College London, ⟨d.kipping@ucl.ac.uk⟩

M. KISSLER-PATIG, European Southern Observatory, ⟨mkissler@eso.org⟩

L. KOECHLIN, LATT - Observatoire Midi Pyrénées, ⟨lkaurent.koechlin@ast.obs-mip.fr⟩

D. KONTOPOULOS, Aristotle University of Thessaloniki, ⟨dimitris-jp@hotmail.com⟩

T. KOTANI, ISAS/JAXA, ⟨tkotani@ir.isas.jaxa.jp⟩

A. LABEYRIE, Collège de France, ⟨antoine.labeyrie@obs-azur.fr⟩

P.-O. LAGAGE, CEA-Saclay, ⟨pierre-olivier.lagage@cea.fr⟩

H. LAMMER, Austrian Academy of Sciences, ⟨helmut.lammer@oeaw.ac.at⟩

D. LATHAM, Harvard-Smithsonian Center for Astrophysics, ⟨dlatham@cfa.harvard.edu⟩

M. LATTANZI, INAF - Osservatorio Astronomico di Torino, ⟨lattanzi@oato.inaf.it⟩

P. LAWSON, NASA – JPL, ⟨Peter.R.Lawson@jpl.nasa.gov⟩

R. LEE, NASA – JPL, ⟨roger.a.lee@jpl.nasa.gov⟩

A. LÉGER, IAS, ⟨Alain.Leger@ias.fr⟩

M. LEITZINGER, University of Graz, ⟨martin.leitzinger@uni-graz.at⟩
P. LÉNA, Université Paris Diderot / Observatoire de Paris,
⟨pierre.lena@obspm.fr⟩
D. LIN, University of California Santa Cruz / KIAA Peking University,
⟨lin@ucolick.org⟩
J. LINSKY, JILA/University of Colorado / NIST, ⟨jlinsky@jila.colorado.edu⟩
R. LISEAU, Chalmers University of Technology, ⟨Rene.Liseau@chalmers.se⟩
A. LO, Northrop Grumman Coporation, ⟨amy.lo@ngc.com⟩
N. LODIEU, IAC, Tenerife, ⟨nlodieu@iac.es⟩
M. LOPEZ-MORALES, Department of Terrestrial Magnetism/Carnegie
Institution of Washington, ⟨mercedes@dtm.ciw.edu⟩
R. LYON, NASA – GSFC, ⟨Richard.G.Lyon@nasa.gov⟩
S. MAHADEVAN, Pennsylvania State University, ⟨suvrath@astro.ufl.edu⟩
W. MAJID, NASA – JPL / Caltech, ⟨walid.majid@jpl.nasa.gov⟩
F. MALBET, Laboratoire d'Astrophysique de Grenoble / CNRS / UJF,
⟨Fabien.Malbet@obs.ujf-grenoble.fr⟩
J. MALDONADO, Universidad Autonoma de Madrid,
⟨jesus.maldonado@uam.es⟩
G. MANDUSHEV, Lowell Observatory, ⟨gmand@lowell.edu⟩
R. MANGINA, NASA – JPL / Caltech, ⟨rmangina@jpl.nasa.gov⟩
J. MARCAIDE, Univ. Valéncia, ⟨j.m.marcaide@uv.es⟩
E. MARTIN, CSIC-INTA Centro de Astrobiologia, ⟨ege@iac.es⟩
G. MARTIN, LAOG, ⟨guillermo.martin@obs.ujf-grenoble.fr⟩
S. MARTIN, NASA – JPL, ⟨stefan.r.martin@jpl.nasa.gov⟩
J. MARTIN RODRÍGUEZ, IEEC, ⟨jmartin@am.ub.es⟩
F. MARTINACHE, Subaru Telescope, ⟨frantz@naoj.org⟩
T. MATSUO, NAOJ / NASA – JPL, ⟨Taro.Matsuo@jpl.nasa.gov⟩
M.-C. MAUREL, ANBioPhy - UPMC-CNRS FRE3207,
⟨marie-christine.maurel@upmc.fr⟩
M. MCCAUGHREAN, ESA – ESTEC, ⟨mjm@rssd.esa.int⟩
S. MESCHIARI, University of California Santa Cruz, ⟨smeschia@ucolick.org⟩
M. MEYER, ETH Zurich, ⟨mmeyer@phys.ethz.ch⟩
G. MICELA, INAF OAPA, ⟨giusi@astropa.unipa.it⟩
E. MILLOUR, LMD, ⟨ehouarn.millour@lmd.jussieu.fr⟩
J. MIRALDA-ESCUDE, ICREA/University of Barcelona,
⟨jordi.miralda@gmail.com⟩
P. MONTAÑÉS-RODRÍGUEZ, IAC, ⟨pmr@iac.es⟩
D. MONTES, Universidad Complutense de Madrid, ⟨dmg@astrax.fis.ucm.es⟩
A. MOORHEAD, University of Florida, ⟨altheam@astro.ufl.edu⟩

Participants

J.-C. MORALES, Institut d'Estudis Espacials de Catalunya,
⟨morales@ieec.uab.es⟩

N. MURAKAMI, Hokkaido University, ⟨nmurakami@eng.hokudai.ac.jp⟩

T. NAKAGAWA, ISAS/JAXA, ⟨nakagawa@ir.isas.jaxa.jp⟩

N. NARITA, National Astronomical Observatory of Japan,
⟨norio.narita@nao.ac.jp⟩

J. NISHIKAWA, National Astronomical Observatory of Japan,
⟨jun.nishikawa@nao.ac.jp⟩

P. ODERT, Institute of Physics/IGAM, University of Graz,
⟨petra.odert@uni-graz.at⟩

M. OLLIVIER, Institut d'Astrophysique Spatiale d'Orsay,
⟨marc.ollivier@ias.u-psud.fr⟩

I. PAGANO, INAF / Catania Astrophysical Observatory, ⟨ipa@oact.inaf.it⟩

E. PALLE, Instituto de Astrofisica de Canarias, ⟨epalle@iac.es⟩

F. PEPE, Observatoire Astronomique de l'Université de Genève,
⟨Francesco.Pepe@unige.ch⟩

G. PIOTTO, Dipartimento di Astronomia, Universita' di Padova,
⟨giampaolo.piotto@unipd.it⟩

E. PODLEWSKA, Institute of Physics, University of Szczecin,
⟨edytap@univ.szczecin.pl⟩

R. POLIDAN, Northrop Grumman Aerospace Systems,
⟨ron.polidan@ngc.com⟩

D. POLLACCO, Queens University Belfast, ⟨d.pollacco@qub.ac.uk⟩

M. POSTMAN, Space Telescope Science Institute, ⟨postman@stsci.edu⟩

A. QUIRRENBACH, Landessternwarte, ⟨A.Quirrenbach@lsw.uni-heidelberg.de⟩

F. RANTAKYRO, Gemini, ⟨frantaky@gemini.edu⟩

L. RENOUARD, Astrium, ⟨laurent.renouard@astrium.eads.net⟩

I. RIBAS, Institut de Ciències de l'Espai (IEEC/CSIC), ⟨iribas@ieec.uab.es⟩

S. RINEHART , NASA – GSFC, ⟨Stephen.A.Rinehart@nasa.gov⟩

A. ROBERGE, NASA – GSFC, ⟨Aki.Roberge@nasa.gov⟩

F. RODLER, IAC, ⟨frodler@iac.es⟩

D. ROIG, IEEC, ⟨roig@ieec.uab.es⟩

B. ROJAS-AYALA, Cornell University, ⟨babs@astro.cornell.edu⟩

D. ROUAN, LESIA / Observatoire de Paris, ⟨daniel.rouan@obspm.fr⟩

B. SAMUEL, IAS, ⟨benjamin.samuel@ias.u-psud.fr⟩

J. SANZ FORCADA, Centro de Astrobiologa, ⟨jsanz@laeff.inta.es⟩

D. SASSELOV, Harvard University, ⟨sasselov@cfa.harvard.edu⟩

S. SCUDERI, INAF - Osservatorio Astrofisico di Catania,
⟨scuderi@oact.inaf.it⟩

J. SCHNEIDER, CNRS-Observatoire de Paris, ⟨jean.schneider@obspm.fr⟩

C. Schnupp, Max Planck Institute for Astronomy, ⟨schnupp@mpia.de⟩

P. Schuller, Institut d'Astrophysique Spatiale,
⟨peter.schuller@ias.u-psud.fr⟩

R. Schwarz, Institute for Astronomy / University of Vienna,
⟨schwarz@astro.univie.ac.at⟩

F. Selsis, Laboratoire d'Astrophysique de Bordeaux,
⟨selsis@obs.u-bordeaux1.fr⟩

M. Shao, NASA – JPL / Caltech, ⟨michael.shao@jpl.nasa.gov⟩

H. Shibai, Osaka University, ⟨shibai@ess.sci.osaka-u.ac.jp⟩

D. Sing, IAP, ⟨hs1136@gmail.com⟩

I. Snellen, Leiden Observatory, ⟨snellen@strw.leidenuniv.nl⟩

R. Soummer, Space Telescope Science Institute, ⟨soummer@stsci.edu⟩

A. Sozzetti, INAF - Osservatorio Astronomico di Torino,
⟨sozzetti@oato.inaf.it⟩

W. Sparks, Space Telescope Science Institute, ⟨sparks@stsci.edu⟩

D. Spiegel, Princeton University, ⟨dsp@astro.princeton.edu⟩

D. Stam, SRON Netherlands Institute for Space Research,
⟨d.m.stam@sron.nl⟩

T. Sumi, Nagoya University, ⟨sumi@stelab.nagoya-u.ac.jp⟩

M. Swain, NASA – JPL, ⟨Mark.R.Swain@jpl.nasa.gov⟩

M. Tamura, NAOJ, ⟨motohide.tamura@nao.ac.jp⟩

H. Terada, Subaru Telescope / NAOJ, ⟨terada@subaru.naoj.org⟩

G. Tinetti, University College London, ⟨g.tinetti@ucl.ac.uk⟩

W. Traub, NASA – JPL, ⟨wtraub@jpl.nasa.gov⟩

J. Trauger, NASA – JPL / Caltech, ⟨john.trauger@jpl.nasa.gov⟩

J. Trigo-Rodriguez, Institut de Ciències de l'Espai (IEEC/CSIC),
⟨trigo@ieec.uab.es⟩

G. van Belle, European Southern Observatory, ⟨gvanbell@eso.org⟩

M. Vasquez, Institut für Methodik der Fernerkundung/ Deutsches Zentrum fr
Luft- und Raumfahrt e.V. (DLR), ⟨mayte.vazquez@dlr.de⟩

E. Villaver, Universidad Autonoma de Madrid, ⟨eva.villaver@uam.es⟩

G. Vladilo, Osservatorio Astronomico di Trieste / INAF,
⟨vladilo@oats.inaf.it⟩

K. von Braun, Caltech, ⟨kaspar@caltech.edu⟩

A. Vosteen, TNO Science and Industry, ⟨amir.vosteen@tno.nl⟩

U. Wehmeier, NASA – JPL, ⟨kurtman00@gmail.com⟩

J. Weingrill, Space Research Institute, Austrian Academy of Sciences,
⟨joerg.weingrill@oeaw.ac.at⟩

G. White, Open University / Rutherford Appleton Laboratory,
⟨glenn.white@stfc.ac.uk⟩

M. WITTKOWSKI, European Southern Observatory, ⟨mwittkow@eso.org⟩
R. WORDSWORTH, Laboratoire de Meteorologie Dynamique, ⟨rwlmd@lmd.jussieu.fr⟩
G. WUCHTERL, Thüringer Landessternwarte, ⟨gwuchterl@TLS-Tautenburg.de⟩
N. YAITSKOVA, European Southern Observatory, ⟨nyaitsko@eso.org⟩
N. YOSHITSUGU, Kobe University, ⟨yoshi@kobe-u.ac.jp⟩
O. ZAKHOZHAY, Main Astronomical Observatory of National Academy of Sciences of Ukraine, ⟨zkholga@mail.ru⟩
P. ZARKA, LESIA / Observatoire de Paris, ⟨philippe.zarka@obspm.fr⟩
H. ZINNECKER, Astrophysical Institute Potsdam, ⟨hzinnecker@aip.de⟩

xxix

Conference participants outside the meeting venue,
the CosmoCaixa Museum of Science in Barcelona, Spain.

xxx

Pathways Towards Habitable Planets meeting organizers and sponsors.

Session I
Welcome

Pathways Towards Habitable Planets
ASP Conference Series, Vol. 430, 2010
Vincent Coudé du Foresto, Dawn M. Gelino, and Ignasi Ribas, eds.

How to Find a Habitable Planet

J. F. Kasting
Department of Geosciences, The Pennsylvania State University, University Park, PA, USA

Abstract. Over 400 extrasolar planets have now been discovered by ground-based methods, especially the radial velocity (RV) method. Most of these planets are much bigger than Earth, and only a handful of them are rocky planets that could conceivably harbor life. Within the next few years, we may be able to identify Earth-sized planets within the habitable zones around M stars by looking for transits, and RV techniques may even enable us to find Earths around more Sun-like (F-G-K) stars. To characterize these planets, though, we are likely to need new space-based telescopes that operate either in the visible/near-infrared or in the thermal-IR. Some conceptions of what these telescopes might look like and what we might learn from them are discussed.

1. Introduction

To date (December 2009), some 405 extrasolar planets have been reported (http://exoplanet.eu/), with most of them found around nearby stars using the radial velocity (RV) method. That is the good news for planet hunters and astrobiologists. The bad news is that nearly all of these planets are thought to be gas or ice giants that lack solid or liquid surfaces on which life might originate and evolve. Although Carl Sagan once suggested that bubble-like creatures might hover in Jupiter's clouds, originating life in such an unstable environment would be difficult, if not impossible. Rocky planets like Earth are much more likely abodes for life.

A second requirement for detectable life is that it must be able to modify a planet's atmosphere in ways that might be observed remotely. Earth provides an example of such a planet: Organisms, specifically photosynthetic algae, plants, and cyanobacteria, have created an atmosphere rich in O_2 that is very different from the predominantly CO_2 atmospheres on neighboring Mars and Venus. What allows this to happen on Earth is the presence of abundant liquid water on its surface, which provides a home for photosynthetic organisms. Liquid water is also an essential requirement for life as we know it. Clever biochemists have speculated that life might develop in other types of fluids, e.g., liquid CH_4 (NRC 2007). Even if this is possible, though, such life would be so different from us that we might not be able to recognize its existence from afar. For this reason, we henceforth limit our discussion to the conventional circumstellar habitable zone (HZ), the region around a star within which a planet can maintain liquid water on its surface. Although this might seem overly restrictive, planets falling into this category may exist around many, or even most, stars. So, making this assumption should not seriously impede our search for life elsewhere in the Galaxy.

2. Habitable Zones around Stars

The concept of the circumstellar HZ was introduced by Shapley (1953), Strughold (1953), and Huang (1959). Hart (1978, 1979) extended this idea by defining the continuously habitable zone (or CHZ) as the region that remains habitable for some finite period of time as the star evolves and increases in luminosity. Hart attempted to estimate the width of both the HZ and the CHZ around the Sun and other stars. He did so by performing time-dependent numerical simulations of Earth-like planets at different orbital distances around their parent stars. Hart's conclusions were extremely pessimistic. He found that the 4.6-Gyr CHZ around the Sun was itself rather narrow, 0.95-1.01 AU, and that CHZs around other main sequence stars were either narrower or completely nonexistent. If this conclusion was correct, Earth might be the only habitable planet in the entire Galaxy.

Theories of the habitable zone evolved, partly as a consequence of studies of the so-called "faint young Sun" problem on Earth. The Sun was roughly 30 percent less luminous when the Solar System formed 4.6 billion years (b.y.) ago (Gough 1981); yet Earth has had life since $\sim$3.5 b.y. ago and liquid water since 4.4 b.y. ago (Valley et al. 2002). At least part of the solution to this problem is that early Earth probably had a much larger atmospheric greenhouse effect, caused by the presence of high concentrations of CO_2 (Walker et al. 1981) and perhaps CH_4, as well (Kiehl & Dickinson 1987; Haqq-Misra et al. 2008). CO_2 concentrations would have been enhanced because of a strong negative feedback in the carbonate-silicate cycle, which controls atmospheric CO_2 concentrations over long time scales. As Earth's surface cools, silicate weathering slows down, thereby reducing the rate at which CO_2 can be sequestered into carbonate rocks. Steady state can be achieved only when liquid water is present, thereby ensuring that early Earth remained within the liquid water regime.

This same feedback implies that if Earth had formed further from the Sun, it should have developed a dense CO_2 atmosphere that would have kept its surface warm. Thus, the outer edge of the habitable zone should be farther out than Hart calculated, probably somewhere around 1.7–2.0 AU (Kasting et al. 1993; Forget & Pierrehumbert 1997). Hence, the HZ should be relatively wide, and the 4.6-Gyr CHZ around the Sun should extend from $\sim$0.95 AU to $\sim$1.4 AU. If terrestrial planets around other stars exist, and if they are distributed in orbital distance as they are in our own Solar System, the chances of finding another habitable planet appear to be pretty good.

3. The Gliese 581 System

Although none of the exoplanets discovered so far is thought to be habitable, two candidates have come extremely close. Both of them orbit the star Gliese 581. Gliese 581 is an M3V star with 0.31 times the mass of the Sun and 0.0135 times its luminosity. It has 4 known planets, b-e, whose properties are listed in Table 1. These planets were all discovered using the RV method (Udry et al.

Table 1. The Gliese 581 system

Planet	Mass ($M_\oplus$)	Semi-major axis (AU)
b	$30.4 \geq M \geq 15.65$	0.04
c	$10.4 \geq M \geq 5.36$	0.07
d	$13.8 \geq M \geq 7.09$	0.22
e	$3.1 \geq M \geq 1.94$	0.03

2007), and so the lower limits on planetary mass (in Earth masses) are derived in the usual manner. The upper mass limits are derived from considerations of orbital stability (Mayor et al. 2009).

The two planets that are closest to being within the habitable zone are Gliese 581c and d. Planet c is most likely rocky, given that its mass is between 5 and 10 Earth masses. But the flux incident on planet c is about 30 percent greater than that hitting Venus today (von Bloh et al. 2007; Selsis et al. 2007), so planet c is unlikely to be habitable. Planet d is a better candidate, although it may or may not be rocky, as its upper mass limit is close to that of Uranus (about 14 Earth masses). Its estimated semi-major axis, 0.22 AU, has come down somewhat since it was first reported; hence, the stellar flux available at the planet's orbit has increased to about 28 percent of that hitting the Earth (1365 W/m^2). This is slightly higher than the "maximum greenhouse" limit for a slightly bluer M0 star (Kasting et al. 1993, Table III); hence, it could very well be habitable if it is indeed a rocky planet.

4. Searching for Habitable Planets by using Transits

The discovery of the Gliese 581 system has demonstrated that it should already be possible to search for potentially habitable planets around M stars. One may be able to do this using the RV technique, particularly if that technique can be extended to the near-IR, where M stars are at their brightest. But the most promising technique is to look for transits (Deming et al. 2009). Finding potentially habitable planets around M stars is much easier than finding them around F-G-K stars because the stars themselves are smaller, allowing a transiting Earth-sized planet to produce a strong signal, and because the HZ is close to the star, increasing the probability that a transit will occur. (The transit probability is equal to the radius of the star divided by the radius of the planet's orbit.) The odds of seeing Earth around the Sun are only 1 in 200. For a late M star, the HZ is at $\sim$0.03 AU and the stellar diameter is about 0.1 times that of the Sun; hence, the transit probability increases to about 1 in 60.

If transiting M-star planets can be located within the next few years, it may be possible to study their atmospheres spectroscopically with the James Webb Space Telecope (JWST), scheduled for launch in 2014. The idea would be to employ secondary eclipse spectroscopy, as has been done already for hot Jupiter-type exoplanets using the Spitzer Space Telescope. In this technique, one measures the infrared emission from the combined star plus planet while the planet is visible, then subtracts the spectrum obtained when the planet is behind the star. Whether or not this can yield useful information about terrestrial type planets around M stars remains to be seen, but it is an intriguing possibility.

Other transiting potentially habitable planets may be discovered by NASA's Kepler Space Telescope (http://kepler.nasa.gov/), which has been taking data since mid-summer 2009. Kepler is staring at a patch of the Milky Way, monitoring the brightness of ~100,000 stars. Kepler can do photometry to about 1 part in 10^5; hence, given enough time (about 3 years), Kepler should be able to find Earth-sized planets around solar-type stars.

5. Astrometric Exoplanet Searches

While Kepler may tell us the frequency of Earth-sized planets, it is not likely to find any nearby candidates. That is because the average distance to stars in the Kepler field of view is about 600 pc. To find Earth-like planets around nearby stars, we need another method. The traditional RV method may or may not work. Earth's RV signal on the Sun is about 9 cm s^{-1}. The current RV sensitivity threshold is about 1 m s^{-1} for most stars, but reaches as low as 20 cm s^{-1} for the very best targets (Mayor et al. 2009). Hence, given enough observing time, bright stars could potentially be searched for planets using this method. The caveat is that stellar variability may make it difficult to pull the signal out of the noise. Nevertheless, one hopes that ground-based astronomers will actively pursue this problem. A dedicated 8-10 m ground-based telescope would distinctly improve the chances for success with this method.

An alternative way of searching nearby stars for exoplanets is to do accurate space-based astrometry. Astrometry, of course, is the measurement of a star's position on the sky. Stars move as they orbit, along with their planets, around the barycenter of the planetary system. Earth is lighter than the Sun by a factor of $\sim 3 \cdot 10^{-6}$; hence, it moves the Sun by a distance of $3 \cdot 10^{-6}$ AU. Viewed from a distance of 10 pc, this would appear as a shift of 0.3 μas in the Sun's position. An angle that small cannot be measured from the ground; however, it can be measured from space using optical interferometry (Unwin et al. 2008). NASA's proposed Space Interferometry Mission-Lite (SIM-Lite) (http://planetquest.jpl.nasa.gov/SIM/index.cfm) should be able to do this, were it to be funded.

A question of considerable interest to exoplanet researchers is whether the stellar target list that SIM-Lite might obtain would enhance the return from direct imaging missions discussed below. The joint NASA/NSF Exoplanet Task Force (ExoPTF) reasoned that having a target list might be a significant advantage for direct imaging missions, particularly those which involve an occulter. However, detailed studies of at least one such mission have not supported this conclusion (Savransky et al. 2009). This question should probably be studied by an unbiased panel before decisions are made as to whether such an astrometric precursor mission would be useful.

Regardless of whether an astrometric mission is needed as a precursor, it would be useful to fly one at some point, as these measurements yield the true mass of the planet, not just a lower limit. A planet's mass affects its habitability by determining whether it can retain an atmosphere and whether it can generate enough internal heat to drive plate tectonics. The actual mass might also

How to Find a Habitable Planet 7

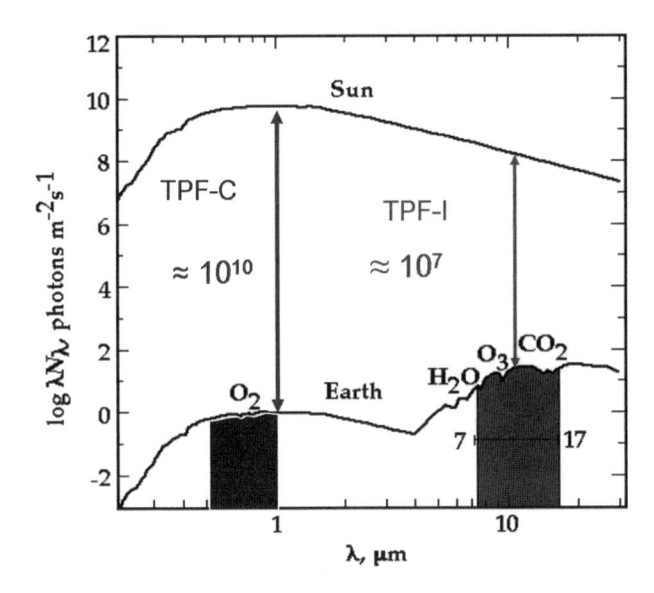

Figure 1. Photon flux from the Sun and the Earth, as viewed from a distance of 10 pc. (Courtesy C. Beichman, NASA JPL)

be obtained from RV in combination with direct imaging, although the errors involved are potentially large and (to my knowledge) unstudied. This may be the best argument for launching SIM-Lite.

6. Direct Space-based Imaging of Earth-like Exoplanets

Ultimately, we want to image extrasolar planets directly and obtain spectra of their atmospheres. Such spectra should be more detailed than those obtained from transits, because the stellar background would be largely or totally subtracted. Separating light from the planet from that of its parent star is difficult because the planet is much dimmer than the star and very close to it. The problems depend on which spectral region is being used, as illustrated by Figure 1. The figure shows calculated radiative fluxes from the Sun and from Earth, as viewed from a distance of 10 pc. The Sun's spectrum appears smooth at this low spectral resolution: it is approximately that of a 5780 K blackbody. By contrast, Earth's spectrum shows two distinct humps, along with several pronounced wiggles. At visible/near-IR wavelengths ($\lambda < 4$ μm), Earth's flux parallels that from the Sun, because Earth is being viewed in reflected sunlight. At longer wavelengths, Earth's flux increases while the Sun's flux declines, because here one is observing Earth's own emitted thermal-IR radiation.

The question of which wavelength to observe at appears straightforward at first glance. The contrast ratio in brightness between the star and the planet is about 10^{10} in the visible, compared to "only" 10^7 in the thermal-IR. Consequently, NASA's initial plans (and ESA's postponed Darwin mission) focused on the latter wavelength region. The disadvantage of observing in the thermal-IR is that the required telescope aperture is significantly larger than in the visible. Because of diffraction by the telescope mirror, the smallest angle, θ, that can be resolved is given by:

$$\theta \simeq \frac{n\lambda}{D} \qquad (1)$$

Here, λ is wavelength, D is the mirror diameter, and n is a number that depends on the characteristics of the optical system. Straightforward calculation shows that for mid-visible wavelengths ($\lambda = 0.5$ μm), a telescope with $n = 4$ would need to be 4 m in diameter to resolve Earth around the Sun from a distance of 10 pc. To get a spectrum out to 1 μm, the required telescope diameter would be twice that, or 8 m. Although large, a monolithic mirror of this diameter could be accommodated by existing launch vehicles. NASA's proposed Terrestrial Planet Finder-Coronagraph (TPF-C) mission (Levine et al. 2006) employed an 8×3.5 m elliptical mirror that could be stacked lengthwise in the cargo bay of a Delta-IVH rocket.

At thermal-IR wavelengths, the required telescope aperture is much larger. Working at 10 μm with the same value for n would require a mirror diameter of 80 m, which is clearly beyond our present space-based capabilities. One can get around this problem, though, by linking several smaller telescopes together and using them as an *interferometer*. ESA's proposed Darwin mission would consist of four ~3 m telescopes, each carried by a separate spacecraft (http://www.darwin.rl.ac.uk/). The beams from these four telescopes would be combined at a fifth spacecraft to do the interferometry. NASA has discussed a similar mission called Terrestrial Planet Finder-Interferometer, or TPF-I. A mission of this sort would probably be expensive, as it would involve multiple spacecraft and cooled telescopes, so hopefully it would end up being a cooperative international venture.

Most U.S. exoplanet scientists think that some sort of visible/near-IR (and possibly UV) direct imaging mission should fly first. This mission itself comes in two flavors: *1)* an internal coronagraph similar to TPF-C, or *2)* an external occulter similar to the proposed New Worlds Observer mission (http://newworlds.colorado.edu/). The occulter concept is relatively new, but it has garnered significant attention because the requirements on the telescope mirror are much less severe. The mirror for TPF-O, as NASA now calls it, would need only to be approximately as smooth as that of the Hubble Space Telescope. The TPF-C mirror, by contrast, would need to be roughly 20 times smoother – a feat that could only be achieved by using a deformable mirror.

A TPF occulter mission would create its own problems, because the occulter (a 50–70 m diameter flower-shaped disk) would need to be accurately positioned at a distance of ~50,000 km from the telescope, and it would need to be repositioned as one moved from one target star to another. Thus, it might be useful to know the desired target stars in advance, as noted in the previous section.

How to Find a Habitable Planet 9

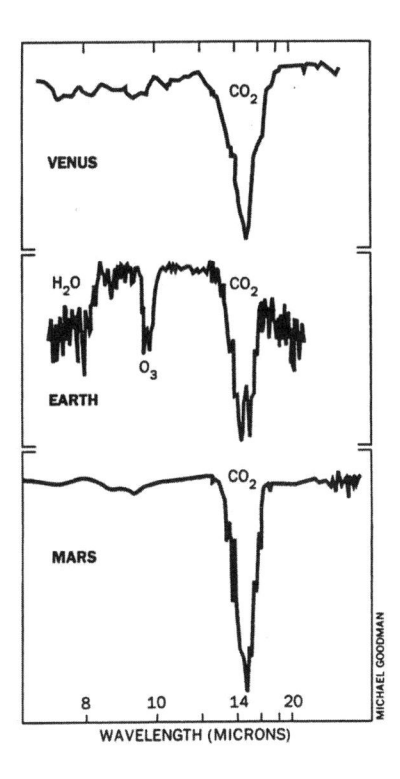

Figure 2. Thermal-IR spectra of Venus, Earth, and Mars. (Courtesy of R. Hanel, NASA Goddard)

7. Spectral Indicators of Planetary Habitability and Life

If one or more of these direct imaging missions can be launched, then the prospects for discovering other Earth-like exoplanets are good. One may at the same time be able to determine whether these planets are actually inhabited. Figure 2 shows low-resolution spectra of Venus, Earth, and Mars, as they might be observed by Darwin or TPF-I. For Venus and Mars, only one strong spectral feature is seen: the 15-μm band of CO_2. For Earth, the 15-μm band is still prominent, despite the fact that the concentration of CO_2 in Earth's atmosphere is only ~380 ppmv. But on Earth one can also see the edge of the 6.3-μm rotation-vibration band of H_2O, along with part of the pure rotation band of H_2O, which extends longward of 13 μm. This is gas-phase H_2O, not liquid H_2O, and so it suggests (but does not prove!) that liquid water is present on Earth's surface.

An even more interesting feature seen in Figure 2 is the 9.6-μm band of O_3 (ozone). Ozone is formed photochemically from O_2, and nearly all of Earth's O_2 has been produced by photosynthesis. Hence, under most circumstances, detection of an O_3 or O_2 signal in an exoplanet atmosphere would be strong

10   Kasting

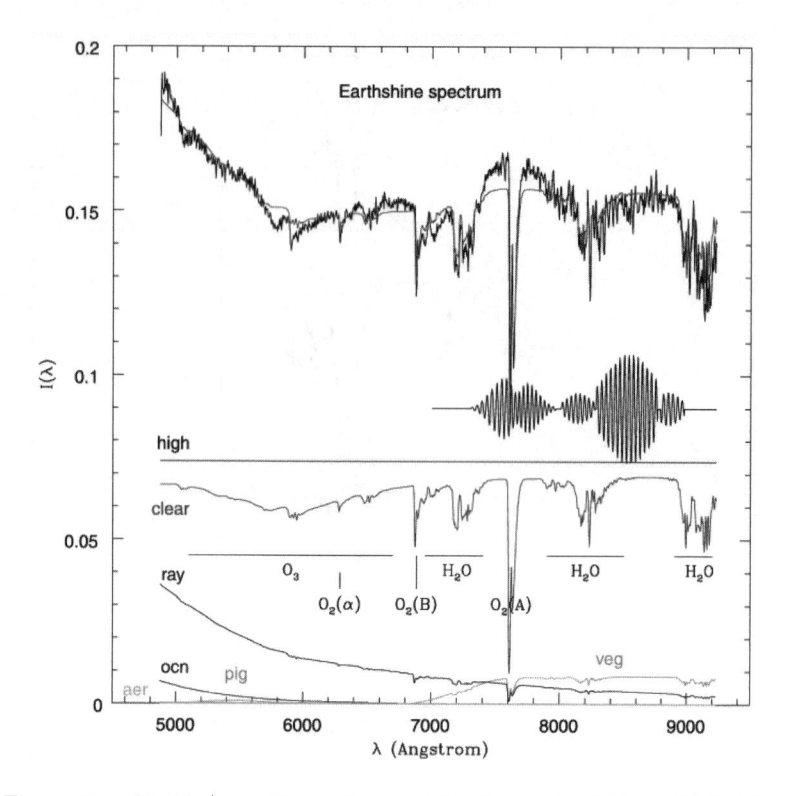

Figure 3. Visible/near-IR spectrum of Earth obtained from Earthshine measurements (from Woolf et al. 2002).

evidence for the existence of life. That said, possible exceptions to this statement, so-called "false positives" for life, have already been identified. The most obvious of these is a runaway greenhouse atmosphere, like that which may have occurred on early Venus (Kasting 1997; Des Marais et al. 2002). If hydrogen was lost quickly, and if much of the oxygen was retained, O_2 could conceivably build up to high levels on such a planet, without any help from biology. A frozen planet, like Mars but with about twice the mass so that it did not lose oxygen to space, is another known false positive (ibid.). Planets that are within their star's habitable zone, and which are large enough to support volcanism, are not thought to be subject to these particular ambiguities.

At visible/near-IR wavelengths, the spectral features that could be observed on Earth are different (Figure 3), but they too provide useful information. H_2O has three strong bands in this region, centered at 0.72, 0.82, and 0.94 μm. H_2O is actually considered to be easier to measure in the visible than in the thermal-IR because the absorption bands are narrower. Also seen in Figure 3 is the strong 0.76-μm "A" band of O_2. Detection of O_2 in the near-IR would have the same implications as detection of O_3 in the thermal-IR.

Finally, suppose that one flew both types of direct imaging missions, visible and thermal-IR, and that one found an exoplanet that exhibited signs of life. What could one do to confirm this result? The answer to this question was suggested over 40 years ago in two papers that appeared almost simultaneously in Nature magazine. In the first, Joshua Lederberg suggested that one should look for evidence of extreme thermodynamic disequilibrium (Lederberg 1965). All planetary atmospheres are in disequilibrium to some extent, because of the competing influences of the local thermal environment and incident stellar UV radiation. But Earth's atmosphere is clearly farther out of equilibrium than are the atmospheres of Venus or Mars. Lovelock (1965) made this argument more concrete by suggesting that the simultaneous presence of O_2 and a reduced gas, such as CH_4 or N_2O, would be strong evidence for life. O_2 and CH_4 in Earth's atmosphere are more than 20 orders of magnitude out of thermodynamic equilibrium – a phenomenon that is made possible only by the fact that they are both actively produced by organisms. CH_4 in Earth's atmosphere would be difficult to observe remotely, as its concentration is only ~1.7 ppmv. But a really big space telescope, with capabilities in the near-IR, should be able to see both CH_4 and N_2O, along with O_2 (Sagan et al. 1993). So, a "Life Finder" mission of this sort could provide even stronger evidence for extraterrestrial life.

8. Conclusions

The study of extrasolar planets has already progressed from infancy to childhood. We are now on the verge of entering the adult stage, where the focus is shifting towards terrestrial planets and life. A number of ambitious space missions are either being constructed or have been proposed to further this development, including JWST, SIM-Lite, TPF-C (or -O), and TPF-I/Darwin. We should do everything possible to get these missions approved and to search the nearby solar neighborhood for Earth-like planets and for life.

References

Deming, D., et al. 2009, PASP, 121, 952
Des Marais, D. J., et al. 2002, Astrobiology, 2, 153
Forget, F., & Pierrehumbert, R. T. 1997, Sci, 278, 1273
Gough, D. O. 1981, Solar Phys., 74, 21
Haqq-Misra, J. D., Domagal-Goldman, S. D., Kasting, P. J., & Kasting, J. F. 2008, Astrobiol., 8, 1127
Hart, M. H. 1978, Icarus, 33, 23
Hart, M. H. 1979, Icarus, 37, 351
Huang, S. S. 1959, Amer. Scientist, 47, 397
Kasting, J. F. 1997, Origins of Life, 27, 291
Kasting, J. F., Whitmire, D. P., & Reynolds, R. T. 1993, Icarus, 101, 108
Kiehl, J. T., & Dickinson, R. E. 1987, JGR, 92, 2991
Lederberg, J. 1965, Nat, 207, 9
Levine, M., Shaklan, S., Traub, W., Kasting, J., & Dressler, A. 2006, http://planetquest.jpl.nasa.gov/TPF/STDT_Report_Final_Ex2FF86A.pdf
Lovelock, J. E. 1965, Nat, 207, 568
Lovelock, J. E. 1965, Nat, 207, 568
Mayor, M., et al. 2009, A&A, 507, 487

NRC 2007, The Limits of Organic Life in Planetary Systems (Washington DC: National Academies Press)
Sagan, C., Thompson, W. R., Carlson, R., Gurnett, D., & Hord, C. 1993, Nat, 365, 715
Savransky, D., Kasdin, N. J., Cady, E. 2009, arXiv:0903.4915v1
Selsis, F., Kasting, J. F., Paillet, J., Levrard, B., Ribas, I., & Delfosse, X. 2007, A&A, 476, 1373
Shapley, H. 1953, Climatic Change: Evidence, Causes, and Effects (Cambridge, MA: Harvard Univ. Press)
Strughold, H. 1953, The Green and Red Planet (Albuquerque: Univ. of New Mexico Press)
Udry, S., et al. 2007, A&A, 469, L43
Unwin, S. C., et al. 2008, PASP, 120, 38
Valley, J. W., Peck, W. H., King, E. M., & Wilde, S. A. 2002, Geol., 30, 351
von Bloh, W., Bounama, C., Cuntz, M., & Franck, S. 2007, A&A, 476, 1365
Walker, J. C. G., Hays, P. B., & Kasting, J. F. 1981, JGR, 86, 9775
Woolf, N. J., Smith, P. S., Traub, W. A., & Jucks, K. W. 2002, ApJ, 574, 430

Session II

Reports from Prospective Exercises

Pathways Towards Habitable Planets
ASP Conference Series, Vol. 430, 2010
Vincent Coudé du Foresto, Dawn M. Gelino, and Ignasi Ribas, eds.

The Blue Dots Initiative and Roadmapping Exercise

V. Coudé du Foresto for the Blue Dots participants
Observatoire de Paris – LESIA, Meudon, Paris, France

Abstract. The Blue Dots initiative (a grassroot effort to build a scientific community in Europe around the exoplanet theme) is introduced. The Blue Dots activities include the elaboration of a roadmap towards the spectroscopic characterization of habitable exoplanets, a summary of which is presented here. While the roadmap will need to be updated regularly, it is expected that the methodology developed within Blue Dots will provide a durable framework for the elaboration of future revisions.

1. Introduction

The Blue Dots initiative was created in 2008 to contribute towards building a community in Europe around the exoplanet theme, and to converge towards a strategy enabling a more coherent approach to Calls for Proposals in ground- and space-based projects. The scope of the initiative is science-oriented and not restricted to a particular detection technique.

The intiative gathers more than 180 scientists, mostly located in Europe, with additional participation from the US, Japan, and India. Participants are organized in different working groups covering the relevant science themes (Targets and their Environments, Formation and Evolution of Planetary Systems, Habitability Criteria, Observation of Planetary Atmospheres) and methods (Single Aperture Imaging, Multiple Aperture Imaging, Microlensing, Modelling Habitable Planets, Radial Velocities, Astrometry, Transits). The reports from these working groups were presented at this Conference and are published in Part 3 of these Proceedings. Most of the other material produced by the initiative can be found on its web site at http://www.blue-dots.net.

2. Elements for a Roadmap

The initial Blue Dots activity consisted of preparing a roadmap towards the detection and characterization of habitable exoplanets, recognizing that this ambitious goal will require several intermediate steps.

To achieve this a framework for discussion had to be created and several underlying principles were applied. First, science questions should drive the roadmap – techniques should be seen as tools to address these questions. Specific missions should be introduced as late as possible in the process if convergence is to be sought. We also felt it was important to clearly identify the points of consensus and the matters of debate. Points of consensus provide an opportunity for the community to send out a common message.

Scientific debates are obviously healthy and useful when they can be organized in a way to help clarify the issues. Four of these debates were reproduced in this Conference as special Panel Sessions (a synthetic summary is published for three of them in Part 6 of these Proceedings), and cover what were found to be cornerstone questions in the roadmapping exercise:

1. What is the complementarity between astrometry and radial velocity (RV) to discover habitable planets?

2. Is spectroscopic characterization of the atmosphere of telluric exoplanets possible by transit spectroscopy?

3. Do we need to solve the exozodi issue? If yes, how best to solve it?

4. How to consolidate efforts within the community and the related agencies?

A roadmap should also find a way to get around the "pathfinder dilemma" which could be expressed this way: it is clear that the goal of spectroscopic characterization of the atmosphere of habitable planets, in search of biomarkers, will ultimately require one or more very ambitious and innovative missions. Those cannot meet the feasibility criteria as currently established by the space agencies, whose current trend is to select missions which are both low-risk and with an immediate science return. And the immediate science return of a more affordable demonstrator (one that would retire the risk on the bigger mission) does not necessarily meet the agency standards.

The frame of discussion required the creation of homogeneous grids that would cover the science that can can/should be achieved, and the methods that can be employed to achieve it. These are presented below.

2.1. Science Potential Levels

Following the lines of the step-by-step approach mentioned above, for a given class of objects the **science potential level** (SPL) of a technique is defined by its capacity:

* * To carry out a statistical study of objects in a given class;
* ** To designate targets in the solar neighborhood for spectroscopic follow-up study;
* *** To carry out a spectroscopic characterization of the object.

2.2. Target Classes

These classes are not meant to categorize objects according to their physical nature – rather they group objects of similar detection difficulty. The five classes identified are (by order of increasing detection difficulty):

1. **Hot giant planets**: These planets can be hot either because they are close to their host stars and highly irradiated, or because they are young;

2. **Other giant planets**: These planets include the warm and cold gaseous giants, down to Neptune size;

3. **Hot telluric planets** that might be hot, young or slightly more massive than the Earth (super-Earth), not necessarily located in the habitable zone neither around specific stars;

4. **Telluric planets in the habitable zone of M-type stars**;

5. **Telluric planets in the habitable zone of solar-type stars**.

2.3. Observing Methods

The methods employed for detecting and characterizing exoplanets can be classified in six large families, which are represented in the corresponding Blue Dots working groups:

- The **microlensing** method which consists in monitoring the photometry of distant stars in order to detect microlensing events.

- The **transit photometry** method which relies on measuring the relative change of the photometry of the star due to a primary eclipse (the planet transiting in front of the stellar photosphere) or to a secondary eclipse (the planet disappearing behind the star).

- The **radial velocities** method (RV) which relies on measuring the Doppler shift of star spectra with high precision in order to detect the reflex motion due to the presence of one or several planets. This method also includes timing techniques.

- The **astrometry** method whether in narrow-angle or globally consists in measuring the relative position of stars in order to detect the reflex motion due to the presence of orbiting planets.

- The **single aperture imaging** (SAI) technique includes all types of coronagraphic methods, including external occulters as well as imaging techniques using Fresnel lenses, in order to separate the direct light of the planets from the stellar light which is usually hidden.

- The **multiple aperture imaging** (MAI) technique uses interferometric nulling, hypertelescopes, etc. to extract the direct light from the planet and to some extent the stellar light.

It is clearly understood that these methods are in many ways interrelated and complementary, inasmuch as observables obtained by one technique can often be interpreted only with the help of additional information from another technique. The reader is referred to the Blue Dots report for more details on each method, its current state of the art, the prospective performance, and the required R&D efforts needed to achieve them.

2.4. Project Scales

Projects related to the different detection methods are listed and discussed in the Blue Dots report. It is recognized that those projects, which can be ground-based instruments or facilities, or spaceborne missions, can be of different scales. These are symbolized by different letter codes:

- **E** code: Existing, or already programmed efforts on an existing facility;

- **G** code: Projects which are not yet funded but whose effort is equivalent to that of a ground based instrument on an extremely large telescope ($\simeq$30 M euro, 5-10 years);

- **M** code: Projects whose effort is comparable to an ESA M-class mission ($\simeq$450 M euro);

- **L** code: Flagship projects corresponding to an ESA L-class mission ($\simeq$650 M euro, 10-15 years), or even larger projects (**XL**, $\geq$1 G euro) carried out in worldwide collaboration where ESA's participation would be an L-size mission in itself. A typical time frame for such projects is 20+ years.

3. Producing Synthetic Grids

With this approach, it is possible to produce a grid that provides a synthetic view of the potential of each family of detection methods for different classes of exoplanets (Figure 1). Another way to present the information is to produce a timeline grid (Figure 2) which presents the progression in observation capacity for each family of methods, as a function of the increasing size of the projects.

4. Analysis

In the context of the goal of Blue Dots (spectroscopic characterization of habitable telluric exoplanets), a roadmap should be represented by the path of least effort from the current state of the art (the E-coded boxes, located mostly in the upper left of the grid of Figure 1) to where lies the goal of Blue Dots (SPL *** in the last two columns, or in the last column if one is only concerned with earth twins around solar twins).

One should first note the successes of the discipline and recognize that the E-coded boxes encompass some major achievements: spectroscopic characterization (SPL***) has already been achieved on one class of exoplanets (hot giants), and a few telluric planets, albeit non habitable, have already been identified

	Planet classes					
Methods	Hot Giant Planets (young or hot)	Other Giant Planets (same as in Solar System)	Hot Terrestrial Planets (hot, young or super-Earth)	Telluric Planet in habitable zone around M-dwarfs	Telluric Planet in habitable zone around solar-type stars
Microlensing	N/A	★ E	★ E	N/A	★ M
Transits	★★★ E	☆☆ G	★★★ M	★★ / ☆☆☆ M	★ E
Radial velocities	★★ E	★★ E	★★ E	★★ G	☆☆ G
Astrometry	★★ L	★★ E	★★ M	N/A	★★ L
V imaging / coronagraphy (SAI)	★★★ E	★★★ G	★★★ L	N/A	★★★ L
IR imaging / nulling (MAI)	★★★ G	N/A	★★★ L	★★★ L	★★★ L

Figure 1. Science potential levels (*: statistics, **: identification, ***: spectral characterization, N/A when non applicable) of each family of detection methods for different classes of exoplanets.

Methods	SPL	Project classes			
		Existing	Ground-based	M-class in space	L-XL class in space
Microlensing	★	Giants	N/A	HZ telluric solar stars	N/A
Transits	★★/☆☆☆	Close giants	Other giants	All other terrestrials	N/A
Radial velocities	★★/☆☆	Giant, close telluric	Habitable telluric	N/A	N/A
Astrometry	★★	Giants	N/A	Young telluric planets	Young + telluric in HZ
V imaging / coronagraphy (SAI)	★★★	Young giants	Far giants	N/A	Telluric planets
IR imaging / nulling (MAI)	★★★	N/A	Hot giants	N/A	All others

Figure 2. Timelines for various families of detection methods.

(SPL✶✶). If one considers the youth of our field (the first exoplanets were identified less than 15 years ago), this is quite remarkable indeed, and if the trends continue it would give all reasons to be optimistic for the future.

One should also note that not every method is relevant for every class of object. This means that a roadmap necessarily has to rely on a portfolio of methods in order achieve the Blue Dots goal. Likewise, the timeline grid (Figure 2) shows that no single method has relevant projects at all scale levels. So, while some methods may be extremely productive now, other techniques need to be developed in order to take the relay when they are needed.

"Least effort" in the roadmap means that priority should be given, at each step towards the Blue Dots goal, to the method which enables to achieve it with the projects of the lowest scale.

Applying this principle leads to the conclusion that SPL✶✶ on habitable exoplanets should be carried out preferably from the ground by radial velocity. This is contingent to the acceptance that RV techniques are indeed capable of identifying telluric habitable exoplanets (first cornerstone question mentioned above). Likewise, spectroscopic characterization (SPL✶✶✶) of habitable planets around M stars should be attempted by transit photometry if it is indeed possible (second cornerstone question). If it is, this means that the goal of Blue Dots, at least in a select sample of targets, can be achieved with a medium term (M coded) project. If not, and in any case for the solar type stars, spectroscopic characterization will require a flagship mission (L, and most likely XL sized) which will involve either an infrared interferometer, or an imager in the visible (and obviously preferentially both). It is not possible at this stage to choose between the two options, but work should be continued on collecting the elements that will help make the decision including the following:

- Pursue the study of the exozodi issue to see how it impacts the detectability of habitable exoplanets in both cases;

- Comparative system study for the two concepts in order to be able to compare costs with an equivalent maturity level ;

- Pursue the identification of biomarkers and assess their detectability both in the infrared and the visible/near IR range.

Conclusion

What emerges from the current exercise is that in several instances it might be possible to fast track the science towards the Blue Dots goal with lower scale projects (e.g. RV for detection, transit spectroscopy for characterization) if one selects a subset of targets (biased and incomplete sample) that happens to be more accessible to some techniques. Achieving the same potential level on a complete sample, however, will require projects of a larger scale, including some for which broad international collaboration will be needed. Nevertheless, the fast track approach on a subset of targets seems to be one way to get around the "pathfinder dilemma".

No roadmap is carved in stone forever, and this is especially true with such a vibrant field as exoplanet science. It is likely to be revised every few years as the technology improves and new (possibly unexpected) results pour in. Blue Dots will continue to provide a forum and a framework to discuss these updates.

Acknowledgments. The material produced by Blue Dots comes from the contribution of all of the Blue Dots participants, and most notably its core team and working group coordinators which include Ignasi Ribas, Hans Zinnecker, Sebastian Wolf, Franck Selsis, Charles Cockell, Lisa Kaltenegger, Giovanna Tinetti, Anthony Boccaletti, Marc Ollivier, Jean-Philippe Beaulieu, Nuno Santos, Damien Segransan, Fabien Malbet, Alessandro Sozzeti, Ewa Szuszkiewicz, Helmut Lammer, Christoph Keller, Gerard van Belle, Szymon Gladysz, Chas Beichman, Hiroshi Shibai, Maxim Khodachenko, Gang Zhao, Abhijit Chakraborty, as well as $\simeq$180 other scientists who participate at some level in the initiative.

Exoplanets Forum 2008

W. A. Traub,[1] P. R. Lawson,[1] S. C. Unwin,[1] M. W. Muterspaugh,[2] R. Soummer,[3] W. C. Danchi,[4] P. Hinz,[5] B. S. Gaudi,[6] G. Torres,[7] D. Deming,[4] J. Lazio,[8] and A. Dressler[9]

Abstract. The *Exoplanets Forum 2008* meeting led to a book *Exoplanet Community Report* with about 180 authors from the exoplanet community. This book describes eight technique-oriented methods for detecting and characterizing exoplanets, with an emphasis on space missions. The topics are astrometry, optical imaging, infrared imaging, exozodiacal disks, microlensing, radial velocity, transits, and magnetospheric emission. Several of these techniques have counterparts in proposed space mission studies carried out under the Astrophysics Strategic Mission Concept Studies (ASMCS) program in 2008-9, and the exoplanet-related Astro2010 studies in 2009. The Forum meeting, through the vehicle of its resulting book, provides a snapshot of the science and potential missions for exoplanets during this period, and as such should provide a relevant reference for several years to come.

1. Introduction

Exoplanets Forum 2008 was an open meeting with over 180 participants from the exoplanet science research community in the US and abroad, held on 29-30 May 2008 in Pasadena, California. The purpose of the meeting was to examine eight technique-oriented methods for detection and characterization of exoplanets. Most of these methods require space missions.

In addition to this 2-day face-to-face meeting, the participants also met electronically from April through October 2008, in chapter-writing sub-groups. The resulting book *Exoplanet Community Report* is available in print and online (Lawson 2009).

---

[1] Jet Propulsion Laboratory, California Institute of Technology, M/S 301-451, 4800 Oak Grove Dr., Pasadena, CA, USA

[2] Tennessee State Univ. Nashville, TN, USA

[3] Space Telescope Science Institute, Baltimore, MD, USA

[4] NASA Goddard Space Flight Center, Greenbelt, MD, USA

[5] Univ. of Arizona, Tucson, AZ, USA

[6] The Ohio State Univ., Columbus, OH, USA

[7] Smithsonian Astrophysical Observatory, Cambridge, MA, USA

[8] Naval Research Laboratory, Washington, DC, USA

[9] Carnegie Institution of Washington, Pasadena, CA, USA

Each of the eight chapters summarizes the issues, scientific potential, state-of-the-art, and technology needed for the respective observational technique. The broad-ranging discussion at the Forum allowed the scientists and engineers working in each area to prepare their own chapter in the context of the full scope of many goals and techniques. Thus, inherent in each chapter are cross-references of how the various approaches complement and compete with each other.

Our tempered optimism in each area springs from our belief that exoplanet science is the most exciting new area in astrophysics, for scientists and the public alike. We believe that US leadership in exoplanet science is essential for major advances in this field. We also believe that the public expects no less.

2. Frame of Reference

The *Exoplanets Forum 2008* meeting, organized by the Exoplanet Exploration Program (ExEP), formerly the Navigator Program, was the third in a series on the science and technology of exoplanet missions for Earth-like planets: the first was the Navigator Program Forum 2006, on Earth-Like Exoplanets, at the US Naval Observatory, on 8-9 May 2006; and the second was the Navigator Program Forum 2007, on Small and Mid-Scale Exoplanet Space Missions, at NASA Ames, on 17-18 May 2007. The Forum meetings are designed to provide a thread of continuity over the years, as scientists and agencies continue their technological and funding struggles toward the ultimate goal of finding and characterizing nearby Earth-like planets. This process was initiated most famously by the early proposals to move directly to that goal, the Terrestrial Planet Finder Interferometer (TPF-I) and the Terrestrial Planet Finder Coronagraph (TPF-C), for both of which the scientific support is unwavering, but for which the funding support has not yet materialized.

Meanwhile, committees of scientists continue to meet to consider the science, technology, and political appeal of searching for Earth twins. These include: the ExoPTF(see below); the Exoplanet Analysis Committee (ExoPAG) in the US, starting in 2010; the Astro2010 Astronomy and Astrophysics Decadal Survey, including exoplanets, in the US, from 2008-2010; and the ExoPlanetary Roadmap Advisory Team (EPRAT) in Europe, in 2008-2010.

3. Recommendations from the ExoPTF

The Exoplanet Task Force (ExoPTF) (2008) was a blue-ribbon panel of exoplanet scientists, convened to study the field and make recommendations. The ExoPTF was chartered by the National Science Foundation (NSF) and NASA through the Astronomy and Astrophysics Advisory Committee (AAAC) "to advise NSF and NASA on the future of the ground-based and space-based search for and study of exoplanets, planetary systems, Earth-like planets and habitable environments around other stars."

The final ExoPTF Report became public one week before the Exoplanet Forum meeting, but preliminary versions of the report were available well before then. Therefore, the participants were able to discuss scientific goals and techniques in the context of the ExoPTF Report. The exoplanet community acknowledges and appreciates the substantial efforts of the ExoPTF. A majority of the exoplanet community agrees with the substance of the ExoPTF Report.

4. **Astrometry**

Micro-arcsecond astrometry is the only probe-scale technique that can detect Earth-twins around as many as 100 nearby stars, independent of orbital inclination or exozodi brightness. Because no other technique can do this, the ExoPTF recommended the near-term launch of a space mission for micro-arcsec astrometry. The science rationale for this mission is two-fold. (1) If Earth-twins are detectable, then many other planets, with larger astrometric signatures, will be detectable, so the Earth-twin goal stands as a useful benchmark of instrument performance. (2) The detection of potential Earth-twins is of great value as existence proofs of planets, as design optimizers for follow-up characterization missions, and as search optimizers. Combining ground-based radial velocity (RV) with space-based astrometry is a powerful detection strategy, because RV aids astrometry in extracting masses and orbits from the extremes of short- and long-period planet signatures in multi-planet systems. Once nearby planets are identified, and their masses and orbits known, then follow-up can be planned to measure exozodi brightness and to characterize the planets through imaging observations or transits. The technology for such a mission has been developed by NASA and is now complete. The astrometry group strongly endorses the major finding of the ExoPTF that astrometry is a critical part of the search for nearby Earth-like planets.

5. **Optical Imaging**

A visible-wavelength coronagraph, internal or external, is the only technique capable of observing the spectrum of a benchmark Earth-twin deep enough in its atmosphere to probe the lower troposphere and surface to assess habitability and to search for direct signs of life. A small coronagraph could measure the brightness, colors, and spectra of large planets outside the habitable zone (HZ) around nearby stars, possibly a few terrestrial planets in the HZ, and the zodi brightness distribution in the same regions. A large coronagraph can measure closer to the parent star, and it will have better spatial resolution, so it can characterize more giant as well as terrestrial planets, including more Earth-twins in the HZ. If either type of coronagraph is preceded by a successful astrometry mission, then the coronagraph design and observing efficiency could be optimized. Coronagraph technology is well advanced, and the technological requirements are well understood. The key technologies for a large-scale mission are identical to those for a small-scale one, so all such work on a small mission is a sound investment for a future large one.

6. **Infrared Imaging**

A mid-infrared mission would enable the detection of biosignatures of Earth-like exoplanets around more than 150 nearby stars. The mid-infrared spectral region is attractive for characterizing exoplanets because contrast with the parent star brightness is more favorable than in the visible (10 million vs. 10 billion), and because mid-infrared light probes deep into a planet's troposphere. The mid-infrared offers access to several strong molecular features that are key signs of life, and also provides a measure of the effective temperature and size of a planet. Taken together, an infrared mission plus a visible one would provide a nearly full picture of a planet, including signs of life; with mass from an astrometric mission, we would have a virtually complete picture. The technology needed for a large formation-flying mission is similar to that for a small connected-element one (e.g., cryogenics and detectors), with the addition of formation-flying technology. The technology is now in hand to implement a probe-scale mission; starlight suppression has even been demonstrated to meet the requirements of a flagship mission.

7. Exozodiacal Disks

From the viewpoint of direct imaging of exoplanets in the visible or infrared, exozodi dust disks can be both good and bad. An exozodi disk is good if it has structures (cleared regions or resonant clumps) that suggest the gravitational presence of planets, however it is bad if the dust fills the instrumental field of view with brightness that swamps the signal from a planet. Unfortunately, it takes very little dust to compete with the light from a planet: an Earth-twin signal is roughly equal to a 0.1-AU patch of solar system-twin zodi, in the visible or infrared. Current one-sigma limits of detection, in units of the solar system brightness, are a few hundred using the Spitzer Space Telescope, about one hundred with the Keck Interferometer (KI), and about 10 expected from the Large Binocular Telescope Interferometer (LBTI). A small coronagraph or small interferometer in space could reach the sensitivity required to detect the glow at the level of our own solar system.

8. Microlensing

Gravitational focusing of light from a distant star by a closer star (and planet system) is known as microlensing, and has already produced several dramatic exoplanet detections. The statistics of these detections suggest that low-mass planets are more common than high-mass ones, an extremely important result in the context of the search for Earth analogs. Microlensing is typically sensitive to planets around distant (several kpc) stars, and is therefore most valuable as a statistical sampling tool. It is not an effective method for detecting planets around nearby stars. Owing to the scarceness of microlensing events, more ground-based telescopes are needed; a space telescope dedicated to the purpose would substantially improve the statistics. An attractive feature of this approach is that no new technology is needed.

9. Radial Velocity

Radial velocity (RV) detects planets by measuring a parent star's periodic line-of-sight velocity change due to its orbit around a common center of mass, using the Doppler shift of the star's spectral lines. The method is sensitive to walking-speed motions of a star, and promises to reach crawling-speed sensitivity in a few years. The RV method is responsible for most of the 400-plus exoplanets detected to date. Ground-based telescopes are adequate for this task, although more telescopes, and additional stable spectrographs, would be very welcome. The main new technology is the laser comb spectrum, which can provide many more calibration points across a spectrum than traditional laboratory sources. In combination with a space microarcsec astrometry mission, RV holds good promise for helping unscramble long-period multi-planet signals, and thus detecting Earth-twins.

10. Transits

The transit technique refers to observations of decreased brightness of a star system as a planet passes in front of or behind the star. Specifically it includes the spectral information that comes from starlight transmitted through the planet's atmospheric annulus during a primary transit and reflected or radiated light from the planet's full disk during a secondary transit. It can also include the phase effect from reflected or radiated light during its orbital motion. The method works in the visible as well as thermal infrared. The transit method gives us the planet's diameter, and the absorption spectrum of its upper atmosphere. The transit technique is especially useful for short-period planets because these are close to their star and have a greater chance of transiting as well as more transits. The full power of the transit technique is limited to a relatively small fraction of planetary systems, but similar combined-light techniques can be applied to many close-in planets, even those that do not transit. Transits are also most useful for giant planets, owing to their large opaque cross sections and relatively large atmospheric annulus areas; transits of Earth-sized planets across solar-type stars are quite weak, but super-Earths transiting M dwarf stars, for example, can produce signals large enough to be detected even from the ground with small telescopes. The Kepler mission will be especially important for generating a measure of the frequency of Earth-size planets at separations out to the HZ, and the frequency of larger-size planets. The main technology need is for greater stability in ground and space instruments.

11. Magnetospheric Emission

Planets in the solar system produce radio-frequency waves by radiation from electrons spiraling along the planet's magnetic field lines, and it is anticipated that exoplanets will broadcast similar radio waves. The radio luminosity is proportional to the solar wind power incident on a planet, so is expected to be largest for active stars. Since these stars are difficult to observe with astrometry or RV, magnetospheric emissions may offer a path to detecting planets in these systems. There are a number of ground-based instruments under construction that promise significant improvements in sensitivity. The main technology needs are the development of improved radio frequency interference (RFI) avoidance and excision. Ground observations will ultimately be limited by a combination of the Earth's ionosphere and RFI, however antennas on the far side of the Moon could have large areas and very low RFI.

12. Overarching Science Goals

The goal of exoplanet science is to extend our picture of the formation of planets and the conditions that can lead to life on planets. This is a relatively new scientific endeavor, but it is clearly the logical next step for astrophysics, given our natural curiosity about the origins of life on Earth. We already know that more than 400 relatively massive planets exist around nearby stars; these were discovered using RV, microlensing, transits, direct imaging, and astrometry. There is broad support in the community for continuing with these techniques. Beyond these, the Forum discussion highlighted some of the major steps that the exoplanet community hopes to take within the next decade or two, likely in this order: (1) Measurement of the frequency of low-mass exoplanets, i.e., the mass distribution function. (2) Searches for all types of exoplanets, from terrestrial mass to gas giants, around nearby stars. Probe-scale missions can likely fulfill this need. A leading technique appears to be a space-based astrometric mission, working with ground-based RV. (3) Measurements of exozodi disk brightnesses to levels approaching that of the solar system. Ground-based interferometers will pioneer this area, but a space mission is needed to reach faint (solar system level) exozodis. (4) Characterization and search for signs of life on nearby exoplanets, especially Earth-twins. The only viable techniques are coronagraphs and interferometers in space. Probe-scale missions can explore this area, but flagship-scale missions will be absolutely necessary for characterizing nearby Earth-twins.

13. Summary Statement

The exoplanet community's top priority is that a line of probe-class missions for exoplanets be established, leading to a flagship mission at the earliest opportunity.

Acknowledgments. Part of the research described in this paper was carried out at the Jet Propulsion Laboratory, California Institute of Technology, under a contract with the National Aeronautics and Space Administration. ©2009. All rights reserved.

References

Lawson, P.R., Traub, W.A., & Unwin, S.C. 2009, editors, Exoplanet Community Report, JPL Publication 09-3 3/09, 252 pp., available in print, and online at http://exep.jpl.nasa.gov/documents/ExoplanetCommunityReport.pdf

Session III

Working Group Reviews

Multi-Aperture Imaging of Extrasolar Planetary Systems

O. Absil
Département d'Astrophysique, Géophysique & Océanographie, Université de Liège, 17 Allée du Six Août, Sart Tilman, Belgium

Abstract. In this paper, we review the various ways in which an infrared stellar interferometer can be used to perform direct detection of extrasolar planetary systems. We first review the techniques based on classical stellar interferometry, where (complex) visibilities are measured, and then describe how higher dynamic ranges can be achieved with nulling interferometry. The application of nulling interferometry to the study of exozodiacal discs and extrasolar planets is then discussed and illustrated with a few examples.

1. Introduction

When considering the direct imaging of extrasolar planetary systems, one is faced with two main challenges: the small angular separation and the high contrast between exoplanets and their host stars. If the goal is to characterise the mid-infrared emission of exoplanets in the habitable zone of nearby stars, the typical angular resolution of 50 mas required to resolve a Sun-Earth system at 20 pc leads to an impractical aperture size of 40 m. The only option is then stellar interferometry, which synthesises the resolving power of a larger aperture by using multiple telescopes separated by an appropriate distance called the interferometric baseline B. The associated angular resolution equals $\lambda/2B$ (to be compare with λ/D for a single aperture of diameter D). The need for interferometry to image planetary systems in the infrared was recognised already in the late 70s (Bracewell 1978). Its application to the search and characterisation of habitable worlds was proposed 15 years later by Léger et al. (1996).

Reaching the appropriate angular resolution to separate the signals of the planet from that of its host star is only half of the solution, and the high dynamic range still needs to be addressed. In this review, we discuss the various techniques that can be implemented with infrared interferometers to reach the required dynamic range to image extrasolar planetary systems of various kinds (from dusty discs and hot giant planets down to Earth-like planets).

2. Classical Stellar Interferometry

The main quantity that is measured by a (two-telescope) stellar interferometer is the complex visibility of the interference fringes recorded at the detector. The complex visibility of a high-contrast binary system, given by the sum of the complex visibilities of the two objects, can be approximated as follows:

$$\mathcal{V} \simeq (1-r)\mathcal{V}_\star + r\exp(i2\pi(u\Delta\alpha + v\Delta\beta)) \quad (1)$$

with $\mathcal{V}_\star$ the complex visibility of the central star, u and v the Fourier plane coordinates associated to the baseline vector **B** ($u = B_x/\lambda$, $v = B_y/\lambda$), r the flux ratio between the planet and the star, and α and β the cartesian angular coordinates of the planet relative to the star in the sky plane. In this expression, we have assumed that $r \ll 1$ and that the amplitude of the complex visibility of the sole planet is equal to 1 (completely unresolved object). This expression shows that the presence of a faint companion around the target star affects both the amplitude $V = |\mathcal{V}|$ and the phase $\phi = \arg(\mathcal{V})$ of the interference fringes, as follows:

$$V^2 \simeq V_\star^2 - 2r(V_\star - \cos(2\pi(u\Delta\alpha + v\Delta\beta))) \quad (2)$$

$$\phi \simeq \frac{r\sin(2\pi(u\Delta\alpha + v\Delta\beta))}{V_\star} \quad (3)$$

where we have assumed that the star is point-symmetric (so that $\mathcal{V}_\star$ is real). In both cases, the magnitude of the planet-related effect is directly proportional to the flux ratio between the planet and the star.

With the existing, state-of-the-art ground-based interferometers, the typical accuracy that can be reached on the squared visibility is of the order of 10^{-2} on bright unresolved stars (see e.g., Kervella et al. 2004). This level of performance does not allow extrasolar planets to be detected, even in the most favourable cases. Therefore, planet detection techniques mostly rely on phase measurements. The main problem with interferometric phases is that they are corrupted by atmospheric turbulence: the random piston between two apertures induces a time-varying phase that adds to the astrophysical phase. Furthermore, making an absolute measurement of the fringe phase would require a perfect knowledge of the zero optical path difference position in the instrument, which is impractical. Several techniques can however be used to retrieve (part of) the phase information. The main three of them are described below.

2.1. Wavelength-differential Phase

When dispersing the interferometric fringes into several spectral channels, the fringe phase can be measured in one channel with respect to any other channel. Since the various wavelengths have travelled through the same atmosphere and through the same optical train, their phases should be the same unless the astrophysical object has a wavelength-dependent phase. In the case of a star-planet system, the phase is proportional to the flux ratio r, which changes with wavelength since the two bodies have different temperatures and various spectral signatures. By measuring the phase variations across the spectrum, one can thus theoretically reconstruct the variations of the star-planet contrast relative to a reference wavelength.

This method, which was thoroughly described by Vannier et al. (2006), is however not immune to chromatic effects in the Earth atmosphere and in the instrument. In particular, the compensation of the delay between the beams collected by two telescopes is generally done in the ambient (moist) air, whose refraction index changes with wavelength, thereby creating a wavelength-dependent phase. The magnitude of this effect can be predicted if the refraction index of air is modelled with a good enough accuracy. However, the fluctuations of the column density of water vapour ("water vapour seeing") above the two telescopes creates a random chromatic component in the phase difference between the two beams, which is much more difficult to correct.

The most advanced use of this technique for extrasolar planet detection has been performed with the AMBER instrument at the VLTI (see e.g., Millour et al. 2008). In practice, chromatic effects generally prevent the different phase to be measured with an accuracy better than a few 0.01 radians, even when using advanced calibration techniques such as beam commutation. Flux ratios larger than 100:1 are therefore currently out of reach.

2.2. Phase Referencing

Phase referencing consists in observing simultaneously two stars located within the same isoplanatic patch so that their phases can be measured relative to each other. This requires a high precision internal metrology system in order to monitor the non-common path differences between the two detected fringe packets. The final result is an accurate measurement of the relative positions of the two objects on the sky plane at the observing wavelength. If one of the two stars is accompanied by an exoplanet, the photocentre of the system will be slightly shifted towards the planet. Such a shift could potentially be detected, if the position of the sole star was known with a high enough accuracy. This is generally not the case, and phase referencing is therefore not appropriate to directly image an extrasolar planet. On the other hand, by measuring the time evolution of the position of a star with respect to a fixed object, one can evaluate the astrometric shift due the gravitational influence of planets around this star. This method has already been used at the Palomar Testbed Interferometer to search for planets in close binary system (Lane et al. 2004) and will soon be used at the VLTI with the PRIMA instrument (Launhardt et al. 2008).

2.3. Closure Phase

When using three telescopes (or more) at a time, one can exploit the closure phase Φ, which is defined as the argument of the triple product of the complex visibilities measured on a closed triangle of baselines:

$$\Phi = \arg(V_{12}e^{i\phi_{12}}V_{23}e^{i\phi_{23}}V_{31}e^{i\phi_{31}}) \quad (4)$$
$$= \phi_{12} + \phi_{23} + \phi_{31} \quad (5)$$

By construction, this quantity is insensitive to telescope-specific phases, because an additional phase φ_i on a given telescope i appears positively in $\phi_{ij} = (\phi_i + \varphi_i) - \phi_j$ and negatively in $\phi_{ki} = \phi_k - (\phi_i + \varphi_i)$. The intrinsic closure phase of an astrophysical object can thus be measured by adding the phases of the fringes detected on the three baselines of a closed triangle, without being affected by atmospheric or instrument effects (including the chromatic effects that spoil differential phase measurements). Furthermore, one can show that the closure phase differs from zero only for targets that are not point-symmetric, which is precisely the case of high contrast binary systems. In this case, the closure phase can be written as follows assuming the host star to be unresolved:

$$\Phi \simeq r(\sin(2\pi \mathbf{u_{12}} \cdot \mathbf{\Delta}) + \sin(2\pi \mathbf{u_{23}} \cdot \mathbf{\Delta}) + \sin(2\pi \mathbf{u_{31}} \cdot \mathbf{\Delta})) \quad (6)$$

with $\mathbf{u_{ij}} = (u_{ij}, v_{ij})$ and $\mathbf{\Delta} = (\Delta\alpha, \Delta\beta)$. The closure phase is thus proportional to the flux ratio r, and its sensitivity to faint companions can potentially be significantly increased if the host star is resolved by at least one baseline of the triangle, as described by Chelli et al. (2009). Exoplanet host stars are however generally too small to be resolved with the current generation of interferometers.

Closure phase measurements are nowadays performed routinely by several interferometers around the world. Attempts to detect extrasolar planets with this method have been carried out with the MIRC instrument at the CHARA array (Zhao et al. 2008), and with the AMBER instrument at the VLTI. In the best cases, the closure phase accuracy reaches $\sim 0.1°$, allowing low-mass companions with contrast up to 600:1 to be detected at 1σ. It is expected that improvements in the accuracy of closure phase measurements could soon provide the first direct detection of a hot extrasolar giant planet with interferometry. The combination of spectral dispersion with closure phase measurements (as done in AMBER) could potentially allow the detection of molecular bands in the exoplanet's atmosphere, such as CH_4 or CO features in the infrared K band.

3. Nulling Interferometry

The main limitation to the capabilities of stellar interferometry in the detection of extrasolar planets resides in the high dynamic range that must be achieved. The previous section has shown that contrasts of 1000:1 and higher are difficult to reach with the current generation of ground-based arrays. One avenue to improve the dynamic range of interferometers is to use the undulatory nature of light to perform a destructive interference on the blinding stellar light. When overlapping the light beams collected by two telescopes on a balanced beam splitter, one can tune their respective phases by means of phase-shifting devices so that all the on-axis stellar light is sent to only one of the two complementary outputs of the beam splitter (see Figure 1). Such a configuration happens when a phase shift of π radians is maintained between the two input beams. This is in contrast with classical interferometry, where the optical path difference between the beams takes a range of values (either in a pupil plane or an image plane) in order to record a complete interferogram, or at least a part of it.

A two-telescope nulling interferometer is characterised by its transmission map $T_\lambda(\theta, \alpha)$, displayed in Figure 1, which results from the fringe pattern produced by the interference between the two beams:

$$T_\lambda(\theta, \alpha) = 2P_\lambda(\theta, \alpha) \sin^2\left(\pi \frac{B\theta}{\lambda} \cos\alpha\right) \qquad (7)$$

where θ and α are respectively the radial and polar angular coordinates with respect to the optical axis, B the interferometer baseline (whose direction defines $\alpha = 0$), D the telescope diameter, λ the wavelength, and where $P_\lambda(\theta, \alpha)$ represents a field-of-view taper function resulting from the size of the collecting apertures and from the particular design of the instrument. For small values of θ ($\ll \lambda/B$), one can see that the interferometer transmission is proportional to θ^2, so that the central part of the dark fringe is parabolic.

The final detection can be done either in a pupil or in an image plane. In the former case, a single-pixel detector is sufficient to record the total flux in the output pupil, emanating from all the sources in the diffraction-limited field-of-view. In the latter, an image similar to that of a single telescope is formed, except

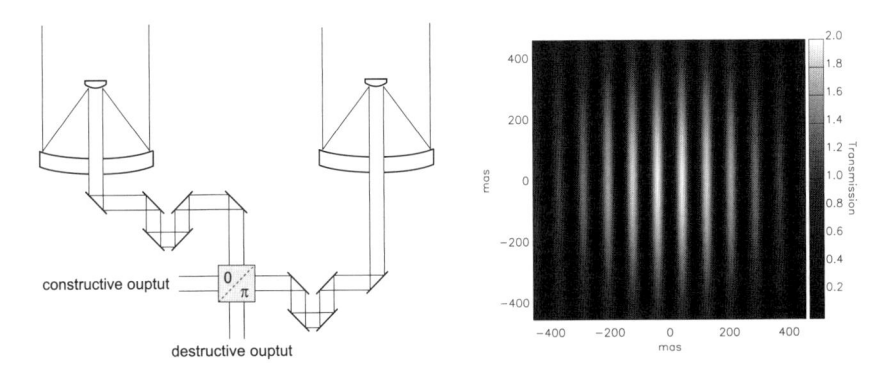

Figure 1. *Left*: Principle of a two-telescope nulling interferometer. The beam-combining system produces a destructive interference by applying a π phase shift to one of the two input beams and by superposing them in a coaxial way. *Right*: Associated transmission map, showing the parts of the field that are transmitted (bright stripes) and those that are blocked (dark stripes, including the central dark fringe) by the interference process.

that the relative contribution of each source is affected by the interferometer's intensity response at its location. In any case, no fringes are formed, and the final output generally consists in a single value: the total intensity in the diffraction-limited field-of-view. The final output flux then writes:

$$F(\lambda) = \int_\alpha \int_\theta 2P_\lambda(\theta, \alpha) \left(\sin^2(\pi \frac{B\theta}{\lambda} \cos\alpha) I_c(\theta, \alpha, \lambda) + I_i(\theta, \alpha, \lambda) \right) \theta \, d\theta \, d\alpha \quad (8)$$

with $I_c(\theta, \alpha, \lambda)$ and $I_i(\theta, \alpha, \lambda)$ the intensity distributions of the coherent and incoherent sources located within the fied-of-view. The nulling ratio N is then defined as the fraction of stellar light that makes it to the destructive output due to its finite angular diameter $\theta_\star$. Assuming the stellar angular diameter to be small compared to the angular resolution λ/B of the interferometer, it can be shown that the nulling ratio reduces to:

$$N \simeq \frac{\pi^2}{16} \left(\frac{B\theta_\star}{\lambda} \right)^2 \quad (9)$$

In principle, nulling interferometry can be generalised to any number of telescopes, provided that the phases introduced in each beam result in a destructive interference on the optical axis of the interferometer (i.e., for a zero optical path difference). In particular, using more that two telescopes at the same time allows deeper nulls to be achieved, with the central transmission proportional to higher powers of θ than the θ^2 proposed by a two-telescope interferometer.

3.1. Two-telescope Nullers: Exozodi Finders

Because all the sources located within the coherent field-of-view of a single aperture contribute (with various transmission factors) to the destructive output of the interferometer, a simple two-telescope nulling interferometer can only detect the sources that are producing the largest contribution. Stellar leakage, related to the finite size of the stellar photosphere, is generally one of the main contributors. However, its expression is analytical (Equation 9), so that its contribution can be subtracted *a posteriori* if the stellar diameter θ_* is known with a good enough accuracy. Another major contribution to the flux detected at the destructive output is the incoherent background, produced by the thermal emission of the Earth atmosphere and of the instrument itself. Since this emission is uniform across the field-of-view, its contribution can be suppressed e.g., by rotating the baseline of the interferometer and subtracting two measurements.

Once the star and background contributions have been subtracted, the remaining contribution comes from the immediate environment of the star, i.e., the planetary system in our case. The largest source of infrared emission in extrasolar planetary systems generally comes from the cloud of interplanetary dust, which represents the primordial material from which planets are formed in the case of protoplanetary discs, or results from the collisional activity of larger bodies in the case of exozodiacal discs (the extrasolar equivalent of the zodiacal disc of dust grains surrounding our Sun). Owing to their brightness, such discs are generally the main targets of two-telescope nullers. Hot giant planets could also be detected in a few cases.

Until now, only two ground-based nulling instruments have been producing scientific results, in particular in the field of extrasolar planetary systems. The BLINC nulling camera has been used since 1998 with two sub-apertures of the 6.5-m MMT at Mount Hopkins, Arizona. The observations of bright main sequence stars in the N band have resulted in 3σ upper limits on the dust density of exozodiacal discs ranging between 220 and 10,000 zodi (1 zodi being the luminosity of our own zodiacal cloud as seen from outside the solar system, see Liu et al. 2009). The second nulling instrument is the Keck Interferometer Nuller (KIN, Colavita et al. 2009), which is combining the lights of the two 10-m Keck telescopes in the N band on top of Mauna Kea, Hawaii. In order to subtract the background from the scientific measurements, this instrument divides the pupils of the Keck telescopes into two halves, produces two (equivalent) nulled outputs on the 85-m Keck baseline using two pairs of half pupils, and then combines the two nulled outputs with a time-varying phase shift in order to modulate the coherent signal contributing to the nulled outputs. Synchronous demodulation can the be used to remove the incoherent background. This process is similar to the phase chopping technique discussed in the next section and illustrated in Figure 2. KIN reaches a typical sensitivity of 300 zodi for nearby main sequence stars, and has been performing a survey for exozodiacal discs on a sample of about 40 stars. The first results are currently being published (see first paper by Stark et al. 2009).

Several projects of ground-based or space-based two-telescope nulling interferometers are currently being considered. The ALADDIN project (Absil et al. 2007) aims at performing deep nulling in the L band on the high Antarctic plateau, where the atmospheric conditions (turbulence, background, transparency) are well suited to such observations. Pegase and FKSI (Defrère et al. 2008) are two space mission projects with similar goals and designs, except that Pegase is based on a free flying flotilla of three spacecrafts, while FKSI relies on a structurally connected interferometer with a 12-m baseline. The expected performance of these projects are compared to the KIN sensitivity in Figure 3.

Multi-Aperture Imaging 35

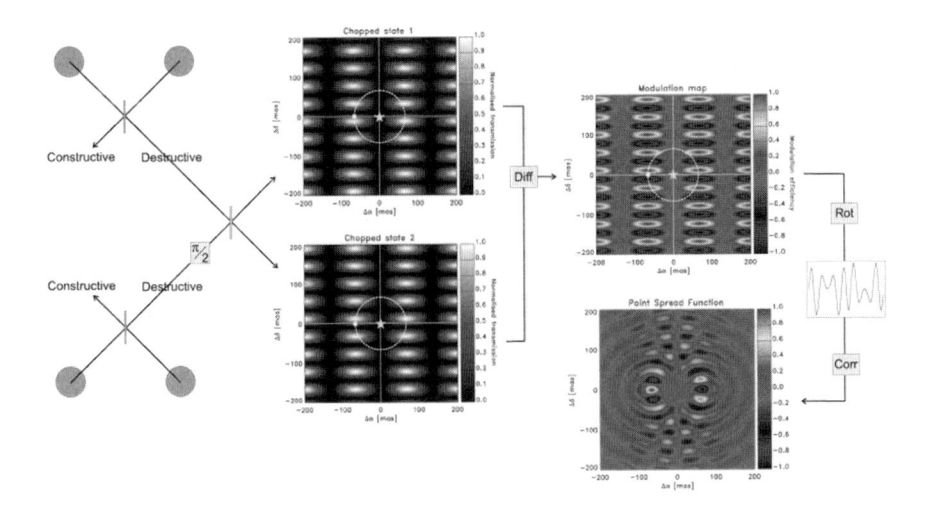

Figure 2. Principle of phase chopping, illustrated for the X-array configuration. Combining the beams with different phases produces two conjugated transmission maps (or chop states), which are used to produce the chopped response. Array rotation then locates the planet by cross-correlation of the modulated chopped signal with a template function.

3.2. Multi-telescope Nullers: Planet Imagers

Because planets (and in particular Earth-like planets) are generally not the brightest component in extrasolar planetary systems, a simple two-telescope nulling interferometer is mostly inappropriate for their detection. An additional subtraction technique must be used to get rid of the exozodiacal disc. A few techniques have been considered so far, mostly relying on the fact that the intensity distribution of the exozodi is point-symmetric while the planet is an off-axis point-like source. The most promising of these techniques is phase chopping. Its principle is to synthesise two different transmission maps with the same telescope array, by applying different phase shifts in the beam combination process. In order to remove all point-symmetric sources in the field-of-view (star, background, exozodiacal disk) while keeping the planetary signal, the two maps must be linked to each other by point symmetry, but have no point symmetry themselves. This phase chopping technique can be implemented in various ways, and is now an essential part of the future space-based planet finding missions such as the Darwin and TPF-I projects that have been considered by ESA and NASA respectively during the last decade. Based on a free-flying array of four 2-m class telescopes, these missions are capable of detecting and spectroscopically characterising the thermal emission from Earth-like planets located in the habitable zone of nearby solar-type stars. The interested reader is referred to the recent review by Cockell et al. (2009) for a detailed description of these missions, and to Defrère et al. (2010) for a description of the mission performance and of the influence of exozodiacal light on this performance.

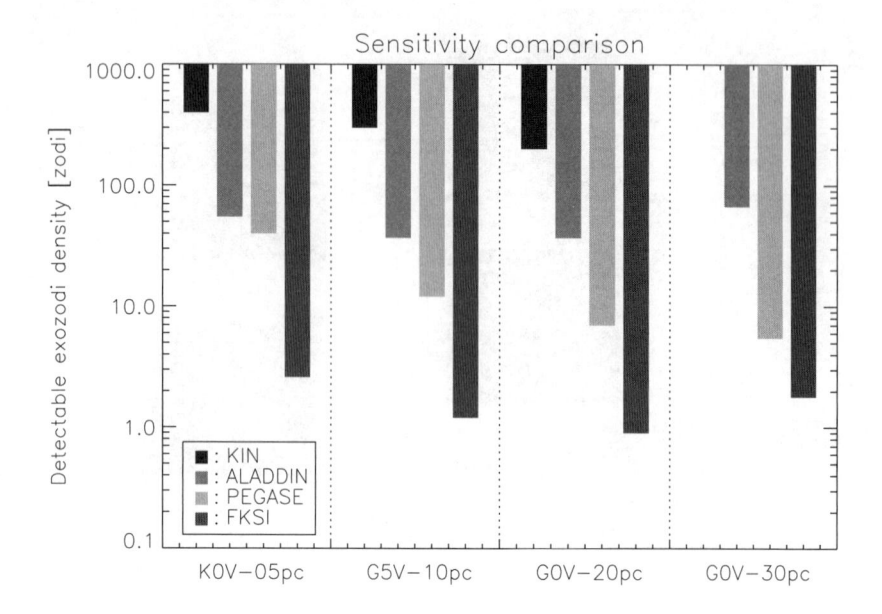

Figure 3. Measured or simulated performance of various ground- and space-based nulling interferometers, in terms of the smallest exozodiacal dust density that can be detected at 3σ around various solar-type stars.

Acknowledgments. The author acknowledges the financial support from an F.R.S.-FNRS postdoctoral fellowship, and from the Communauté Française de Belgique (ARC – Académie universitaire Wallonie-Europe).

References

Absil, O., Coudé du Foresto, V., Barillot, M., & Swain, M. 2007, A&A, 475, 1185
Bracewell, R. 1978, Nat, 274, 780
Chelli, A., Duvert, G., Malbet, F., & Kern, P. 2009, A&A, 498, 321
Cockell, C., Herbst, T., Léger, A., et al. 2009, Exp. Astron., 23, 435
Colavita, M., Serabyn, E., Millan-Gabet, R., et al. 2009, PASP, 121, 1120
Defrère, D., Absil, O., Coudé du Foresto, V., Danchi, W., & den Hartog, R. 2008, A&A, 490, 435
Defrère, D., Absil, O., den Hartog, R., Hanot, C., & Stark, C. 2010, A&A, 509, 9
Kervella, P., Ségransan, D., & Coudé du Foresto, V. 2004, A&A, 425, 1161
Lane, B. & Muterspaugh, M. 2004, ApJ, 601, 1129
Launhardt, R., Queloz, D., Henning, T., et al. 2008, SPIE, 7013, 2I
Léger, A., Mariotti, J.-M., Mennesson, B., et al. 1996, Icarus, 123, 249
Liu, W., Hinz, P., Hoffman, W., Brusa, G., Miller, D. & Kenworthy, A. 2009, ApJ, 693, 1500
Millour, F., Petrov, R., Vannier, M. & Kraus, S. 2008, SPIE, 7013, 1G
Stark, C., Kuchner, M., Traub, W., Monnier, J. D., et al. 2009, ApJ, 703, 1188
Vannier, M., Petrov, R., Lopez, B. & Millour, F. 2006, MNRAS, 367, 825
Zhao, M., Monnier, J. D., ten Brummelaar, T., Pedretti, E. & Thureau, N. D. 2008, SPIE, 7013, 1K

Pathways Towards Habitable Planets
ASP Conference Series, Vol. 430, 2010
Vincent Coudé du Foresto, Dawn M. Gelino, and Ignasi Ribas, eds.

Scientific Output of Single Aperture Imaging of Exoplanets

R. Gratton,[1] M. Bonavita,[1] S. Desidera,[1] A. Boccaletti,[2] M. Kasper,[3] and F. Kerber[3]

[1] INAF-Osservatorio Astronomico di Padova, Vicolo dell'Osservatorio 5, Padova Italy

[2] LESIA, Observatoire de Paris-Meudon, Meudon, France

[3] European Southern Observatory, Karl Schwarzschildstrasse 2, Garching bei München, Germany

Abstract. We discuss the scientific output expected from single aperture imaging of exoplanets by comparing expected detections from different ground-based and space instruments, either in construction or proposed for the future. Images of a few young giant planets have been already obtained, and many more are expected with planet finders on 8 meter class telescopes or JWST. Direct detections of Neptune-like and rocky planets require a new generation of instruments, either specialized satellites or extremely large telescopes from the ground. We emphasize that ELT's, and in particular the E-ELT, may allow the characterization of planets detectable with indirect techniques, like RV and astrometry, allowing an important step forward in planetary science.

1. Introduction

Direct imaging is not only the technique that most easily captures the attention of the general public, but it also provides a wealth of information about exoplanets. First, it might be the fastest way to detect them: a single image is in principle enough, although practically confirmation of candidates might be not easy. Even multiple systems, which require a huge number of observations from indirect techniques, can be solved at once by imaging. Second, images taken at various epochs might provide the planet orbit, if short enough. Finally and most importantly, detailed characterization of planets is possible, using photometric, spectroscopic or polarimetric techniques.

However, direct imaging of exoplanets is very difficult, due to the very large contrast between the star and the planet, and the very small apparent separation: it was then for a long time a dream. However, this is not true anymore. Table 1 lists the planets already discovered by direct imaging (from the Extrasolar Encyclopaedia[1]), along with some of their properties. As can be seen, detection was made possible because some of the following circumstances (or all of them) hold: (i) the planet-star mass ratio is small (the contrast is

---
[1] http://exoplanet.eu/

Table 1. Exoplanets discovered by direct imaging.

Planet	Planet Mass (M_J)	Star Mass ($M_\odot$)	Separation (AU)	Separation (arcsec)	Age (Myr)
2M1207 b	4	0.025	46	0.88	8
CT Cha b	17		440	2.67	
USco CTIO 108 b	14	0.057	670	4.62	5.5
AB Pic b	13.5		275	6.03	30
GQ Lup b	21.5	0.7	103	0.74	1
β Pic b	8	1.8	8	0.41	12
HR8799 b	7	1.5	68	1.73	60
HR8799 c	10	1.5	38	0.96	60
HR8799 d	10	1.5	24	0.61	60
Fomalhaut b	<3	2.06	115	14.94	200

lower); (ii) young age (the planet is brighter); or (iii) large physical/angular separation between the planet and the star.

The concerns related to the large contrast and small separation determine the characteristics of the instrumentation required to image exoplanets. In the optical and near infrared (NIR), all but the very youngest and massive planets shine due to reflected starlight. Contrast is expected to be very large: e.g. the most luminous solar system planets (Venus, Jupiter, and Earth) have luminosities of about 10^{-9} that of the Sun or less. The contrast is much more favourable in the mid-IR, where thermal emission from the planets may be 10^{-4} (or even more, for very young and massive planets) that of the star. However, background is a serious issue at these wavelengths, making these observations as difficult as those at shorter wavelengths.

2. Projects for the Next Decade

Several instruments specifically designed to image extra-solar planets are under construction or in design. These instruments will provide a wealth of new data about exoplanets, and ultimately we hope that direct imaging will allow us to answer fundamental questions about planetary systems. In this talk, we will review some of the expected output from these instruments, focusing on only those instruments that use single apertures; instruments based on interferometric approaches will be discussed in (Absil 2010). From the various existing projects, we selected only a few which appeared to us more likely to be realized in the mid-term for our analysis.

(1) A number of planet finders are already available or under construction for ground-based 8m telescopes. These include Hi-Ciao at Subaru (Tamara et al. 2006), SPHERE at VLT (Beuzit et al. 2008), and GPI at Gemini South (Macintosh et al. 2008), the two last being more ambitious and likely more performing. Both of them are foreseen to be operative in 2011. Designs are similar, including an Extreme AO systems, a coronagraphic module, and an Integral Field Spectrograph (IFS) working in the Y-J-H bands. SPHERE will also include IRDIS, which allows differential imaging over a wider field of view and long slit spectroscopy, and a high precision differential polarimeter (ZIMPOL). Performances

of SPHERE and GPI are expected to be similar, although GPI should reach a deeper contrast (due to a further stage in the control of static aberrations), while SPHERE should allow observations of slightly fainter sources (being still limited at targets with $I < 10$).

(2) JWST, expected to launch in 2014, will provide facilities for imaging exosolar planets in the NIR (< 5 μm: NIRCAM/TFI[2]) and in the mid IR (> 5 μm: MIRI[3]). Due to the smaller telescope size and lesser control of the wavefront errors, NIRCAM will not allow higher contrasts and smaller inner field of view with respect to SPHERE and GPI; however, it will not be limited to bright targets, and will then have an important niche for nearby extreme M stars. MIRI will open a new window, outperforming by far existing ground-based mid-IR imagers.

(3) A third group of instruments includes 1.5 m class space coronagraphs, forerunners of the much more ambitious TPF. There are many projects proposed (PECO: Guyon et al. (2008); EPIC: Clampin et al. (2006), Lyon et al. (2008); ACCESS: Trauger et al. (2008); SEE-COAST: Schneider et al. (2009)). None of these projects are presently funded, but if approved they can likely become operative before the end of the next decade. Without entering into the details of this competition, expected performances from these various instruments are quite similar each other, defining the same science niche.

(4) Lastly, there are three projects for extremely large telescopes that are being actively pursued by the North American (TMT and GMT) and European (E-ELT) astronomical communities. While not yet approved, these programs appear feasible in 8-10 years from now. All these projects consider instruments for the direct imaging of exoplanets, either in the near-IR (EPICS at E-ELT: Kaspar et al. (2008); PFI at TMT, HRCAM at GMT) or in the mid-IR (METIS at E-ELT: Brandl et al. (2008); MIRES at TMT; MIISE at GMT). E-ELT, with its 42 m diameter is the most ambitious among these projects.

Table 2 lists the main properties of these projects.

3. Which Planets Can be Observed in the Next Decade?

We used the MESS code (Bonavita et al. 2009) to analyze the discovery niches for each of the instruments considered in the previous section. MESS is a Monte Carlo code that compares expected properties of a population of exoplanets with detection limits for different instruments, using various techniques (imaging, RV, astrometry, transits). To this purpose, MESS considers stellar parameters (mass, distance, age, etc.) from samples of real stars, in the present case either a sample of about 1200 young stars (age < 500 Myr, distance < 100 pc), or a sample of about 600 stars (distance < 20 pc). These samples were prepared for the science case of SPHERE and are then limited in magnitude at $I < 10$. For each star in these samples, a number of planets (5 in our application) characterized by a mass and period are randomly generated using the distributions from Cumming et al. (2008), generated following the observed RV distribution, extrapolated

---

[2]http://ircamera.as.arizona.edu/nircam/

[3]http://www.roe.ac.uk/ukatc/consortium/miri/index.html

Table 2. Instruments for direct imaging of exoplanets considered in this analysis.

Instrument	Contrast	Wavelength (μm)	IWA (arcsec)	Year
\multicolumn{5}{c}{8 m ground-based telescopes}				
VLT-SPHERE Gemini-GPI	10^{-7}	0.9-1.7	0.08	2011
\multicolumn{5}{c}{JWST}				
NIRCAM	10^{-5}	2.1-4.6	0.30	2014
MIRI	10^{-4}	5-25	0.35	
\multicolumn{5}{c}{1.5 m Space Coronagraphs}				
	$10^{-9} - 10^{-10}$	0.3-1.3	0.08	?
\multicolumn{5}{c}{ELT's class instruments}				
E-ELT-EPICS	$10^{-8} - 10^{-9}$	0.9-1.7	0.03	> 2018
E-ELT-METIS	10^{-5}	2.5-20	0.08	

up to periods corresponding to a semi major axis of 40 AU (for 1 $M_\odot$); we assume that both planet period and mass scale with the stellar mass. For each of these planets, we assumed an orbit eccentricity extracted randomly from a uniform distribution in the range 0.0-0.6. All remaining orbital elements (including inclination) are randomly generated using uniform distributions. This set of parameters allows the definition of the circumstances of observation, and a number of planet properties and observables: semi-major axis (AU) and projected separation (arcsec), evaluated using distance and mass of each star; planetary radius estimated following the approach of Fortney et al. (2007); planet effective temperature, using the models by Sudarsky et al. (2003); intrinsic luminosity obtained using the models by Baraffe et al. (2003); reflected luminosity in visible and NEAR-IR (V, H, K, L band) obtained scaling the Jupiter luminosity with planet semi-major axis and radius; intrinsic luminosity in mid-IR (11.4 μm) from a black body emission; RV semi-amplitude; astrometric signature.

Contrast (in the appropriate wavelength range), RV semi-amplitude and astrometric signature for each planet in the population can be compared with detection limit curves for direct imaging, RV, and astrometry (GAIA). Planets for which the expected observables are above limits are potentially detected with the different techniques.

Figure 1 compares the planets expected to be discovered by the various instruments considered in the present review. Different colours are used for warm giants, cold giants, Neptune-like, and rocky planets, respectively. Table 3 summarizes the results. Ground-based 8 m telescopes and JWST will allow us, by the middle of the next decade, to have a quite clear picture of the outer parts of the planetary systems, not explored by the indirect techniques. Not only the architecture of the systems can be determined, but also photometry and spectroscopy of these planets can be obtained, providing information about the physics of their atmospheres. Most observations will be focused on young systems, providing crucial information about the formation phases, and data about the disk-planet interactions.

Single Aperture Imaging of Exoplanets 41

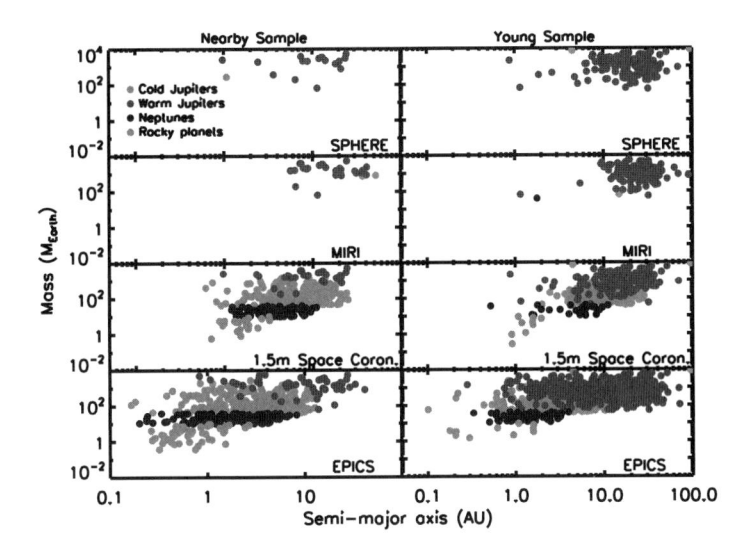

Figure 1. Planets expected to be discovered by SPHERE (representative of planet finders on 8 m class ground-based telescopes), JWST-MIRI, 1.5m space coronagraphs, and EPICS/E-ELT (representative of 30-40 m class telescopes) in the mass vs separation plane. Different grey shades are used for warm giant, cold giant, Neptune-like, and rocky planets respectively (see key in upper left). Plots at left are obtained considering the nearby sample, those on the right with the young star sample.

At the end of the decade, 1.5 m space coronagraphs and ground-based ELT will considerably increase the range of masses and separations of the imaged planets. Some tens of Neptune-like and even a few rocky planets will become detectable. With some luck, some planets in the habitable zone might even be discovered.

4. Spectroscopy and Atmosphere Composition

Imagers equipped with IFSs or long-slit spectrographs will allow the measurement of low-resolution spectra of exosolar planets, information that can now

Table 3. Summary of expected detections from imagers in the next decade.

Instrument	Year	Young Giants	Old Giants	Neptunes Planets	Rocky Planets	Habit. Planets
Ground based 8 m	2011	tens	few			
JWST	2014	tens	few			
1.5m Space Coro.	?	tens	tens	tens	few	??
ELTs	>2018	hundreds	hundreds	tens	few	??

only be obtained for transiting systems. This will help to probe the atmospheric composition. Bands of different species are accessible in the spectral ranges observed by the various instruments. CO, O_2, NH_3, CO_2, CH_4 and H_2O can be observed by NIR instruments (Sphere, GPI, NIRCAM, EPICS) in the next few years for giant planets, and possibly at the end of the decade for Neptune-like and rocky planets. Additional species, including O_3, H_2S, PH_3, C_2H_2 and H_3^+ have prominent bands in the mid-IR, observable with MIRI and METIS; such observations will be limited to giant planets due to the detection limits of these instruments. Finally, space coronagraphs working at visible wavelengths will also allow observations of several molecules, including life signalling O_2, possibly for rocky planets. However, the rocky planets discovered by such instruments will likely be too cold for hosting life. In spite of these limitations, all this spectroscopic information, coupled with additional polarimetric observations (crucial to determine the height where haze or clouds form), will allow a much better understanding of planetary atmospheres, and will pave the road for ambitious projects like TPF.

Spectra at higher resolution than in standard set-up for planet detection will allow a more detailed characterization (for planets detected with high enough S/N). Some science goals include identification of spectral features, determination of physical parameters (temperature, gravity, chemical composition), cloud formation processes and their variation with time (e.g. for planets in eccentric orbits). Various resolutions, from R=3000 to R=20000 are considered, with the higher resolution modes being more suitable on ELT's (see Claudi et al. 2010): we expect hundreds of exoplanets observable at medium resolution, and some tens at the higher resolution. Planet radial velocities could also be determined. This will be useful to constrain the planetary orbit if only visual measurements are available, and the planet-star mass ratio even based on small time baseline; or to detect binary planets, if any. Finally, planet rotational velocity could also be determined (for reference, Jupiter equatorial rotation velocity is 12.6 km/s); field T dwarfs typically rotate faster (30-50 km/s: McLean et al. 2007; Zapatero Osorio et al. 2006). R=20,000 corresponds to FWHM=15 km/s, and R=3000 to 100 km/s. There is then the possibility of measuring rotational velocity similar to that of Jupiter. This is very interesting if coupled with photometric rotational modulation, providing the planet radius is independent of luminosity, or the inclination of rotational axis over the orbital plane.

5. Synergies with Other Techniques

In this last section we comment on synergies of imaging with other techniques. Those with dynamical methods are interesting, because these last allow determining the planet masses, eliminating the degeneracy with age. Until the middle of the next decade, exoplanets discovered using ground-based 8 m telescopes or JWST will likely be at large distances from the star. Periods of these planets are long (tens of years or even more), and only limited information about them could be obtained from dynamical methods: dynamical masses will then be determined only in a few favourable cases. However, the situation will change completely if 1.5 m class space coronagraphs are available, and even more with ELT's. The discovery space for EPICS at E-ELT overlaps well with those from

Single Aperture Imaging of Exoplanets 43

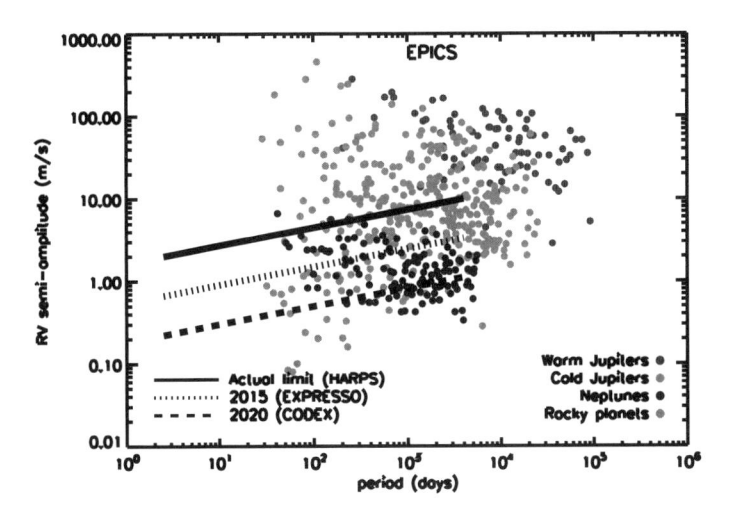

Figure 2. Planets expected to be detected by EPICS (nearby sample) in the RV signal vs. period plane, compared with detection limits for RV instruments (HARPS, ESPRESSO, and CODEX). The grey scale code is the same as Figure 1.

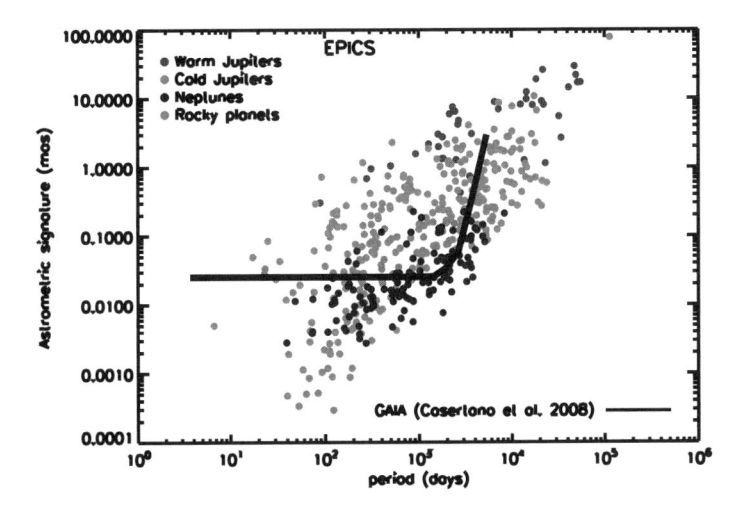

Figure 3. Planets expected to be detected by EPICS (nearby sample) in the astrometric signal vs. period plane, compared with detection limits for astrometric satellites GAIA and SIM-LITE. Grey scale code is the same as in Figure 1.

radial velocity (RV) instruments (HARPS at ESO 3.6m telescope, ESPRESSO at VLT, and especially CODEX at E-ELT: see Figure 2), so that we may expect to get both spectra and masses for a large number of targets. The discovery space of EPICS overlaps also well with that of GAIA (Casertano et al. 2008) (see Figure 3), which will likely detect hundreds, or even thousands of giant planets from astrometric signatures. GAIA is on track for launch in 2011 and data will be available before the end of the decade. However, astrometric detection of rocky planets requires dedicated observations of a limited number of targets, as will be possible if the SIM-LITE mission (Goullioud et al. 2008) is approved and launched.

Finally, there is also some potential overlap with transit missions such as PLATO, an ESA Cosmic Vision proposed mission. 10 M_{Earth} exoplanets around K and M dwarfs with V=8.5-10 (bright end of PLATO) can be detected also with EPICS. For K dwarfs, planets in the habitable zones are detectable. Availability of planet spectrum from EPICS and radius from PLATO will be relevant for the physical study of the planets. For G and F stars (and K and M dwarfs as well) planets at separations larger than that accessible to PLATO can be detected, allowing the study of the outer planetary system of PLATO targets (see talk by Claudi et al. 2010).

In summary, enormous progress can be expected in the next decade: we will be moving from the pure discovery of existence to characterisation of planets. This will be a basic step toward the discovery of habitable planets.

References

Absil, O., 2010, this volume
Baraffe, I., Chabrier, G., & Barman, T. S. 2003, A&A, 402, 701
Beuzit, J.-L., Feldt, M., Dohlen, K., et al. 2008, SPIE, 7014, 41
Bonavita, M., Desidera, S., Gratton, R.G., et al. 2009, in preparation
Brandl, B. R., Lenze, R., Pantin, E., et al. 2008, SPIE, 7014, 55
Casertano, S., Lattanzi, M. G., Sozzetti, A. et al. 2008, A&A, 482, 699
Clampin, M., Melnick, G., Lyon, R., et al. 2006, SPIE, 6265, 38
Claudi, R., et al. 2010, this volume
Cumming, A., Butler, R. P., Marcy, G. W., Vogt, S. S., Wright, J. T., & Fischer, D. A. 2008, PASP, 120, 531
Fortney, J. J., Marley, M. S., & Barnes, J. W. 2007, ApJ, 659, 1661
Goullioud, R., Catanzarite, J. H., Dekens, F. G., Shao, M., & Marr, J. C. 2008, SPIE, 7013, 151
Guyon, O., Angel, J. R. P., Backman, D., et al. 2008, SPIE, 7010, 59
Kasper, M., Beuzit, J.-L., Verinaud, C. et al. 2008, SPIE, 7015, 46
Lyon, R. G., Clampin, M., Melnick, G., Tolls, V., Woodruff, R., & Vasudevan, G. 2008, SPIE, 7010, 118
Macintosh, B. A., Graham, J. R., Palmer, D. W., et al. 2008, SPIE, 7015, 31
McLean, I. S., Prato, L., McGovern, M. R., Burgasser, A. J., Kirkpatrick, J. D., Rice, E. L., & Kim, S. S. 2007, ApJ, 658, 1217
Schneider, J., Boccaletti, A., Mawet, D., et al. 2009, Exp. Astron. 23, 357
Sudarsky, D., Burrows, A., & Hubeny, I., 2003, ApJ, 588, 1121
Tamura, M. & Hodapp, K., Takami, H., et al. 2006, SPIE, 6269, 28
Trauger, J., Stapelfeldt, K., Traub, W., et al. 2008, SPIE, 7010, 69
Zapatero Osorio, M. R., Martin, E. L., & Bouy, H. 2006, ApJ, 647, 1405

Pathways Towards Habitable Planets
ASP Conference Series, Vol. 430, 2010
Vincent Coudé du Foresto, Dawn M. Gelino, and Ignasi Ribas, eds.

Blue Dots Team Transits Working Group Review

A. Sozzetti,[1] C. Afonso,[2] R. Alonso,[3] D. L. Blank,[4] C. Catala,[5] H. Deeg,[6] J. L. Grenfell,[7] C. Hellier,[8] D. W. Latham,[9] D. Minniti,[10] F. Pont,[11] and H. Rauer[12]

Abstract. Transiting planet systems offer an unique opportunity to observationally constrain proposed models of the interiors (radius, composition) and atmospheres (chemistry, dynamics) of extrasolar planets. The spectacular successes of ground-based transit surveys (more than 60 transiting systems known to-date) and the host of multi-wavelength, spectro-photometric follow-up studies, carried out in particular by HST and Spitzer, have paved the way to the next generation of transit search projects, which are currently ongoing (CoRoT, Kepler), or planned. The possibility of detecting and characterizing transiting Earth-sized planets in the habitable zone of their parent stars appears tantalizingly close. In this contribution we briefly review the power of the transit technique for characterization of extrasolar planets, summarize the state of the art of both ground-based and space-borne transit search programs, and illustrate how the science of planetary transits fits within the Blue Dots perspective.

1. Introduction

Within the Blue-Dots Team (BDT) initiative (http://www.blue-dots.net/), the primary goal of the Transits Working Group (TWG) is to gauge the potential and limitations of transit photometry (and follow-up measurements) as a tool to detect and characterize extrasolar planets, while emphasizing its complemen-

---

[1]INAF - Osservatorio Astronomico di Torino, Strada Osservatorio 20, Pino Torinese, Italy

[2]Max Planck Institute for Astronomy, Königstuhl 17, Heidelberg, Germany

[3]Observatoire de Genéve, Université de Genève, 51 Ch. des Maillettes, Sauverny, Switzerland

[4]Centre for Astronomy, School of Mathematical & Physical Sciences, James Cook University, Townsville, Australia

[5]LESIA, Observatoire de Paris, 5 place Jules Janssen, Meudon, France

[6]Instituto de Astrofisica de Canarias, La Laguna, Tenerife, Spain

[7]Technische Universität Berlin, Hardenbergstr. 36, Berlin, Germany

[8]Astrophysics Group, Keele University, Staffordshire, UK

[9]Harvard-Smithsonian Center for Astrophysics, 60 Garden Street, Cambridge, MA, USA

[10]Dept. of Astronomy, Pontificia Universidad Católica de Chile, Casilla 306, Santiago, Chile

[11]School of Physics, University of Exeter, Exeter, UK

[12]Institute of Planetary Research (DLR), Rutherfordstr. 2, Berlin, Germany

tarity with other techniques. Following the "grid approach" agreed upon within the BDT, the mapping is to be performed as a function of depth of the science investigation, project scale, and detectable exoplanet class. In this review of the TWG activities, we first describe the main observables accessible by means of photometric transits. We then focus on the host of follow-up techniques that can be utilized to deepen our understanding and characterization of transiting systems, and briefly touch upon some key science highlights, while keeping in mind the primary difficulties and limitations inherent to the transit technique when applied to planet detection. Next, a summary of the present-day and future projects devoted to detection and characterization of transiting planets, both from the ground and in space, is presented. Finally, we use a 'grid approach' to properly gauge how the scientific prospects of photometric transits and follow-up techniques fit within the BDT perspective.

2. Planet Detection with Transit Photometry

2.1. Observables

The primary observable of the transit technique is the periodic decrease of stellar brightness of a target, when a planet moves across the stellar disk. The magnitude of the eclipse depth is defined as $\Delta F/F_0 = (R_p/R_\star)^2$, where F_0 is the measured out-of-transit flux, and R_p and $R_\star$ are the planet and primary radius, respectively. Transits require a stringent condition of observability, i.e., the planetary orbit must be (almost) perpendicular to the plane of the sky. The geometric probability of a transit, assuming random orientation of the planetary orbit, is: $P_{tr} = 0.0045(1\,\mathrm{AU}/a)((R_\star + R_p)/R_\odot)((1 + e\cos(\pi/2 - \varpi))/(1 - e^2))$, where a is the orbital semi-major axis, e the eccentricity, and ϖ the argument of periastron passage (e.g., Charbonneau et al. 2007). The duration of a central transit is $\tau = 13(R_\star/R_\odot)(P/1\,\mathrm{yr})^{(1/3)}(M_\odot/M_\star)^{(1/3)}$ hr, with a reduction for non-central transits by a factor $\sqrt{1 - b^2/R_\star^2}$, where $b = (a/R_\star)\cos i$ is the impact parameter for orbital inclination i (e.g., Seager & Mallén Ornelas 2003). The transit method allows the determination of parameters that are not accessible to Doppler spectroscopy, such as the ratio of radii, the orbital inclination, and the stellar limb darkening. When combined with available radial-velocity (RV) observations, actual mass and radius estimates for the planet can be derived, provided reasonable guesses for the primary mass and radius can be obtained.

2.2. False Positives Reconnaissance

The typical transiting system configurations known to-date (for a comprehensive list see http://exoplanet.eu) encompass Jovian planets orbiting solar-type stars on a few-day orbits, resulting in eclipse depths of 0.3 − 3% and transit durations of 1.5 − 4 hr. A great variety of stellar systems can reproduce such signals in terms of depth and duration. These include grazing eclipsing binaries, large stars eclipsed by small stars, and 'blends' consisting of faint eclipsing binaries whose light is diluted by a third, brighter star (e.g., Brown 2003). Typically, such impostors constitute over 95% of the detected transit-like signals in wide-field photometric surveys datasets. Extensive campaigns of follow-up observations of transit candidates must then be undertaken in order to ascertain

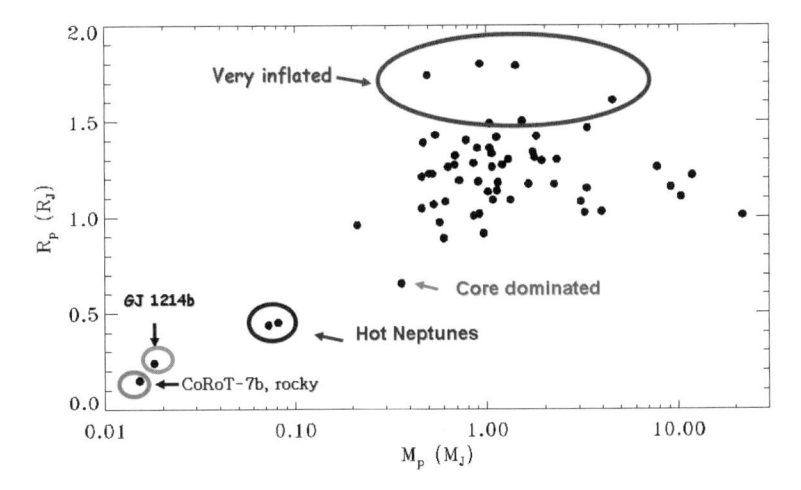

Figure 1. Masses and radii for the sample of transiting planets known as of December 2009.

the likely nature of the system. High-quality light curves and moderate precision ($\approx$ km s^{-1}), low signal-to-noise ratio (SNR) spectroscopic measurements are usually gathered to deepen the understanding of the primary, the consistency of the transit shape with that produced by a planet and, via detailed modeling of the combined datasets (e.g., Torres et al. 2004), to rule out often subtle blend configurations. Only if the candidate passes all the above tests, does one resort to use 10-m class telescopes and high-resolution, high-precision Doppler measurements to determine the actual spectroscopic orbit (e.g., Mandushev et al. 2005). The plague of false positives contamination is however rather diminished for ground-based transit surveys targeting cool, nearby M-dwarf stars (Nutzman & Charbonneau 2008). This holds true also for space-borne transit programs, with very high photometric precision which allows to reveal many of the stellar companions via very shallow secondary eclipses, and/or ellipsoidal variations out of transit. Giant stars can be excluded beforehand by using a target catalog, such as the Kepler Input Catalog or Gaia data (as it's envisioned for PLATO). Background eclipsing binaries can instead be identified efficiently if precision astrometry can be performed on the photometric times series themselves, to measure centroid shifts due to variability-induced movers (Wielen 1996). This technique is currently being applied with considerable success to Kepler data (Monet et al. 2010).

2.3. Transiting Systems Highlights

Sixty-two transiting planets of main-sequence stars are known today. The direct measure of their masses and radii, and thus densities and surface gravities, puts fundamental constraints on proposed models of their physical structure (Charbonneau et al. 2007, and references therein). Figure 1 shows the M_p-R_p relation for the known transiting systems. Strongly irradiated planets cover a range of almost three orders of magnitude in mass and more than one order of magnitude in radius. The variety of inferred structural properties is posing a great challenge to evolutionary models of their interiors (e.g., Baraffe et al. 2008; Valencia et al. 2007; Miller et al. 2009). Among the most interesting systems found by photometric transit surveys are *a*) those containing super-massive ($M_p \approx 7-13$ M_J) hot Jupiters such as HAT-P-2b (Bakos et al. 2007), WASP-14b (Joshi et al. 2009), XO-3b (Johns-Krull et al. 2008), and particularly WASP-18b (Hellier et al. 2009), the first planet to be likely tidally disrupted on a short timescale, *b*) very inflated ($\sim 1.8\ R_J$) Jovian planets such as WASP-12b (Hebb et al. 2009), TrES-4 (Mandushev et al. 2007), and WASP-17b (Anderson et al. 2010), *c*) the tilted, most eccentric planet, HD 80606b (Winn et al. 2009, and references therein), *d*) the first transiting planet in a multiple system, HAT-P-13b (Bakos et al. 2009), *e*) the first transiting low-mass brown dwarf, CoRot-3b (Deleuil et al. 2008), and *f*) the first transiting super-Earth, CoRot-7b (Léger et al. 2009), itself a member of a multiple-planet system (Queloz et al. 2009).

3. Transiting Planet Characterization: Follow-up Techniques

When the primary is sufficiently bright (see below), a host of follow-up photometric and spectroscopic measurements can be carried out over a large range of wavelengths, to deepen the characterization of the physical and dynamical properties of transiting systems (Charbonneau et al. 2007). At visible wavelengths, long-term, high-cadence, high-precision photometric monitoring can allow the detection of additional components in a system (not necessarily transiting) via the transit time variation method (Holman & Murray 2005), while RV measurements collected during transit offer the opportunity to determine the degree of alignment between the stellar spin and the orbital axis of the planet (e.g., Winn et al. 2009). These data are powerful diagnostics for models of orbital migration and tidal evolution of planetary systems (e.g., Fabrycky & Winn 2009, and references therein). The technique of transmission spectroscopy opens the way to measurements of specific elements seen in absorption in the planet's atmosphere (Charbonneau et al. 2002), including water (Tinetti et al. 2007). At infrared wavelengths, mostly thanks to the Spitzer Space Telescope, the photometric and spectroscopic monitoring over a wide range of planetary phases, and particularly during secondary eclipse, has allowed the detailed study of strongly-irradiated atmospheres of a few planets, with successful detection of the planet's thermal emission (Charbonneau et al. 2005) and characterization of the longitudinal temperature distribution (e.g., Knutson et al. 2007; Charbonneau et al. 2007). The direct measurements of exoplanets' atmospheric compositions and temperature profiles, atmospheric dynamics, and phase light curves are key inputs for models of atmospheric physics, chemistry, and dynamics (Burrows et al. 2008).

4. Ground-based and Space-borne Projects

The success of (wide-field) ground-based photometric transit surveys bears upon two distinct approaches, the first adopting moderate-size telescopes to search relatively faint stars, the second utilizing small-size instrumentation for searches around brighter targets (see Figure 2). Most of the projects have focused on high-cadence, visible-band photometry of tens of thousands of stars, and only recently near-infrared filters have been contemplated. Solar-type stars have been so far the main focus of all searches, but in recent times the MEarth project (Nutzman & Charbonneau 2008; and also Damasso et al., this volume) and the UKIRT WF-CAM Transit Survey (WTS) program (http://star.herts.ac.uk/RoPACS/) have identified as targets nearby, bright M dwarfs and fainter, more distant low-mass stars, respectively. The MEarth project has recently announced the first detection of a transiting super-Earth companion to a late M-dwarf star (Charbonneau et al. 2009). Transit discovery programs typically achieve photometric precisions of 3-5 mmag. The best-case performances of $\sim$ 1-2 mmag are mostly obtained with dedicated follow-up programs at 1-2m class telescopes.

A photometric precision of $\sim$ 3 mmag is enough to detect Jupiter- and Saturn-sized companions in transit across the disk of solar-type stars, or $2-4$ $R_\oplus$ planets transiting M dwarfs (see the definition of $\Delta F/F_0$ above). If the goal becomes the detection of transits of Earth-sized planets around solar-type primaries, it is necessary to go to space, in order to achieve $0.0001 - 0.00001$ mag photometric accuracy. The CoRoT satellite (Baglin et al. 2009), the Kepler mission (Borucki et al. 2009; Monet et al. 2010), and the PLATO mission, currently under study by ESA (Catala 2009; Catala et al., this volume), have been designed to reach the above performances (we direct the reader to the above contributions for details on the science). The first transiting super-Earth of a solar-type star was recently announced by the CoRoT team (Léger et al. 2009; see also Rouan et al., this volume).

As mentioned above, the spectroscopic characterization of transiting systems has been carried out primarily from space, by HST and Spitzer. In the future, the prospects for detailed atmospheric characterization of transiting planets will rely on the James Webb Space Telescope (JWST), the SPICA satellite, and the proposed THESIS concept. We refer to the contributions by Clampin and Enya in this volume and by Swain et al. (2009) for details on these projects and their potential.

4.1. The Star's the Limit

It is not uncommon to believe that main-sequence, solar-type stars astrophysics is a solved problem, for practical purposes. In reality, when it comes to transiting planetary systems, the knowledge of the central star is oftentimes the limit for the accurate determination of the most sensitive planetary parameters. The precise characterization of transiting planets is intimately connected to the accurate determination of a large set of stellar properties (activity levels, age, rotation, mass, radius, limb darkening, and composition). Some of them (activity, rotation) can critically limit the possibility of successfully determining the spectroscopic orbit of the detected planets. Others (mass, radius, age) are strongly model-dependent quantities, and the correct evaluation of their uncertainties is not trivial (e.g., Brown 2009). Furthermore, precise measurements of the stellar characteristics become increasingly more challenging for fainter targets.

Survey	Location	Apert.(mm)	CCD	FOV (deg^2)	Range(mag)	Scale(")	Since	Nr. stars	Filters
OGLE[a]	Las Campanas	1300	8K×8K	0.34		0.26	1992	>10^5	UBVRI
APT[b]	Australia	500	2K×2K	6	9.4	10-15	1995		B,V,R,I
Vulcan[c]	Lick Obs.	120	4K×4K	49	<13		1999	6000	V, R
STARE[d] (TrES)	Tenerife	99	2K×2K	32		10.8	1999	>24000	B,V,R
ASAS-3[e]	Las Campanas	2×71, 250, 50	2×2K × 2K	64, 4.8, 936			2002		V,I
SuperWasp[f]	S. Africa, La Palma	2×8×111	2K×2K	16×61	<13	13.7	2002	>100K	
BEST[g]	OHP	195	2K × 2K	9.6	10-14	5.5	2002	100K	clear
XO[h]	Haleakla	2 × 110	1K×1K	51.84	12	25.4	2003	>100K/year	400-700 nm
WHAT[i]	Wise Obs.	110	2K×2K	67.24	10-14	14	2004	15000	I
HATNet[j]	Hawaii, FLWO	6×110	2K×2K	67	I<14	14	2003	96K	I
VulcanSouth[k]	Antarctic	200	4K×4K				2004-2005		600-700nm
SLEUTH[l](TrES)	Palomar	100	2K×2K	36	<14	10	2003	10000	r',g,i,z
PSST[m](TrES)	Arizona	100	2K×2K	36	10-13		2004	4000-12000	B,V,R, VR
BEST II[n]	Armazones	250	4K × 4K	2.8°	10-16	1.5	2007	100K	clear
TEST[p]	Tautenburg	300	4K × 4K	4.8	10-15	2	2007	50000	(UBVI)R
ASTEP-South[q]	Antarctic	100	4K × 4K				2008		
MEarth[r]	FLWO	2×400	2K × 2K	0.18	<9	0.75	2008	4131	
PANSTARRS[s]	Haleakla	4×1800	1.4bil pix.	49	<24	0.3	ongoing	6000/night	g,r,i,y
VISTA-ROPACS[t]	Paranal	4000	8K × 8K			0.339	ongoing		Z,Y,J,H,K_s
ASTEP	Antarctic	400					2010		
PASS[u]	Antarctic	all sky		5.5-10.5				250K	
ICE-T[v]	Antarctic	2×600		65			2012	1.3M	yes
OmegaTrans[w]	Paranal	2600	16K×16K	1	13.5-17.5	0.26		200K	R

Figure 2. Ground-based transit surveys summary table.

	1	2	3	4	5
	Giant (close / young)	Giants (others)	Telluric (others)	Telluric HZ (M)	Telluric HZ (others)
Transit	3	2 ?	3 ?	3	1

Figure 3. Science potential of the family of techniques for transit detection and characterization for different classes of exoplanets. Existing, or already programmed efforts, are coded in light grey. Projects whose effort is comparable to an ESA M-class mission are coded in dark grey. Numbers in the grey boxes indicate the following: (1) programs capable of carrying out a statistical study of objects in a given class, (2) projects able to designate targets in the solar neighborhood for spectroscopic follow-up studies, and (3) large-scale programs with the goal of carrying out spectroscopic characterization of the objects.

As discussed in Section 2.2., follow-up observations of transiting planet candidates can be very time consuming (the RV campaign for CoRoT-7 required over 100 spectra, a total of more than 70 hrs of observing time distributed over 4 months). For CoRoT, Kepler, and PLATO confirmation via RV measurements may not even be feasible below a certain radius size, depending on spectral type (for reference, the semi-amplitude of the RV motion induced by the Earth at 1 AU on the Sun is ~ 9 cm s^{-1}, way below the currently best-achievable precision of 50-100 cm s^{-1} with the HARPS spectrograph). Devices for ultra-stable wavelength calibration such as laser combs can in principle allow to push towards < 10 cm s^{-1} precision (Li et al. 2008), provided the star cooperates (Walker 2008; Makarov et al. 2009). Achievable SNR for a given host's spectral type make also complicated the problem of atmospheric characterization of transiting rocky planets via transmission spectroscopic or secondary-eclipse observations (Kaltenegger & Traub 2009; Deming et al. 2009).

Given the above issues, it is clear that bright ($V \leq 12$) stars are the privileged targets for transit searches. The challenge is then designing a survey capable of covering large areas of the sky to maximize the yield of good targets. This is planned for PLATO, thanks to its step & stare mode (Catala et al., this volume) and, in all-sky fashion, by TESS (Deming et al. 2009).

5. The BDT Perspective

As discussed in detail by Coudé du Foresto et al. (this volume), the BDT has devised a strategy to gauge the interplay between families of techniques, project scale, scientific potential, and detectable exoplanet class within the context of the multi-step approach recognized as necessary in order to reach the final goal of detection and characterization of terrestrial, habitable planets. Following the "grid methodology" agreed upon within the BDT, and outlined in detail by Coudé du Foresto et al. (this volume), the TWG has attempted to gauge the potential and limitations of transit photometry (and follow-up measurements) as a tool to detect and characterize extrasolar planets, while emphasizing its complementarity with other techniques. Overall, our preliminary conclusion is that transit-discovery observations can crucially contribute to statistical studies of planetary systems (science potential 1) and to identify systems suitable for follow-up (science potential 2). Follow-up observations of known transiting systems have the potential to achieve their full spectroscopic characterization (science potential 3). However, the full potential of the technique might not be realized for all classes of extrasolar planets encompassed in this exercise. In particular (see Figure 3):

Hot giant planets: Ground-based, wide-field transit surveys with typical photometric accuracy < 0.01 mag, have detected several tens of hot Jupiters. The ongoing CoRoT mission is also providing many detections of close-in giants, and the prospects with Kepler are very encouraging too. The Spitzer and Hubble Space Telescopes have been utilized as follow-up tools for the (broad-band) spectral characterization of several hot Jupiters at visible, near, and mid IR wavelengths, with several molecules identified.

Giant planets at large orbital radii: CoRoT and Kepler can achieve an accuracy of $10^{-4} - 10^{-5}$ mag, respectively, in the visible. They will provide a census of transiting giant planets out to 1 AU based on $\sim 10^5$ targets. The proposed TESS all-sky survey (~ 2012) could achieve a photometric precision similar to that of CoRoT, and could provide a census of transiting giants with periods up to several tens of days around bright stars. While not truly all-sky survey, the PLATO mission (~ 2017) could achieve, on a statistically significant sample of bright stars, a photometric precision exceeding Kepler's. It will be sensitive to Jupiters on orbital periods similar to those accessible by Kepler. Statistical information on the rate of occurrence of longer-period giant planets will also be collected by ongoing and upcoming large-scale ground-based surveys, such as LSST and PANSTARRS (see Figure 2). For the sample of relatively bright stars, several very efficient space- and ground-based facilities will become available in the near future for spectroscopically characterizing the discovered planets. These include JWST and SPICA (and possibly THESIS) for infrared photometry and spectroscopy, and the ELTs for high-resolution spectroscopy in the optical and near IR.

Telluric planets in and out of the habitable zone of M dwarfs and solar-type stars: CoRoT and TESS have the potential to detect super-Earth planets around all targets, and at a range of orbital radii, including the habitable zone of low-mass stars (CoRoT has recently announced its first detection). Kepler has the potential to provide the first statistically sound estimate of $\eta_\oplus$. The ultra-high-precision photometry delivered by PLATO will also allow the detection of Earth-sized planets in the habitable zone of F-G-K targets. As for low-mass stars, the ongoing ground-based MEarth cluster of telescopes is optimized to search for transiting super-Earths in the habitable zone of nearby M dwarfs, while the WTS survey will target a large sample of low-mass stars, searching for transiting rocky planets with periods of a few days. Theoretical studies are now maturing, which can predict the range and strength of spectral fingerprints of terrestrial, habitable planets (e.g., Grenfell et al. 2007; Kaltenegger et al. 2009). The proposed SPICA mission, JWST, and the THESIS concept will be able to perform spectral characterization (broad bands, spectra) in the near- and mid-IR possibly down to telluric planets in the HZ of bright M dwarfs. The proposed SIMPLE instrument for the E-ELT would also be able to perform transmission spectroscopy of low-mass planets transiting M dwarfs.

6. Summary

On the "bright" side, transit photometry allows the characterization of the bulk composition of a planet, and it identifies systems suitable for atmospheric characterization. Such studies are more difficult to undertake for non-transiting systems. In order to fully exploit the potential of this technique (and follow-up measurements), there does not seem to be a clear need for a facility devoted to planetary transits which would require investments on the order of an ESA flagship (L-class) mission. On the other side, this technique requires large amounts of follow-up work, and the stellar host can often be the limiting factor in the precision with which the crucial physical parameters of the planets are determined.

The relevant technology for transit detection of terrestrial-type planets is already available. Ongoing and future programs have the potential to nail the occurrence rate of habitable planets around main-sequence stellar hosts, and, provided some degree of further technological development, characterize those around stars with favorable spectral type.

Acknowledgments. A.S. gratefully acknowledges financial support from the Italian Space Agency through ASI contract I/037/08/0 (Gaia Mission - The Italian Participation to DPAC).

References

Anderson, D. R., et al. 2010, ApJ 709, 159
Baglin, A., et al. 2009, in IAU Symp. 253, Transiting Planets, eds. F. Pont, D. D. Sasselov & M. J. Holman, (Cambridge: Cambridge University Press), 71
Bakos, G.A., et al. 2007, ApJ, 670, 826
Bakos, G.A., et al. 2009, ApJ, 707, 446
Baraffe, I., Chabrier, G., & Barman, T. 2008, A&A, 482, 315
Borucki, W., et al. 2009, in IAU Symp. 253, Transiting Planets, eds. F. Pont, D. D. Sasselov & M. J. Holman, (Cambridge: Cambridge University Press), 289
Brown, T.M. 2003, ApJ, 593, L125
Brown, T.M. 2009, ApJ submitted
Burrows, A., Budaj, J., & Hubeny, I. 2008, ApJ, 678, 1436
Catala, C. 2009, Experimental Astronomy, 23, 329
Charbonneau, D., Brown, T.M., Noyes, R.W., & Gilliland, R.L. 2002, ApJ, 568, 377
Charbonneau, D., et al. 2005, ApJ, 626, 523
Charbonneau, D., et al. 2007, in Protostars and Planets V, eds. B. Reipurth, D. Jewitt, & K. Keil, (University of Arizona Press : Tucson), 701
Charbonneau, D., et al. 2009, Nat, 462, 891
Deleuil, M., et al. 2008, A&A, 491, 889
Deming, D., et al. 2009, PASP, 121, 952
Fabrycky, D.C., & Winn, J.N. 2009, ApJ, 696, 1230
Grenfell, J.L. et al. 2007, Planet. Space Sci., 55, 661
Hebb, L., et al. 2009, ApJ, 693, 1920
Hellier, C., et al. 2009, Nat, 460, 1098
Holman, M.J., & Murray, N.W. 2005, Sci, 307, 1288
Johns-Krull, C.M., et al. 2008, ApJ, 677, 657
Joshi, Y.C, et al. 2009, MNRAS, 392, 1532
Kaltenegger, L., et al. 2009, Astrobiology, in press (arXiv:0906.2263)
Kaltenegger, L., & Traub, W.A. 2009, ApJ, 698, 519
Knutson, H.A., et al. 2007, Nat, 447, 183

Léger, A., et al. 2009, A&A, 506, 287
Li, C.-H., et al. 2008, Nat, 452, 610
Makarov, V.V., et al. 2009, ApJ, 707, L73
Mandushev, G., et al. 2005, ApJ, 621, 1061
Mandushev, G., et al. 2007, ApJ, 667, L195
Miller, N., Fortney, J.J., & Jackson, B. 2009, ApJ, 702, 1413
Monet, D.G., et al. 2010, ApJ, submitted (arXiv:1001.0305)
Nutzman, P., & Charbonneau, D. 2008, PASP, 120, 317
Queloz, D., et al. 2009, A&A, 506, 303
Seager, S. & Mallén-Ornelas, G. 2003, ApJ, 585, 1038
Swain, M.R., et al., 2009, Astro2010: The Astronomy and Astrophysics Decadal Survey, Technology Development Papers, 61
Tinetti, G., et al. 2007, Nat, 448, 169
Torres, G., Konacki, M., Sasselov, D.D., & Jha, S. 2004, ApJ, 614, 979
Valencia, D., Sasselov, D.D., & O'Connell, R.J. 2007, ApJ, 665, 1413
Walker, G. 2008, Nat, 452, 538
Wielen, R. 1996, A&A, 314, 679
Winn, J.N., et al. 2009, ApJ, 703, 2091

ております
Pathways Towards Habitable Planets
ASP Conference Series, Vol. 430, 2010
Vincent Coudé du Foresto, Dawn M. Gelino, and Ignasi Ribas, eds.

Habitability of Planets

F. Forget and R. D. Wordsworth

Laboratoire de Météorologie Dynamique, IPSL, Paris, France

Abstract. What makes a planet suitable for life and its evolution? This question has been discussed for years and we are slowly making progress. Liquid water remains the key criterion for habitability. It can exist in the interior of a variety of planetary bodies, but it is usually assumed that liquid water at the surface, interacting with rocks and light, is necessary for the emergence of a life able to modify its environment and evolve. The climatic conditions allowing surface liquid water have been studied with global mean 1-D models which have defined the "classical habitable zone" (Kasting et al. 1993). Key issues are now to better understand the geophysical feedback that seems to be necessary to control atmospheric evolution and maintain a continuously suitable climate (is this unique to the Earth?), as well as exploring with accuracy climate regimes that could locally allow liquid water. For this last purpose a new generation of 3-D climate models based on universal equations and tested on bodies in the solar system is now available.

1. Introduction: Habitability and Liquid Water

What makes a planet suitable for life?

With only our planet as a viable example, and only one kind of life to define the necessary ingredients, addressing this question requires enormous scientific extrapolations and some trust in purely theoretical studies.

Obviously, the answer depends on the kind of life that we want to consider. Life as we know it always uses carbon-based molecules with liquid water as a solvent, with no exceptions. In fact, our experience on Earth has told us that the requirement for life is liquid water, regardless of mean temperature and pressure (Brack 1993). Living organisms can exist and thrive in almost any conditions on Earth if liquid water is available (Rothschild & Mancinelli 2001). Conversely, no creatures can "live" (i.e., have metabolic activity) without liquid water. One can speculate on forms of life based, say, on liquid ammonia, condensed methane or even plasma ions interactions. However, exploring the wide field of modern chemistry and challenging the most open-minded chemists reveals that with our present knowledge it is difficult to imagine any alternative chemistry approaching the combination of diversity, versatility and rapidity afforded by liquid water-based biochemistry. This results from the unique characteristics of water as a liquid solvent (a large dipole moment, the capability to form hydrogen bonds, to stabilize macromolecules, to orient hydrophobic-hydrophilic molecules, etc.).

Within this context, the primary definition of habitability is the presence of liquid water. This may be narrow-minded, but if optimistic conclusions can be reached with such a focus, then whatever we have ignored will only serve to broaden the biological arena (Sagan 1996). On the other hand, it can be argued that liquid water by itself may not be sufficient and that a few other elements and a source of energy (chemical gradient or light) are necessary to support life forms. This is further discussed in Section 2. However, the discovery of a very large variety of extremophiles on Earth in recent years suggests that just about every chemical gradient imaginable can support some sort of life (Lammer et al. 2009).

Water is abundant in our galaxy, e.g., Cernichara & Crovisier (2005) and is expected to be part of the initial inventory of terrestrial planets. In practice, the primary difficulty to get liquid water is thus to be in the right range of temperature and pressure. Pressures must be significantly higher than the triple point (near 6.1 mb). Temperatures should range between the freezing point (0 °C, or lower with dissolved salts) and the boiling point, which depends on the pressure.

2. Four Classes of Habitable Planets

In their review of the factors which are important for the evolution of habitable Earth-like planets, Lammer et al. (2009) proposed a classification of four liquid-water habitat types which we find very useful to structure the scientific debate on habitability. We propose here a slightly simplified version of these classes:

Class I habitats represent Earth-like analog planets where stellar and geophysical conditions allow water to be available at the surface, along with sunlight. Light is important because the most productive natural way of powering an organism is by either using sunlight via photo-synthesis, or by digesting something that does (at least for life as we know it). On Earth, even most subsurface ecosystems derive their energy from photosynthesis. The deep-sea vent communities derive energy from the reaction of H_2S from the vent with O_2 from the ambient seawater. However the O_2 comes from surface photosynthesis, so these ecosystems are ultimately also dependent on it. Only three ecosystems completely independent of photosynthesis have been reported, all of which have limited metabolisms (McKay et al. 2008).

Class II habitats include bodies which initially enjoy Earth-like conditions, but do not keep their ability to sustain liquid water on their surface due to stellar or geophysical conditions. Mars, and possibly Venus are example of this class. On such planets, it is reasonable to assume that life could start, and that this life could potentially migrate to the limited habitats left once the planet is no longer able to sustain liquid water on its surface. On Mars, for instance, deep subsurface aquifers are considered to be potential sites for remnant life, while on Venus it is speculated that some exotic lifeforms could be present in the liquid cloud droplets of the upper atmosphere.

Class III habitats are planetary bodies where water oceans exist below the surface, and where the oceans can interact directly with a silicate-rich core. Such a situation can be expected on water-rich planets located too far from their star to allow surface liquid water, but on which subsurface water is in liquid form because of the geothermal heat. An example of such an environment is given by Europa, one of Jupiter's satellites, which has only about a hundredth of Earth's mass and almost no atmosphere, but which is strongly heated by internal deformation resulting from tidal forces. In such worlds, not only is light not available as an energy source, but the organic material brought by meteorites (thought to have been necessary to start life in some scenarios) may not easily reach the liquid water. Nevertheless, interaction with silicates and hydrothermal activity, also thought to be important for the origin of life, are possible.

Class IV habitats are very water-rich worlds which have liquid water oceans or reservoirs lying above a solid ice layer. Indeed, even if most planets are expected to possess a silicate core covered by a water layer, if this layer is thick enough, water at its base will be in solid phase (ice polymorphs) because of the high pressure. Ganymede, Callisto are likely examples of this class. Their oceans are thought to be enclosed between thick ice layers. In such conditions, the emergence of life may be very difficult because the necessary ingredient for life will be likely completely diluted. Lammer et al. (2009) considered the lack of rocky substrate such a severe constraint that they also put the "Ocean planets" with the ocean lying over a thick ice layer in Class IV, even if the water was liquid at the surface and therefore exposed to light and meteoritic inputs.

3. The Classical Habitable Zone

Considering these four classifications, it seems hard to imagine higher life forms as we know them populating anything but a Class I habitable planet (Lammer et al. 2009). Furthermore, if a planet can only harbor life below its surface, the biosphere would not likely modify the whole planetary environment in an observable way (Rosing 2005). Detecting its presence on an exoplanet would thus be impossible. Within that context, the concept of *habitable* exoplanets is usually limited to surface habitability, and the term "habitable zone" is usually defined as the range of orbital distances within which worlds can maintain liquid water on their surfaces.

The key reference on the subject remains the masterly work of Kasting et al. (1993) (see references therein for previous studies). An up-to-date description of the habitable zone is also available from Selsis et al. (2007) in the framework of their assessment of the habitability of the planets around the star Gl581.

All the studies which now define the "classical habitable zone" are based on 1-D climate modelling, which assesses the habitability of an entire planet by calculating the global average conditions using a single atmospheric column illuminated by the global averaged flux. Projects aiming at improving these calculations in 3-D are described in Section 5.

3.1. Inner Edge of the Habitable Zone

The classical inner edge of the habitable zone is the distance where surface water is completely vaporized or where the warm atmospheric conditions allows water to reach the upper atmosphere. There, it can be rapidly dissociated by ultraviolet radiation, with the hydrogen lost to space (the Earth currently keeps its water thanks to the cold-trapping of water at the tropopause). As thus defined, the inner edge may not be very far inside Earth's current orbit because of a destabilizing mechanism called the "runaway" greenhouse effect: if a planet with liquid water on its surface is "moved" toward the sun, its surface warms, increasing the amount of water vapor in the atmosphere. This water vapor strongly enhances the greenhouse effect, which tends to further warm the surface. On the basis of simple 1-D model calculations, Kasting (1988) found that on an Earth-like planet around the Sun, oceans would completely vaporize at 0.84 Astronomical Units (AU). However, he also showed that the stratosphere would become completely saturated by water vapor at only 0.95 AU, quickly leading the loss of all water. Clearly, this "water-loss" limit is the one of primary physical concern on the inner edge of the habitable zone. A lot of uncertainties exist, and the 0.95 AU limit can be considered to be conservative, mostly because clouds feedbacks are ignored (Kasting et al. 1993). Assuming that clouds may protect a planet by raising its albedo up to 80% (this approximately corresponds to continuous and thick water cloud cover), a habitable planet at about 0.5 AU from the sun is conceivable. This is an extreme value: physical processes able to maintain liquid water at, say, 0.4 AU are hard to imagine.

3.2. Outer Edge of the Habitable Zone

The classical outer edge of the habitable zone is the limit outside which water is completely frozen on the planet surface. Estimating this limit with a classical model of Earth climate with a present-day atmosphere would suggest that this limit is very close to the current Earth orbit because of strong positive feedbacks on the temperature related to the process of "runaway glaciation": a lower solar flux decreases the surface temperatures, and thus increases the snow and ice cover, leading to higher surface albedos which tend to further decrease the surface temperature (Sellers 1969; Gerard et al. 1992; Longdoz & Francois 1997).

In reality, on Earth there is a long-term stabilization of the surface temperature and CO_2 level due to the carbonate-silicate cycle (Walker 1981). This may be the case on other planets, assuming that they are geologically active and continuously outgas or recycle CO_2, and that carbonates form in the presence of surface liquid water. Consequently, CO_2 accumulates until the geological source is balanced by the liquid water sink, which ensures the presence of liquid water (this key assumption is discussed in Section 4.).

Within this context, one can define the outer edge of the habitable zone as the limit where a realistic atmosphere - in terms of composition and thermal structure - can keep its surface warm enough for liquid water. The most likely greenhouse gases on a habitable planet are CO_2 and of course H_2O. Other gases like NH_3 or CH_4 are possible in a reducing atmosphere, but they are rapidly photodissociated so that they must be shielded from solar UV (Sagan & Chyba 1997) or produced by a continuous source or a recycling process (Kasting 1997). It turns out that a thick CO_2 atmosphere may be one of the most efficient solutions for keeping a planet warm. This is not only due to the properties of the CO_2 gas itself. In fact, the greenhouse effect of a purely gaseous atmosphere is limited, and in particular adding more and more greenhouse gas to keep a planet warm does not work indefinitely. The infrared opacity tends to saturate, while the absorbed solar energy decrease because of the increase of the albedo by Rayleigh scattering. If we consider a cloud-free CO_2 atmosphere (with a water pressure fixed by temperature), the classical habitable zone outer edge is at 1.67 AU from the present Sun (with a CO_2 pressure of about 8 bar; Kasting et al. 1993). Recent work suggests that this value is too large, because the CO_2 gas opacity was probably overestimated in the model of Kasting et al.

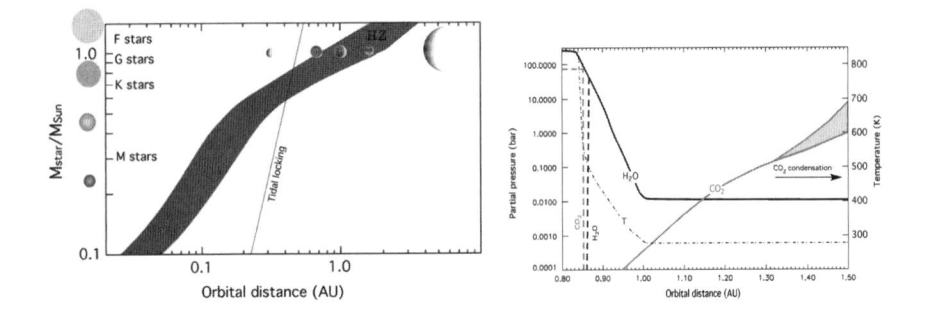

Figure 1. The classical habitable zone (left) and the main atmospheric composition (right) of an Earth-analog atmosphere as a function of distance from its host star. The classical habitable zone theory assumes that geophysical cycles will adjust the atmospheric CO_2 content and its greenhouse effect to compensate for weaker radiation flux when distant from the star. Without that, the habitable zone would be here a thin line rather than the thick swath shown. On the right, the dashed-dotted line represents the surface temperature of the planet and the dashed lines correspond to the inner edge of the habitable zone. The grey zone corresponds to the incertitude related to the effects of CO_2 ice clouds (see text). Figure from Kaltenegger & Selsis (2007) and Lammer et al. (2009) with data from Kasting et al. (1993) and Forget & Pierrehumber (1997).

(1993) (see Wordsworth et al. this issue). However, taking into account the radiative effects of the CO_2 ice clouds, which tend to form in such thick CO_2 atmospheres allows further increases in the warming of the surface thanks to a cloud "scattering greenhouse effect" (Forget & Pierrehumber 1997). Taking into account this process, the outer edge of the habitable zone has been extended as far as 2.5 AU. However, more realistic simulations that spatially resolve the formation and effects of the clouds are required to confirm this value. For now, the value of 2.5 AU may be considered an optimistic upper limit.

3.3. Around Other Stars

To first order, the limits given above for the solar system can be extrapolated to planets orbiting other stars by scaling the orbital distance to the same stellar luminosity, which strongly depends on the stellar mass (Figure 1). However, **stars smaller than the Sun** with low effective temperature emit their peak radiation at longer wavelengths, where the radiation is less reflected by the atmospheric Rayleigh scattering and a water-rich atmosphere is more absorbant. In such conditions the planet is more efficiently heated. The edges of the habitable zone are shifted accordingly (Kasting et al. 1993) (Figure 1). In fact, small M stars with masses 0.1 – 0.5 times the mass of the sun are particularly interesting, because they constitute approximately 75% of the stellar population in our Galaxy, and have a negligible evolution in 10 Gyr. Thus, their "continuously habitable zone" is identical to their initial habitable zone. Terrestrial planets around M stars are also easier to detect! However, estimating their habitability requires us to address several exotic problems such as tidal resonance/locking (in the extreme 1:1 case, this means that one side of the planet will always face the star), active stellar flarings, and the related atmospheric escape (Tarter et al. 2007; Buccino et al. 2007; Joshi 2003; Selsis et al. 2007).

Stars larger than the Sun are much less numerous in the Galaxy, and have a shorter lifetime. If one assumes that, say, 2 Gyr are necessary for complex organisms to form (and build radiotelescopes), only stars with masses less than 1.5 solar mass can be considered. A large stellar mass also affects the radiation output from the star, which is emitted at shorter wavelengths. The stellar light is more readily reflected by an atmosphere, and the habitable zone is shifted accordingly.

4. Habitability and Geologic Activity

The classical theory of habitability and the current definition of the habitable zone relies on the assumption that there is long-term stabilization of the surface temperature and CO_2 level due to the carbonate-silicate cycle. Without this stabilization, Earth would not be habitable, and the habitable zone would be severely limited in size. On Earth, the slow increase of solar fluxes has always been compensated by a decrease in the greenhouse effect, and accidental excursions of the climate toward global glaciation (e.g., Hoffman et al. (1998)) are thought to have been counterbalanced by the CO_2 greenhouse effect without the interruption of life.

The key process allowing the carbonate-silicate cycle on Earth, and more generally the long-term recycling of atmospheric components chemically trapped at the surface, is plate tectonics. This is a very peculiar regime induced by the convection in the mantle, which results from the geothermal heat gradient and surface cooling. How likely is the existence of plate tectonics elsewhere? Is the geophysical stabilization of the climate necessary to maintain life a rare phenomenon? In the solar system, Earth plate tectonics is unique and its origin not well understood. Other terrestrial planet or satellites are characterized by a single "rigid lid" plate surrounding the planet, and this may be the default regime on exosolar terrestrial planets. Plate tectonics is a complicated process that primarily requires lithospheric failure, deformation and subduction (The lithosphere is the "rigid layer" forming the plates that includes the crust and the uppermost mantle). To enable plate tectonics, two conditions have been suggested : 1) Mantle convective stresses large enough to overcome lithospheric resistance to allow plate braking and 2) Plates denser (i.e., colder) than the underlying asthenosphere, to drive plate subduction. On planets smaller than the Earth (e.g., Mars), the rapid interior cooling corresponds to a weak convection stress and a thick lithosphere, and no plate tectonics is expected to be maintained in the long term. On larger planets (i.e., "Super-Earths"), available studies have reached very different views. On the one hand, in their theoretical study entitled "Inevitability of Plate tectonics and Super-Earths", Valencia et al. (2007) showed that, as the planetary mass increases, convection should be more vigorous, making the lithosphere thinner (and therefore reducing lithospheric strengh), while increasing the convective stresses (owing to larger velocities in the mantle). Such conditions should lead to plate tectonics. On the other hand, on the basis of numerical mantle convection simulations, O'Neill & Lenardic

(2007) showed that increasing the planetary radius acts to decrease the ratio of convective stresses to lithospheric resistance. They concluded that super-sized Earths are likely to be in an "episodic or stagnant" lid regime rather than in a plate tectonics regime. Who is right? In fact, the thermo-tectonic evolution of terrestrial planets is a complex combination of phenomena, which has not yet been accurately modeled (D. Breuer, private communication). For instance, the two models mentioned above did not take into account the fact that in super-Earths, the very high internal pressure increases the viscosity near the core-mantle boundary, creating a "low lid" which may decrease the convection and reduce the ability of plate tectonics (Stamenkovic et al. 2009). The effect of size on plate density and subduction has not yet been studied in detail. What these studies highlight is the possibility that the Earth may be very "lucky" to be in an exact size range (within a few percent) that allows for plate tectonics. Venus, which is about the size of the Earth but does not exhibit plate tectonics, shows that the Earth case may be rare, and that many factors control the phenomenon. On Venus, for instance, it is thought that the mantle is drier than on Earth, and that consequently it is more viscous and the lithosphere thicker (Nimmo & McKenzie 1998).

5. Habitability on Local Scales using 3-D Climate Models

So far, nearly all studies of habitability have been performed with simple 1-D steady-state radiative convective models that simulate the global mean conditions. Exceptions to this rule have either been parameterised energy-balance models (EBMs) that study the change in surface temperature with latitude only (Williams & Kasting 1997; Spiegel et al. 2008), or three-dimensional simulations with greatly simplified radiative transfer schemes (Joshi 2003).

5.1. 3-D Models in the Solar System

The climates of most planets within our solar system can now be predicted using 3-D time-marching global climate models (GCMs). A full GCM can be considered as a "planet simulator" that aims to simulate the complete environment on the basis of universal equations only. These models have initially been developed for Earth, as atmosphere numerical weather prediction models (designed to predict the weather a few days in advance) and global climate models (design to fully simulate the climate system and its long term evolution). Such models are now used for countless applications, including tracer transport, coupling with the oceans or the geological CO_2 cycles, photochemistry, data assimilation to build data-derived climate database, etc. Because they are almost entirely built on physical equations (rather than empirical parameters), several teams around the world have been able to succesfully adapt them to other terrestrial planets or satellites. For instance, our team at Laboratoire de Meteorologie Dynamique has adapted the "LMDZ" Earth model to Mars (Hourdin et al. 1993; Forget et al. 1999), Titan (Hourdin et al. 1995), Venus (Lebonnois et al. 2009) and soon Triton and Pluto. These models are used to predict and simulate the volatile cycles, atmospheric photochemistry, clouds and aerosols, past climates, and more.

5.2. Habitability Studies

Relatively complicated 3-D models such as these have several (inter-related) advantages that make them very attractive for future habitability studies. First, they allow simulation of *local* habitability conditions due to e.g., the diurnal and seasonal cycles, which lets us investigate the meaning of the habitable zone more precisely than is possible in a globally averaged simulation. They also help us to better understand the distribution and impact of clouds, which are of central importance to both the inner and outer edges of the habitable zone, as discussed earlier. Finally, 3-D models allow predictions of the poleward and/or nightside transport of energy by the atmosphere and, in principle, the oceans. This is necessary to assess if the planetary water or a CO_2 atmosphere will collapse on the night side of a tidally locked planet, or at the poles of a planet with low obliquity.

5.3. Building a 3-D Model for Exoplanets

In practice, GCMs simulate (a) the motion of the atmosphere, including heat and tracer transport on the basis of the Navier-Stokes equations, (b) the heating and cooling of the atmosphere and surface by solar and thermal radiation (i.e., the radiative transfer), (c) the storage and diffusion of heat in the subsurface, (d) the mixing of subgrid-scale turbulence and convection and (e) the formation, transport and radiative effects of any clouds and aerosols that may be present. Additional levels of complexity may include ice formation / sublimation, interaction with oceans, and even the effects of vegetation and the biosphere. (a), (c), and (d) are almost universal processes. We have learned from studying the solar system that the corresponding parameterizations can be applied without changes to most terrestrial planets. To simulate the 3-D climates on a new planet, the real challenge is to develop a radiative transfer code fast enough for 3-D simulations and versatile enough to model any atmospheric cocktail or thick atmosphere accurately.

We are currently developing such a GCM, which by construction will be general enough to study a wide range of possible habitability scenarios. The model uses the correlated-k method for the radiative transfer, which allows us to calculate the climatic effects of any combination of greenhouse gases accurately, as long as line-by-line data exists over the pressure and temperature range of interest. Due to its importance in habitability studies, we have paid particular attention to the radiative properties of dense CO_2, using new experimental data to produce a more accurate parameterisation of the continuum absorption in the infra-red. The radiative effects of clouds, aerosols and Rayleigh scattering are included through the scheme of Toon et al. (1989), and other processes such as ice formation / sublimation already implemented in the LMDZ Mars GCM are also accounted for.

Development of the model is ongoing, but preliminary studies for the exoplanet Gliese 581d have already begun to yield interesting results. As the planet is extremely close to the outer edge of its system's habitable zone, Selsis et al. (2007) and others have suggested that it could be at least locally habitable. So far, 3-D modelling has shown that due to its eccentric orbit and likely tidally locked rotation, dramatic differences in insolation across the surface can cause local temperatures that are temporarily well above the freezing point of water

Habitability of Planets 63

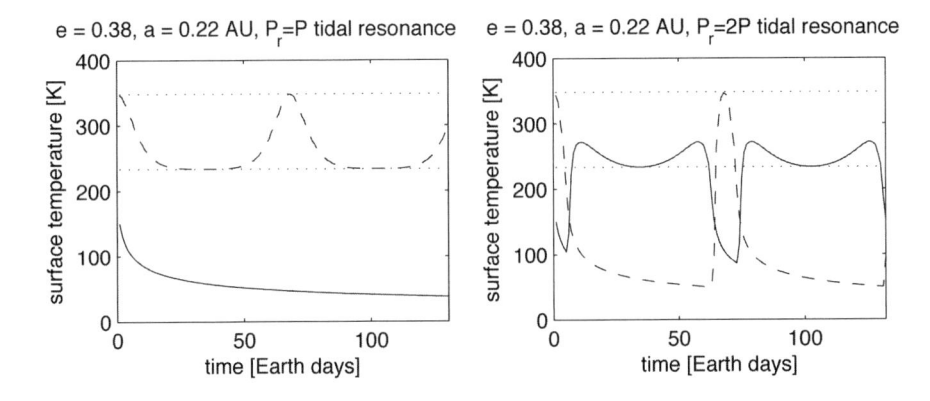

Figure 2. Snapshot surface temperature on the "light" (dashed line) and "dark" (solid line) sides of a hypothetical tidally locked Gliese 581d with minimal atmosphere and surface albedo $A = 0.2$. In both cases the surface heat capacity is 1×10^4 J m^{-2} K^{-1}.

(Figure 2). Thick CO_2 atmospheres increase the global surface temperatures significantly by both traditional greenhouse warming and the CO_2 cloud scattering effect. However, they are extremely vulnerable to collapse via CO_2 condensation in the coldest regions of the planet. As the temperature of CO_2 condensation increases with pressure, the stability of such an atmosphere hangs in a delicate balance between warming and collapse, which can only be accurately investigated using a 3-D climate model.

In the future, we plan to study these outer-edge habitability cases in more detail, as well as making several further improvements to the model. The most important of these will be the addition of a water cycle, including the radiative effects of H_2O and formation and transport of clouds. As well as increasing the realism of the outer edge calculations, this should allow us to investigate "inner edge" cases such as the primitive atmosphere of Venus, in which thick water clouds are expected to play a critical role.

References

Brack, A. 1993, Orig. of Life and Evol. Biosphere, 23, 3
Buccino, A. P., Lemarchand, G. A., & Mauas, P. J. D. 2007, Icarus, 192, 582
Cernicharo, J. & Crovisier, J. 2005, Space Sci.Rev., 119, 29
Forget, F. et al. 1999, JGR, 104, 24155
Forget, F. & Pierrehumbert, R. T. 1997, Sci, 278, 1273
Gerard, J.-C., Hauglustaine, D. A. & Francois, L. M. 1992, Palaeo3., 97, 133
Hoffman, P. F., Kaufman, A. J., Halverson, G. P., & Schrag, D. P. 1998, Sci, 281, 1342
Hourdin, F., Le Van, P., Forget, F., & Talagrand, O. 1993, J. Atmos. Sci., 50, 3625
Hourdin, F., Talagrand, O., Sadourny, R., Régis, C., Gautier, D., & McKay, C. P. 1995, Icarus, 117, 358
Joshi, M. 2003, Astrobiology, 3, 415
Kaltenegger, L. & Selsis, F. 2007 in Extrasolar Planets: Formation, Detection, and Dynamics, ed. R. Dvorak, (New York: Wiley), 79
Kasting, J. F. 1998, Icarus, 74, 472

Kasting, J. F. 1997, Sci, 276, 1213
Kasting, J. F., Whitmire, D. P., & Reynolds, R. T. 1993, Icarus, 101, 108
Lammer, H. et al. 2009, Astron. Astrophys. Rev., 17, 181
Lebonnois, S., Hourdin, F., Eymet, V., Crespin, A., Fournier, R., & Forget, F. 2009, JGR (Planets), in press.
Longdoz, B. & Francois, L. M. 1997, Global and Planetary Change, 14, 97
McKay, C. P., Porco, C. C., Altheide, T., Davis, W. L., & Kral, T. A. 2008, Astrobiology, 8, 909
Nimmo, F. & McKenzie, D. 1998, Ann. Rev. Earth Planet. Sci., 26, 23
O'Neill, C. & Lenardic, A. 2007, GRL, 34, 19204
Rosing, M. T. 2005, Int. J. Astrobio., 4, 9
Rothschild, L. J. & Mancinelli, R. L. 2001, Nat, 409, 1092
Sagan, C. 1996, in Circumstellar Habitable Zones, ed. L. R. Doyle, (Menlo Park, CA: Travis House Publications), 3
Sagan, C. & C. Chyba, C. 1997, Sci, 276, 1217
Sellers, W. D. 1969, J. Appl. Met., 8, 392
Selsis, F., Kasting, J. F., Levrard, B., Paillet, J., Ribas, I., & Delfosse, X. 2007, A&A, 476, 1373
Spiegel, D. S., Menou, K., & Scharf, C. A., 2008, ApJ, 681, 1609
Stamenkovic, V., Noack, L., & Breuer, D. 2009, in European Planetary Science Congress Abstracts, 4, 372
Tarter, J. C., et al. 2007, Astrobiology, 7, 30
Toon, O. B., McKay, C. P., Ackerman, T. P. & Santhanam, K. 1989, JGR, 94, 16287
Valencia, D., O'Connell, R. J., &Sasselov, D. D. 2007, ApJ, 670, L45
Walker, J. C. G., Hays, P. B. & Kasting, J. F. 1981, JGR, 86, 9776
Williams, D. M. & Kasting, J. F. 1997, Icarus, 129, 254

Pathways Towards Habitable Planets
ASP Conference Series, Vol. 430, 2010
Vincent Coudé du Foresto, Dawn M. Gelino, and Ignasi Ribas, eds.

Characterizing Exoplanet Atmospheres and Surfaces

T. Encrenaz

LESIA, Observatoire de Paris, CNRS, Paris, France

Abstract. After more than twelve years of successful campaigns leading to the detection of several hundreds of exoplanets, it is now possible to determine, for some of them, the spectroscopic properties of their atmospheres. In the present paper, our knowledge of solar system planets is used to try to extrapolate what kind of atmosphere can be expected for a given exoplanet, on the basis of its mass, its asterocentric distance, and the spectral type of its host star. This simple classification leads to three main categories of exoplanets: (1) the rocky planets (less than 10 terrestrial masses and within the snow line), (2) the icy planets (below 10 terrestrial masses and beyond the snow line), and (3) the giant planets (above 10 terrestrial masses). It is shown that the thermal spectrum of an exoplanet critically depends upon the existence (or the absence) of a stratosphere, which itself depends upon the exoplanet's composition. The icy and giant planets (Jupiter-, Neptune-, or Titan-type) are expected to have a stratosphere as a result of the methane photodissociation. Rocky planets (Mars- and Venus-type) are expected to show no stratosphere except in the case of the presence of ozone (Earth-type). Typical spectra are shown for each class of object and a discussion is given about the resolving power needed for detecting atmospheric species and/or surface features.

1. Introduction

Since the first detection of an exoplanet around a solar-type star in 1995, about 400 exoplanets have been detected. Most of them have been found by velocimetry, and the lower mass limit is presently below ten terrestrial masses (M_E). This threshold value is critical because, below this value, the gravity field of the object is not sufficient to capture the protosolar gas, mostly composed of hydrogen and helium. The nature of the exoplanet (rocky or icy, depending upon its temperature) is thus expected to be very different from the giant exoplanets. In the case of the solar system, giant planets are believed to have formed beyond the snow line (i.e., the line of water condensation, at a temperature of about 180-200 K) from the accretion of an icy core larger than 10-15 M_E, and the subsequent capture of the protosolar gas (Mizuno 1980). At smaller heliocentric distances, rocky planets formed from the accretion of silicates and metals but the available mass was not sufficient to build objects of more than 1 M_E.

A major result of the exoplanets' exploration has been the detection of a large number of giant exoplanets in the immediate vicinity of their parent star. This situation, very different from the case of the solar system, is usually explained in terms of migration: the giant exoplanet, formed beyond the snow line in a cold environment, interacts with the protoplanetary disk and migrates toward the star. In the case of the solar system, there is also evidence of a moderate migration of the giant planets which has played a major role in the dynamical history of the outer solar system (Morbidelli et al. 2005).

In addition to velocimetry, the transit method has provided us with the detection of several tens of exoplanets. Coupled with velocimetric measurements, this method leads to the determination of the radius, mass and density of the object. Also, spectroscopic measurements of the exoplanet's atmosphere can be achieved, in some cases, during the primary transit (when the planet transits in front of the star) and the secondary transit (when the planet is behind the star). These observations have become possible, in particular in the near-infrared with the HST, and in the mid- and far-infrared with the Spitzer satellite (Tinetti 2010). Such observations will develop in the future with increasing capabilities in detection limit and resolving power.

The purpose of this paper is to provide reasonable guesses about the possible nature of the atmospheres and surfaces of exoplanets, on the basis of a few main parameters: the mass, the asterocentric distance and the spectral type of the host star. This analysis is based on an extrapolation of what is known in the solar system. The main limitation of this analysis is that it does not take into account possible migration effects. Still, it is hoped that the simple classification proposed here may be of some use as a first guess, in the absence of any other information. On the basis of the proposed classification, typical spectra of exoplanets will be shown, and information will be derived upon the spectral resolving power required for detecting specific atmospheric and/or surface species.

2. Spectroscopy of an Exoplanet

As for solar system objects, the spectrum of an exoplanet is characterized by two main components: the reflected stellar light and the thermal emission, corresponding to the absorbed part of the stellar flux. The ratio between the two components depends upon the planet's albedo (i.e. the fraction of reflected stellar light); in the solar system, a typical value of the planetary albedo is about 0.3. In the case of solar system planets, the maximum of the thermal component (in frequency scale) ranges from 7 μm (Mercury) to 100 μm (Uranus and Neptune). In the case of hot Jupiters, it peaks in the near-infrared (at 4 μm for an effective temperature of 1250 K).

In the case of the reflected component, planetary atmospheric features are observed in absorption in front of the stellar continuum. The thermal emission is different, as it mostly depends upon the thermal profile of the emitting atmospheric region. If the temperature gradient is negative (as in planetary tropospheres), lines are observed in absorption because the line center is formed at a higher (and thus colder) level than the wings. In contrast, if the gradient is positive, the lines are observed in emission, as in planetary stratospheres.

A third mechanism is also observed in planetary spectra. Fluorescence emissions can be observed in the UV, visible and near-IR ranges. A stellar photon is absorbed by a planetary species (atom, radical or molecule) and re-emitted, either at the same frequency (resonant scattering) or through a cascade at longer wavelengths. Typical emissions of this kind are the H Lyman α line at 1216 A, N_2 and H_2 bands also in the UV, and CH_4 in the near-infrared.

While the fluorescence mechanism usually allows the probing of atoms and radicals in upper atmospheric regions, the infrared range is best suited for probing exoplanets' neutral atmospheres, as most molecules exhibit their rotational bands and their vibration-rotation bands at wavelengths typically higher than 3 μm.

3. The Solar System: A Planetary Zoo

Within their classes (terrestrial or giant planets), solar system planets show an extreme variety in atmospheric parameters (temperature and pressure, composition). The same diversity is observed among the outer satellites. Still, it is possible to build a simple classification in three main categories, considering mainly their mass and their temperature.

(1) Rocky planets are small and formed within the snow line. Mars and Venus have similar atmospheric compositions (CO_2, N_2) but very different surface pressures (90 bar for Venus, 0.006 bar for Mars) and temperatures (730 K for Venus, about 220 K for Mars). The Earth is an exception because of the emergence of life and the presence of oxygen. Mars and Venus have no stratosphere but the Earth has, due to the photolysis of O_2 and the presence of the ozone layer.

(2) Icy "planets" are small and formed beyond the snow line. They are actually outer satellites but can be considered as possible representatives of a class of exoplanets. There are three objects of this kind (Titan, Triton and Pluto), all characterized with a (N_2, CH_4) atmosphere (with some CO), and with very different surface pressures (1.5 bar for Titan, a few microbars for Triton and Pluto). The photodissociation of CH_4 and the formation of many hydrocarbons leads to the presence of a a stratosphere.

(3) Giant planets were formed (and still are) beyond the snow line. They have a mass larger than 10 M_E and have captured the protosolar gas (mostly hydrogen and helium) around their core. There are often classified in two subcategories, depending upon their masses. Jupiter (318 M_E) and Saturn (90 M_E) are mostly composed of protoplanetary gas and are called the gaseous giants. Uranus (14 M_E) and Neptune (17 M_E) are mostly made of their icy core and are called the icy giants. But all four giant planets have a (H_2, CH_4) atmosphere and thus a stratosphere. Their spectra are comparable, with the difference that in Uranus and Neptune, the temperature is cold enough for many minor species to condense (NH_3, PH_3, H_2O...).

For completeness, we should also quote the bare objects of the solar system. They are mainly of two categories, the refractory objects (Mercury, Moon, asteroids) and the icy objects (outer satellites, comets and trans-neptunian objects).

4. What Kind of Exoplanets Can We Expect?

On the basis of our solar system classification, let us try to extrapolate the possible nature of an exoplanet's atmosphere and surface, on the basis of its mass and its asterocentric distance. Let us first consider a solar-type star. We first estimate the exoplanet's effective temperature on the basis of its distance to the star. The effective temperature is defined as the temperature of the blackbody

Te (K)	1200	850	460	220	120	50
Stellar distance (AU) (solar-type star)	0.05	0.1	0.3	1.5	5.0	20.0

```
Small Exoplanet       <    ROCKY PLANETS      >       <ICY PLANETS >
                              (WARM)                     (COLD)
(0.01 - 10 M_E)  No atmosphere    Atmosphere          Atmosphere
                 (Mercury-type)   N_2, CO_2, CO, H_2  N_2, CH_4(+CO)
                                  (Mars-Venus type)   hydrocarbons, nitriles
                                  if O_2 -> O_3 (Earth-type)   (Titan-type)
                                       STRATOSPHERE        STRATOSPHERE


Giant Exoplanet   <PEGASIDES>        < GASEOUS GIANTS >   <ICY GIANTS>
                    (HOT)                 (WARM)             (COLD)
(10 - 1000 M_E)  Atmosphere           Atmosphere          Atmosphere
                 H_2,CO,N_2,H_2O      H_2,CH_4,NH_3,H_2O  H_2,CH_4
                                      hydrocarbons        hydrocarbons
                                      (Jupiter-type)      (Neptune-type)
                                      STRATOSPHERE        STRATOSPHERE
```

Figure 1. What kind of exoplanets can we expect?

which emits the same quantity of absorbed stellar flux. The following equation is used:

$$[F^*/D^2](1-a) = 4\sigma T_e^4 \qquad (1)$$

where F^* is the stellar flux, D the distance between the exoplanet and its star, a is the planetary albedo, σ the Stefan constant and T_e the effective temperature of the exoplanet. Note that this equation corresponds to a fast-rotating planet (like the Earth or the giant planets), which radiates over the entire sphere the stellar flux which it collects on the day side; if the exoplanet had a slow rotation (like Venus or Mercury), the factor of 4 would be replaced by a factor of 2. We assume for the albedo a value of 0.3, typical of solar system planets.

Figure 1 summarizes the various classes of exoplanets as a function of their mass and their stellar distances, and the type of atmosphere which can be foreseen. These classes are described below.

(1) Small exoplanets (M ≤ 10 M_E) show two sub-classes:

- *the rocky exoplanets* (warm), with temperatures higher than about 200 K (coresponding to stellar distances of about 3-4 AU). Near the star they have no atmosphere (Mercury-type). At larger distances (the frontier depending upon their mass), they are expected to have a (CO_2, N_2) atmosphere with no stratosphere (Mars, Venus-type). The presence of an ozone stratospheric layer could be considered as a possible evidence for life (Earth-type).

- *the icy exoplanets* (cold), with temperatures lower than 180 K, beyond about 4 AU. Like Titan, Triton and Pluto, their atmospheres would be presumably made of N_2 and CH_4, with possibly CO. The presence of nitrogen is expected because NH_3 (the main alternate nitrogen-bearing molecule) is very easily transformed into N_2 (by photodissociation in particular) while the opposite reaction is very difficult. As a result of this atmospheric composition, stratospheric hydrocarbons and nitriles are expected to be present.

Table 1. Relationship between the temperature and the asterocentric distance as a function of the stellar type

Te (K)	1200	850	460	273	220	120	50
A (T = 10000 K)	0.15	0.3	0.9	3.0	4.5	15.0	60.0
F (T = 7000 K)	0.08	0.16	0.5	1.6	2.4	8.0	32.0
G (T = 5700 K)	0.05	0.1	0.3	1.0	1.5	5.0	20.0
K (T = 4200 K)	-	0.04	0.12	0.4	0.6	2.0	8.0
M (T = 3200 K)	-	-	0.04	0.14	0.2	0.4	1.4

(2) Giant exoplanets ($10\ M_E \leq M \leq 1000\ M_E$)
In the solar system, giant planets are found beyond the snow line. However we know that hot Jupiters are found in the close vicinity of their parent stars, probably as a result of migration. We thus consider three sub-classes:
- *Hot Jupiters*: Numerous exoplanets of this kind have been detected in the close vicinity of their host stars. Thermochemical calculations allow us to get a first estimate of their possible atmospheric composition. In addition to hydrogen and helium, N_2, CO and H_2O are expected to be found within 0.05 AU; H_2, CO, NH_3 and H_2O are likely to be found between 0.05 and 0.1 AU; H_2, NH_3, CH_4 and H_2O are expected to be found beyond 0.1 AU (Goukenleuque et al. 2000). In the latter case the presence of a stratosphere is expected due to the presence of methane.
- *Gaseous giants*: Their composition could be comparable with the ones of Jupiter and Saturn, at stellar distances above a few AUs. The temperature (above 80 K) is sufficient for a few tropospheric species to be detectable (PH_3, GeH_4, AsH_3...) in addition to H_2 and CH_4. Again, a stratosphere is expected.
- *Icy giants*: At asterocentric distances above about 15 AU, the temperature is cold enough for most of the minor species to condense at detectable levels (at a pressure level of 0.1 – 1 bar), including methane. The amount of saturated methane above the tropopause is still sufficient for CH_4 to be photodissociated, leading to the formation of a stratosphere where several hydrocarbons are present (C_2H_2, C_2H_6).

How does this classification evolve in the case of a non-solar type star? The relationship between the exoplanet's effective temperature and its stellar distance is changed. For A, F, G, K and M stars, stellar temperatures are typically 10000 K, 7000 K, 5700 K, 4200 K and 3200 K respectively. As a result, the snow line is shifted toward the inside for low-mass stars (K and M types) and outside for massive stars (A and F types). Table 1 gives the relationship between the exoplanet's effective temperature and its stellar distance for the various stellar types. Using both Figure 1 and Table 1, it is possible, to first order, to estimate the basic composition of an exoplanet's atmosphere, knowing its mass and its asterocentric distance.

There are several limitations to keep in mind regarding the above classification; a few of them are listed below. First, the possible effect of migration within a planetary disk may considerably modify the simple "static" view described above. In addition, other planetary parameters may influence the radia-

70   Encrenaz

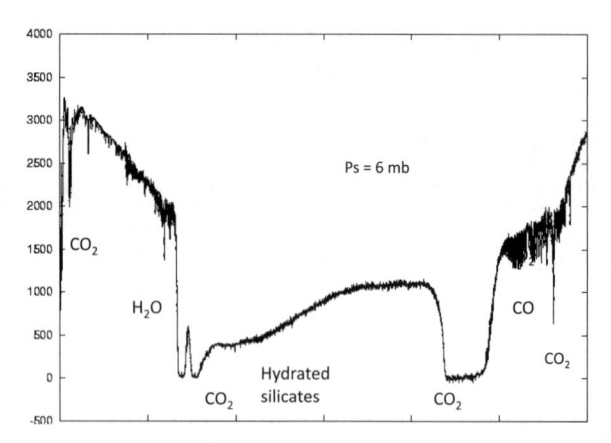

Figure 2. The spectrum of Mars, as observed by ISO-SWS. The spectrum is dominated by CO_2 absorptions. Adapted from Lellouch et al. (2000).

tive balance between the stellar flux and the planetary emission: (1) the value of the albedo; (2) the rotation period of the planet; (3) a possible greenhouse effect as in the case of Venus, which depends upon the atmospheric content in greenhouse gases (CO_2, H_2O, CH_4); (4) the cloud structure which also influences the albedo; (5) the eccentricity and the obliquity of the planet which may induce seasonal changes and atmospheric dynamics; (6) the magnetic field which, if present, may affect the long-term evolution of the atmosphere as it tends to prevent the atmospheric escape.

5. Small Rocky Exoplanets

5.1. Spectral Signatures

The spectrum of a small rocky exoplanet is characterized by signatures of CO_2, CO and H_2O. Water vapor is by far the most active spectroscopic agent. In the case of Mars (and also Venus above the clouds), the H_2O content is very low (less than a percent) and the CO_2 signatures dominate (Figure 2), while water vapor dominates the terrestrial spectrum (Figure 3). Solid signatures can be detected at 1.0 and 2.0 μm (silicates and ferric oxides), and 3.2 – 3.5 μm (hydrated silicates). Water ice has characteristic signatures at 1.5, 2.0, and 3.1 μm.

Another surface signature of potential interest is the Red Vegetation Edge (RVE). This sharp slope in the near-infrared spectrum of the terrestrial surface is due to the presence of chlorophyll and has been considered as a possible biomarker on exoplanets. Attempts have been made to measure this signature on Earthshine observations (Seager et al. 2005). The interpretation is difficult, however, due to the presence of the water ice cloud cover which partly hides the terrestrial surface, at a level of 20 – 30 percent. The same problem might also take place in the case of exoplanets' observations.

The dark side of a hot rocky exoplanet (like Venus) shows an interesting near-infrared spectrum. With a surface temperature above 700 K, the thermal

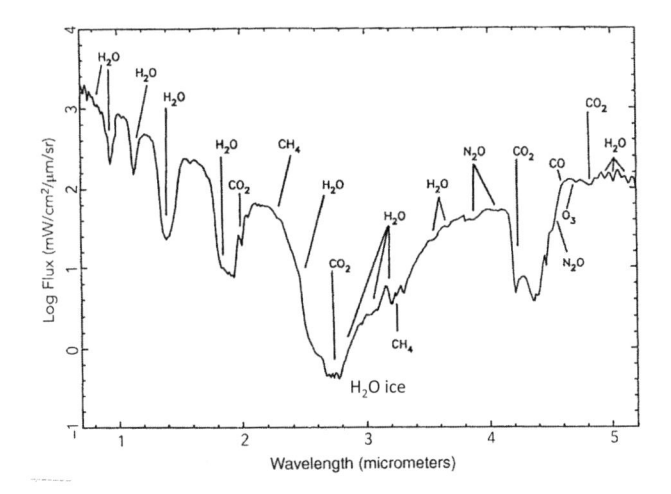

Characterizing Exoplanet Atmospheres and Surfaces   71

Figure 3. The near-infrared spectrum of the Earth, as observed by the Galileo probe during its flyby in Dec. 1990. Water vapor signatures dominate the spectrum. After Drossart et al. (1993).

emission coming from the surface and the lower troposphere is strong enough to be detectable even at short wavelengths, down to 1 μm, between the strong CO_2 bands (Carlson et al. 1991). With enough sensitivity, such kind of spectrum could be possibly detectable on rocky exoplanets in the future, before and after primary transits.

The thermal emission of a rocky planet is dominated by the strong CO_2 band at 15 μm. On Mars and Venus, the core of the band is in absorption (because of the absence of stratosphere), but, in the case of Mars, the wings can appear in emission if the surface temperature is colder than the atmosphere (Figure 4). The solid signatures to be searched for in this spectral range are H_2O ice around 12 μm (700 – 900 cm^{-1}) and silicates around 9 μm (1000 – 1200 cm^{-1}).

On Earth, the thermal spectrum shows the band of ozone at 9.5 μm and, more marginally, the CH_4 band at 7.7 μm (Figure 5). A low resolving power (R = 3) is sufficient for detecting the CO_2 band. The detection of O_3 on an Earth-like exoplanet would require a resolving power of 10, while R= 50 is needed to detect the CH_4 band.

5.2. Variations of the Spectrum with the Stellar Type and the Stellar Distance

Figure 6 shows the variation of a Mars-type spectrum between 1 and 5 μm, as a function of (a) the stellar type and (b) the asterocentric distance. It can be seen that, for F stars, the thermal component is expected to dominate for $\lambda \geq 2.5$ μm (still assuming an albedo of 0.3). In the case of a solar-type star, at small distances (D $\leq$ 0.1 AU), the reflected component dominates.

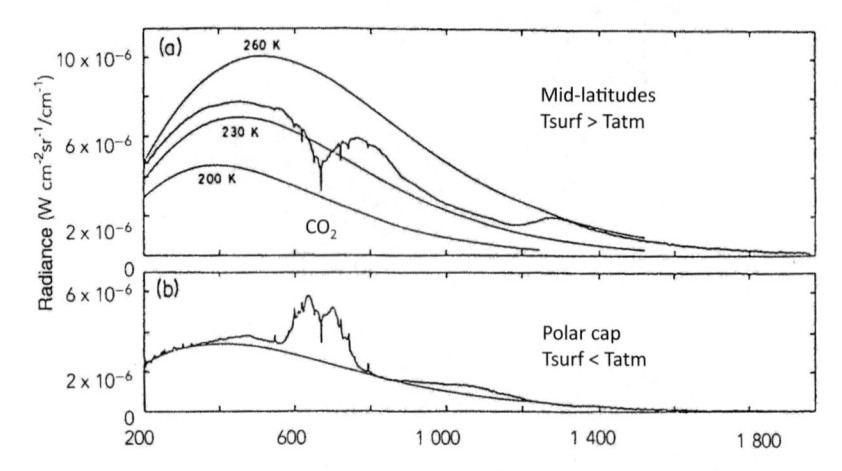

Figure 4. The thermal spectrum of Mars measured by Mariner 9 at mid latitudes (top) and over a polar cap (bottom). As a result, the shape of the CO_2 band at 15 μm is drastically different. The figure is adapted from Hanel et al. (1992).

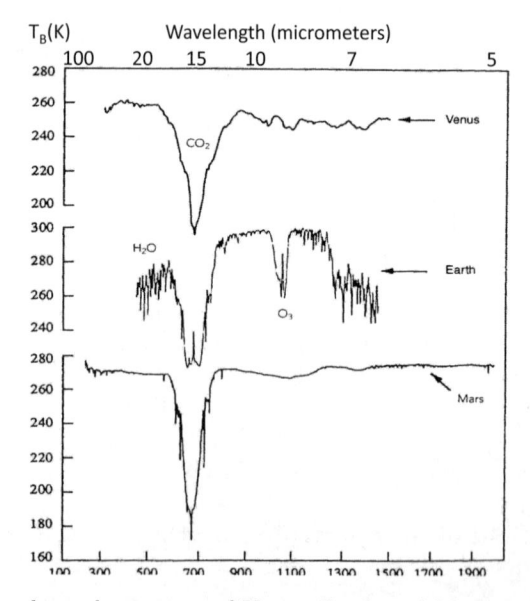

Figure 5. The thermal spectrum of Venus, Earth and Mars expressed in brightness temperatures. It can be seen that on Earth, the CO_2 and O_3 bands show an emission core because of the presence of a stratosphere. After Hanel et al. (1992).

Characterizing Exoplanet Atmospheres and Surfaces 73

(a) Variation with the stellar type (b) Variation with the heliocentric distance
(D = 1 UA) (solar-type star)
Radiance (μwatt/cm^{2}sr/μm)

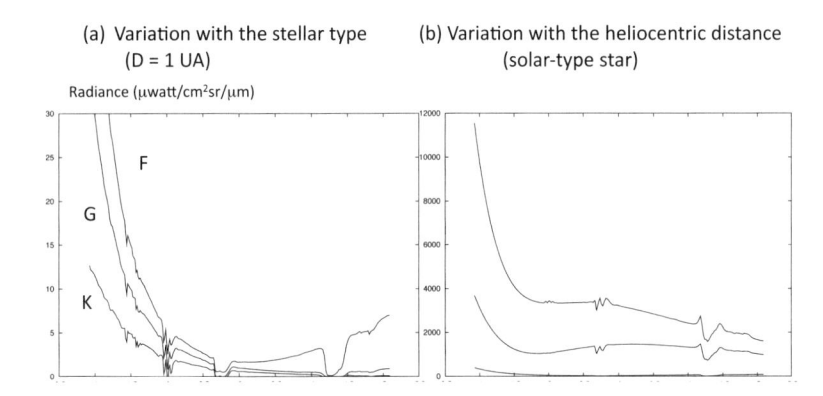

Figure 6. Left: (a) Variations of a Mars-type spectrum as a function of the stellar type. From top to bottom: : F-type star (T = 7000 K), G-type star (T = 5700 K) and K-type star (T = 4200 K). Corresponding values of Te are 356 K, 273 K and 174 K respectively. Right: (b) Variation of the same spectrum as a function of the stellar distance, in the case of a solar-type star. From top to bottom: D = 0.1, 0.3 and 1.0 AU.

6. Icy and Gaseous Exoplanets

Although very different in nature, icy and gaseous exoplanets (except for hot Jupiters) could show comparable near-infrared spectra (Larson 1980), because methane, in all cases, is the dominant spectroscopic agent in spite of its low mixing ratio (a few percent at most). The situation would be different for Pegasides, because of the expected presence of water vapor. Hydrogen, water and ammonia may also be expected in the spectrum of a hot Jupiter.

In terms of solid signatures, ices can be searched for at the surface of icy planets (H_2O, CH_4, CO, N_2...) while condensates of H_2O, CH_4, hydrocarbons (C_2H_2, C_2H_6...), NH_3, NH_4SH or H_2S could be present in the clouds of the gaseous planets.

There is a striking difference between the thermal spectra of Jupiter and Saturn (Figure 7). Still, as observed from another star, these two giant planets could be considered as similar. What is the reason for such differences? It is probably connected to the stronger vertical mixing of Saturn, which possibly comes from a stronger internal energy source. As a result, PH_3, a disequilibrium species, is more abundant; in contrast, NH_3, which shows a strong absorption band in Jupiter, is absent in Saturn's spectrum because it is condensed in clouds, as a result of Saturn's lower temperature. In addition, emission and absorption features are present on both spectra, and their interpretation depends upon the thermal atmospheric structure. This example illustrates how it may be difficult to interpret thermal spectra of gaseous exoplanets when these data become available.

What is the resolving power needed to separate the atmospheric species of a giant exoplanet? In the case of a Jupiter-type planet, $R \geq 150$ is required for detecting CH_4 and $R \geq 100$ is required for detecting NH_3. The detection of hydrocarbons is easier: only $R \geq 20$ is needed for detecting C_2H_6. In the case of

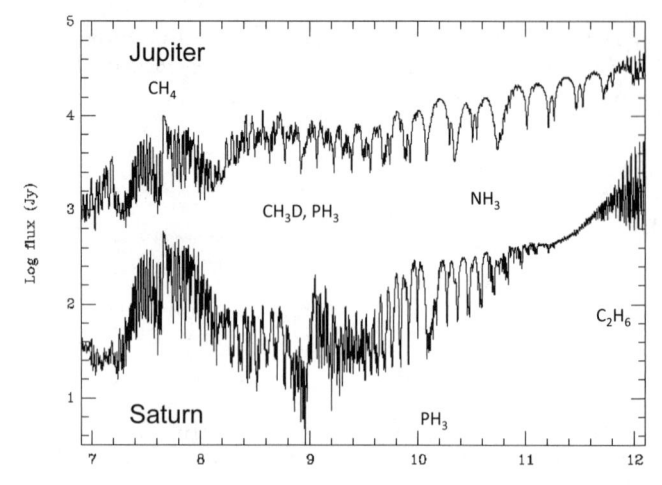

Figure 7. The thermal spectra of Jupiter and Saturn, measured by ISO-SWS. Both spectra are combinations of emission and absorption features. The emission features are due to stratospheric CH_4 (at 7.7 µm) and C_2H_6 (at 12 µm). Other features are tropospheric absorptions (in particular NH_3 in Jupiter and PH_3 in Saturn. Adapted from Encrenaz et al. (2003).

a small icy exoplanet or an icy giant exoplanet, the detection of CH_4 and other hydrocarbons is made easier by the absence of other species (Figure 8). $R \geq 5$ is sufficient to detect hydrocarbons (C_2H_2-C_2H_6); $R \geq 10$ allows the detection of CH_4.

7. Conclusions

Starting from our knowledge of solar system planets, a simple classification has been proposed to try to extrapolate the possible nature of the exoplanets' atmospheres, on the basis of their mass, their asterocentric distance and the stellar type of their host star. Three main classes are defined: the small rocky planets (temperature above about 200 K), the small icy planets (colder temperature) and the giant planets (mass above 10 terrestrial masses).

Using this simple classification, it is possible to constrain, to first order, the possible properties of their spectra, both in the reflected and the spectral ranges. Thermal spectra, in particular, strongly depend upon the presence or the absence of a stratosphere, and the knowledge of the thermal structure is essential for their interpretation. No stratosphere is expected in the case of rocky planets, except if oxygen is present; a stratosphere is expected in all the other cases.

In the reflected range, the spectrum of rocky planets should be dominated by CO_2 and H_2O, while CH_4 is expected to dominate the spectra of icy and giant planets. The near-infrared spectra of hot Jupiters could show the signatures of H_2, H_2O, CH_4, and possibly CO, CO_2 and NH_3. In the thermal range, a resolving power higher than 10 is typically required for the identification of major gaseous

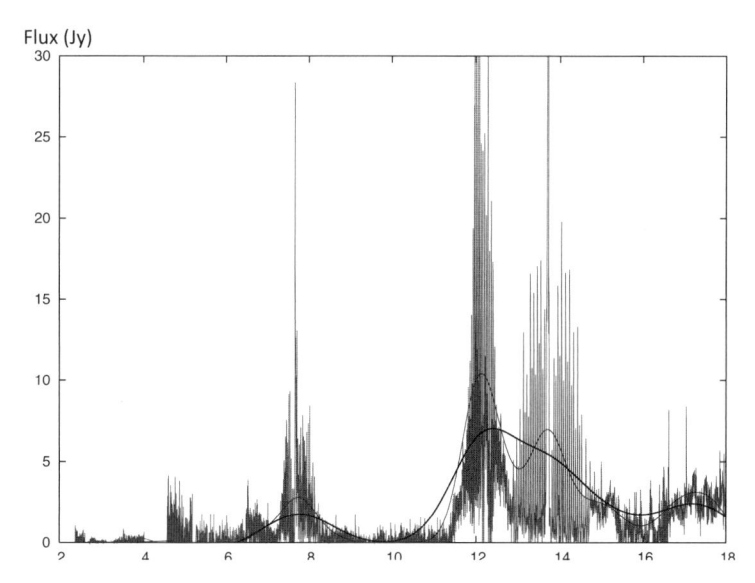

Characterizing Exoplanet Atmospheres and Surfaces

Figure 8. The thermal spectrum of Neptune, measured by ISO-SWS. The emission features are due to stratospheric CH_4 (at 7.7 μm), C_2H_6 (at 12 μm) and C_2H_2 (at 14 μm). The spectrum is convolved with R = 5 (thin line) and R = 10 (thick line).

and solid signatures; as an exception, hydrocarbons (C_2H_2, C_2H_6) should be detectable with R ≥ 5, and thus easier to detect than methane itself.

References

Carlson, R. W. et al. 1991, Sci, 253, 1541
Drossart, P. et al. 1993, Plan. Space Sci., 41, 551
Encrenaz, T., Drossart, P., Orton, G., Feuchtgruber, H., Lellouch, E., & Atreya, S. K. 2003, Plan. Space Sci., 51, 89
Goukenleuque, C., Bézard, B., & Lellouch, E. 2000, in ASP Conf. Ser. 212, From Giant Planets to Cool Stars, eds. C. Griffith and M. Marley, (San Francisco: ASP), 242
Hanel, R. A., Conrath, B. J., Jennings, D. E., & Samuelson, R. E. 1992, Exploration of the Solar System by Infrared Remote Sensing, (Cambridge: Cambridge University Press)
Larson, H. P. 1980, ARA&A, 18, 43
Lellouch, E. et al. 2000, Plan. Space Sci., 48, 1393
Mizuno, H. 1980, Prog. Theor. Phys., 64, 544
Morbidelli, A., Levison, H. F., Tsiganis, K., & Gomes, R. 2005, Nat, 435, 462
Seager, S., Turner, E. L., Schafer, J., & Ford, E. B. 2005, Astrobiology, 5, 372
Tinetti, G. 2010, this volume

Pathways Towards Habitable Planets
ASP Conference Series, Vol. 430, 2010
Vincent Coudé du Foresto, Dawn M. Gelino, and Ignasi Ribas, eds.

Habitable Planets: Targets and their Environments

M. Güdel

ETH Zurich, Institute of Astronomy, Zurich, Switzerland

Abstract. The environment in which a planet resides is largely controlled by the host star, not least by its magnetic activity. The resulting interplay between "the environment" and a planet may include magnetic interactions, interactions between stellar radiation (optical, UV, X-rays) and planetary atmospheres, and interactions between stellar particle fluxes (high-energy particles, stellar winds) and planetary magnetospheres.

1. Introduction

Habitability of planets is most significantly determined by the radiation environment defined by the host star. Kasting et al. (1993) defines habitability based on the criterion of the potential presence of liquid water on a planet bearing an Earth-like atmosphere; the inner edge of the habitable zone (HZ) is defined by loss of water as a consequence of photolysis and hydrogen escape, and the outer edge is determined by the formation of CO_2 clouds cooling the planet by the increased albedo, leading to runaway glaciation. Of course, the exact location of the HZ depends on the atmospheric composition, in particular the presence of greenhouse gases such as CO_2 or CH_4.

Further conditions contribute to the potential habitability or inhospitality of planets. Specifically, the presence of plate tectonics has been recognized to be of prime importance for climate stabilization through the carbonate-silicate cycle (Walker et al. 1981; Kasting & Catling 2003). Also, habitable zones vary on evolutionary time scales of the host star, with the slow increase of a main-sequence star's luminosity leading to increasing radii of the HZ. For example, the zero-age main-sequence Sun's bolometric luminosity was about 30% lower than the present-day Sun's (Sackmann & Boothroyd 2003), implying less insolation of the young planetary atmospheres in our solar system (Sagan & Mullen 1972).

However, actual habitable conditions on planets also depend on a number of additional properties of the host star not related to the relatively stable optical output of the solar surface. Planetary atmospheres are particularly susceptible to short-wavelength/high-energy radiation from the outer layers of the host stars, induced by magnetic activity. Further, late-type main-sequence stars are assumed to continuously lose mass in ionized, magnetized stellar winds that strongly interact with planetary atmospheres or magnetospheres. In contrast to the optical output from a main-sequence star, the short-wavelength radiation is subject to very strong variations on evolutionary time scales, on time scales related to so-called activity cycles, on daily to monthly time scales reflecting the changes in the magnetic configuration on the star, and on hourly time scales related to flares. In turn, these variations are a function of stellar mass. The combination of the optical luminosity (determining the "classical" HZ for the potential presence of liquid water) and the magnetic activity behavior of the star (driven by an internal dynamo and being a strong function of stellar age) leads to a number of important constraints for habitable planets. The following sections summarize issues related to stellar magnetic activity relevant to the habitability of planets, and discuss implications for the conditions for life.

2. The Radiation Environment in Stellar Evolution

The Sun has significantly brightened during its evolution on the main sequence, starting on the zero-age main sequence at a level of approximately 70% of its present-day bolometric luminosity (Sackmann & Boothroyd 2003). The lower luminosity of the young Sun poses a problem for our understanding of the evolution of young planetary atmospheres as both Earth and Mars should have been in deep freeze during the first few 100 Myr (Sagan & Mullen 1972; Kasting & Catling 2003). In contrast, geological records indicate a warmer climate and the presence of liquid water on both planets at ages younger than 1 Gyr. Although greenhouse gases may significantly elevate atmospheric temperatures, even a high-pressure CO_2 rich atmosphere on Mars would not have produced a sufficiently mild climate (Kasting 1991; Sackmann & Boothroyd 2003). However, upper atmospheres of planets are much more susceptible to changes in the short wavelength (ultraviolet [UV], far-ultraviolet [FUV], extreme-ultraviolet [EUV], and X-ray) radiation due to stellar magnetic activity. Opposite to the bolometric luminosity of a star, its magnetically induced high-energy luminosity decreases with time as the internal dynamo decays as a consequence of stellar spin-down by angular momentum loss. The short-wavelength decay is much more dramatic than the optical increase, the X-ray luminosity decreasing over three orders of magnitude during the main-sequence life of a star; the radiation also becomes much softer on evolutionary time scales (e.g., Güdel et al. 1997). The decay time scale of magnetic activity is a function of stellar mass. While the X-ray luminosity of a solar-mass star drops by a factor of ≈100 during the first Gyr, stars of spectral type M may reside at a high, saturated magnetic activity level during this entire period, after which a rather slow decay sets in (West et al. 2008).

The decay laws are also a function of wavelength: shorter wavelength radiation decays by larger factors than longer wavelength radiation. Ribas et al. (2005) studied the long-term evolution of radiation less than 1700 Å to construct a comprehensive model of the spectral evolution of the "Sun in Time". The integrated irradiances correlate tightly with the stellar rotation period or age, the relations being represented by power laws, as illustrated in Figure 1. From band-integrated luminosities and individual lines (see Telleschi et al. 2005 for X-ray lines) one finds the following decay laws for chromospheric, transition region UV, transition region FUV, EUV (including the softest X-rays at 20–100 Å), and soft X-rays (1–20 Å), where we also use the rotation law given by Ayres (1997), namely $\Omega \propto P^{-1} \propto t^{-0.6 \pm 0.1}$ for a solar analog (P is

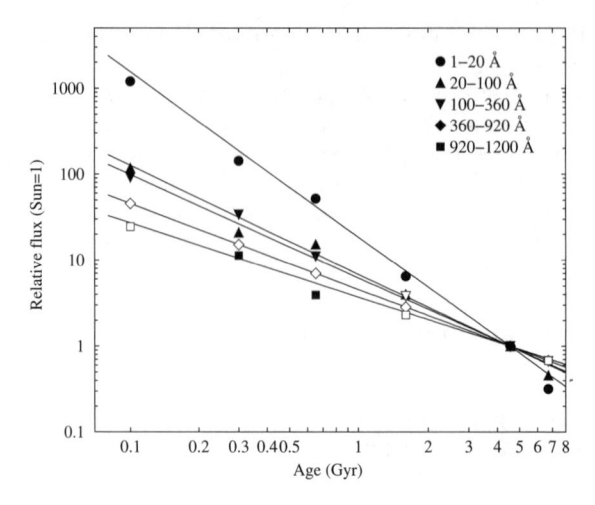

Figure 1. Power-law decays in time for various spectral ranges, normalized to the present-day solar flux. Note that the hardest emission decays fastest (from Ribas et al. 2005, reproduced by permission of AAS).

the stellar rotation period, Ω the angular rotation rate, and t the stellar age):

$$L_{\rm ch} \propto P^{-1.25\pm0.15} \propto t^{-0.75\pm0.1} \quad (1)$$

$$L_{\rm UV} \propto P^{-1.60\pm0.15} \propto t^{-1.0\ \pm0.1} \quad (2)$$

$$L_{\rm FUV} \propto P^{-1.40} \propto t^{-0.85} \quad (3)$$

$$L_{\rm EUV} \propto P^{-2.0} \propto t^{-1.2} \quad (4)$$

$$L_{\rm X} \propto P^{-3.2} \propto t^{-1.9} \quad (5)$$

The heating of planetary upper atmospheres (thermospheres) and the generation of ionospheres is primarily due to the solar EUV-to-X-ray (XUV) radiation. As the mean free path in the outer atmosphere (in the exospheric layer) becomes sufficiently small, the lighter, heated species may escape into space if their thermal velocity exceeds the gravitational escape velocity (e.g., Kulikov et al. 2007). For sufficiently high gas temperatures, the upper atmosphere may move off hydrodynamically (see, e.g., Watson et al. 1981).

Atmospheric evaporation induced by short-wavelength heating may have been responsible for the loss of a large water reservoir initially present on Venus. A water ocean would evaporate as a consequence of the strong insolation, inducing a "runaway greenhouse" in which water vapor would become the major constituent of the atmosphere (Ingersoll 1969). Water vapor was then photodissociated in the upper atmosphere by *enhanced* solar EUV irradiation (Kasting & Pollack 1983), followed by the escape of hydrogen into space (Watson et al. 1981; Kasting & Pollack 1983). Detailed model calculations suggest that the amount of water in the present terrestrial ocean could have escaped from Venus in only 50 Myr if the H_2O mass fraction (mass mixing ratio) was at least 0.46, and the XUV flux was 70–100 times the present values (Kulikov et al. 2007). Evaporative losses may also have been relevant for young Earth and Mars although the protecting magnetosphere of Earth (and some magnetism also on Mars) may partly account for a different scenario compared to Venus, and hydrodynamic blow-off may not have occurred (Lammer et al. 2003).

Atmospheric evaporation driven by short-wavelength irradiation should then be very important for exoplanets in close orbit around their host stars. Penz et al. (2008) considered Roche-lobe effects to the hydrodynamic loss, and used realistic heating efficiencies and IR cooling terms for HD 209458b, a Jupiter-sized planet in orbit around an old, inactive solar analog. The atmospheric temperatures in their model reach ≈ 8000 K at a distance of $1.5\,R_{\text{planet}}$ which leads to blow-off due to gravitational effects by the Roche lobe. These authors obtained atmospheric loss rates of 3.5×10^{10} g s^{-1}. Assuming 100 times higher XUV flux at an age of 0.1 Gyr, the loss rate may exceed 10^{12} g s^{-1}, leading to an integrated hydrogen mass loss of 1.8–4.4% of the present mass over the entire main sequence (MS) life time of the host star. This loss rate is not leading to substantial alterations in the planet but may still play a significant role in the evolution of its atmosphere. Also, higher evaporation fractions are possible for planets orbiting closer to the host star.

In this context, the very slow evolution of magnetic activity in M dwarfs is potentially important for planets in their habitable zones. The high L_X (relative to L_{bol} defining the habitable zone) affects planetary atmospheres (and potentially the presence of water) for a much longer time than for a more massive star (e.g., Scalo et al. 2007, see below).

3. High-energy Environments

Apart from (quasi-)steady short-wavelength radiation from magnetically active (chromospheric and coronal) regions on the star, explosive energy release in flares may alter the radiative environment of planets on time scales of minutes to hours, by orders of magnitude in flux. Flares are most likely a consequence of magnetic reconnection in the stellar corona. This process leads to both heating and accelerated electrons which travel toward lower layers where they impact in the chromosphere, thus heating this material and evaporating it into the corona where high levels of X-rays are emitted. Some flares, in particular the most energetic examples, are accompanied by "Coronal Mass Ejections" (CMEs) that join the stellar wind at high velocities.

Extreme flare events may be effective in modifying planetary atmospheres. Schaefer et al. (2000) identified, regardless of main-sequence age, "superflares" for solar-like stars with total radiated energies of order $10^{35} - 10^{38}$ erg (in X-rays or optical bands). Their recurrence time must be long (decades to centuries), but the irradiation by some of these flares may exceed the total irradiation input by the *whole* star for maybe an hour. The result could be temporary excess heating, break-up of the ionosphere, and build-up of nitrous oxides at high altitudes. Nitrous oxides in turn destroy ozone; Schaefer et al. (2000) estimate that an event with 10^{36} erg of ionizing energy results in 80% of ozone loss for more than a year, thus increasing the ultraviolet irradiation of the planetary surface from normal stellar emission. Again, the prolonged activity of M dwarfs makes this an important issue for planetary habitability.

However, the class of large flares may, on average and on long time scales, not be energetically important for the stellar XUV output. Many flares are too small to be recognized individually in disk-integrated light curves, but these are the most frequent events. Both in the case of the Sun and active stars, X-ray and EUV studies show that flares are distributed in energy according to a power law,

$$\frac{dN}{dE} = kE^{-\alpha} \qquad (6)$$

where dN is the number of flares per unit time with a total energy in the interval $[E, E + dE]$, and k is a constant. If $\alpha \geq 2$, then the energy integration (for a given time interval, $\int_{E_{\min}}^{E_{\max}} E[dN/dE]dE$) diverges for $E_{\min} \to 0$, that is, if the power law is extrapolated to small flare energies, a lower cut-off is required for the power-law distribution. On the other hand, any energy release power is possible depending on the value of $E_{\min}$.

Specifically, for young, magnetically active G–M dwarf stars including pre-main sequence objects, monitoring studies in the EUV and X-ray bands found α mostly in the range of 2–2.5 for G–M dwarfs (Audard et al. 2000; Kashyap et al. 2002; Güdel et al. 2003; Stelzer et al. 2007), supporting the view that moderate flares are the dominant heating source for these active coronae, and therefore the dominant contributors to short-wavelength radiation. Supporting evidence comes from a *linear* correlation between the rate of EUV flares (above a given threshold) and the time-averaged X-ray luminosity for late-type stars (Audard et al. 2000), and probably also for T Tauri stars (Stelzer et al. 2007).

Why is it important that short-wavelength radiation is due to flares rather than a more steady source? One important aspect for the evolution of planetary atmospheres is the short-term fluctuation of the input energy. Although details are not well studied, fluctuating XUV radiation may drive planetary atmospheres significantly out of equilibrium as a consequence of variable photochemical reactions (e.g., Scalo et al. 2007).

Further, although the soft X-ray output of a frequently flaring corona may be nearly equivalent to a constant radiator, the spectrum of the former is different as it should be accompanied by a tail of non-thermal X-rays above 10 keV (Güdel 2009), as a result of the bombardment of the chromospheric layers by accelerated electrons. Flare-produced gamma-rays and X-rays will usually not propagate to the surface of planets with any substantial atmospheres, but a fraction of the hard radiation can be reprocessed and re-emitted as UV light that showers the planetary surface, perhaps at biologically relevant doses (Smith et al. 2004). This secondary radiation may involve damaging effects on living cells, but may also act as an evolutionary driver.

4. Stellar Winds and Non-thermal Escape

While the solar wind has been well investigated in situ, relatively little is known about such winds from other solar-like stars. None of these winds has so far been detected directly (e.g., from their bremsstrahlung, charge exchange, etc.), but interesting upper limits to the mass loss rate have been derived from corresponding searches (e.g., $\dot{M} < 1.7 \times 10^{-11}\, M_\odot$ yr^{-1} for the *evolved* subgiant Procyon, Drake et al. 1993; $\dot{M} < 7 \times 10^{-12}\, M_\odot$ yr^{-1} or $\dot{M} < 3 \times 10^{-13}\, M_\odot$ yr^{-1}

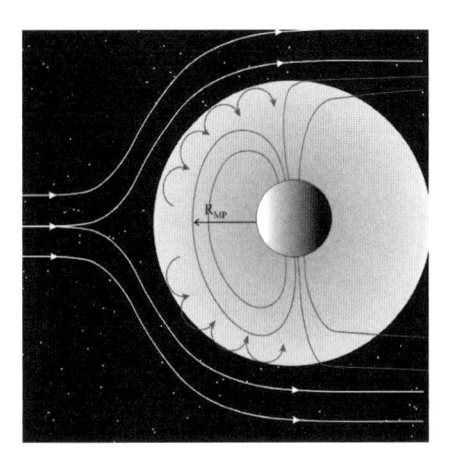

Figure 2. Interaction between a stellar wind or a coronal mass ejection with a moderate planetary magnetosphere but extended (heated) upper atmosphere. The atmosphere above the compressed magnetospause can be eroded (from Lammer et al. 2007).

for the old M dwarf Proxima Centauri, Lim et al. 1996 and Wargelin & Drake 2002; $\dot{M} < 4 - 5 \times 10^{-11} M_\odot$ yr^{-1} for nearby active solar analogs, Gaidos et al. 2000). The upper limits reported by Gaidos et al. (2000) for solar analogs imply a maximum mass loss during solar history of $\lesssim 6\%$ of $M_\odot$.

Wood et al. (2002, 2005) discuss a promising indirect approach making use of Lyα absorption in so-called "astrospheres". The latter are formed where the outflowing stellar wind collides with the interstellar medium. The heliosphere is permeated by interstellar H I with $T \approx (2-4) \times 10^4$ K (Wood et al. 2002). Much of this gas is piled up between the heliopause and the bow shock, forming the so-called "hydrogen wall" that can be detected as an absorption signature in the Lyα line. The measured absorption depths in Lyα are compared to hydrodynamic model simulations (Wood et al. 2002, 2005). The amount of astrospheric absorption should scale with the wind ram pressure, $P_w \propto \dot{M}_w V_w$, where V_w is the (unknown) wind velocity (Wood & Linsky 1998). The latter is usually assumed to be the same as the solar wind speed. From nearby stars, the following relations are found for the mass loss rate ($\dot{M}_w$) (Wood et al. 2005),

$$\dot{M}_w \propto F_X^{1.34\pm0.18} \quad (7)$$
$$\dot{M}_w \propto t^{-2.33\pm0.55} \quad (8)$$

Stellar ionized (and magnetized) winds interact with the upper planetary atmosphere/ionospheres and their magnetospheres (if present). Such interactions further deteriorate the atmospheres as ions are carried away. Important mechanisms include ion pick-up, sputtering, ionospheric outflow, and also dissociative recombination; the ionizing source for the upper planetary atmosphere is the EUV and X-ray emission from the Sun (for a summary see, e.g., Chassefière & Leblanc 2004; Lundin et al. 2007). Such "non-thermal" processes are thought to have fundamentally altered the atmospheres of Venus, Earth, and

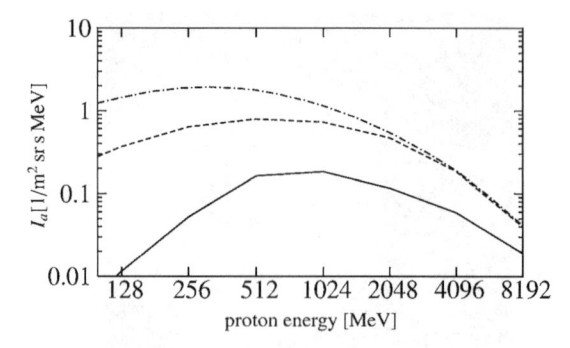

Figure 3. Energy spectrum of cosmic rays reaching the top of the atmosphere for present-day Earth (solid), a tidally locked, Earth-like planet at 0.2 AU around a $0.5 M_\odot$ star (dashed), compared to the incident spectrum outside the magnetosphere (dash-dotted; from Grießmeier et al. 2005).

Mars, leading to strong loss of oxygen (Lammer et al. 2006) and therefore water (after photodissociation of water vapor in the upper atmosphere) from young Venus and Mars. Clearly, a strong magnetosphere as in the case of the Earth shields the lower atmospheric layers; such shielding may have been important in retaining much of the terrestrial water. The amount of shielding obviously depends on the strength of the magnetosphere but also on the strength of the wind and its ability to compress the magnetosphere (Figure 2).

"Hot Jupiters" or planets in the close-in "habitable zones" around low-mass M dwarfs should be in tidally locked rotation. The slow rotation produces comparatively small magnetic moments (through dynamo action), and the ram pressure of the stellar wind can compress the magnetosphere sufficiently to expose the upper atmosphere; this effect is enhanced if the upper atmosphere is heated and expanded by strong XUV irradiation. Magnetic protection of the atmosphere is thus strongly reduced (Grießmeier et al. 2005; Lammer et al. 2007), leading to increased erosion, while the cosmic ray flux at the top of the atmosphere will be enhanced given the weaker and smaller magnetosphere (Figure 3, Grießmeier et al. 2005).

We already alluded to coronal mass ejections in the previous section. If such CMEs are ejected at a rate of several per day, they will essentially act like an enhanced solar wind, which will further compress planetary magnetospheres and erode the upper atmospheres (Khodachenko et al. 2007). The combination of enhanced exospheric heating by EUV irradiation with consequent exospheric expansion and a CME "wind" could induce atmospheric losses of up to tens or hundreds of bars for close-in planets. This may again be particularly important for planets in the habitable zones around M dwarfs (Lammer et al. 2007).

References

Audard, M., Güdel, M., Drake, J. J., & Kashyap, V. L. 2000, ApJ, 541, 396
Ayres, T. R. 1997, JGR, 102, 1641
Chassefière, E. & Leblanc, F. 2004, P&SS, 52, 1039
Drake, S. A., Simon, T., & Brown, A. 1993, ApJ, 406, 247

Gaidos, E.J., Güdel, M., & Blake, G.A. 2000, GRL, 27, 501
Güdel, M. 2009, in AIP Conf. Ser. 1126, Simbol-X: Focusing on the Hard X-Ray Universe, ed. J. Rodriguez, (AIP), 341
Güdel, M., Audard, M., Kashyap, V. L., Drake, J. J., & Guinan, E. F. 2003, ApJ, 582, 423
Güdel, M., Guinan, E. F., & Skinner, S. L. 1997, ApJ, 483, 947
Grießmeier, J.-M., Stadelmann, A., Motschmann, U., Belisheva, N. K., Lammer, H., & Biernat, H. K. 2005, Astrobiology, 5, 587
Ingersoll, A. P. 1969, JAtS, 26, 1191
Kashyap, V. L., Drake, J. J., Güdel, M., & Audard, M. 2002, ApJ, 580, 1118
Kasting, J. F. & Pollack, J. B. 1983, Icarus, 53, 479
Kasting, J. F., Whitmire, D. P., & Reynolds, R. T. 1993, Icarus, 101, 108
Kasting, J. F. 1991, Icarus, 94, 1
Kasting, J. F. & Catling, D. 2003, ARA&A, 41, 429
Khodachenko, M. L., et al. 2007, Astrobiology, 7, 167
Kulikov, Y. N., et al. 2007, SSRv, 129, 207
Lammer, H., et al. 2006, P&SS, 54, 1445
Lammer, H., et al. 2003, Icarus, 165, 9
Lammer, H., et al. 2007, Astrobiology, 7, 185
Lim, J., White, S. M., & Slee, O. B. 1996, ApJ, 460, 976
Lundin, R., Lammer, H., & Ribas, I. 2007, SSRv, 129, 245
Penz, T., et al. 2008, P&SS, 56, 1260
Ribas, I., Guinan, E.F., Güdel, M., & Audard, M. 2005, ApJ, 622, 680
Sackmann, I.-J., & Boothroyd, A. I. 2003, ApJ, 583, 1024
Sagan, C., & Mullen, G. 1972, Sci, 177, 52
Scalo, J., et al. 2007, Astrobiology, 7, 85
Schaefer, B. E., King, J. R., & Deliyannis, C. P. 2000, ApJ, 529, 1026
Smith, D. S., Scalo, J., & Wheeler, J. C. 2004, Icarus, 171, 229
Stelzer, B., et al. 2007, A&A, 468, 463
Telleschi, A., Güdel, M., Briggs, K., Audard, M., Ness, J.-U., & Skinner, S. L. 2005, ApJ, 622, 653
Walker, J. C. G., Hays, P. B., & Kasting, J. F. 1981, JGR, 86, 9776
Wargelin, B. J. & Drake, J. J. 2002, ApJ, 578, 503
Watson, A. J., Donahue, T. M., & Walker, J. C. C. 1981, Icarus, 48, 150
West, A. A., et al. 2008, AJ, 135, 785
Wood, B. E., & Linsky, J. L. 1998, ApJ, 492, 788
Wood, B. E., Müller, H.-R., Zank, G. P., & Linsky, J. L. 2002, ApJ, 574, 412
Wood, B. E., Müller, H.-R., Zank, G. P., Linsky, J. L., & Redfield, S. 2005, ApJ, 628, L143

Pathways Towards Habitable Planets
ASP Conference Series, Vol. 430, 2010
Vincent Coudé du Foresto, Dawn M. Gelino, and Ignasi Ribas, eds.

Review from the Blue Dots Astrometry Working Group

F. Malbet,[1] A. Sozzetti,[2] P. Lazorenko,[4] R. Launhardt,[3] D. Ségransan,[5] F. Delplancke,[6] N. Elias,[7] M. W. Muterspaugh,[8] A. Quirrenbach,[9] S. Reffert,[9] and G. van Belle[6]

Abstract. The astrometry technique is an important tool for detecting and characterizing exoplanets of different types. In this review, the different projects which are either operating, in construction or in discussion, are presented and their performance is discussed in the framework of the Blue Dots study. We investigate the sensitivity of astrometry to different sources of noise and we show that astrometry is a key technique in the path of discovering and characterizing new types of planets including the very challenging category of Earth-like planets orbiting the habitable zone of solar-type stars.

1. Introduction

In the Blue Dots study (see Coudè du Foresto et al. in this volume) which aims at preparing a road map towards the detection and characterization of habitable exoplanets, precise stellar astrometry is one of the methods identified to detect exoplanets and characterize them. The goals of the Astrometry Working Group within the Blue Dots initiative are to identify the type of exoplanets that astrometry can detect and characterize so that the Blue Dot table can be filled in with the astrometry prospects, to investigate the limitations of astrometry and finally to emphasize the complementarity of astrometry with the other techniques. The reader is invited to read a deeper review on the subject by Sozzetti (2009) which is an essential resource for this work.

---

[1]Universit de Grenoble J. Fourier/CNRS, Laboratoire d'Astrophysique de Grenoble, BP 53, Grenoble, France

[2]Osservatorio Astronomico di Torino INAF, Strada Osservatorio 20, Pino Torinese, Italy

[3]Max-Planck-Institut fr Astronomie, Knigstuhl 17, Heidelberg, Germany

[4]Main Astronomical Observatory, 27 Acad. Zabolotnoho St, UA Kiev, Ukraine

[5]Observatoire astronomique de l'Universit de Genve, Sauverny, Versoix, Switzerland

[6]European Southern Observatory, Karl-Schwarzschild-Str. 2, Garching bei Mnchen, Germany

[7]National Radio Astronomy Observatory, 1003 Lopezville Road, Socorro, NM, USA

[8]Tennessee State University, 3500 John A. Merritt Blvd, Nashville, TN, USA

[9]Zentrum fr Astronomie der Universitt Heidelberg - Landessternwarte, Knigstuhl 12, Heidelberg, Germany

Table 1. Expected astrometric signal for different flavors of planetary systems.

	Giants planets			Telluric planets		
Type of planet	Classical jupiter	Young jupiter	Hot jupiter	Hot super-Earth	Earth in HZ	Earth in HZ
Stellar spectral type	G2	G2	G2	M	G2	M
M_P (M_{Earth})	300	300	300	5	1	1
M_P ($M_{Jupiter}$)	1	1	1	0.02	0.003	0.003
a_P (AU)	5	5	0.1	0.1	1	0.28
P (yr)	11	11	0.03	0.05	1	0.2
P (d)	4084	4084	12	17	365	82
M_* (M_{Sun})	1	1	1	0.45	1	0.45
d (pc)	10	150	10	2.5	10	9
Astrometric signal (in μas)	495	33	10	1	0.3	0.2

2. How Can Astrometry Detect Exoplanets?

The motion of a star projected onto the plane of sky is a combination of three types of apparent motion: parallax which is the apparent motion due to the change of perspective of the observer (mainly the location of the Earth during one year), the proper motion which is the motion of the star+planets system in the galaxy, and the reflex motion due to the presence of orbits. A precise orbit determination unravels the presence of planets of different masses, only if one is able to subtract the effect of parallax and proper motion.

It is generally assumed that a minimal signal-to-noise ratio of 5-6 on the reflex motion of the star is required to detect a planet. In this case, astrometric measurements of star motion yield the period P and the planetary mass M_P, but also the six parameters of the orbit: the semi-major axes a_P, the inclination of the orbit i, the eccentricity of the orbit e, the longitude of the ascending node Ω, the argument of periapsis ω and the mean anomaly ν at epoch T_o.

The contribution of a planet to the reflex motion of its host star is given by the following formula:

$$\Delta\alpha = 0.33 \left(\frac{a_P}{1\,\text{AU}}\right) \left(\frac{M_P}{1\,\text{M}_{\text{Earth}}}\right) \left(\frac{M_*}{1\,\text{M}_{\text{Sun}}}\right)^{-1} \left(\frac{d}{10\,\text{pc}}\right)^{-1} \mu\text{as} \qquad (1)$$

Typical numbers corresponding to various flavors of planetary systems are given in Table 1. There are almost 4 orders of magnitude between a Jupiter in a Solar-like planetary system located at 10 pc which gives a signal of the order of 500 μas and an Earth-like planet in its habitable zone which gives a signal of 0.3μas. Astrometry, unlike some other methods, is best suited when looking to nearby sources.

There are a variety of techniques to accurately measure the astrometric motion of stars, i.e., to accurately measure the positions of stars on the sky plane. These techniques can lead to wide-angle or narrow-angle observations between stars, using relative or absolute measurements, with local or global strategy. Atmospheric turbulence is an important limitation in performing astrometry and therefore there are both ground-based and spaceborne astrometric projects.

3. Main Astrometric Projects

We have listed five main types of astrometric projects in the domain of exoplanet detection and characterization. We have sorted them by accuracy level and number of potential targets.

3.1. Large Ground-based Telescopes

For many years starting in the 1960s, narrow-field astrometry was frequently tried for finding exoplanets. These attempts resulted in various discoveries, but none were confirmed (Barnard's star, Lalande 21185). These failures, caused by insufficient precision and systematic errors of the photographic technique, cast doubt on the use of astrometry for exoplanet studies. Besides, narrow-field astrometry was considered to be limited by a ~ 1 mas precision due to the atmospheric image motion which came from the expression $\varepsilon \sim \theta/D^{2/3}$ derived by Lindegren (1980) for the r.m.s. of the image motion ε in the measurement of a distance θ between a pair of stars on a telescope of diameter D. Recently, Lazorenko (2002) has found that for the reference field represented by a grid of stars, the power law is $\varepsilon \sim \theta^{11/6}/D^{3/2}$ with an improvement with D. However, the above expression is asymptotic and refers to the very narrow field mode of observations requiring use of very large telescopes. In practice, this opens a way to a $< 100\,\mu$as astrometry in a field of view limited to $\theta \leq 1'$ with integration times of a few minutes only, yet sufficient for the detection of massive exoplanets around nearby stars. This challenging precision, of course, assumes a proper handling of many other effects, in particular related to a highly complicated shape of star profiles at the sub-pixel level and to the unfixed, floating position of reference stars due to the differential color refraction, proper motion, parallax, etc. Incorrect treatment of these effects degrades precision to ~ 1 mas. Currently, astrometry is used to search for exoplanets at three telescopes:

- The STEPS instrument installed on the 5-m Palomar telescope (Pravdo et al. 2004) with astrometric precision of 1 mas per a single series of 20-30 exposures is used for exoplanet search around 30 late M stars since 1997. In 2009, this resulted in the first astrometric discovery of the giant exoplanet VB 10b around a main-sequence star (Pravdo & Shaklan 2009), although it is still debated (Zapatero Osorio et al. 2009; Bean et al. 2010).

- The CAPSCam camera on the 2.5-m Las Campanas telescope has an estimated astrometric precision of $300\,\mu$as/h. The search program includes about 100 late M, L, and T dwarfs to be observed for 10 years (Boss et al. 2009).

- The FORS2 camera installed on one of the 8-m VLT telescopes was tested in a series of theoretical and observational studies which revealed its long-term astrometric precision of $50 - 100\,\mu$as at time scales from a few days to a few years (Lazorenko et al. 2009). Science observations started in 2009 with the VB 10b as a program object and include 14 nearby L dwarfs.

Precision narrow-field astrometry is applicable to the targets of 12–17 mag at galactic latitudes up to $15 - 30°$ allowing for a sufficient number of reference stars. Astrometry enables the detection of $\sim \mathrm{M_{Jupiter}}$ planets at nearby low-mass red stars and brown dwarfs with orbital periods of ≥ 1 year. We will take the range $0.3 - 1$ mas as a conservative number for the accuracy of large telescope astrometry.

3.2. Hubble Space Telescope / Fine Guidance Sensors

Relative, narrow-angle astrometry from space has been performed so far with the *Fine Guidance Sensors* aboard the *Hubble Space Telescope* (HST/FGS). For HST/FGS astrometry with respect to a set of reference objects near the target (within the 5×5 arcsec instantaneous field of view of FGS), $1 - 2$ mas single measurement precision down to $m_V \sim 16$ has been demonstrated (see Benedict et al. 1994, 1999) using an ad hoc calibration and data reduction procedures to remove a variety of random and systematic error sources from the astrometric reference frame (spacecraft jitter, temperature variations and temperature-induced changes in the secondary mirror position, constant and time-dependent optical field angle distortions, orbit drifts, lateral color corrections). The limiting factor is the spacecraft jitter. A single-measurement precision below $0.5 - 1$ mas is out of reach for HST/FGS. The first undisputed value of the actual mass of a Doppler-detected planet was obtained by Benedict et al. (2002) who derived the perturbation size, inclination angle, and mass of the outer companion in the multiple-planet system GJ 876 from a combined fit to HST/FGS astrometry and high-precision radial-velocities. Six other objects are under observation.

3.3. Ground-based Dual Star Interferometry

Long-baseline optical/infrared interferometry is an important technique to obtain high-precision astrometry measurements (Shao & Colavita 1992). The idea is to operate an interferometer equipped with a dual instrument observing two stars simultaneously so that the optical delay between the fringes can be accurately measured. The first observations have been achieved by the Palomar Testbed Interferometer (PTI, Colavita et al. 1999) reaching $\sim 100\,\mu$as short-term accuracy for $30''$ binaries (Lane et al. 2000) and $20 - 50\,\mu$as for sub-arcsec binaries (Lane & Muterspaugh 2004) within the Palomar High-precision Astrometric Search for Exoplanet Systems (PHASES) program.

Two projects are underway: the *ASTrometric and phase-Referenced Astronomy* (ASTRA: Pott et al. 2008) on the Keck Interferometer and *Phase-Referenced Imaging and Micro-arcsecond Astrometry* (PRIMA: Delplancke 2008) at the VLT interferometer. These two projects are designed to perform narrow-angle interferometric astrometry at an accuracy better than $100\,\mu$as with two telescopes of a target and one reference star separated by up to $1'$. However, this is a challenging technique requiring optical path difference errors less than 5 nm.

The Exoplanet Search with PRIMA (ESPRI) Consortium (Launhardt et al. 2008) plans to carry out a three-fold observing program focused on the astrometric characterization of known radial velocity planets within ≤ 200 pc from the Sun, the astrometric detection of low-mass planets around nearby stars of any spectral type within ≈ 15 pc from the Sun, and the search for massive planets orbiting young stars with ages in the range $5 - 300$ Myr within ~ 100 pc from the Sun. The target list includes ~ 100 stars with good references.

3.4. Space-borne Global Astrometric Survey: Gaia

In its all-sky survey, Gaia, due to launch in Spring 2012, will monitor the astrometric positions of all point sources in the magnitude range 6 − 20 mag, a database encompassing $\sim 10^9$ objects. Using the continuous scanning principle first adopted for Hipparcos, Gaia will determine positions, proper motions, and parallaxes for all objects, with end-of-mission precision between 6 μas at $V = 6$ mag and 200 μas at $V = 20$ mag and an averaged 80 transits per object in 5 years.

Gaia astrometry, complemented by on-board spectrophotometry and radial velocity information, will have the precision necessary to quantify the early formation, and subsequent dynamical, chemical and star formation evolution of the Milky Way Galaxy. One of the relevant areas in which the Gaia observations will have great impact is the astrophysics of planetary systems (e.g. Casertano et al. 2008), in particular when seen as a complement to other techniques for planet detection and characterization (e.g. Sozzetti 2009).

Using Galaxy models, our current knowledge of exoplanet frequencies, and Gaia's estimated precision of $\sim 10\,\mu$as on bright targets ($V < 13$), Casertano et al. (2008) have shown that Gaia will measure actual masses and orbital parameters for possibly thousands of giant planets and determine the degree of coplanarity of possibly hundreds of multiple-planet systems. Gaia will be sensitive as far as the nearest star-forming regions for systems with massive giant planets on $1 \leq a \leq 4$ AU orbits around solar-type hosts and out to 30 pc for Saturn-mass planets with similar orbital semi-major axes around late-type stars.

Gaia holds the promise to make crucial contributions to many aspects of planetary systems astrophysics, in combination with present-day and future extrasolar planet search programs. Gaia data over the next decade will allow us to (a) significantly refine of our understanding of the statistical properties of extrasolar planets, (b) carry out crucial tests of theoretical models of gas giant planet formation and migration, (c) achieve key improvements in our comprehension of important aspects of the formation and dynamical evolution of multiple-planet systems, (d) provide important contributions to the understanding of direct detections of giant extrasolar planets, and (e) collect essential supplementary information to optimize the target lists of future observatories aimed at the direct detection and spectroscopic characterization of terrestrial, habitable planets in the vicinity of the Sun.

3.5. Space-based Astrometric Observatory: SIM Lite

The *Space Interferometry Mission* is a spaceborne instrument (Marr et al. 2008) which would carry out astrometry to micro-arc second precision on the visible light from a large sample of stars in our galaxy and search for earth-like planets around nearby stars (Unwin et al. 2008). SIM Lite is an alternative concept for SIM, and it is the current proposed implementation for SIM for NASA program on Search for Earth-like planets and life. The SIM Lite instrument is an optical interferometer with a baseline of 6 meters, including a guiding interferometer and a guiding siderostat for spacecraft pointing and a science interferometer to perform high accuracy astrometric measurements on target stars. The primary objective for SIM Lite is to search 65 nearby stars for exoplanets of masses down to one Earth mass, in the habitable zone. SIM Lite is designed to deliver better than 1 µas narrow-angle astrometry in 1.5 hr integration time on bright targets ($m_V \leq 7$) and moderately fainter references ($m_v \simeq 9-10$) (Goullioud et al. 2008). An accuracy on the position of the delay lines of a few tens of picometers with a 6-m baseline must be achieved (Zhai et al. 2008). Furthermore, a positional stability of internal optical path lengths of ~ 10 nm is required, in order to ensure maintenance of the fringe visibility.

SIM Lite has planned for three planetary system surveys. (1) The *deep planetary survey* will focus on fewer than one hundred nearby stars of the main sequence within 10 parsecs from the Sun. The main objective is to identify planetary systems with Earth-like planets in the habitable zone around these Sun-like stars. This deep survey requires the highest possible astrometric accuracy, below the single micro-arc-second. This accuracy is achieved by multiple visits to the target stars with 12 to 60 chops per visits in order to lower the instrument accuracy down to the level required for detection of one Earth mass planet. A 60 chop visit to a magnitude 6 target will require less than two hours of observation time. The target would then be observed several hundred times during the lifetime of the mission. (2) The *broad planetary survey* will study more than one thousand sets of stars of many types including O, B, A, F, binary with a reduced astrometric accuracy of 5 µas in order to cover the diversity of planetary systems and to increase our knowledge on the nature and the evolution of planetary systems in their full variety. This accuracy can be achieved by short visits to the target stars. To accurately determine the orbital parameters of the planets, 100 visits per target star will be scheduled over the 5 year mission. (3) The *young planetary system survey* will observe fewer than one hundred nearby solar type stars with ages below 100 millions years within 100 parsecs from the Sun with the aim to understand the frequency of Jupiter-mass planets and the early dynamical evolution of planetary systems. This survey can be conducted with a reduced astrometric accuracy of 4 µas achieved with 4-chops visits requiring about eight minutes of observation time for a $V \sim 11$ young star.

4. Limitations and Performance

Using astrometry to detect and characterize extra-solar planetary systems is challenging in many aspects. In this part we focus on the main issues that astrometry will have to face especially in the field of Earth-mass planets.

4.1. Influence of Giant Planets on Earth-like Planets Detection

Correct determination of the astrometric orbits of planetary systems involves highly non-linear orbital fitting procedures, with a large number of model parameters. Particular attention will have to be devoted, for example, to the assessment of the relative robustness and reliability of different procedures for orbital fits (and associated uncertainties). The quality of the coplanarity analysis for multiple systems has to be measured against the achieved single-measurement precision and available redundancy in the observations. Correctly identifying signals with amplitude close to the measurement uncertainties is a challenge, particularly in the presence of larger signals induced by other companions and/or sources of astrophysical noise of comparable magnitude, like Jupiter-like planets for telluric ones (500 µas signal compared to 0.3 µas signal in the case of a Solar System located at 10 pc).

All these issues could have a significant impact on the capability of Gaia and SIM Lite to detect and characterize planetary systems. Double-blind test campaigns have been carried out to estimate the potential of both Gaia and SIM Lite for detecting and measuring planetary systems (Casertano et al. 2008; Traub et al. 2009). The Gaia *Data Processing and Analysis Consortium* (DPAC) is developing the modeling of the astrometric signals produced by planetary systems, implementing multiple robust procedures for astrometric orbit fitting (such as Markov Chain Monte Carlo and genetic algorithms) and in order to determine the degree of dynamical stability of multiple-component systems. Similar work is also underway for SIM Lite (see Traub et al. in this volume) which shows that detection completeness is 90% for all planets reaching 95% for terrestrial planets in the habitable zone with a reliability of 97% and 100%, respectively.

4.2. Impact of Stellar Activity

In the domain of terrestrial planets, the expected signal is so small (0.3µas for an Earth at 10 pc) that one must take into account the effect of spots on the surface of the target stars. These spots will introduce a spurious signal, the level of which should be evaluated all the more because these spots can last for several weeks. This phenomenon also impacts the other detection techniques, like transits and radial velocity measurements. Several papers have addressed this issue in the case of astrometry, at least for the most sensitive instruments, Gaia and SIM Lite (Eriksson & Lindegren 2007; Desort et al. 2009; Makarov et al. 2009).

The general conclusion is that, unlike the radial-velocity case, µas-level astrometry is significantly less affected by the above astrophysical noise sources. For example, a Sun-like star inclined at $i = 90°$ at 10 pc is predicted to have a jitter of 0.087 µas in its astrometric position along the equator, below the expected signal of an Earth at a level of 0.3 µas, and 0.38 m/s in radial velocities above the expected signal from a Earth at 0.09 m/s. If the presence of spots due to stellar activity is the ultimate limiting factor for planet detection, then the sensitivity of SIM Lite to Earth-like planets in habitable zones is about an order of magnitude higher that the sensitivity of prospective ultra-precise radial velocity observations of nearby stars.

4.3. Performance and Mission Scales

Table 2 summarizes the various projects presented in Section 3. with their operational status or when they will happen, the expected accuracy, and their main targets. We have also indicated their scale in the *Blue Dots* terminology (see Coudé du Foresto et al. in this volume). Ground-based facilities are categorized in existing low cost projects, whereas Gaia is in the M-type mission class and SIM Lite in the L-type mission class, even though if approved, it will fly and operate earlier than some ground-based instruments (like spectrographs on extremely large telescopes).

Table 2. Summary of the astrometry mission scopes and their scales.

Type of astrometry mission	Ground-based large telescope large field imaging	Ground-based optical interferometry differential narrow-angle astrometry	Space-borne wide-angle astrometry survey	Space-borne narrow angle astrometry
Project	Palomar/STEPS, VLT/FORS	VLTI/PRIMA (ESO), KI/ASTRA (NSF/NASA)	Gaia (ESA)	SIM-Lite (NASA)
Status	in operation	commissioning	launch date: T2/2012	end of phase B
Availability	now	2011	2012	2015-2017?
Accuracy	0.3-1 mas	10-50 μas	25 μas	0.2 μas
Exoplanet targets	Giant planets around M stars / Neptunes around brown dwarfs	Giant planets around solar-type stars / Neptunes around M stars / Young planetary systems	Giant planets around several 10^5s of stars of different types	Earth in HZ of F-G stars / survey of 1000 systems around different stars / Young planetary systems
Mission scale	Existing	Existing	~M-type mission	~L-type mission

5. Conclusion: Specificity of Astrometry in the Exoplanet Field

This review has allowed us to show that astrometry can detect and characterize new categories of exoplanets especially in the Solar neighborhood. It encompasses techniques ranging from ground-based instruments limited to the largest planets to spaceborne missions allowing Earth-like planets to be detected in the habitable zone around solar-type stars. Astrometry will obtain full characterization of orbits which (a) gives us masses and, (b) tells us where and when to look with spectroscopic characterization missions.

Astrometry is mostly immune to stellar activity even at the signal level due to Earth-like planets, and therefore is well positioned for identifying and characterizing planets for future direct detection and spectroscopic follow-up projects like DARWIN/TPF-I or TPF-C.

Astrometry is a difficult and challenging technique, yet it may be the only way to detect and characterize Earth-like planets in the habitable zone of solar-type stars. Last but not the least, astrometry projects are underway or technically ready and should soon contribute significantly to the exoplanet field.

Acknowledgments. We would like to thank S. Shaklan, S. Pravdo, F. Mignard, M. Shao, J. Catanzarite, and V. Makarov for providing material for the review presentation in Barcelona meeting.

References

Bean, J. L., Seifahrt, A., Hartman, H., et al. 2010, ApJ, 711, L19
Benedict, G. F., McArthur, B., Chappell, D. W., et al. 1999, AJ, 118, 1086
Benedict, G. F., McArthur, B., Nelan, E., et al. 1994, PASP, 106, 327
Benedict, G. F., McArthur, B. E., Forveille, T., et al. 2002, ApJ, 581, L115
Boss, A. P., Weinberger, A. J., Anglada-Escudé, G., et al. 2009, PASP, 121, 1218
Casertano, S., Lattanzi, M. G., Sozzetti, A., et al. 2008, A&A, 482, 699

Colavita, M. M., Wallace, J. K., Hines, B. E., et al. 1999, ApJ, 510, 505
Delplancke, F. 2008, New Astronomy Review, 52, 199
Desort, M., Lagrange, A., Galland, F., et al. 2009, A&A, 506, 1469
Eriksson, U. & Lindegren, L. 2007, A&A, 476, 1389
Goullioud, R., Catanzarite, J. H., Dekens, F. G., Shao, M., & Marr, IV, J. C. 2008, SPIE, 7013, E151
Lane, B. F., Kuchner, M. J., Boden, A. F., Creech-Eakman, M., & Kulkarni, S. R. 2000, Nat, 407, 485
Lane, B. F. & Muterspaugh, M. W. 2004, ApJ, 601, 1129
Launhardt, R., Queloz, D., Henning, T., et al. 2008, SPIE, 7013, E76
Lazorenko, P. F. 2002, A&A, 382, 1125
Lazorenko, P. F., Mayor, M., Dominik, M., et al. 2009, A&A, 505, 903
Lindegren, L. 1980, A&A, 89, 41
Makarov, V. V., Beichman, C. A., Catanzarite, J. H., et al. 2009, ApJ, 707, 1, L73
Marr-IV, J. C., Shao, M., & Goullioud, R. 2008, SPIE, 7013, E79
Pott, J., Woillez, J., Akeson, R. L., et al. 2008, (arXiv: 0811.2264)
Pravdo, S. H. & Shaklan, S. B. 2009, ApJ, 700, 623
Pravdo, S. H., Shaklan, S. B., Henry, T., & Benedict, G. F. 2004, ApJ, 617, 1323
Shao, M. & Colavita, M. M. 1992, A&A, 262, 353
Sozzetti, A. 2009, in Extrasolar Planets in Multi-Body Systems: Theory and Observations (Torun, Poland, August 25-29, 2008), (arXiv:0902.2063)
Traub, W. A., Beichman, C., Boden, A. F., et al. 2009, (arXiv: 0904.0822)
Unwin, S. C., Shao, M., Tanner, A. M., et al. 2008, PASP, 120, 38
Zapatero Osorio, M. R., Martín, E. L., del Burgo, C., et al. 2009, A&A, 505, L5
Zhai, C., Yu, J., Shao, M., et al. 2008, SPIE, 7013, E157

The (Un)Lonely Planet Guide: Formation and Evolution of Planetary Systems from a "Blue Dots" Perspective

M. R. Meyer

Institute for Astronomy, ETH, Zürich, Switzerland

Abstract. In this contribution I summarize some recent successes, and focus on remaining challenges, in understanding the formation and evolution of planetary systems in the context of the Blue Dots initiative. Because our understanding is incomplete, we cannot yet articulate a design reference mission engineering matrix suitable for an exploration mission where success is defined as obtaining a spectrum of a potentially habitable world around a nearby star. However, as progress accelerates, we can identify observational programs that would address fundamental scientific questions through hypothesis testing such that the null result is interesting.

1. Introduction

Planet formation is a complex process, yet one that can capture the imagination of a hurried public in a busy place (Figure 1). In our own solar system, we can identify several possible outcomes such as the terrestrial planets at orbital radii inside the asteroid belt, the gas giants Jupiter and Saturn, the ice giants Uranus and Neptune, objects in the Edgeworth-Kuiper Belt, and comets. Without entering into debate concerning "what is a planet" suffice it to say that there are several flavors of planet formation. And this diversity is expanded greatly when we consider the range of exoplanets discovered around other stars reported elsewhere in this volume. These planets are thought to have formed in circumstellar disks of gas and dust which initially surrounded the young Sun and indeed most sun-like stars [1].

What are the initial conditions for planet formation? These are constrained from observations that span the electromagnetic spectrum of primordial disks around young pre-main sequence stars in nearby star-forming regions. Disk masses are inferred from (often) optically-thin millimeter wave emission to be between 0.001-0.1 $M_\odot$ (assuming standard gas-to-dust ratios). Observed mass surface density profiles range from $\Sigma \sim r^{-p}$ with $0.0 < p < 1.0$ (rather than p = 1.5 which is often assumed; see Andrews et al. 2009 and Isella et al. 2009). It is generally accepted that disks exhibit inner holes, variable gas to dust ratios, as well as hydrostatic flaring (Pinte et al. 2008; Cortes et al. 2009). Increasingly complicated models are needed to calculate in a self-consistent way their thermal and geometrical structure, not to mention dynamics (Dullemond et al. 2007). These gas-rich disks exhibit large ranges of other properties as well including outer radii, accretion rates, and even lifetimes. It remains to be

---

[1]Even the major satellites of the giant planets likely formed from circumplanetary disks.

94   Meyer

Figure 1. An interested public in the Zürich Hauptbahnhof surveying ESA's plans to search for terrestrial planets around Sun-like stars. The author fielded questions such as "how do you define life?" and "what sorts of planets are habitable?" for which he had no good answers. There was also a general admonition not to "oversell" future missions to the public.

demonstrated whether this diversity in initial conditions is responsible for the diversity observed in extra-solar planets (Meyer, 2009).

What should we expect from theory regarding the products of the planet formation process? The basic ideas for collisional growth of planetesimals were outlined by Safronov in the late 1960's. Refinements by Wetherill (1991) and many others are roughly consistent with observations (see below). Planet formation should proceed more quickly around stars of higher mass due to both higher disk mass surface densities and shorter orbital timescales. Yet disks around more massive stars appear to dissipate more quickly. It is not yet clear which process wins out in planet formation.

In what follows, I briefly review some successes but concentrate on failures in our understanding of planet formation in the hope of motivating the work needed to define near-term missions as well as realizing our ultimate aspirations within the Blue Dots initiative.

2. Evolution of Gas Disks

All Sun–like stars probably form with some sort of circumstellar disk surrounding them (see for example Figure 2). These disks are bins for storing angular

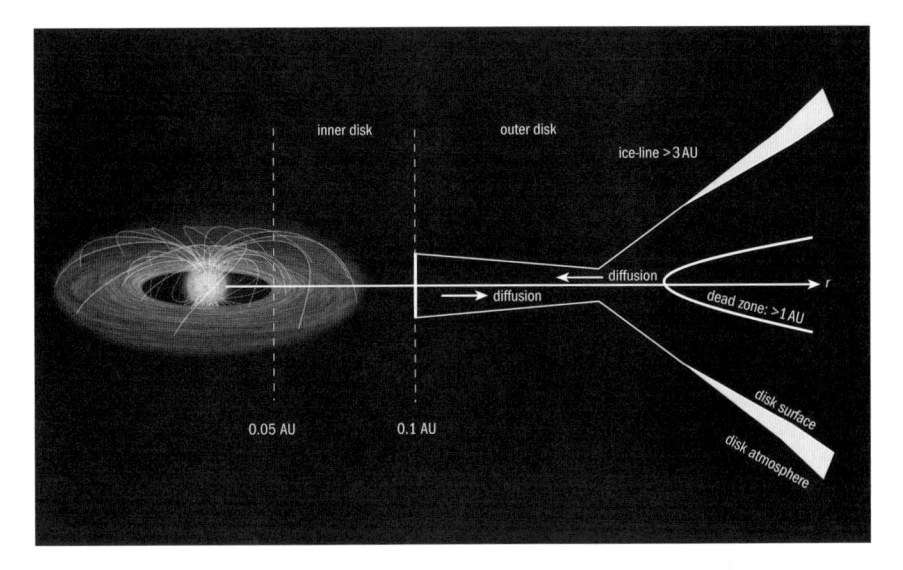

Figure 2. Schematic of a circumstellar accretion disk adapted from Dullemond et al. (2001) incorporating artwork of R. Hurt (NASA/JPL) as it appeared in Meyer (2009b).

momentum and provide the initial conditions for planet formation. The "typical" lifetime of such disks is 1–3 Myr with large dispersion (Mamajek 2009). In this section we discuss the evolution of solids, the formation of gas giant planets, aspects of disk chemistry, as well as the final stages of disk dissipation.

2.1. Draining of Solids

Theorists who study planet formation create their own problems. One of the current challenges is to understand how to preserve solid materials in gas-rich disks from which the cores of gas giants as well as terrestrial planets form. A potentially serious loss mechanism is due to the gas drag suffered by bodies in Keplerian orbits embedded in a gas disk that is partly supported by thermal pressure. Such bodies would orbit faster than the gas, suffer a head-wind, and spiral into the star. For example, a meter-sized body at 1 AU suffering this effect would be lost in less than one century (e.g. Weidenschilling, 1977). But as pressure gradients take, they can also give. Others have investigated whether density enhancements or vortices expected in a turbulent accretion disk could serve to concentrate particles. Johansen et al. (2007) suggest these effects could lead to gravitationally unstable regions capable of producing large bodies (100 km) very quickly. If bodies can grow faster than they migrate, they can reach a size where the effect is not important and are saved. Further work is needed to explore observations that would demonstrate empirically the existence of the problem as well as the range of mechanisms which can solve it. Perhaps thermal/chemical discontinuities in the disk (like the ice line) are those places where solids can collect and speed-up planet formation (e.g. Kretke & Lin 2007).

Another thorny issue is Type I migration where lunar mass and larger bodies launch spiral waves in a gas–rich disk. Numerous torques permit the proto-planet to exchange angular momentum with the gas and migrate. Calculations assuming a previously favored mass surface density profile $\Sigma(r) \sim r^{-3/2}$ led to the conclusion that the co-rotation torques are small, in agreement with linear theory. In this case the inward migration rate is very rapid, leading to concerns about the viability of planet formation that depends on a healthy population of these embryos. Nelson (2005) has explored whether scattering of migrating protoplanets off of density enhancements in the inner accretion disk could save these building blocks. Recent calculations by Paardekooper & Papaloizou (2009) have pointed out that co-rotation torques are non-zero for a range of relevant surface density profiles calling into question the ubiquity of rapid inward migration. Models that produce planets favor migration rates an order of magnitude smaller than the early estimates (Ida & Lin 2008).

Current models of solar system formation start with a disk of gas and dust that cannot be much more than 10% the mass of the young Sun. With a standard gas to dust ratio that implies no more than a few hundred Earth-masses of material in heavy elements. Because the current solar system has ∼100 Earth-masses of solids, any methods of removing solids cannot have been extremely efficient.

2.2. Giant Planet Formation

Recent reviews of giant planet formation models can be found in Durisen et al. (2007) and Lissauer & Stevenson (2007). To say that there is great controversy among experts regarding how giant planets form is an overstatement. Two mechanisms proposed gather most of the attention primarily because the differences in the predictions appear to be stark in at least two important respects. In the theory of gravitational fragmentation, no core (of several Earth-masses in heavy elements) is needed while in the core accretion theory, there is a core. Further, gravitational fragmentation should operate primarily at large radii where the disk is cooler while core accretion would seem to favor formation at smaller radii (perhaps at the ice line). The expected differences in planet composition are murky at best (Helled & Schubert 2009). Nonetheless, many researchers favor the core accretion theory, in part because of the expected correlation of gas giants with host system metallicity (Santos et al. 2001; Fischer & Valenti 2005; Sozzetti et al. 2009). Yet the planets we can most easily image directly are those at large radii (e.g. Marois et al. 2008, Kalas et al. 2008, and Lagrange et al. 2009). It would appear likely that planets can form through gravitational instability though this is probably not the main pathway for gas giant planet formation (Boley, 2009; Rafikov, 2009). And yet, they move. The physics of Type II migration differs from Type I in that the proto–planet in question is large enough to open a gap in the disk. The net effect is inward migration towards the star on the viscous timescale. Evidence for this comes from multi-planet systems where one planet appears to have been caught in resonance with an outer more massive planet during migration (e.g., Kley et al. 2005).

2.3. Disk Chemistry

Sophisticated models for the temperature and density structure of circumstellar disks are starting points to investigate chemical processes within those disks and ultimately, the composition of planets that form within them (Woitke et al. 2009). Heroic work in analyzing spectra from the Spitzer Space Telescope has revealed a number of molecular species including many emission features due to water vapor (Salyk et al. 2008; Carr & Najita 2008). Near-infrared echelle spectra obtained with 6-10 meter telescopes can provide velocity as well as spatial information (Pontoppidan et al. 2008). Longer wavelength observations with Herschel and ground-based millimeter-wave telescopes trace cooler gas. Combining these techniques could in principle yield molecular abundances as a function of disk radius. Discontinuities in these abundance gradients could shed light on special radii in the disk where physical processes relevant to planet formation occur. For example, compared to the Sun, the terrestrial planets in the solar system are underabundant in carbon with respect to silicon by a factor of x 20. If a significant fraction of the carbon from the interstellar medium (ISM) is delivered in the form of hydrogenated amorphous grains (HACs or even PAHs), then some process that transmutes this carbon to another form must be identified to explain why these grains do not contribute significantly to the bulk composition of the terrestrial planets. Gail (2002) suggests that combustion reactions at temperatures above 800 K could decrease the solid carbon abundance substantially. Lee et al. (2010) have provided an alternate explanation. Whatever the solution, chemistry matters in determining the structure, geometry, and evolution of the disk and ultimately the composition of forming planets.

Molecular abundances containing volatile species can also place constraints on the radial migration of icy solids into the inner disk (e.g. Ciesla & Cuzzi 2006). If Type I migration is responsible for transporting C–N–O rich icy planetesimals into regions where temperatures are high enough to liberate these species into the gas phase, we may see novel chemical abundances as a result. In a pioneering study using chemistry as a probe of disk conditions, Pascucci et al. (2009) report differences in the C_2H_2/HCN emission line ratios between the disks surrounding solar-mass T Tauri stars and young brown dwarfs. They speculate that the observations can be explained with a model where molecular nitrogen, photo-dissociated by UV radiation, drives HCN production more rapidly around higher mass stars with stronger accretion luminosity. These studies are in their infancy but hold great promise for unlocking some secrets of planet formation.

2.4. The Final Stages

There is a general concensus that gas-rich disks around Sun-like stars dissipate their inner disks on timescales of 1-10 Myr with a concomitant cessation of accretion onto the star (see Meyer 2009 and references therein). Further there is growing evidence that gas-rich disks persist longer around stars of low mass (cf. Mamajek 2009), although there is some controversy concerning the duration of this transition phase. Currie et al. (2009) argue for an extended transition phase of millions of years while Luhman et al. (2010) derive a transition phase of $< 10^6$ yrs consistent with several previous studies. Estimates are made of the fraction of objects observed "in transition" (from optically-thick to optically-thin) as well as the mean age of the sample. The duration of the evolutionary phase is derived from the product. Part of the discrepancy can be traced to the normalization of the fraction to the whole population versus the number of stars with detected disks[2]. More substantive issues concern which types of objects are defined as "in transition" and the fact that different regions sampled appear to yield different results. Some studies find disks that are either: i) optically-thick at all wavelengths; ii) thick at long wavelengths and optically-thin at shorter wavelengths; or iii) thin at all wavelengths. Other researchers find evidence for a continuum of disk spectral energy distributions with some exhibiting very modest excess emission over a range of wavelengths perhaps indicating flatter disk geometries (or "homologously depleted" disks of lower mass). Self-consistent analysis of a number of data sets could resolve apparent contradictions.

The issue is one of substance as models of disk dispersal make different predictions concerning the duration of this phase of evolution. New observations with the Herschel Space Telescope of atomic fine-structure lines of [CII], [OI], and [NII] are sensitive probes of remnant gas at large radii. UV and x-ray radiation ionizes and photo-evaporates a thin layer of exposed material on the disk surface (Gorti & Hollenbach 2008; Ercolano et al. 2009). PACS observations should be sensitive to fractions of an Earth mass in remnant disk gas around nearby young stars (e.g. Augereau et al. 2008). The ability to observe the last gasps of these disks as they dissipate could be crucial to understand the formation of the ice giants Uranus and Neptune in our own solar system.

3. Rocky Bodies: Fire and Ice

As the gas goes away we turn our attention to remnant dust created in collisions of larger parent bodies. Indeed T Tauri disks themselves may already bear witness to collisions of kilometer-sized planetesimals as the dust that dominates the opacity we see may be second-generation rather than pristine ISM grains. Nonetheless, in the gas-poor debris phase, observing this dust helps us to trace the evolution of planetesimal belts which in turn build rocky planets from the material left over after the end of the gas-rich accretion disk phase (e.g. Meyer 2009 and references therein).

3.1. Tracing the Formation of Terrestrial Planets

It is certainly tempting to link dust generated in collisions in debris disks with processes responsible for planet formation. An observational starting point is to determine the frequency of infrared excess emission as a function of stellar age and compare the results to recent models of planet formation (e.g. Siegler et al. 2007; Currie et al. 2008). Fortunately for observers, some modeling groups take the extra step of estimating this expected infrared excess as a function of input assumptions facilitating comparisons to the data (e.g. Kenyon & Bromley 2008). Yet a flux measured at a single wavelength can be explained in many ways. Meyer et al. (2008) interpreted the evolution of 24 micron excess emission

---

[2]Thought experiments regarding the possible outcomes under a range of assumptions suggest that the former is preferred.

observed towards Sun-like stars in terms of a warm dust model related to the formation of terrestrial planets. Detectable mid-IR emission decreases substantially around Sun-like stars at ages > 300 Myr. However, it is not clear whether this evolution is in the dust temperature or in the overall dust luminosity. Carpenter et al. (2009) were able to test this using spectro-photometry from 5-30 microns and found that the dust was cooler than typically assumed. This implies planetesimal belts located beyond 3 AU with a mode in the distribution of inner radii of 10 AU. There is no evidence for temperature evolution and modest evidence for luminosity evolution of the dust for ages 3-300 Myr.

Occasionally one detects objects whose IR excess is so large, it cannot be explained with any reasonable quasi-equilibrium debris disk model (Wyatt et al. 2007). In these cases one invokes the uncomfortable hypothesis that the phenomena is short lived. Because we observed only the product of the duration and frequency of the events, it is difficult to interpret the statistics in the context of planet formation through giant impacts (Nagasawa et al. 2007). While the frequency of detectable mid-IR excess emission around Sun-like stars is about 10-20% for ages between 3-300 Myr, fewer than 10% of these exhibit evidence for transient emission. If we hypothesize that every sun-like star has 1-3 terrestrial planets, and each is built through a series of 3-10 giant collisions, we should expect to see dozens of collisions during the epoch of terrestrial planet formation (10-100 Myr). If each collision produces a cascade of debris that persists for 10^5 years, the observed statistic of 1-3% can be explained. Note that a duration of 10^5 years is much longer than the expected lifetime of small dust grains (Wyatt, 2008), so an extended phase of dust generation would be required.

What are the expected observational signatures of such giant collisions? Some transient systems exhibit unusual spectral features due to amorphous silica thought to be produced only in high velocity impacts (e.g. Rhee et al. 2008)[3]. Lisse et al. (2009) suggest that the Spitzer spectra of HD 172555 contains evidence for both silica and SiO gas in the system. Such emission could be produced by a giant impact in which some fraction of the impactor was vaporized. In the simulations of Canup (2004) about 5×10^{-3} Earth-masses of hot gas are released in the impact of a Mars-sized body with the proto-Earth. If the vapor condensed into micron sized grains, this would result in mid-IR emission 100× above the Spitzer detection limit. However such dust might only persist for 10^3 years (around stars of age 10^7) before removal by radiation pressure. Unless subsequent collisions prolong the dust generation phase (not to mention the expected duration of gas emission), such objects are expected to be very rare. SiO gas emission has only been claimed for a single object (and awaits ground-based confirmation).

Could we detect directly a forming terrestrial planet, the result of such a giant impact? Zahnle et al. (2007) review the geophysical consequences of the giant impact hypothesis for the evolution of the early Earth. Pahlevan & Stevenson (2007) predict the existence of a large magma disk (>1500 K!) of a few Earth radii that would persist for thousands of years (Figure 3). An object of such high temperature and large solid angle would be easily detectable (if resolved from the host star). Finally we note that Miller-Ricci et al. (2009)

---

[3]Silica is also observed toward a handful of T Tauri stars (Sargent et al. 2009).

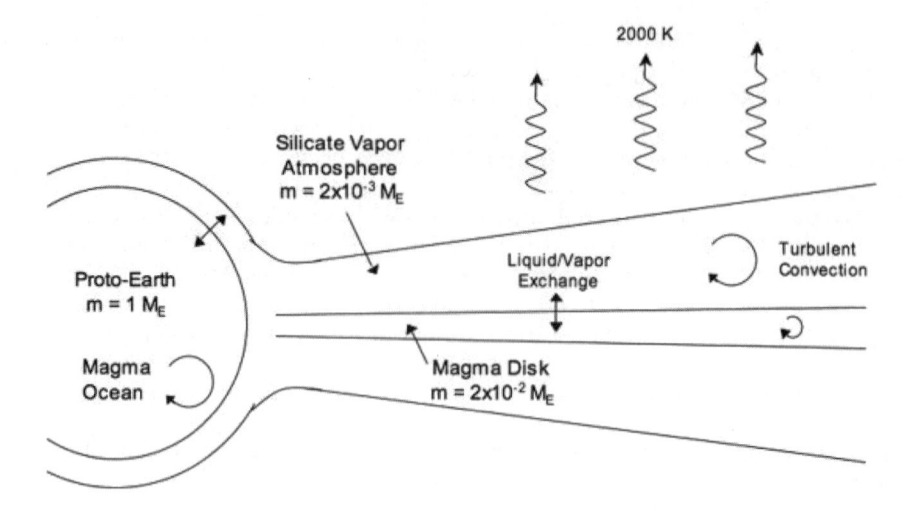

Figure 3. Schematic of the magma disk surrounding the proto-Earth after the Moon-forming impact. Such a large hot disk would be easy to detect around near-by stars (from Pahlevan & Stevenson 2007).

have explored whether we could directly detect molten Earths or super-Earths. Current technology would enable us to detect such bodies beyond 30 AU around stars in typical search samples, but they appear to be rare. New instruments that employ novel diffraction suppression strategies on existing 6-10 meter telescopes (e.g. Kenworthy et al. 2007) as well as the next generation ELTs present exciting possibilities.

3.2. Dynamics = Composition

Physical models of circumstellar disks inform calculations of disk chemistry telling us what can be where at a given time. Fluid dynamics governing the motions of solids in gas disks can move things around. But so can gravity alone after the dispersal of the gas. Raymond et al. (2004) assessed what fraction of volatiles (such as water) could end up as part of forming terrestrial planets built through collisions of planetesimals that originate from a variety of locations. The basic idea is that static temperature considerations (such as the location of the ice-line) can tell you where water-rich planetesimals might be located in a disk. And the dynamics of N-body systems can help you assess where they end up. A general feature of N-body calculations of collisions capable of forming terrestrial planets within a few AU of Sun-like stars is that their eccentricities and inclinations are generally too high. O'Brien et al. (2006) have explored dynamical friction from swarms of small bodies to dampen the eccentricities with some success (see also Morishima et al. 2008).

Bond et al. (2010) have taken this approach one step further combining the results of a time-dependent accretion disk model with a chemical condensation sequence model to predict the composition of planetesimals as a function of radius in the disk (assuming no bulk migration of solids). Convolving the output of these calculations with a dynamical code that tracks what fraction of each planet formed from the initial annulus in the disk, one can estimate the bulk elemental abundances of the planets. The approach, though simplistic, does a remarkably good job for the abundances of many elements in the Earth and Venus (excluding volatiles). Future work exploring similar models as a function of stellar mass, stellar composition, comparing results from other codes, and including the effects of solid migration could provide additional insight. A potentially observable test of such models could be the amount of heavy elements deposited in the stellar photosphere after the radiative core of the Sun-like star develops. Models can predict the amount of solid material lost into the star from collisions of rocky planetesimals. These calculations can be compared to the dispersion in heavy element abundances observed in young clusters such as the Pleiades. Observations reported by Wilden et al. (2002) indicate that fewer than 5 Earth-masses of refractory elements could have been accreted in a differential sense. This limit approaches the predictions from models and future work could provide more stringent constraints. Finally we note that, contrary to pollution scenarios, Melendez et al. (2009) find evidence that the Sun is depleted in refractory elements compared to other Sun–like stars, perhaps offering another tool to pre–select "systems like our own" for closer scrutiny.

3.3. Evolution of Outer Belts

Wyatt (2008) and references therein recently review the evolution of debris disks around stars as a function of stellar type. In general one can say that debris disks (of higher mass) appear more common around A stars than around G stars or M dwarfs (Su et al. 2006; Trilling et al. 2008; Carpenter et al. 2009; and Gautier et al. 2007). There is some evidence for luminosity evolution in that older stars sport fainter debris though there is a large dispersion in most observed quantities. One striking feature of the data reported for Sun-like stars from the FEPS project (Carpenter et al. 2009) is that inner disk hole sizes are typically 10 AU, even for stars 3-30 Myr old. Morales et al. (2009) also report large inner holes for A star samples. Whatever clears these debris disks out beyond 10 AU (perhaps efficient planet formation) happens relatively fast. A healthy fraction of these debris disks cannot be fitted with single temperature models implying debris over a range of radii. It remains to be seen whether models of extended debris (like a modified T Tauri disk model; Hillenbrand et al. 2008) fit as well, worse, or better than models with two distinct inner and outer belts (like scaled versions of the asteroid and Kuiper belts; Su et al. 2009).

It is clear that Spitzer observations have detected only the tip of the iceberg regarding dusty debris around Sun-like stars. Bryden et al. (2006) estimate that the data are consistent with the median of Sun-like stars having a debris signature within × 10 of our own (see also Greaves et al. 2010). Recent models for the evolution our own solar system evolution invoke migration of the giant planets as a key feature about 900 Myr after formation (Morbidelli et al. 2009). Jupiter and Saturn migrate inward, and Uranus and Neptune migrate outward, wreaking havoc on both the asteroid and Kuiper belts in the process. This dynamical depletion of dust-producing parent bodies is perhaps the main reason why our system is so faint today (Booth et al., 2009; Meyer et al. 2007). Other planetary system architectures might be expected to undergo comparable dynamical rearrangements at times determined by their configurations.

Is there a connection between outer debris disks traced by far-IR/mm-wave emission and extra-solar planets detected via other means? Moro-Martin et al. (2007) searched for a correlation based in disk samples from the Spitzer Space Telescope and found none (see also Bryden et al. 2009). Yet there are some very notable exceptions: HD 69830, eps Eri, Beta Pic, Fomalhaut, and HR 8799. Su et al. (2009) point out the unique spectral energy distribution of HR 8799 may hint at disk sculpting by planets. Imaging such disks in scattered light (e.g. SPHERE and GPI) , and in thermal emission (Herschel, ALMA, JWST) will provide fundamental constraints on models needed to make accurate estimates of their physical properties.

3.4. Are Planetary Systems Common?

Radial velocity surveys suggest that 8.5% of Sun-like stars possess planets with masses $> 0.3 M_{JUP}$ within 3 AU (Cumming et al. 2008). Extrapolations of those results beyond about 30 AU are inconsistent with existing null results (e.g. Nielsen & Close, 2009). The next generation of imaging surveys will test the prediction that $\sim 20\%$ of Sun-like stars have gas giants inside of 20 AU (e.g. Cumming et al. 2008). There are already hints from on-going RV work as well as micro-lensing statistics that lower mass (3-10 Earth-mass) planets are even more common (Gould et al. 2010; Mayor et al. 2009). If we take these results as confirmation of the planetary population synthesis models, we can speculate that terrestrial planets will be more common still (Mordasini et al. 2009; Ida & Lin 2004). Finally we note the recent work of Wright et al. (2009) who point out that at least 28 % of known planetary systems contain multiple planets (like our own).

And if terrestrial planets are so common, can we detect them? Above, we explored a high-risk scenario for detecting hot terrestrial planets during their collisional formation. Absil et al. (this volume) present observations of reflected light from inner zodiacal dust emission which could present challenges for the direct detection of terrestrial planets. The Herschel Space Telescope as well as the LBT Interferometer will be powerful tools to explore cool as well as warm debris down to levels $\times$ 10 that inferred for our planetary systems (Hinz 2010). The case for a 1-2 meter space-based visible coronagraph capable of detecting debris down to solar system levels seems clear. Such a facility could also detect Earth-like planets around a handful of nearby stars as well as characterize dozens to hundreds of super-Earths and gas giants in visible reflected light (e.g. Guyon et al. this volume). The joint JAXA/ESA mission SPICA could also play a crucial role in planning for a future space-based mission (Goicoechea et al. this volume). Finally, we note that the next generation of extremely large telescopes might also take the first image of a terrestrial planet around a nearby star (Kasper et al. 2009).

4. Implications

One thing is very clear from studies concerning the formation and evolution of planetary systems: all disk properties exhibit great dispersion at a given time. Yet some key parameters appear to be a function of star mass (like initial disk mass and gas-disk lifetime). While the draining of solids represents serious challenges to current theories of planet formation, these physical mechanisms may also present opportunities for planetesimal growth. Regarding the formation of giant planets, some lines of evidence favor the core accretion theory, yet it seems clear that gravitational instability might also play a role. It may be that hybrid approaches are needed where gravitational instabilities collect solid material in gas rich disks enhancing the prospects for the core accretion scenario. Available observational constraints (meager though they are) are consistent with terrestrial planets being very common. We may even be lucky to catch some in the process of collisional formation!

Many open questions remain. Does planet formation favor dynamically dense systems? Perhaps all planetary systems are packed in the sense of being stable only on timescales comparable to their current age (Laskar 1994; Barnes & Greenberg 2007). If so, it is those systems lacking circumstellar debris that could harbor the richest planetary architectures. Is planet formation in multiple star systems different than around single stars? We are only beginning to tackle this question from theory (Quintana et al. 2007) and observations (Eggenberger 2009). How does planet formation vary from one star forming environment to another? Studies of disk evolution as a function of cluster environment are ongoing (e.g. Mann & Williams, 2009). The presence of massive stars (likely in a rich cluster) could have important effects on the planet-forming potential of disks found around nearby sun-like stars (Adams et al. 2006).

If we could perform an on-line search to get directions for a "Pathway Toward Habitable Planets" the results might read something like this: 1) Straight AHEAD: with medium class missions capable of answering key questions in the next decade; 2) 'RIGHT': strong exoplanet science cases for a variety of ELT instrumentation; 3) Don't be LEFT behind: invest now in diverse technologies for a flagship mission in order to save money when final design concepts are needed; 4) WATCH for: differences between exploration and scientific hypothesis testing in developing level one requirements. Will you learn something from the null result? Such an investment should teach us something fundamental about the formation and evolution of planetary systems; and 5) Don't STOP: until you reach your goal of imaging and taking spectra of terrestrial planets around nearby Sun-like stars!

Acknowledgments. I would like to thank the organizers of the Blue Dots initiative for staging such a stimulating conference in an inspiring setting as well as colleagues on the Working Group for the Formation and Evolution of Planetary Sytems for sharing their ideas. Special thanks to I. Ribas for his patience and advice in preparing this manuscript and H. Zinnecker for providing comments on a draft version.

References

Adams, F. C., Proszkow, E. M., Fatuzzo, M., & Myers, P. C. 2006, ApJ, 641, 504
Andrews, S. M., et al. 2009, ApJ, 700, 1502
Augereau, J.-C., et al. 2008, SF2A-2008, 443 (http://proc.sf2a.asso.fr)
Barnes, R., & Greenberg, R. 2007, ApJ, 665, L67
Boley, A. C. 2009, ApJ, 695,
Bond, J. C., Lauretta, D. S., & O'Brien, D. P. 2010, Icarus, 205, 321 L53
Booth, M., et al. 2009, MNRAS, 399, 385

Bryden, G., et al. 2006, ApJ, 636, 1098
Bryden, G., et al. 2009, ApJ, 705, 1226
Canup, R. M. 2004, ARA&A, 42, 441
Carr, J. S., & Najita, J. R. 2008, Science, 319, 1504
Carpenter, J. M., et al. 2009, ApJS, 181, 197
Cortes, S. R., et al. 2009, ApJ, 697, 1305
Ciesla, F. J., & Cuzzi, J. N. 2006, Icarus, 181, 178
Cumming, A., et al. 2008, PASP, 120, 531
Currie, T., Lada, C. J., Plavchan, P., Robitaille, T. P., Irwin, J., & Kenyon, S. J. 2009, ApJ, 698, 1
Currie, T., Kenyon, S. J., Balog, Z., Rieke, G., Bragg, A., & Bromley, B. 2008, ApJ, 672, 558
Dullemond, C. P., Hollenbach, D., Kamp, I., & D'Alessio, P. 2007, Protostars and Planets V, eds. B. Reipurth, D. Jewitt, & K. Keil, (Tucson: Univ. of Arizona) 555
Dullemond, C. P., Dominik, C., & Natta, A. 2001, ApJ, 560, 957
Durisen, R. H., et al. 2007, Protostars and Planets V, eds. B. Reipurth, D. Jewitt, & K. Keil, (Tucson: Univ. of Arizona) 607
Eggenberger, A. 2009, arXiv:0910.3332
Ercolano, B., Clarke, C. J., & Drake, J. J. 2009, ApJ, 699, 1639
Fischer, D. A., & Valenti, J. 2005, ApJ, 622, 1102
Gail, H.-P. 2002, A&A, 390, 253
Gautier, T. N., III, et al. 2007, ApJ, 667, 527
Gorti, U., & Hollenbach, D. 2008, ApJ, 683, 287
Gould, A., et al. 2010, arXiv:1001.0572
Greaves, J. S., & Wyatt, M. C. 2010, MNRAS, 322
Helled, R., & Schubert, G. 2009, ApJ, 697, 1256
Hillenbrand, L. A., et al. 2008, ApJ, 677, 630
Hinz, P. M. 2009, American Institute of Physics Conference Series, 1158, 313
Ida, S., & Lin, D. N. C. 2008, ApJ, 673, 487
Ida, S., & Lin, D. N. C. 2004, ApJ, 604, 388
Isella, A., Carpenter, J. M., & Sargent, A. I. 2009, ApJ, 701, 260
Johansen, A., et al. 2007, Nat, 448, 1022
Kalas, P., et al. 2008, Science, 322, 1345
Kasper, M., et al. 2010, 1st AO4ELT conf. - Adaptative Optics for Extremely Large Telescopes, Paris, France, eds Y. Clénet, J.-M. Conan, Th. Fusco, and G. Rousset (EDP Sciences) 2010, id.02009,
Kenworthy, M. A., at al. 2007, ApJ, 660, 762
Kenyon, S. J., & Bromley, B. C. 2008, ApJS, 179, 451
Kley, W., Lee, M. H., Murray, N., & Peale, S. J. 2005, A&A, 437, 727
Kretke, K. A., & Lin, D. N. C. 2007, ApJ, 664, L55
Lagrange, A.-M., et al. 2009, A&A, 493, L21
Laskar, J. 1994, A&A, 287, L9
Lee, J.-E., Bergin, E. A., & Nomura, H. 2010, ApJ, 710, L21
Lissauer, J. J., & Stevenson, D. J. 2007, Protostars and Planets V, eds. B. Reipurth, D. Jewitt, & K. Keil, (Tucson: Univ. of Arizona) 591
Lisse, C. M., et al. 2009, ApJ, 701, 2019
Luhman, K. L., et al. 2010, ApJS, 186, 111
Mamajek, E. E. 2009, American Institute of Physics Conference Series, 1158, 3
Mann, R. K., & Williams, J. P. 2009, ApJ, 694, L36
Marois, C.,et al. 2008, Science, 322, 1348
Mayor, M., et al. 2009, A&A, 493, 639
Meyer, M. R. 2009a, IAU Symposium, 258, 111
Meyer, M., 2009, Physics World, 22, 26
Meyer, M. R., et al. 2008, ApJ, 673, L181

Miller-Ricci, E., Meyer, M. R., Seager, S., & Elkins-Tanton, L. 2009, ApJ, 704, 770
Morbidelli, A., et al. 2009, A&A, 507, 1041
Mordasini, C., Alibert, Y., Benz, W., & Naef, D. 2009, A&A, 501, 1161
Morales, F. Y., et al. 2009, ApJ, 699, 1067
Moro-Martín, A., et al. 2007, ApJ, 658, 1312
Nagasawa, M., et al. 2007, Protostars and Planets V, eds. B. Reipurth, D. Jewitt, & K. Keil, (Tucson: Univ. of Arizona) 639
Nelson, R. P. 2005, A&A, 443, 1067
Nielsen, E. L., & Close, L. M. 2009, arXiv:0909.4531
O'Brien, D. P., Morbidelli, A., & Levison, H. F. 2006, Icarus, 184, 39
Paardekooper, S.-J., & Papaloizou, J. C. B. 2009, MNRAS, 394, 2283
Pahlevan, K., & Stevenson, D. J. 2007, Earth and Planetary Science Letters, 262, 438
Pascucci, I., Apai, D., Luhman, K., Henning, T., Bouwman, J., Meyer, M. R., Lahuis, F., & Natta, A. 2009, ApJ, 696, 143
Pinte, C., et al. 2008, A&A, 489, 633
Pontoppidan, K. M., Blake, G. A., van Dishoeck, E. F., Smette, A., Ireland, M. J., & Brown, J. 2008, ApJ, 684, 1323
Quintana, E. V., Adams, F. C., Lissauer, J. J., & Chambers, J. E. 2007, ApJ, 660, 807
Rafikov, R. R. 2009, ApJ, 704, 281
Raymond, S. N., Quinn, T., & Lunine, J. I. 2004, Icarus, 168, 1
Rhee, J. H., Song, I., & Zuckerman, B. 2008, ApJ, 675, 777
Salyk, C., Pontoppidan, K. M., Blake, G. A., Lahuis, F., van Dishoeck, E. F., & Evans, N. J., II 2008, ApJ, 676, L49
Santos, N. C., Israelian, G., & Mayor, M. 2001, A&A, 373, 1019
Sargent, B. A., et al. 2009, ApJ, 690, 1193
Siegler, N., Muzerolle, J., Young, E. T., Rieke, G. H., Mamajek, E. E., Trilling, D. E., Gorlova, N., & Su, K. Y. L. 2007, ApJ, 654, 580
Su, K. Y. L., et al. 2009, ApJ, 705, 314
Su, K. Y. L., et al. 2006, ApJ, 653, 675
Trilling, D. E., et al. 2008, ApJ, 674, 1086
Weidenschilling, S. J. 1977, MNRAS, 180, 57
Wetherill, G. W. 1991, Science, 253, 535
Wilden, B. S., Jones, B. F., Lin, D. N. C., & Soderblom, D. R. 2002, AJ, 124, 2799
Winnick, R. A., Demarque, P., Basu, S., & Guenther, D. B. 2002, ApJ, 576, 1075
Woitke, P., Kamp, I., & Thi, W.-F. 2009, A&A, 501, 383
Wyatt, M. C. 2008, ARA&A, 46, 339
Wyatt, M. C., Smith, R., Greaves, J. S., Beichman, C. A., Bryden, G., & Lisse, C. M. 2007, ApJ, 658, 569
Zahnle, K., Arndt, N., Cockell, C., Halliday, A., Nisbet, E., Selsis, F., & Sleep, N. H. 2007, Space Science Reviews, 129, 35

The organizers gather for the opening session.

Session IV
Astrophysical Themes

Pathways Towards Habitable Planets
ASP Conference Series, Vol. 430, 2010
Vincent Coudé du Foresto, Dawn M. Gelino, and Ignasi Ribas, eds.

General Milankovitch Cycles

D. S. Spiegel,[1] S. N. Raymond,[2] C. D. Dressing,[1] C. A. Scharf,[3,4] J. L. Mitchell,[5] and K. Menou[4]

[1] *Department of Astrophysical Sciences, Princeton University, Peyton Hall, Princeton, NJ, USA*

[2] *Center for Astrophysics and Space Astronomy, University of Colorado, UCB 389, Boulder CO, USA*

[3] *Columbia Astrobiology Center, Columbia Astrophysics Laboratory, Columbia University, 550 West 120th Street, New York, NY, USA*

[4] *Department of Astronomy, Columbia University, 550 West 120th Street, New York, NY, USA*

[5] *Institute for Advanced Study, School of Natural Sciences, 1 Einstein Drive, Princeton, NJ, USA*

Abstract. The Earth is thought to have gone through at least one globally frozen, "snowball" state in the last billion years that it presumably exited after several million years of buildup of greenhouse gases when the ice-cover shut off the carbonate-silicate cycle. Extrasolar terrestrial planets with the capacity to host life might fall into similar snowball states. Here we show that if a terrestrial planet has a giant companion on a sufficiently eccentric orbit, it can undergo Milankovitch-like oscillations of eccentricity of great enough magnitude to melt out of a snowball state.

1. Introduction

Even very mild astronomical forcings can have dramatic influence on the Earth's climate. Although the orbital eccentricity varies between ~0 and only ~0.06, and the axial tilt, or obliquity, between 22.1° and 24.5°, these slight periodic changes are sufficient to help drive the Earth into ice ages at regular intervals. Milankovitch (1941) recognized and articulated this possibility in his astronomical theory of the ice ages. Specifically, Milankovitch posited a causal connection between three astronomical cycles (precession – 23 kyr period, and variation of both obliquity and eccentricity – 41 kyr and 100 kyr periods, respectively) and the onset of glaciation/deglaciation. Though much remains to be discovered about the Milankovitch cycles, they are now generally acknowledged to have been the dominant factor governing the ice ages of the last several million years (Berger 1975).

The nonzero (but, at just 0.04, nearly zero) eccentricity of Jupiter's orbit is the primary driver of the Earth's eccentricity Milankovitch cycle. Were Jupiter's eccentricity greater, it would drive larger amplitude variations of the Earth's eccentricity. This same mechanism might be operating in other solar systems.

Among the more than 400 currently known extrasolar planets, there are many that have masses comparable to Jupiter's and that are on highly eccentric orbits; ~20% of the known exoplanets have eccentricities greater than 0.4, including such extreme values as 0.93 and 0.97 (HD 20782b: O'Toole et al. 2009; HD 80606: Naef et al. 2001).[1] Furthermore, tantalizing evidence suggests that lower mass terrestrial planets might be even more numerous than the giant planets that are easier to detect (O'Toole et al. 2009). Therefore, it seems highly likely that many terrestrial planets in our galaxy experience exaggerated versions of the Earth's eccentricity Milankovitch cycle.

These kinds of cycles could have dramatic influence on life that requires liquid water. Since seminal work by Dole (1964) and Hart (1979), a variety of theoretical investigations have examined the possible climatic habitability of terrestrial exoplanets. Kasting et al. (1993) emphasized that the habitability of an exoplanet depends on the properties of the host star. Several authors have considered how a planet's climatic habitability depends on the properties of the planet, as well (Williams & Kasting 1997; Williams & Pollard 2002; Selsis et al. 2007; Spiegel et al. 2008, 2009). In particular, two recent papers have focused on the climatic effect of orbital eccentricity. Williams & Pollard (2003) used a general circulation climate model to address the question of how the Earth's climate would be affected by a more eccentric orbit. A companion paper to this one (Dressing et al. 2009; hereafter D09) uses an energy balance climate model to explore the combined influences of eccentricity and obliquity on the climates of terrestrial exoplanets with generic surface geography. A more eccentric orbit both accentuates the difference between stellar irradiation at periastron and at apoastron, and increases the annually averaged irradiation (in proportion to $(1-e^2)^{-1/2}$, where e is eccentricity). Thus, periodic oscillations of eccentricity will cause concomitant oscillations of both the degree of seasonal extremes and of the total amount of starlight incident on the planet in each annual cycle. Since these oscillations depend on gravitational perturbations from other companion objects, the present paper can be thought of as examining how a terrestrial planet's climatic habitability depends not just on its star, not just on its own intrinsic properties, but also on the properties of the planetary system in which it resides.

There is evidence that, at some point in the last billion years, Earth went through a "Snowball Earth" state in which it was fully (or almost fully) covered with snow and ice (see the review by Hoffman & Schrag 2002 and references therein; Williams 1993 and Williams et al. 1998 suggest an alternative interpretation of the data). The high albedo of ice gives rise to a positive feedback loop in which decreasing surface temperatures lead to greater ice-cover and therefore to further net cooling. As a result, the existence of a low-temperature equilibrium climate might be a generic feature of water-rich terrestrial planets, and such planets might have a tendency to enter snowball states. The ice-albedo feedback makes it quite difficult for a planet to recover from such a state (Pierrehumbert 2005). In temperate conditions, the Earth's carbonate-silicate weathering cycle acts as a "chemical thermostat" that tends to prevent surface temperatures from straying too far from the freezing point of water (Zeebe & Caldeira 2008). A

---

[1] See http://exoplanet.eu

snowball state would interrupt this cycle. The standard explanation of how the Earth might have exited its snowball state is that this interruption of the weathering cycle would have allowed CO_2 to build up to concentrations approaching ~1 bar over 10^6-10^7 years, at which point the greenhouse effect would have been sufficient to melt the ice-cover and restore temperate conditions (Hoffman & Schrag 2002). However, an exoplanet in a snowball state that is undergoing a large excitation of its eccentricity might be able to melt out of its globally frozen state in significantly less time, depending on the magnitude of the eccentricity variations and on other properties of the planet. Exploring this possibility is the primary focus of this paper.

2. Climate Model

In lieu of full, self-consistent modeling of atmospheric dynamics and radiative transfer, we use nearly the same 1-dimensional time-dependent energy balance model (EBM) that we have used in previous studies of exoplanet climatic habitability (Spiegel et al. 2008, hereafter SMS08; Spiegel et al. 2009, hereafter SMS09; see references therein for a justification of this model in habitability studies). The present work's model has a slight modification to handle starting in a snowball state. Our model treats the redistribution of heat as a diffusive process, and it (or close variants) has been used by other authors both in the context of exoplanet habitability (Williams & Kasting 1997, hereafter WK97) and in other contexts, including studies of both Martian climate and the Earth's climate (North et al. 1981). The response of this model to nonzero eccentricity is explored extensively in a companion paper (D09). Here, we provide a brief synopsis of the model.

In accord with WK97 and our own previous work, we use the following Equation for planet surface temperatures as a function of time and location:

$$C\frac{\partial T}{\partial t} - \frac{\partial}{\partial x}\left(D(1-x^2)\frac{\partial T}{\partial x}\right) = S(1-A) - I, \qquad (1)$$

where $x \equiv \sin\lambda$ and λ is latitude. The values of heat capacity C and effective diffusivity D, and the albedo and infrared cooling functions A and I are as described in SMS08 and SMS09. We briefly recapitulate. The heat capacity has the following values over land (C_l), ice (C_o), and ocean (C_o): $C_l = 5.25 \times 10^9$ erg cm^{-2} K^{-1}, $C_i = 9.2C_l$ when $263 < T < 273$ K and $2.0C_l$ when $T \leq 263$ K, and $C_o = 40C_l$. The diffusivity $D = 5.394 \times 10^2$ erg cm^{-2} s^{-1} K^{-1}. The albedo is A_2 of SMS08 and SMS09: $A[T] = 0.525 - 0.245\tanh[(T-268\text{ K})/5\text{ K}]$. We adopt two different functional forms for the infrared cooling. The first is I_2 of SMS08 and SMS09: $I_2[T] = \sigma T^4 / \{1 + 0.5925(T/273\text{ K})^3\}$, where $\sigma = 5.67 \times 10^{-5}$ erm cm^{-2} s^{-1} K^{-4} is the Stefan-Boltzmann constant. The second is I_{WK97} of SMS09, which depends not only on T but on p_{CO_2}, the CO_2 partial pressure. This function, presented in the appendix of WK97, is a parameterized fit (third order in T and fourth order in $\ln[p_{CO_2}]$) to a radiative-convective model (Kasting 1991). We solve Equation (1) on a grid of 145 points equally spaced in latitude. We use a time-implicit numerical scheme and an adaptive time-step, as described in SMS08 and in Hameury et al. (1998). Obliquity is varied from 0° to 90°, and initial temperature is set to be 100 K.

2.1. Modeling a Cold Start

We place a frozen planet in a variety of pre-set orbits in order to explore the capacity for the orbital configuration (specifically, the semimajor axis, the eccentricity, and the obliquity) to thaw such a planet. In previous work using this model, we have set "hot start" initial conditions in which the initial temperature is set far above the freezing point of water. In modeling the melting of a Snowball Earth planet, we are in a different regime. We have previously implicitly assumed that the latent heat involved in melting ice and freezing water is negligible. But this assumption is no longer valid in the case of a snowball planet that might have a sheet of ice that is a kilometer or more thick.

We here adopt a simple algorithm to handle the melting of an ice-sheet.[2] At a given location, if the temperature has never reached 273 K, the albedo is set to its maximum value (0.77), to indicate that the surface is still covered with snow and ice. When the local temperature reaches 273 K, further excess of radiant/diffusive energy of ΔE is treated as going into changing the thickness of the local ice sheet by Δh cm, where $\Delta h \approx -\Delta E/(3 \times 10^9$ erg). During such periods of melting, the albedo reduces to 80% of its maximum value ($A = 0.616$) because snow and ice are less shiny when melting than when fully frozen (Køltzow 2007). Consider a patch of surface that has just been melting but, because of the onset of winter, now receives less incident flux than its radiant and diffusive losses. First, the summer's melt-water re-freezes (ΔE is negative, so Δh is positive) while the temperature remains fixed at 273 K. Once all melt-water has re-frozen, subsequent energy deficits result in the temperature decreasing, as specified by Equation (1).

Instead of assuming an initial ice thickness and integrating until it has melted, we recognize that if a portion of a model planet, during a given year, receives more incident radiant energy than its radiant and diffusive losses, then it will continue to receive an excess of energy in all following years. Therefore, we define a patch of surface as having melted once the ice on it has been "melting" continuously for an entire model year, where "melting" means either actually melting or refreezing the summer's melt while the temperature is fixed at 273 K. We integrate for 130 model years, or until a portion of surface has melted and thereby restored a nonzero habitability fraction (see SMS08) to the planet. If an initially frozen model planet's surface has partially melted within the integration time, we note this model's orbital configuration as one that could melt a planet out of a snowball state.

3. Source of Earth-like Exoplanet Eccentricities

What would cause an Earth-like planet to have a large eccentricity? Models of planet formation and interactions with gaseous protoplanetary disks predict that planets should form and remain on near-circular orbits (e.g., Tanaka & Ward 2004). The low eccentricities of the Solar System's terrestrial and giant planets

---

[2]See Semtner (1976) for a more sophisticated model of the vertical diffusion of heat through an ice sheet.

fit this picture, but the wide eccentricity distribution of the known extrasolar giant planets provides a dramatic contrast (Wright et al. 2009).

Once a planetary system is formed and any short-term instabilities have occurred, subsequent evolution is dominated by dynamical interactions between the planets (as well as remnant planetesimals). The Earth's eccentricity oscillates between almost zero and about 0.06 in a complicated fashion with a $\sim 100{,}000$ year periodicity (Laskar 1988). The dynamics of the Earth's orbit are controlled by resonant secular forcing from the other planets and undergo chaotic evolution on long timescales (Laskar 1989). We expect a large diversity of the orbits of extra-solar Earth-like planets. For the purposes of this study, we are most interested in the eccentricity evolution of such planets, in terms of both the amplitude of oscillation and the period of oscillation.

Given the importance of resonances and the potential for chaos, it is impossible to understand the detailed dynamical evolution of a planetary system without full knowledge of the system's architecture. However, we can make simple assumptions to get an idea of the range of reasonable eccentricity evolution for Earth-like planets. We therefore consider the simple case of a single Earth analog at 1 AU evolving under the perturbations from a single giant planet. We are interested in the range of variations as well as the systematic dependences on planetary system properties such as the proximity and orbit of the giant planet. To study these simple, two-planet systems we integrate the orbit-averaged rates of change for the planets' eccentricities e and longitudes of pericenter ϖ using the equations of Mardling & Lin (2002) and Mardling (2007). We make the assumption that the planets are coplanar, and evolve a variety of systems for 1 Myr, in each case including an Earth-mass planet at 1 AU. We also neglect the effects of general-relativistic precession.

The amplitude of the Earth-like planet's eccentricity oscillation is a function of the giant planet's semimajor axis and eccentricity but is independent of the giant planet's mass. However, the oscillation frequency (often referred to as the secular frequency) increases linearly with giant planet mass. Not surprisingly, the amplitude of the Earth analog's eccentricity oscillation decreases for lower giant planet eccentricities, and the period of oscillation increases with lower giant planet eccentricities.

The planetary system architecture is another key influence. For example, the following two systems both induce eccentricity oscillations of about 0.1 in the Earth analog, but on very different timescales of 3,000 and 170,000 years: (1) The fast-oscillating case contains a giant planet interior to the Earth-like planet, at 0.5 AU with an eccentricity of 0.1. (2) The slow-oscillating case contains an exterior giant planet, at 5 AU with an eccentricity of 0.4.

We also searched for extreme cases that would be dynamically stable, yet induce very large amplitude oscillations in the Earth's eccentricity. One potential source of large eccentricities comes from a consideration of inclinations called the Kozai mechanism (Kozai 1962). For a star-planet-companion system, where the companion can be stellar or planetary, an inclination between the planet and companion's orbits I larger than $39.2°$ can induce eccentricity (and inclination) oscillations in the planet's orbit (Innanen et al. 1997). For reasonable estimates of the mutual inclination between planetary and companion orbits, the majority of binary stars should induce large eccentricity oscillations in any planets that form around the primary star (Takeda & Rasio 2005).

4. Summary and Conclusions

Using our modified EBM, we searched for orbital configurations that would lead to an ice-covered planet melting out of the snowball state. Dynamical-model and climate-model results will be presented in detail in a paper that is currently in preparation (Spiegel et al. 2009 in prep.). In brief, we found that orbital configurations that are not unlikely could cause a snowball-Earth-analog to melt out by dint of increased eccentricity.

Acknowledgments. We thank Adam Burrows, Ed Turner, and Scott Tremaine for helpful discussions. We acknowledge the use of the Della supercomputer at the TIGRESS High Performance Computing and Visualization Center at Princeton University.

References

Berger, A. 1975, Ciel et Terre, 91, 261
Dole, S. H. 1964, Habitable planets for man (New York, Blaisdell Pub. Co.)
Dressing, C. D., Spiegel, D. S., Raymond, S. et al. 2009, ArXiv 1002.4875
Hameury, J.-M., Menou, K., Dubus, G. et al. 1998, MNRAS, 298, 1048
Hart, M. H. 1979, Icarus, 37, 351
Hoffman, P. F. & Schrag, D. P. 2002, Terra Nova, 14, 129
Innanen, K. A., Zheng, J. Q., Mikkola, S., & Valtonen, M. J. 1997, AJ, 113, 1915
Kasting, J. F. 1991, Icarus, 94, 1
Kasting, J. F., Whitmire, D. P., & Reynolds, R. T. 1993, Icarus, 101, 108
Køltzow, M. 2007, JGR (Atmospheres), 112, 7110
Kozai, Y. 1962, AJ, 67, 591
Laskar, J. 1988, A&A, 198, 341
—. 1989, Nat, 338, 237
Mardling, R. A. 2007, MNRAS, 382, 1768
Mardling, R. A. & Lin, D. N. C. 2002, ApJ, 573, 829
Milankovitch, M. 1941, Kanon der Erdbestrahlung und seine Anwendung auf das Eiszeitenproblem (Belgrade, Mihaila Curcica)
Naef, D., LathaM, D. W., Mayor, M. et al. 2001, A&A, 375, L27
North, G. R., et al. 1981, Rev. Geophys. & SpPhys., 19, 91
O'Toole, S. J., Jones, H. R. A., Tinney, et al. 2009, MNRAS392, 641
Pierrehumbert, R. T. 2005, JGR (Atmospheres), 110, 1111
Selsis, F., Kasting, J. F., Levrard, B., et al. 2007, A&A, 476, 1373
Semtner, Jr., A. J. 1976, J Phys Ocean., 6, 379
Spiegel, D. S., Menou, K., & Scharf, C. A. 2008, ApJ, 681, 1609
—. 2009, ApJ, 691, 596
Takeda, G. & Rasio, F. A. 2005, ApJ, 627, 1001
Tanaka, H. & Ward, W. R. 2004, ApJ, 602, 388
Williams, D. M. & Kasting, J. F. 1997, Icarus, 129, 254
Williams, D. M., Kasting, J. F., & Frakes, L. A. 1998, Nat, 396, 453
Williams, D. M. & Pollard, D. 2002, Int. J. Astrobiology, 1, 61
—. 2003, Int. J. Astrobiology, 2, 1
Williams, G. E. 1993, Earth Science Reviews, 34, 1
Wright, J. T., Upadhyay, S., Marcy, G. W., et al. 2009, ApJ, 693, 1084
Zeebe, R.-E. & Caldeira, K. 2008, Nature Geoscience, 1, 312

Pathways Towards Habitable Planets
ASP Conference Series, Vol. 430, 2010
Vincent Coudé du Foresto, Dawn M. Gelino, and Ignasi Ribas, eds.

Exploring Extrasolar Worlds Today and Tomorrow

G. Tinetti[1] and C. A. Griffith[2]

[1]*Department of Physics and Astronomy, University College London, London WC1E 6BT, UK*

[2]*Lunar and Planetary Laboratory, University of Arizona, Tucson, Arizona, US*

Abstract. More than 440 extrasolar planets have been found since the discovery of 51 Peg b by Mayor & Queloz (1995). The traditional field of planetology has thus expanded its frontiers to include planetary environments not represented in our Solar System. We expect that in the next five years space missions (CoRoT, Kepler, and GAIA) or ground-based detection techniques will both increase exponentially the number of new planets discovered and lower the present limit of a ~1.9 Earth-mass object (Mayor et al. 2009). While the search for an Earth-twin orbiting a Sun-twin has been one of the major goals pursued by the exoplanet community in the past years, the possibility of sounding the atmospheric composition and structure of an increasing sample of exoplanets with current telescopes has opened new opportunities, unthinkable just a few years ago. As a result, it is possible now not only to determine the orbital characteristics of the new bodies, but moreover to study the exotic environments that lie tens of parsecs away from us. The analysis of the starlight not intercepted by the thin atmospheric limb of its planetary companion (transit spectroscopy), or of the light emitted/reflected by the exoplanet itself, will guide our understanding of the atmospheres and the surfaces of these extrasolar worlds in the next few years. We discuss in this manuscript the preliminary results obtained by interpreting current observations of the atmospheres of transiting Gas giants and Neptunes. While the full characterisation of an Earth-twin might require a technological leap, our understanding of large terrestrial planets (so called Super-Earths) orbiting bright, later-type stars is within reach by current space and ground telescopes.

1. Introduction

Half a century ago, the space age began with the launching of the Sputnik. Now at the completion of a fairly detailed study of planets in our own solar system, we are at the dawn of the age of Exoplanets. More than 440 exoplanets, i.e., planets orbiting a star different from our Sun, are now known thanks to indirect detection techniques (Schneider 2010). In the first decade after their initial discovery in 1995 (Mayor & Queloz 1995), the task was to find more and more of these astronomical bodies: the biggest, the smallest, the hottest, the coolest, the system with the most planets in it. In recent years, attention has switched from finding planets to characterising them. Among the variety of exoplanets discovered so far, special attention has been devoted to those planets which transit their parent star, whose presence can be detected by the reduction in the brightness – the extinction – of the central star as the planet passes in front of it (Charbonneau et al. 2000). More than 70 currently identified exoplanets are transiting planets, and for these objects planetary and orbital parameters such as radius, eccentricity, inclination, mass (given by radial velocity combined measurements), are known, allowing first order characterisation on the bulk composition and temperature (gas giant? Neptune type? Terrestrial?). But it is clear that there is great variety even amongst the family of transiting exoplanets. The smallest, CoRoT-7b (Leger et al 2009) and GJ1214b (Charbonneau et al. 2009), have masses of just 0.0151 and 0.0179 M_j. They orbit their parent star at distances of < 0.02 AU. At the other end of the transiting planet distance scale, HD 80606 b orbits its G5 star with a period of more than 100 days and an eccentricity of 0.93. And at the high mass end, WASP-18b and XO-3b have a mass of 10.43 and 11.79 M_j and their atmospheric temperature is likely to be very hot.

Most importantly recent results have been able to demonstrate that for transiting exoplanets orbiting stars brighter than V=12, it is possible to identify the main chemical components in the planet's atmosphere. A stellar occultation (called primary transit) occurs when the light from a star is partially blocked by an intervening body, such as a planet. With this method, we can indirectly observe the thin atmospheric ring surrounding the optically thick disk of the planet while the planet is transiting in front of its parent star (Brown 2001; Charbonneau et al. 2002). In the secondary transit technique, we first observe the combined spectrum of the star and the planet. Then, we take a second measurement of the star alone when the planet disappears behind it: the difference between the two measurements is the planet's own spectrum (Deming et al. 2005; Charbonneau et al. 2005).

In the past three years, key observations with Spitzer and Hubble Space Telescopes have, for the first time, given us real insights into the composition of some of the most unusual exoplanets so far discovered the class known as Hot Jupiters. More specifically, infrared transmission and emission spectroscopy have revealed the presence of the primary carbon and oxygen species such as CH_4, CO_2, CO, and H_2O (Tinetti et al. 2007a; Barman et al. 2008; Swain et al. 2008a; Grillmair et al 2008; Swain et al. 2009a; Beaulieu 2010; Tinetti et al. 2010) and provided constraints for the temperature profiles (Harrington et al 2006; Knutson et al. 2007; Burrows et al. 2007; Swain et al. 2009b), which are coupled to the composition. Today, broad-band or low-resolution spectroscopy from ground- and space-based observatories allow us to:

- determine planetary and orbital parameters
- constrain the albedo
- detect the main molecular species in the hot transiting planets atmosphere
- constrain the horizontal and vertical thermal gradients in the hot exoplanet atmospheres
- constrain the boundary conditions in the upper atmosphere
- detect the presence of clouds or hazes

Relatively high resolution spectroscopy data were recently obtained with ground-based telescopes in the optical (Redfield et al. 2008; Snellen et al. 2008) and the near infrared (NIR) (Swain et al. 2010) confirming alkali metals are present in hot-Jupiter atmospheres and showing non thermal emission processes. These achievements open up enormous possibilities in terms of atmospheric characterisation for the short term future prior to the launch of the next generation of space telescopes (the James Webb Space Telescope, scheduled for launch in 2014) or a dedicated mission (e.g., THESIS, the Terrestrial and Habitable-zone Exoplanet Spectroscopy Infrared Spacecraft, Swain et al. (2009c)).

With current instruments we can already study the atmospheres of more than ten transiting hot-Jupiters and approach the case of hot Neptunes and warm super-Earths transiting later type stars, e.g., GJ436 b and GJ1214 b. Surveys aimed at detecting extrasolar planets are focusing on searches for ever smaller worlds and rocky planets in the habitable zones. While CoRoT and Kepler will increase the statistics of such objects with the ultimate goal of detecting Earth-like planets around G-type stars, transit and radial velocity surveys from the ground (HARPS, MEarth, WFCAM Transit Survey) will actually provide the optimal targets for atmospheric characterisation, in particular super-Earths transiting bright M-dwarfs down to the habitable zone. Feasibility studies show that those objects will be easily studied by JWST and THESIS-like missions (Tessenyi et al. 2010).

2. Retrieving Atmospheric Parameters from Exoplanet Spectra

Within the past year, several efforts to determine the abundances of atmospheric constituents found instead a range of degenerate temperature and composition solutions from the spectra (Swain et al. 2009b; Madhusudhan & Seager 2009; Tinetti et al. 2010), Figure 3 (left). Using an iterative forward model approach for spectral retrieval (Swain et al. 2009b), we evaluated a variety of temperature-pressure $(T - P)$ profiles and molecular mixing ratios. Combining near-infrared spectra with mid-infrared measurements, we find that absorption due to H_2O, CH_4 and CO_2 explains most of the features present in the observed hot-Jupiter spectra (Figures 1, 2, and 3). The additional contribution of CO refines our fit, but we cannot discard the possibility that improved data lists for methane and/or CO_2 would provide the missing opacity. The radiative transfer calculations assume local thermal equilibrium (LTE) conditions – as expected for pressures exceeding 10^{-3} bar that are probed by the infrared spectra – and constant mixing ratios for the molecules.

For emission spectra, we obtain a family of plausible solutions for the molecular abundances and detailed temperature profiles (Figures 2 and 3). Additional observational constraints on the atmospheric temperature structure and composition require either improved wavelength coverage/spectral resolution for the dayside spectrum or a transmission spectrum.

Transmission spectra are less sensitive to the atmospheric temperatures, yet the derived composition at the terminator depends sensitively on the assumed radius. In particular the atmospheric temperature may play an important role in the overall scale height, hence in the amplitude of the spectral signatures, as well as in the molecular absorption coefficients. For most cases a thermochemical equilibrium H_2O mixing ratio of 4.5×10^{-4} (Liang et al. 2003, 2004; Zahnle 2009) provides an excellent match to the data. However, a ~1% difference in the estimate of the planetary radius at the ~1 bar pressure level, would result in a variation of the H_2O abundances by a factor of 10. Transit data at multiple wavelengths are needed to constrain the H_2O abundance (Griffith et al. 2010).

The mixing ratios determined for CH_4, CO and CO_2 depend on the line lists used and on the H_2O mixing ratio (Figure 1).

We note that some of the temperature profiles/molecular mixing ratios consistent with the observations raise the question of whether the dayside atmosphere is in radiative and thermochemical equilibrium. Although advection of heat and/or photochemistry could support departures from radiative and thermochemical equilibrium (Figure 3, right), our present knowledge of most molecular opacities at high temperatures (CH_4, H_2S, C_2H_6 etc.) limits our ability to determine decisively whether this condition is met; thus there is an urgent need for further laboratory studies to obtain high temperature opacities of the most common molecules expected in hot-Jupiter atmospheres.

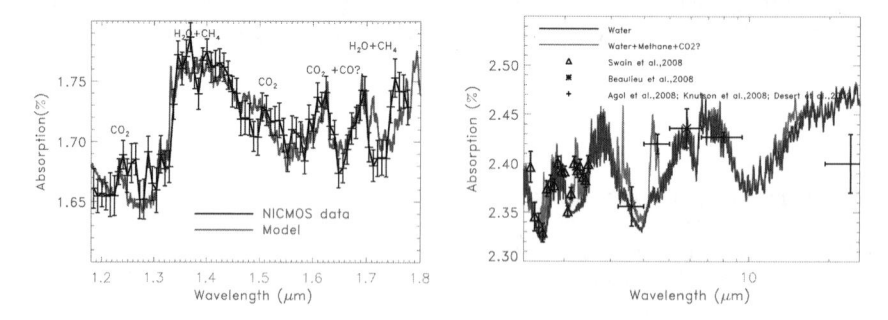

Figure 1. *Left*: transmission spectrum of the hot-Jupiter XO-1b (Tinetti et al. 2010). *Right*: Primary transit photometry and spectroscopy data of HD 189733 b recorded by multiple instruments and different teams. While most of the features can be explained by a combination of water vapour and methane, the IRAC band at 4.5 μm seems to indicate the additional presence of CO2 and/or CO.

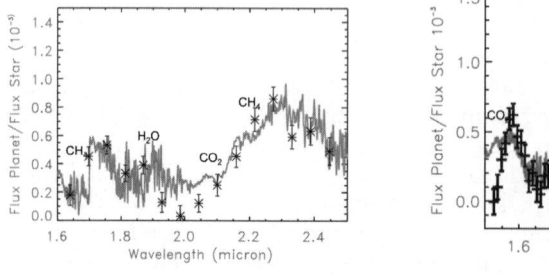

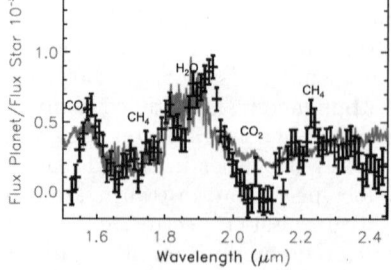

Figure 2. Emission spectra of the hot-Jupiters HD 189733 b (left) and HD 209458 b (right) observed with Hubble-NICMOS (Swain et al. 2009a,b)

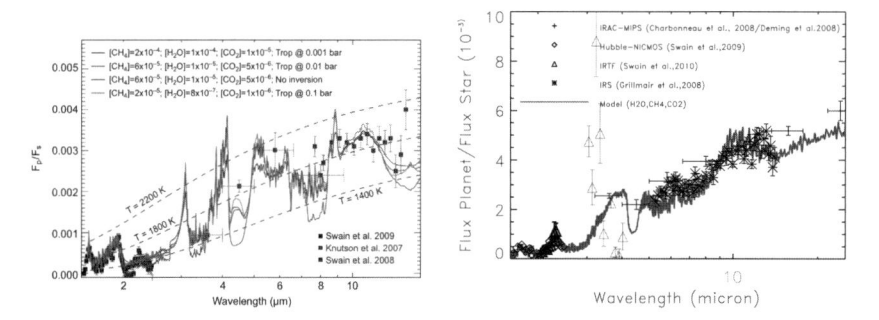

Exploring Extrasolar Worlds Today and Tomorrow 119

Figure 3. Emission photometry and spectroscopy data for two hot-Jupiters. *Left*: HD 209458 b (Swain et al. 2009b). The near-infrared and mid-infrared observations compared to synthetic spectra for four models that illustrate the range of temperature/composition possibilities consistent with the data. For each model case, the molecular abundance of CH_4, H_2O, & CO_2 and the location of the tropopause is given, these serve to illustrate how the combination of molecular opacities and the temperature structure cause significant departures from a purely single-temperature thermal emission spectrum. Note that the mid-infrared data are not contemporaneous with the near-infrared data, and attempting to connect these data sets with a model spectrum is potentially problematic if significant variability is present. *Right*: HD 189733 b (Swain et al. 2010). A radiative transfer model (red) assuming LTE conditions and consistent with the measurements made with Spitzer and Hubble fails to describe the emission structure at 3.1-4.1 μm, and we find no plausible combination of atmospheric parameters that provides a good model of the observations under LTE conditions. The brightness temperature of the 3.25 μm emission feature indicates the likely presence of a non-LTE emission mechanism.

3. The Models

We model the transmission and emission spectra of transiting exoplanets using line by line radiative transfer models which account for the effects of molecular opacities (Tinetti et al. 2005, 2007a,b) and hazes (Griffith et al. 1998, 1991, 2010). In our simulations we include H_2-H_2, H_2O, CH_4, CO, CO_2, NH_3, HCN etc. While the BT2 line list for water (Barber et al. 2006) can be calculated at the appropriate temperatures, the available line lists for methane at high temperature are inadequate to probe the modulations of the atmospheric thermal profile. To cover the spectral range from the visible to the Mid-IR, we have to use multiple line lists for methane, HITRAN 2008 (Rothman et al. 2009), PNNL, and hot-temperature measurements at 800, 1000 and 1273 K (Nassar & Bernath 2003). The Nassar & Bernath (2003) data provide a much better fit to our observations in the region where they overlap with HITRAN 2008. Compared to the results obtained with the Nassar & Bernath (2003) line lists, mixing ratios 10-50 times larger are needed for methane if we use PNNL or HITRAN 2008. The HITRAN 2008 data bank has the advantage of covering the entire spectral range measured by Hubble and Spitzer, with the downside (shared also by the PNNL list) that is based measurements at room temperature, and therefore is quite inadequate to estimate the mixing ratio of methane at the temperatures of interest for hot planets. For CO_2 we use HITEMP (Rothman et al. 2010) and CDSD-1000 (Tashkun et al. 2003), for CO we use HITEMP. The contribution of H_2-H_2 at high temperatures was taken from Borysow, Jorgensen & Fu (2001). The opacity was interpolated to the temperature of each atmospheric layer. As collision-induced absorption scales with the square of the pressure, the H_2-H_2 contribution becomes important for pressures higher than 1 bar. The line shapes of alkali metals are calculated at selected temperatures and interpolated for intermediate values. Their spectral contribution becomes important in the visible-NIR wavelength range (Allard et al. 2003). A most accurate line list for ammonia at high-temperature has recently been accomplished (Yurchenko, Barber & Tennyson 2010).

4. Conclusions

An aspect of exoplanetary science that is both high-impact and cutting-edge is the study of extrasolar planet atmospheres. The ultimate goal is to obtain a high-resolution spectrum of an Earth-like planet, and although such a goal remains lofty, the key intermediate steps towards this end are already being taken with current technology for planets which are more massive and/or warmer than our own Earth. The characterisation of exoplanet atmospheres with current telescopes can be tackled with two main approaches: low resolution spectroscopy, from space using Sptizer and HST or the ground (e.g., NASA-IRTF), and high resolution spectroscopy from the ground (e.g., VLT CRIRES). We can already probe the atmospheric constituents of several giant exoplanets, which orbit very close to their parent star, using transit techniques. The observations can be explained mainly with the combined presence of H_2O, CH_4 and CO_2 in the atmosphere of the planet, but we suggest that CO might also be present. The photometric and spectroscopic emission data observed are consistent with the above composition but a variety of $T-P$ profiles and mixing ratios are possible. Additional observations of transiting hot-Jupiters, especially spectroscopic data, will allow a more thorough classification of this type of planet unknown in our Solar System.

With current telescopes we can also approach the case of hot Neptunes and large terrestrial planets (superEarths) transiting bright later type stars, e.g., GJ 436b or GJ 1214 b. Thanks to CoRoT, Kepler, ground-based transit surveys and the improvements in radial velocity measurements, many rocky planets and, possibly, a few exomoons, are expected to be discovered in the next months/years. Further into the future the James Webb Space Telescope will be the next generation of space telescopes to be online (launch ~2014) and a dedicated mission to characterise transiting exoplanet atmospheres has been recently proposed to NASA and ESA (THESIS). Those observatories will guarantee high spectral resolution from space and the characterisation of smaller/colder targets, allowing us to expand the variety of "characterisable" extrasolar planets down to terrestrial planets and/or the habitable zone of stars cooler than the Sun.

Acknowledgments. We would like to thank Mark Swain, Pieter Deroo, Jean-Philippe Beaulieu, Gautam Vasisht and Ignasi Ribas for their key contribution on observations, Jonathan Tennyson, Bob Barber, Linda Brown and Nicole Allard for their help with data lists. G. T. is supported by a Royal Society University Research Fellowship.

References

Agol E. et al., 2008, in Transiting Planets, IAU Symposium 253, eds. Pont F., Sasselov D., Holman M., 253, 209
Allard, N. F. et al., 2003, A&A, 411, L473
Barber, R. J., Tennyson, J., Harris, G. J. & Tolchenov, R. N., 2006, MNRAS, 368, 1087
Barman T., 2008 ApJL 676, 61
Beaulieu et al., 2008, ApJ, 677, 1343
Beaulieu et al., MNRAS, in press, astroph 0909.0185.
Borysow, A., Jorgensen, U.G., Fu, Y., 2001, JQSRT, 68, 235
Brown, T. M. 2001, ApJ, 553, 1006
Burrows A., et al. 2007, ApJ668, 671
Charbonneau D., Brown T., Latham D., Mayor M., 2000, ApJ, 529, L45
Charbonneau, D., et al., 2002, ApJ, 568, 377
Charbonneau et al., 2005, ApJ, 626, 523
Charbonneau et al., 2009, Nat, 462, 891
Deming, D., et al., 2005, Nat, 434, 740
Desert J.M., Let al., 2009, ApJ, 699, 478
Griffith, C. A., Yelle, R. V. & Marley, M. S., 1998, Sci, 282, 2063
Griffith, C. A., et al., 1991, Icarus, 93, 362
Griffith, C. A., et al., ApJ, submitted
Grillmair, C. J, et al., 2008, Nat, 456, 767
Harrington J., et al., 2006, Sci, 27, 623
Knutson H., et al. 2007, Nat, 447 183
Knutson H., et al., 2009, ApJ, 690, 822
Leger A., et al., 2009, A&A, 506
Liang, M. C., et al., 2003, ApJL, 596, 247
Liang, M.C., et al., 2004, ApJL, 605, 61
Mayor, M., et al., 2009, A&A, 507, 487
Mayor, M., & Queloz, D. 1995, Nat, 378, 355
Madhusudhan, N., & Seager, S., 2009, S., ApJ, 707, 24
Nassar, R., & Bernath, P., 2003, JQSRT, 82, 279
Redfield S., Endl M., Cochran W., & Koesterke L., 2008, ApJ, 673, 87
Rothman, L. S., et al., 2009, JQSRT, 11, 533
Rothman L. S., et al., 2010, JQSRT, in press
Schneider, 2010, http://www.exoplanet.eu
Snellen et al., 2008, A&A, 487, 357
Swain, M. R., et al., 2008a, ApJ, 674, 482
Swain, M. R., Vasisht, G., & Tinetti, G., 2008b, Nat, 452, 329
Swain, M. R., et al., 2009a, ApJ, 960, L114
Swain, M. R., et al., 2009b, ApJ, 704, 1616
Swain, et. al., 2009c, Astro2010: The Astronomy and Astrophysics Decadal Survey, Technology Development Papers, 61
Swain, M. R., et al., 2010, Nat, 463, 637
Tashkun, S. A., et al., 2003, JQSRT, 82 1, 165
Tessenyi M., et al., ApJ, submitted
Tinetti, G., et al., 2005, Astrobiology, 5(4), 461
Tinetti, G., et al., 2007a, ApJ, 654, L99
Tinetti, G., et al., 2007b, Nat, 448, 169
Tinetti, G., et al., 2010, ApJL, 712, L139
Yurchenko S. N., R. J. Barber & J. Tennyson, MNRAS, submitted
Zahnle K.J., et al., 2009, ApJ, 701, L20

Pathways Towards Habitable Planets
ASP Conference Series, Vol. 430, 2010
Vincent Coudé du Foresto, Dawn M. Gelino, and Ignasi Ribas, eds.

Uniformly Derived Orbital Parameters of Exo-planets using EXOFIT

S. T. Balan,[1] G. Lever,[2] and O. Lahav[2]

[1]Astrophysics Group, Cavendish Laboratory, JJ Thomson Avenue, Cambridge, UK

[2]Department of Physics and Astronomy, University College London, Gower Street, London, UK

Abstract. We present the results from a new systematic study of the radial velocity data of more than 200 planets using the Keplerian orbital fitting program EXOFIT. Based on a Bayesian framework, EXOFIT uses Markov Chain Monte Carlo method to simulate the full posterior distribution of the orbital parameters of extrasolar planets. We discuss the disparity in the eccentricity values obtained by EXOFIT with the published values and the possible reasons for the lower estimates of eccentricity obtained by the EXOFIT. The full details of this work, including an online catalogue of exo-planets with the posterior distributions and the radial velocity plots will appears in Balan, Lever and Lahav (in preparation).

1. Introduction

In this article, we introduce a new cataloguing study of extrasolar planets by using the planet search software EXOFIT (Balan & Lahav 2009)[1] and present interesting results from the analysis of more than 200 radial velocity data sets. Rapid expansion of extrasolar planet science has resulted in the introduction of innovative data analysis inference methods. For example, the radial velocity planet search, which has the highest contribution to the planet count, started with traditional data analysis techniques such as periodogram (Lomb 1976; Scargle 1982) search and LM minimisation (Press et al. 1992). A catalogue of extrasolar planets discovered with these techniques has already been published (Butler et al. 2006). In recent years, data analysis based on Bayesian framework has been successfully applied to radial velocity data sets (Gregory 2005; Balan & Lahav 2009). These methods have long enjoyed high popularity in Cosmology; see Trotta 2008 for a general review of Bayesian methods in contemporary cosmology.

By the systematic analysis of the radial velocity data of extrasolar planets we contrast the published estimates of orbital parameters with the ones obtained with EXOFIT. This approach has the unique attribute that all the orbital solutions were analysed by a single program and hence the estimates provide a unique perspective of the extrasolar planet sample.

---

[1]www.star.ucl.ac.uk/~lahav/exofit.html

2. EXOFIT

EXOFIT is a Bayesian data analysis package for extracting orbital parameters of extrasolar planets from radial velocity data. It uses a Markov Chain Monte Carlo (MCMC) method to obtain samples from the posterior distribution of orbital parameters. A brief introduction to Bayesian framework and MCMC is given in EXOFIT User's guide. The output of EXOFIT are the samples from the posterior probability distribution of orbital parameters. Thus we are able to produce more informative inference than mere point estimates, for example mean and standard deviation of the distribution of parameters. At present, EXOFIT can search for either one or two planets in the radial velocity data and it can be easily extended to handle more than two planets or the data from transit photometry. As an illustration, in Figures 1(a) and 1(b), we show the full posterior distribution and the radial velocity plot for HD168443. Note that while the distribution of the eccentricity of the first planet (e1) is symmetric and resembles the Gaussian distribution, the eccentricity of the second planet (e2) is skewed and has plateau-like characteristics.

(a) Posterior distribution   (b) Radial velocity plot

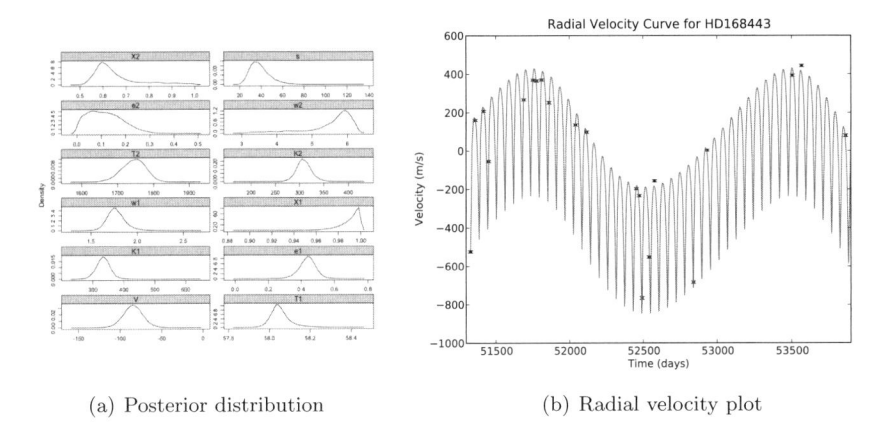

Figure 1. EXOFIT generated 2-planet posterior distribution of the orbital parameters of HD168443 and the corresponding radial velocity plot.

3. Cataloging of Extrasolar Planets

The radial velocity data of extrasolar planets were analysed by using EXOFIT, which provides best estimates of orbital parameters as well as their full posterior distributions. We use the noise factor in the probability model of Balan & Lahav (2009) to decide the best fit model. A higher noise factor implies the presence of an additional signal in the data. At the time of writing this article, we have 209 orbital solutions with 157 single planet systems, 19 two-planet systems and 14 ambiguous systems. The ambiguous systems are those where neither a one-planet nor a two-planet model yields a satisfactory fit to the data. The ambiguous systems are discussed in Section 5.

(a) Mass (b) Semi-major axis

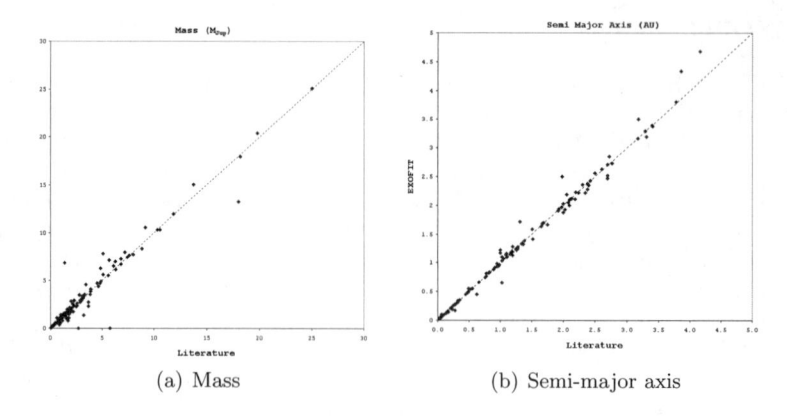

Figure 2. Comparison of values of the mass and the semi-major axis obtained by EXOFIT (vertical axis) and the ones from the literature (horizontal axis).

4. Preliminary Results

The comparison of the best fit orbital parameters obtained by EXOFIT with the published values shows interesting features. We plot a scatter diagram of the published values of mass, semi-major axis, period and eccentricity of extrasolar planets taken from www.exoplanet.eu and the corresponding values obtained by EXOFIT. These plots are shown in Figures 2(a), 2(b), 3(a) and 3(b). It is apparent that, while mass, semi-major axis and orbital period shows little variation, the eccentricity values show higher dispersion[2]. The slope of the eccentricity scatter diagram implies that EXOFIT predicts lower eccentricity solutions for given radial velocity data. We explore the reasons for such a dispersion in the scatter diagram in the next section.

5. Discussion

In this section, we analyse the disagreement between published eccentricity values from www.exoplanet.eu and the ones obtained by EXOFIT. We identify the following reasons for the dispersion in eccentricity values:

1. A different orbital solution (degeneracy) compared to the published one

2. Ambiguous orbital solutions.

A degenerate orbital solution could be the result of higher noise levels in the radial velocity data or of an incomplete coverage of the orbit of the planet. For example, EXOFIT obtains a different orbital solution (period=368 days and eccentricity=0.48) for HD43848 compared to that of Minniti et al. 2009(period=237 days, eccentricity=0.69). The fact that the noise factor values for

---

[2] We caution that a few of the outliers in the eccentricity scatter diagram are due inaccurate data from www.exoplanet.eu

Orbital Parameters of Exo-planets using EXOFIT 125

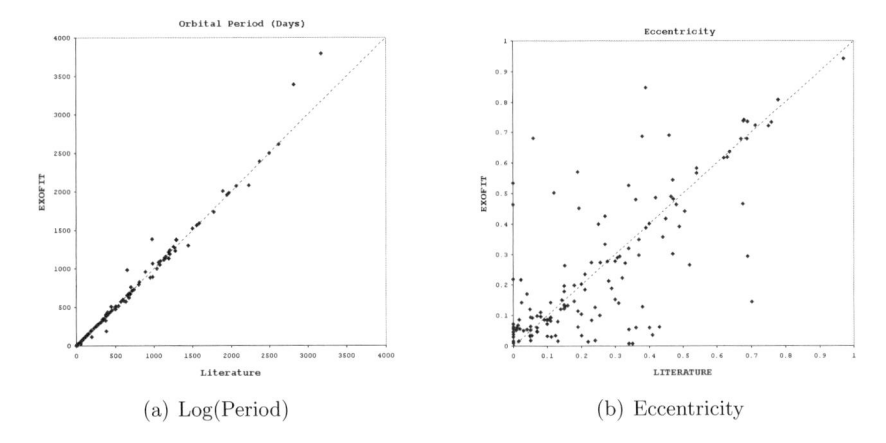

(a) Log(Period)

(b) Eccentricity

Figure 3. Comparison of values of the log(period) and the eccentricity obtained by EXOFIT (vertical axis) and the ones from the literature (horizontal axis).

both the models are similar implies the presence of degeneracy in the orbital solution. The degenerate orbital solutions for HD43848 are shown in Figure 4. Figure 1(a) shows that the posterior distribution of the eccentricity of the second planet in HD50499 has a skewed plateau-like distribution. In this case, it is hard to specify a best estimate for the eccentricity which results in degenerate orbital solutions.

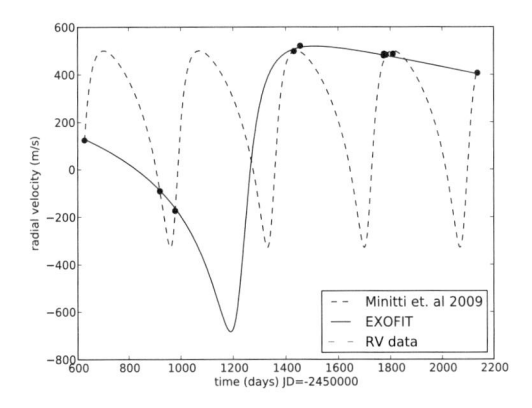

Figure 4. Degenerate orbital solutions of HD43848.

Ambiguous systems are mostly due to the incomplete coverage of the orbital period. As an illustration, we show the orbital solutions for HD50499 and Gl86 in Figures 5(a) and 5(b) respectively. In both cases, there is evidence for the presence of an additional signal. However, a two planet model fits the data poorly and the orbital parameters are loosely constrained. Biased estimates

(a) HD50499 (b) Gl86

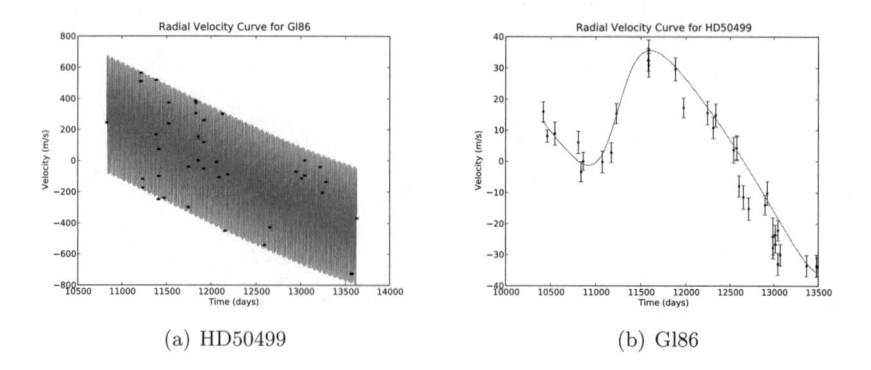

Figure 5. Radial velocity plots of HD50499(right) and Gl86 (left) illustrating the ambiguous orbital solutions where neither a 1-planet nor a 2-planet model yields a satisfactory fit.

are obtained when fitting the data with 1-planet and a linear drift as in the case of HD50499[5(a)] or, with 1-planet and a quadratic drift as in the case of GL86[5(b)].

6. Summary

We have described a new systematic study of extrasolar planets by generating their orbital solutions with the Keplerian orbital fitting program EXOFIT. To date we have analysed the orbital solutions of more than 200 planets. The full details of this work, including an online catalogue, will appear in Balan, Lever, and Lahav (in preparation).

Acknowledgments. SB, GL, and OL would like to thank the organisers of the *Pathways Towards Habitable Planets 2009* for giving them the opportunity to attend the conference. SB is an Issac Newton Student at the Cavendish Laboratory, University of Cambridge. OL acknowledges the support of a Royal Society Wolfson Research Merit Award. SB, GL, and OL thank Paul Gorman, Robert Michaelides, Lisa Menahem, Andrew Strang, Robert C. Clouth, and Milroy Travasso for their help in analysing planetary data sets.

References

Balan, S. T., & Lahav O., 2009, MNRAS, 394, 1936
Butler, R. P., et al., 2006 ApJ, 646, 505
Gregory, P. C. , 2005, ApJ, 631, 1198
Lomb, N. R. , 1976, Ap&SS, 39, 447
Minniti, D., et al., 2009, ApJ, 693, 1424
Press, W. H., Teukolsky, S. A., Vetterling, W. T. and Flannerty, B. P., 1992, Numerical
    Recipes in C: The Art of Scientific Computing, 2nd edn (Cambridge University
    Press, New York)
Scargle, J. D., 1982, ApJ, 263, 835
Trotta, R., 2008, Contemporary Physics, 49, 71

Pathways towards Neptune-mass Planets around Very Low-mass Stars

S. Dreizler,[1] J. Bean,[1] A. Seifahrt,[1,2] H. Hartman,[3] H. Nilsson,[3] G. Wiedemann,[4] A. Reiners,[1] and T. J. Henry[5]

[1] *Institut für Astrophysik, Georg-August-Universität, Friedrich-Hund-Platz 1, Göttingen, Germany*

[2] *Department of Physics, University of California, One Shields Ave., Davis, CA, USA*

[3] *Lund Observatory, Lund University, Box 43, Lund, Sweden*

[4] *Hamburger Sternwarte, Gojenbergsweg 112, Hamburg, Germany*

[5] *Department of Physics and Astronomy, Georgia State University, Atlanta, GA, USA*

Abstract. Radial velocities measured from near-infrared spectra are a potentially powerful tool to search for planets around low-mass stars. The radial velocity precision routinely obtained in the visible can, however, not be achieved in the NIR with existing techniques. In this paper, we describe a method for measuring high-precision radial velocities of a sample of the lowest-mass M dwarfs using CRIRES on the VLT. Our project makes use of a gas cell filled with ammonia to calibrate the instrument response similar to the iodine cell technique that has been used so successfully in the visible. Tests of the method based on the analysis of hundreds of spectra obtained for late M dwarfs over six months demonstrate that precisions of $\sim 5\,\mathrm{m\,s^{-1}}$ are obtainable over long timescales, and precisions better than $3\,\mathrm{m\,s^{-1}}$ can be obtained over timescales up to a week. This allows to search for low-mass planets, i.e., Neptune-mass or even Super-Earth planets around very low-mass stars or sub-stellar objects.

1. Introduction

The vast majority of known exoplanets have been detected with high-precision measurements of stellar radial velocity (RV) variation obtained from spectra in the visible wavelength region. The majority of targets have been solar-type stars, which are brightest at wavelengths shorter than 1 μm as well as rich in deep and sharp spectral lines in the visible suitable for Doppler-shift measurements. Another reason for the success is that visible wavelength spectrograph technology is much more advanced relative to that required for spectrographs operating in other wavelength regions. This includes detector technology and well-established methods for high-precision wavelength calibration. The currently obtainable data quality and wavelength calibration makes it possible to reach RV precisions of $\sim 1\,\mathrm{m\,s^{-1}}$ for many solar-type stars, which is sufficient to detect planets down to a few Earth masses in short-period orbits (e.g., Howard et al. 2009; Bouchy et al. 2009).

The near-infrared (NIR, i.e., $1-5\,\mu$m) is an interesting wavelength region for RV exoplanet studies of low-mass objects that are cooler than ~ 3200 K, which are brightest at wavelengths $\geq 1\,\mu$m. These very low-mass stars are an especially interesting sample of potential planet hosts. They would allow to extend the correlation between the frequency of gas giant planets and stellar masses (Endl et al. 2006; Johnson et al. 2007) down to low-mass host stars. These are the most numerous stars in the Galaxy (Reid et al. 2002; Covey et al. 2008), exhibit a larger dynamical response to orbiting planets, and have closer-in habitable zones (Kasting et al. 1993). Very low-mass stars are an important sample for the possible future study of the atmospheres of habitable planets, e.g., with the *JWST* using the techniques of transit and occultation spectroscopy.

While the lowest-mass stars are out of reach of high-precision spectroscopy at visible wavelengths with current telescopes, their extreme redness means they are reasonably bright at NIR wavelengths and therefore within reach of existing high-resolution NIR spectrographs on 8-10 m class telescopes. Moreover, radial velocities measured from NIR spectra potentially offer the advantage that activity induced RV "jitter" is reduced relative to measurements in the visible. However, the size of this effect depends on unknown properties of active stars like e.g., the temperature contrast between spots and stellar surface (Reiners et al. 2009). The issue of jitter is particularly relevant for low-mass stars because it is well established that a much higher fraction of late-type stars are active than earlier-type stars. In the most comprehensive survey so far, West et al. (2004) found that the fraction of active stars increased from $< 10\%$ at spectral type M3, to 50% at M5, and ultimately 75% at the very bottom of the main sequence.

The advantages offered by NIR radial velocities have been identified before and there has been some previous work to realize them (Martín et al. 2006; Blake et al. 2007; Prato et al. 2008; Huélamo et al. 2008; Seifahrt & Käufl 2008). The precisions achieved range from $300\,\mathrm{m\,s^{-1}}$ down to $\sim 20\,\mathrm{m\,s^{-1}}$. All these previous studies utilized the telluric spectrum imprinted on the data for *in situ* wavelength calibration, although the lines and applied methods varied somewhat. Despite the attention paid to obtaining NIR radial velocities of cool stars, no previous work has achieved a long-term precision on a star other than the Sun (Deming et al. 1987; Deming & Plymate 1994) within an order of magnitude of the same precision that is routinely obtained in the visible.

2. Observing Technique

Long-term precision in RV measurements requires either a long-term stability of the spectrograph or careful monitoring of the instrumental changes. While the former needs to take these science requirements into account at the design phase, the latter method has the advantage that existing spectrographs can be upgraded with an appropriate gas absorption cell. The cell that we have developed for measuring high-precision NIR radial velocities contains ammonia (^{14}NH$_3$). Ammonia is a well established wavelength standard for the NIR (e.g., Urban et al. 1989, and references therein), but to our knowledge it has not been used for calibration of astronomical observations. At room temperatures ammonia is in its gaseous state and exhibits a rich molecular spectrum in the

NIR even with the relatively low column densities possible in a cell to be used at an astronomical observatory. We chose to make observations in a window in the K-band because very low-mass stars exhibit numerous sharp and deep lines in this wavelength region suitable for RV measurements, and ammonia exhibits a number of lines useful for calibration overlapping with the stellar lines. The availability of a window that can be spanned by existing NIR spectrographs and containing a significant number of both calibration lines and stellar lines makes ammonia particular advantageous compared to other possible gases.

The NIR contains a significant number of telluric lines, which was known a priori. Normally, high-precision radial velocity measurements avoid regions containing telluric lines due to the expectation that these lines will exhibit variability on the order of a few $m s^{-1}$. However, the lack of an obvious method for calibrating existing NIR spectrographs means that a more flexible approach is currently called for. We decided that using an ammonia cell in the K-band and accepting the contamination from the telluric lines was a good option considering all the competing issues.

We have implemented the ammonia cell in the high-resolution NIR spectrograph CRIRES (Käufl et al. 2004) at the VLT. The cell is mounted inside the CRIRES "warm optics" box on a carriage that moves the cell in and out of the telescope beam. The cell sits just in front of the Nasmyth focus de-rotator in the converging f/15 beam from the telescope. At this location the cell is in front of all the spectrograph optics, as well as the instrument's integrated AO system. Observations of a star for RV measurements are obtained with the cell in the beam, which causes the absorption lines of the cell to be imprinted on the stellar spectrum. The cell lines, whose position and shapes are well known, serve as a fiducial to precisely establish the wavelength scale and the point spread function of the instrument at the time of the observation independently for each of the obtained spectra during post-processing of the data.

During each observing run since February 2009, we have obtained observations of a portion of a single echelle order spanning a wavelength intervall from 2292 nm to 2350 nm (roughly the K-band), with three gaps of 3.0 – 3.5 nm. The cell does not give enough suitable lines for calibration on the red-most of the four individual detectors, which was skipped in our analysis. For most observations we use a spectrograph entrance slit with a projected width on the sky of $0''.2$, which gives a nominal resolving power $R \approx 100000$. The CRIRES AO system is always utilized to improve the slit throughput. The exposure times are set so that the average S/N will be $\sim$150 pixel^{-1} when possible

The data are reduced separately for each detector and one dimensional spectra are extracted using custom software including bias, dark, as well as non-linear and non-uniform response corrections followed by background subtraction.

In order to measure the relative RV we adopted the iodine cell method pioneered by Butler et al. (1996). The model for fitting the observed spectrum to determine the stellar radial velocity is a composite of the different components that appear in the observed spectrum. In our case, spectra of the gas cell, the star, and the telluric absorption. The reference spectrum of the cell was obtained with the Lund Observatory laboratory FTS using a resolution $R \approx 620000$ at 2.3 μm. The resulting spectrum has a measured signal-to-noise S/N > 700 in the continuum near 2.3 μm. For each star we need a template spectrum which is based on a de-convolved spectrum of the star obtained without the ammonia cell and with the telluric lines divided out using a spectrum of a telluric standard star at a similar airmass. The telluric spectrum is modeled based on spectrum synthesis with a radiative transfer code following exactly the method described by Seifahrt et al. (2009) in order to determine a) the instrumental profile (IP) for the deconvolution following observations for the construction of template spectra and b) as third component in each RV measurement. In addition to the stellar RV, the other free parameters in the model are the wavelength scale and IP of the observed spectrum, the continuum normalization, the water and methane abundance in the model atmosphere used for the telluric spectrum synthesis, and the Doppler-shift of the telluric spectrum.

The parameters that yield the best fit between the model and the observed spectrum are estimated using a standard non-linear least squares algorithm. The adopted uncertainty in the determined RV is estimated from the covariance matrix and multiplied by the square root of the reduced chi-squared of the fit to account for the imperfect modeling. In addition to this error estimate, we also calculate the intrinsic RV content of the data by considering the characteristics of the stellar spectrum using Equation 6 from Butler et al. (1996).

3. First Results

Due to the additional difficulties of NIR high-precision RV measurements it is important to check the performance of our method. In addition to the very low-mass stars we are monitoring as part of our ongoing planet search, we have also frequently observed observed Proxima Centauri (GJ 551) and Barnard's Star (GJ 699) to test our observational and data analysis methodology. These stars are two of the few very low-mass stars for which it is possible to obtain high-precision RV measurements in the visible wavelength range and previous work has shown them to be RV constant at the level of $3\,\mathrm{m\,s^{-1}}$ (Kürster et al. 2003; Endl & Kürster 2008; Zechmeister et al. 2009).

Spectra of Proxima Centauri have been obtained during 17 nights over six months. The radial velocities measured are shown in Figure 1. The derived velocities are completely consistent with the previous findings in that the star does not exhibit large radial velocity variations. The rms of the radial velocities measured for 100 individual spectra obtained over a six night observing campaign is $6.5\,\mathrm{m\,s^{-1}}$, close to the estimated intrinsic RV content of the data of $5.6\,\mathrm{m\,s^{-1}}$ but 30% smaller than our estimated errors of $9.1\,\mathrm{m\,s^{-1}}$. The latter suggests that our error estimation method might be slightly overestimating the uncertainties on this timescale. The dispersion in the radial velocities measured for Proxima Centauri over the full monitoring timespan is $11.7\,\mathrm{m\,s^{-1}}$ for the individual spectra (not shown in Figure 1) and $5.4\,\mathrm{m\,s^{-1}}$ for the binned data. The typical value for our estimated errors over the full timespan is $11.5\,\mathrm{m\,s^{-1}}$, and the expected dispersion in the binned data from these errors is $3.4\,\mathrm{m\,s^{-1}}$.

In order to estimate the noise level on short and long time scales, we examined the rms of the Proxima Centauri radial velocities depending on the number of points binned. Over short timescales the rms of the data nearly follows the expected trend (i.e., the rms goes as $N^{-1/2}$, where N is the number of data points binned). The noise floor over short timescales appears to $\sim 2\,\mathrm{m\,s^{-1}}$. Over long timescales, the rms is a factor of two larger ($\sim 4\,\mathrm{m\,s^{-1}}$) than on short timescales

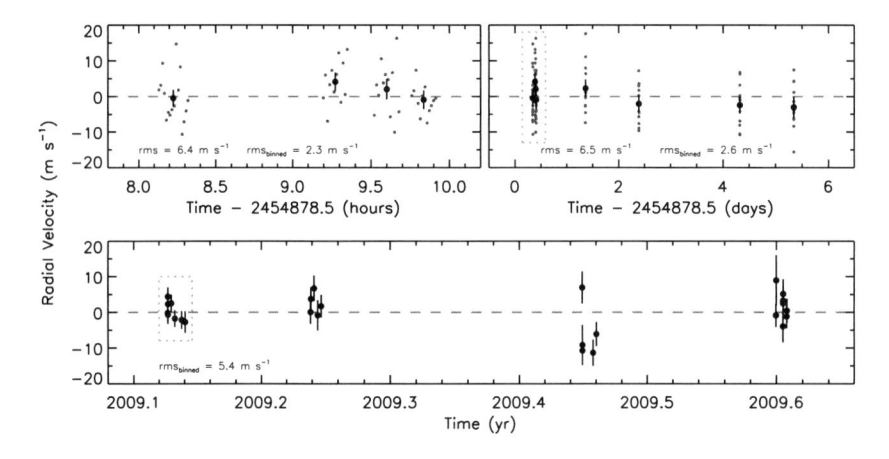

Figure 1. Measured radial velocities for Proxima Centauri. The small grey points are the results for individual exposures, and the large black points are the data binned over 12 minutes. *Top left* A sequence of observations over two hours, *Top right* over the first observing run. The dotted box indicates the data shown in the top left panel. *Bottom* The measured radial velocities obtained over six months (binned points only). The dotted box indicates the data shown in top right panel.

and the radial velocities for Proxima Centauri show significant deviations from the expected reduction in rms with binning. We attribute this to the additional noise arising from the telluric cross-talk resulting from a seasonal movement of the stellar lines relative to the telluric lines.

4. Discussion

The capability of NIR radial velocities studies of low-mass stars and sub-stellar objects for new and potentially very important directions of observational study of exoplanets has been limited up to now due to uncertainties in the calibration of NIR spectra. With an ammonia gas cell at the entrance of CRIRES at the VLT, we have demonstrated the feasibility of long-term RV precisions of $\sim 5\,\mathrm{m\,s^{-1}}$.

It is worthwhile to mention that this result was achieved despite two limitations of the employed instrumental set-up. First, the CRIRES detectors have severe cosmetic blemishes relative to CCDs. Nevertheless, the effects seem to be stable and we have been able to calibrate them out to an acceptable level. Second, our gas cell is not in a temperature stabilized environment. Laboratory tests during the FTS calibration runs demonstrate that calibration down to a few $\mathrm{m\,s^{-1}}$ is possible.

As a result, searching for planets around a much larger number of very low-mass stars is feasible now. Our planet search project includes 31 objects with estimated masses below $0.2\,\mathrm{M_\odot}$, out of which 22 have estimated masses below $0.15\,\mathrm{M_\odot}$. In addition to enabling the search for planets around more low-mass stars, the gas cell and radial velocity measurement algorithm we have developed opens up a new frontier on the search for potentially habitable planets. The orbital period range for a planet in the habitable zone around a star with a mass $M = 0.10\,M_\odot$ would be 3 – 21 days (Selsis et al. 2007). The $5\,\mathrm{m\,s^{-1}}$ precision obtainable with our method corresponds to the velocity semi-amplitudes induced by a $2.5\,M_\oplus$ planet or a $4.6\,M_\oplus$ planet on the inner and outer edges of the habitable zone respectively. Therefore, it should be possible to detect Super-Earth type planets in the habitable zones of very low-mass stars with a reasonable expenditure of observing time on current facilities.

Acknowledgments. This work is based on observations made with ESO Telescopes at the Paranal Observatories under programme ID 182.C-0748. We acknowledge support by the DFG through grants GRK 1351 and RE 1664/4-1, and the BMBF through program 05A0GU2. JLB acknowledges research funding from the European Commissions Seventh Framework Programme as an International Incoming Fellow (grant no. PIFF-GA-2009-234866). AS acknowledges financial support from the NSF under grant AST-0708074. HH acknowledges funding from the Swedish Research Council (VR). HN acknowledges the financial support from the Lund Laser Centre through a Linnaeus grant from the Swedish Research Council (VR). AR received support from the DFG as an Emmy Noether Fellow. The Lund IR-FTS was financed through a grant from the Knut and Alice Wallenberg Foundation.

References

Blake, C. H., Charbonneau, D., White, R. J., Marley, M. S., & Saumon, D. 2007, ApJ, 666, 1198
Bouchy, F., et al. 2009, A&A, 496, 527
Butler, R. P., et al. 1996, PASP, 108, 500
Covey, K. R., et al. 2008, AJ, 136, 1778
Deming, D., Espenak, F., Jennings, D. E., Brault, J. W., & Wagner, J. 1987, ApJ, 316, 771
Deming, D. & Plymate, C. 1994, ApJ, 426, 382
Endl, M., et al. 2006, ApJ, 649, 436
Endl, M. & Kürster, M. 2008, A&A, 488, 1149
Howard, A. W., et al. 2009, ApJ, 696, 75
Huélamo, N., et al. 2008, A&A, 489, L9
Johnson, J. A., et al. 2007, ApJ, 670, 833
Kasting, J. F., Whitmire, D. P., & Reynolds, R. T. 1993, Icarus, 101, 108
Käufl, H.-U., et al. 2004, in SPIE Conference Series, ed. A. F. M. Moorwood & M. Iye, 5492, 1218
Kürster, M., et al. 2003, A&A, 403, 1077
Martín, E. L., Guenther, E., Zapatero Osorio, M. R., Bouy, H., & Wainscoat, R. 2006, ApJ, 644, L75
Prato, L., et al. 2008, ApJ, 687, L103
Reid, I. N., Gizis, J. E., & Hawley, S. L. 2002, AJ, 124, 2721
Reiners, A., et al. 2009, ApJ, 710, 432
Seifahrt, A., & Käufl, H. U. 2008, A&A, 491, 929
Seifahrt, A., et al. 2009, A&A submitted
Selsis, F., et al. 2007, A&A, 476, 1373
Urban, Š., Tu, N., Narahari Rao, K., & Guelachvili, G. 1989, J. Mol. Spec., 133, 312
West, A. A., et al. 2004, AJ, 128, 426
Zechmeister, M., Kürster, M., & Endl, M. 2009, A&A, 505, 859

Pathways Towards Habitable Planets
ASP Conference Series, Vol. 430, 2010
Vincent Coudé du Foresto, Dawn M. Gelino, and Ignasi Ribas, eds.

Tidal Constraints on Planetary Habitability

R. Barnes,[1,2] B. Jackson,[3] R. Greenberg,[4] S. N. Raymond,[2,5] and R. Heller[6]

[1] *Department of Astronomy, University of Washington, Seattle, WA, USA*

[2] *Virtual Planetary Lab*

[3] *NASA Postdoctoral Program Fellow, Planetary Systems Laboratory, Goddard Space Flight Center, Code 693, Greenbelt, MD, USA*

[4] *Lunar and Planetary Laboratory, University of Arizona, Tucson, AZ, USA*

[5] *Center for Astrophysics and Space Astronomy, University of Colorado, UCB 389, Boulder, CO, USA*

[6] *Hamburger Sternwarte, University of Hamburg, Gojenbergsweg 112, Hamburg, Germany*

Abstract. We review how tides may impact the habitability of terrestrial-like planets. If such planets form around low-mass stars, then planets in the circumstellar habitable zone will be close enough to their host stars to experience strong tidal forces. We discuss 1) decay of semi-major axis, 2) circularization of eccentric orbits, 3) evolution toward zero obliquity, 4) fixed rotation rates (not necessarily synchronous), and 5) internal heating. We briefly describe these effects using the example of a 0.25 $M_\odot$ star with a 10 $M_\oplus$ companion. We suggest that the concept of a habitable zone should be modified to include the effects of tides.

1. Introduction

Exoplanet surfaces are probably the best places to look for life beyond the solar system. Remote sensing of these bodies is still in its infancy, and the technology does not yet exist to measure the properties of terrestrial exoplanet atmospheres directly. Indeed, the scale and precision of the engineering required to do so is breathtaking. Given these limitations, a reliable model of habitability is essential in order to maximize the scientific return of future ground- and space-based missions with the capability to remotely detect exoplanet atmospheres.

Here we review one often misunderstood issue: the effect of tides. If the distance between a star and planet is small, $\lesssim 0.1$ AU, the shape of the planet (and star) can become significantly non-spherical. This asymmetry can change the planet's orbital motion from that of spherical planets. Simulating the deviations from the spherical approximation is difficult and uncertain as observations of the solar system, binary stars and exoplanets do not yet provide enough information to distinguish between models. Without firm constraints, qualitatively different models of the planetary response to tides exist. The two most prominent descriptions are the "constant-phase-lag" and "constant-time-lag" models (Greenberg 2009). In the former, the tidal bulge is assumed to lag the perturber by a fixed angle, but in the latter it lags by a fixed time interval. Depending on the mathematical extension in terms of e, the two models may diverge significantly when $e \gtrsim 0.3$. Throughout this review the reader should remember that the presented magnitudes of tidal effects are model-dependent. For more on these differences and the details of tidal models, the reader is referred to recent reviews by Ferraz-Mello et al. (2008) and Heller et al. (2010).

We consider tidal effects in the habitable zone (HZ) model proposed by Barnes et al. (2008) which utilizes the 50% cloud cover HZ of Selsis et al. (2007), but assumes that the orbit averaged flux determines surface temperature (Williams & Pollard 2002). We use the example of a 10 $M_\oplus$ planet orbiting a 0.25 $M_\odot$ star. This choice is arbitrary, but we note that large terrestrial planets orbiting small stars will be preferentially discovered by current detection techniques. This chapter is organized as follows: In section 2 we discuss orbital evolution, in section 3 we describe rotation rates, in section 4 we consider the obliquity, and in section 5 we examine tidal heating.

2. Orbital Evolution

Orbital evolution due to tides should be considered for any potentially habitable world. The asymmetry of the tidal bulge leads to torques which transfer angular momentum between rotations and orbits, and the constant flexing of the planet's figure between pericenter and apocenter dissipates energy inside the planet. These two effects act to circularize and shrink most orbits. In the constant-phase-lag model, the orbits of close-in exoplanets evolve in the following way (Goldreich & Soter 1966; see also Jackson et al. 2009):

$$\frac{da}{dt} = -\Big(\frac{63}{2}\frac{\sqrt{GM_*^3}R_p^5}{m_p Q_p'}e^2 + \frac{9}{2}\frac{\sqrt{G/M_*}R_*^5 m_p}{Q_*'}\Big[1 + \frac{57}{4}e^2\Big]\Big)a^{-11/2} \qquad (1)$$

$$\frac{de}{dt} = -\Big(\frac{63}{4}\frac{\sqrt{GM_*^3}R_p^5}{m_p Q_p'} + \frac{225}{16}\frac{\sqrt{G/M_*}R_*^5 m_p}{Q_*'}\Big)a^{-13/2}e \qquad (2)$$

where a is the semi-major axis, G is Newton's gravitational constant, m_p is the mass of the planet, Q_p' is the planet's tidal dissipation function divided by two-thirds its Love number, Q_*' is the star's tidal dissipation function divided by two-thirds its Love number, R_p is the planet's radius, and R_* is the stellar radius. The Q' values represent the body's response to tidal processes and combines a myriad of internal properties, such as density, equation of state, etc. It is a difficult quantity to measure, so here we use the standard values of $Q_*' = 10^6$ and $Q_p' = 500$ (Mathieu 1994; Mardling & Lin 2002; Jackson et al. 2008a). The first terms in Equations (1 – 2) represent the effects of the tide raised on the planet, the second the tide raised on the star.

Equations (1 – 2) predict a and e decay with time. As tides slowly change a planet's orbit, the planet may move out (through the inner edge) of the habitable zone (HZ). This possibility was considered in Barnes et al. (2008), who showed for some example cases the time for a planet to pass through the inner edge of the HZ. Such sterilizing evolution is most likely to occur for planets with initially large eccentricity near the inner edge of the HZ of low mass stars ($\lesssim 0.3\,\mathrm{M_\odot}$). Even if a planet does not leave the HZ, the circularization of its orbit can require billions of years, potentially affecting the climatic evolution of the planet.

3. Rotation Rates

Planetary rotation rates may be modified by tidal interactions. Although planets may form with a wide range of rotation rates Ω, tidal forces may fix Ω such that no net exchange of rotational and orbital angular momenta occurs during one orbital period. The planet is then said to be "tidally locked," and the rotation rate is "pseudo-synchronous" or in equilibrium. The equilibrium rotation rate in the constant-phase-lag model is

$$\Omega_{eq} = n(1 + \frac{19}{2}e^2) \qquad (3)$$

where n is the mean motion (Goldreich 1966). Note that planets only rotate synchronously (one side always facing the star) if $e = 0$ (the constant-time-lag model makes the same prediction). Therefore, the threat to habitability may have been overstated in the past, as independently pointed out by several recent investigations (Barnes et al. 2008; Ferraz-Mello et al. 2008; Correia et al. 2008). Figure 1 shows the values of the equilibrium rotation period for a 10 $\mathrm{M_\oplus}$ planet orbiting a 0.25 $\mathrm{M_\odot}$ star as a function of a and e.

4. Obliquity

Tidal effects tend to drive obliquities to zero or π. The constant-time-lag model of Levrard et al. (2007; see also Correia & Laskar 2010) found a planet's obliquity ψ changes as

$$\frac{d\psi}{dt} = \frac{\sin(\psi)K_\mathrm{p}}{C_\mathrm{p}\Omega_0 n}\left(\frac{\cos(\psi)\epsilon_1\Omega_0}{n} - 2\epsilon_2\right) \qquad (4)$$

where

$$\epsilon_1 = \frac{1 + 3e^2 + \frac{3}{8}e^4}{(1-e^2)^{9/2}} \qquad (5)$$

$$K_\mathrm{p} = \frac{3}{2}k_{2,\mathrm{p}}\frac{GM_\mathrm{p}^2}{R_\mathrm{p}}\tau_\mathrm{p} n^2\left(\frac{M_\mathrm{s}}{M_\mathrm{p}}\right)^2\left(\frac{R_\mathrm{p}}{a}\right)^6 \qquad (6)$$

$$C_\mathrm{p} = r_{\mathrm{g,p}}^2 M_\mathrm{p} R_\mathrm{p}^2 \qquad (7)$$

and

$$\epsilon_2 = \frac{1 + \frac{15}{2}e^2 + \frac{45}{8}e^4 + \frac{5}{16}e^6}{(1-e^2)^6} \qquad (8)$$

In the preceding equations, $r_\mathrm{g,p}$ ($= 0.5$) is the planet's radius of gyration (a measure of the distribution of matter inside a body), Ω_0 is the initial rotation

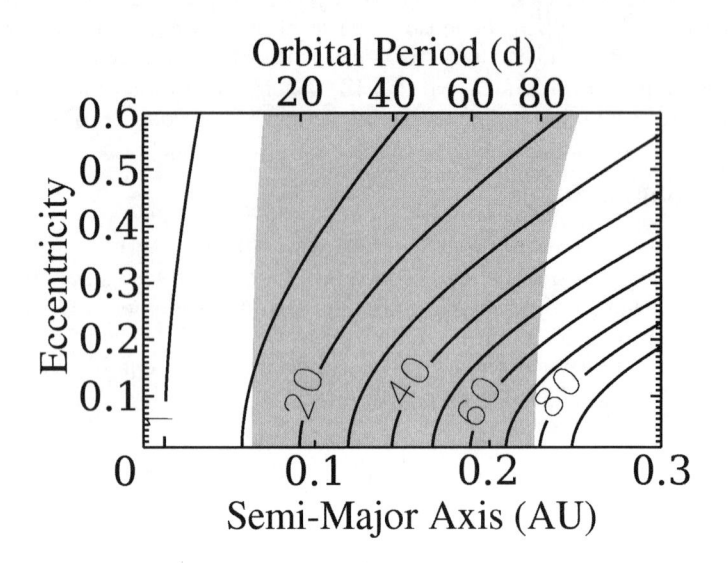

Figure 1. Contours of equilibrium rotation period in days for a 10 $M_\oplus$ planet orbiting a 0.25 $M_\odot$ star. The gray region is the HZ from Barnes et al. (2008).

frequency, and τ_p is the "tidal time lag", which in this constant-time-lag model replaces Q'_p. We assumed $Q'_p = 500$ for the planet at its initial orbital configuration and set $\tau_p = 1/(nQ'_p)$, o initially the planet responds in the same way as in a constant-phase-lag model. In the course of the orbital evolution, τ_p was then fixed while n and Q_p evolved in a self-consistent system of coupled differential equations. In Figure 2 we show the time for a planet with an initial obliquity of 23°.5 to reach 5°, a value which may preclude habitability (F. Selsis, personal communication). However, obliquities may easily be modified by other planets in the system (Atobe et al. 2004; Atobe & Ida 2007) or a satellite (Laskar et al. 1993).

5. Tidal Heating

As a body on an eccentric orbit is continually reshaped due to the varying gravitational field, friction heats the interior. This "tidal heating" is quantified in the constant-phase-lag model as

$$H = \frac{63}{4} \frac{(GM_*)^{3/2} M_* R_p^5}{Q'_p} a^{-15/2} e^2 \qquad (9)$$

(Peale et al. 1979; Jackson 2008b). However, in order to assess the surface effects of tidal heating on a potential biosphere, it is customary to consider the

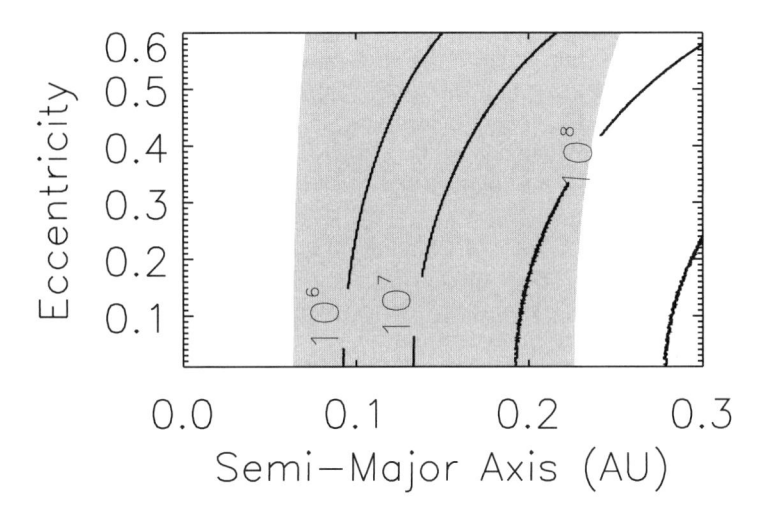

Figure 2. Time in years for a 10 $M_\oplus$ planet orbiting a 0.25 $M_\odot$ star to evolve from an obliquity $\psi = 23\overset{\circ}{.}5$ to $5°$. The gray region is the HZ from Barnes et al. (2008).

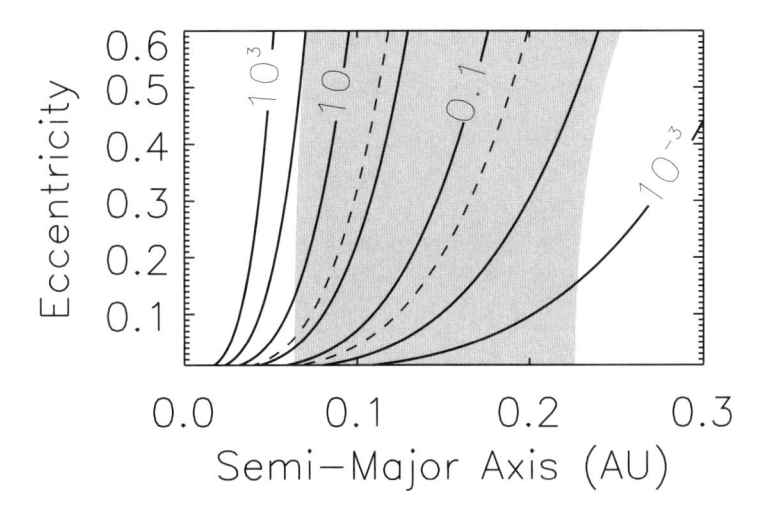

Figure 3. Tidal heating fluxes for a 10 $M_\oplus$ planet orbiting a 0.25 $M_\odot$ star. Contour labels are in W m^{-2}. The dashed contours represent the boundaries of the tidal habitable zone (Jackson et al. 2008c; Barnes et al. 2009b). The gray region is the HZ from Barnes et al. (2008).

heating flux, $h = H/4\pi R_p^2$, through the planetary surface. Jackson et al. (2008c; see also Barnes et al. 2009b) argued that when $h \geq 2$ W m^{-2} (the value for Io; McEwen et al. 2004) or $h \leq 0.04$ W m^{-2} (the limit for plate tectonics; Williams et al. 1997), habitability is less likely. Barnes et al. (2009b) suggested that these limits represent a "tidal habitable zone". In Figure 3 contours of tidal heating are shown for a 10 M$_\oplus$ planet orbiting a 0.25 M$_\odot$ star. The tidal habitable zone is the region between the dashed curves. Note that a and e evolve as prescribed by Equations (1 – 2), and hence the heating fluxes evolve with time as well.

Acknowledgments. RB and SNR acknowledge funding from NASA Astrobiology Institute's Virtual Planetary Laboratory lead team, supported by NASA under Cooperative Agreement No. NNH05ZDA001C. RG acknowledges support from NASA's Planetary Geology and Geophysics program, grant No. NNG05GH65G. BJ is funded by an NPP administered by ORAU. RH is supported by a Ph.D. scholarship of the DFG Graduiertenkolleg 1351 "Extrasolar Planets and their Host Stars".

References

Atobe, K. & Ida, S. 2007, Icarus, 188, 1
Atobe, K., Ida, S., & Ito, T. 2004, Icarus, 168, 223
Barnes, R., Jackson, B., Raymond, S. N., West, A. A., & Greenberg, R. 2009a, ApJ, 695, 1006
Barnes, R., Jackson, B., Greenberg, R., & Raymond, S. N. 2009b, ApJ, 700, L30
Barnes, R., Raymond, S.N., Jackson, B., & Greenberg, R. 2008, Astrobiology, 8, 557
Correia, A. C. M., & Laskar, J. 2010, Icarus, 205, 338
Correia, A. C. M., Levrard, B., & Laskar, J. 2008, A&A, 488, L63
Ferraz-Mello, S., Rodríguez, A., & Hussmann, H. 2008, CeMDA, 101, 171
Goldreich, P. 1966, AJ, 71, 1
Goldreich, P. & Soter, S. 1966, Icarus, 5, 375
Greenberg, R. 2009, ApJ, 698, L42
Heller, R., Jackson, B., Barnes, R., Greenberg, R., & Homeier, D. 2010, A&A, accepted. arXiv:1002.1246
Jackson, B., Greenberg, R., & Barnes, R. 2008a, ApJ, 678, 1396
Jackson, B., Greenberg, R., & Barnes, R. 2008b, ApJ, 681, 1631
Jackson, B., Barnes, R. & Greenberg, R. 2008c, MNRAS, 391, 237
Jackson, B., Barnes, R. & Greenberg, R. 2009, ApJ, 698, 1357
Laskar, J., Joutel, F., & Robutel, P. 1993, Nat, 361, 615
Lainey, V., Arlot, J.-E., Karatekin, O., & van Hoolst, T. 2009, Nat, 459, 957
Levrard, B., Correia, A. C. M., Chabrier, G., Baraffe, I., Selsis, F., & Laskar, J. 2007, A&A, 462, L5
Mardling, R. A. & Lin, D. N. C. 2002, ApJ, 573, 829
Mathieu, R. 1994, ARA&A, 32, 465
McEwen, A. S., Keszthelyi, L. P., Lopes, R., Schenk, P. M., & Spencer, J. R. 2004, in: Jupiter. The planet, satellites and magnetosphere, ed. F. Bagenal, T. E. Dowling, & W. B. McKinnon (Cambridge, UK: Cambridge UP), 307
Peale, S. J., Cassen, P., & Reynolds, R. T. 1979, Sci, 203, 892
Selsis, F., Kasting, J. F., Levrard, B., Paillet, J., Ribas, I., & Delfosse, X. 2007, A&A, 476, 1373
Williams, D. M., Kasting, J. E., & Wade, R. A. 1997, Nat, 385, 234
Williams, D. M. & Pollard, D. 2002, Int. J. Astrobiology, 1, 61

Pathways towards Habitable Moons

D. M. Kipping,[1,2] S. J. Fossey,[2,3] G. Campanella,[4] J. Schneider,[5] and G. Tinetti[2]

[1] *Harvard-Smithsonian Center for Astrophysics, Cambridge, MA, USA*

[2] *University College London, London, UK*

[3] *University of London Observatory, London, UK*

[4] *Queen Mary College, University of London, London, UK*

[5] *LUTH, Observatoire de Paris, Meudon, France*

Abstract. The search for life outside of the Solar System should not be restricted to exclusively planetary bodies; large moons of extrasolar planets may also be common habitable environments throughout the Galaxy. Extrasolar moons, or exomoons, may be detected through transit timing effects induced onto the host planet as a result of mutual gravitational interaction. In particular, transit timing variations (TTV) and transit duration variations (TDV) are predicted to produce a unique exomoon signature, which is not only easily distinguished from other gravitational perturbations, but also provides both the period and mass of an exomoon. Using these timing effects, photometry greater or equal to that of the *Kepler Mission* is readily able to detect habitable-zone exomoons down to 0.2 $M_\oplus$ and could survey up to 25,000 stars for Earth-mass satellites. We discuss future possibilities for spectral retrieval of such bodies and show that transmission spectroscopy with JWST should be able to detect molecular species with ∼ 30 transit events, in the best cases.

1. Introduction

In this conference, named "Pathways Towards Habitable Planets", we are trying to develop a strategy towards detecting life on another planet. However, by only considering *planets* we may be unnecessarily restrictive by excluding another potentially common environment for life to thrive - moons. The possibility of life-bearing moons was addressed by Williams et al. (1997) who noted that a moon of around one third of an Earth mass or larger, residing around a gas giant planet in the habitable zone, would satisfy many of the criteria traditionally placed on habitable planets. For M-dwarf systems, habitable-zone moons have the added bonus that any tidal locking of the planet or the moon would not result in one side of the moon being locked in perpetual darkness.

Whether or not such moons could be dynamically stable remained a relativity open question until Barnes & O'Brien (2002) showed that a habitable-zone Jupiter-like planet could hold onto a $1 M_\oplus$ moon in systems with a host star as small as $M_* \sim 0.4 M_\odot$, and remain dynamically stable for 5 Gyr. In light of these two studies, the potential for habitable moons must be taken as being at least theoretically possible. The formation of a ∼ $1 M_\oplus$ exomoon around a

Jupiter-mass planet is not supported by the current models of the formation of the regular satellites, e.g. see Canup & Ward (2006). However, the possibility of a Jupiter-like planet capturing a terrestrial planet as a moon (an irregular satellite) is a topic on which we could find no detailed studies. Given that such moons would be dynamically stable, and the lessons learnt from the failings of trying to predict exoplanetary systems based on Solar System models prior to the first actual detection, a search for extrasolar moons is not only historically justified but scientifically imperative.

2. Transit Timing Effects due to Extrasolar Moons

2.1. Transit Timing Variation (TTV)

Extrasolar moons do not significantly affect the motion of the host star in a way which would be detectable through radial velocity techniques. It has been proposed that the moons of planets in pulsar systems could be detected through time-of-arrival analysis of the host star's pulses (Lewis et al. 2006), but we can find no publications in the literature proposing an analogous methodology for radial velocity. A discussion of other proposed methods is briefly given by Kipping et al. (2009c), but here we limit ourselves to transiting exoplanet-moon systems. Ostensibly, the most obvious detection method in such cases is the occultation of the star light by the moon itself (Sartoretti & Schneider 1999). However, the position of this signal will appear at different relative positions to the planet for each transit event, with a mean position located directly over the planetary signal. Consequently, we would require a fortunate aligment to make the detection. This point aside, perhaps the most critical objection to detecting exomoons in the same way as exoplanets are found in transit surveys, is that planets can be confirmed through radial velocity whereas moons cannot. CoRoT, HAT-NET and WASP have all reported that the vast majority of their planetary signals are in fact not real planets ($\sim$ 98% for CoRoT) due to the plethora of effects which can mimic such signals (see Almenara et al. 2009; Hartman et al. 2004; Pollacco et al. 2006). Ergo, by the same reasoning, it is imprudent to claim an exomoon can be convincingly detected in this manner.

Sartoretti & Schneider (1999) were the first to propose that exomoons could be detect around transiting exoplanets by looking for transit timing variations (TTV). For a planet-moon system, the two bodies orbit a common center of gravity which itself orbits the host star. Consequently, the planet's tangential position can appear shifted from that expected of a strictly linear ephemeris by a maximal spatial deviation of $a_W = a_S M_S / M_P$, where M_P and M_S are the planetary and satellite masses respectively and a_S is the semi-major axis of the moon's orbit (where we have assumed the simple case of a coplanar, edge-on system with circular orbits). This spatial deviation is observable as a timing deviation in the transit lightcurve where $\delta t \sim a_W / v_P$ where v_P is the tangential velocity of the planet. The effects of planetary and lunar orbital eccentricity are accounted for in the more elaborate derivation of Kipping (2009a) as well reformulating the expression in terms of root-mean-square amplitude to reflect the way such timing signals as proposed to be detected, i.e. by excess scatter.

Pathways towards Habitable Moons 141

$$\delta_{TTV} = \frac{1}{\sqrt{2}} \cdot \frac{a_P^{1/2} a_S M_S M_{PRV}^{-1}}{\sqrt{G(M_* + M_{PRV})}} \cdot \frac{\zeta_T(e_s, \varpi_S)}{\Upsilon(e_P, \varpi_P)} \quad (1)$$

Where ζ_T and Υ are terms absorbing the effects the orbital eccentricity, we reference readers to the original paper for details.

2.2. The Problems with TTV

Although the TTV amplitude has been derived, how would such a signal be detected? One critical problem with exomoon timing signals is that due to dynamical constraints, the orbital period of a moon (P_S) will always be significantly lower than the orbital period of a planet (P_P), but we only see a transit once every planetary period. Kipping (2009a) showed that $P_S \leq P_P \sqrt{\chi^3/3}$ where χ is the orbital distance of the satellite in units of Hill radii. For a prograde exomoon, Domingoes et al. (2006) showed that $\chi_{max} \simeq 0.4895$ meaning $P_S \leq 0.2 P_P$. Consequently, aliasing is unavoidable and a Fourier analysis will reveal a range of possible harmonic frequencies for the lunar period.

If we are unable to provide a singular solution for P_S (or equivalently a_S by re-arrangement through Kepler's Third Law) then equation (2) has no unique solution. Thus a range of possible exomoon masses are possible since the TTV amplitude only gives us $M_S \times a_S$. The problem may be averted if only one harmonic frequency falls within the range of dynamic stability, but short period exomoons will produce numerous possible solutions. Additionally, we have so far only dealt with two free parameters, period and mass, the addition of another unknown, like inclination, makes a unique solution unfeasible.

Finally, another critical problem was that due to the aliasing of frequencies, the TTV signal could in fact appear as a planetary perturber. Holman & Murray (2005) showed that planetary bodies could induce TTV on a transiting planet with a range of frequencies. From the TTV information alone, there is no way to distinguish between the planet or the moon scenario.

2.3. Transit Duration Variation (TDV)

To resolve these problems, Kipping (2009a & 2009b) proposed using transit duration variation (TDV) to break the degeneracy. In the simple coplanar case, a planet's tangential velocity at the moment of transit appears to oscillate around some local mean value as a result of the moon's presence. Changes in velocity will directly affect the transit duration thus leading to TDV.

It is important to understand the differences between TTV and TDV. TTV is caused by changes in the planet's position, which is analogous to the astrometry planet-hunting method, where changes in the star's position imply a companion. TDV is caused by changes in the planet's velocity, which is analogous to the radial velocity planet-hunting method. Like astrometry, TTV is more sensitive to distant companions and scales as mass multiplied by distance, or $M_S a_S$. Like radial velocity, TDV is more sensitive to close-in companions and scales as $M_S a_S^{-1/2}$. Kipping (2009a) showed that:

$$\delta_{TDV} = \sqrt{\frac{a_P}{a_S}} \sqrt{\frac{M_S^2}{M_{PRV}(M_{PRV} + M_*)}} \frac{T}{\sqrt{2}} \frac{\zeta_D(e_s, \varpi_S)}{\Upsilon(e_P, \varpi_P)} \quad (2)$$

Where again ζ_D and Υ are terms absorbing the effects of orbital eccentricity. The scaling of TDV amplitude to $M_S a_S^{-1/2}$ can be seen within this expression.

2.4. The Complementary Nature of TTV and TDV

TDV and TTV are highly complementary due to their different scalings and different phases. Just like simple harmonic motion, the position and velocity of the planet have a $\pi/2$ phase difference (for circular orbits) which provides a highly unambiguous signal. For example, a perturbing planet is not predicted to produce a TDV effect with such a phase difference. Secondly, TDV allows one to detect close-in exomoons which TTV would miss and vice versa meaning the whole parameters space may now be probed.

Furthermore, the different scalings mean that the ratio of the TTV to the TDV amplitude provides the a_S (or equivalently P_S) and M_S uniquely. The derived value of P_S may then be compared to the Fourier spectrum to identify the true orbital frequency and further constrain the system. Any significant deviation between the Fourier frequency and the ratio-derived frequency would indicate orbital eccentricity or inclination.

3. Detectability of Habitable-Zone Exomoons

Despite the theory behind detecting exomoons being now well-established, no detections have been made. This may be primarily due to the fact all of the known transiting exoplanets are short-period or highly eccentric, which dynamically forbid exomoons on Gyr timescales. However, with the launch of *Kepler*, we expect that many long-period transiting planets will shortly be announced. Kipping et al. (2009) evaluated the TTV and TDV signal-to-noise strengths with *Kepler*-class photometry (KCP) for a range of orbital configurations, but fixing the planetary period to always be that of a habitable-zone planet. This was done to limit the parameter space to a more computationally feasible volume and also because such moons would be the most interesting cases, in light of this conference. Details of the methods and assumptions employed can be found in Kipping et al. (2009) and we summarize the key results here for brevity.

- Low-density planets are the ideal hosts for detection as they provide the largest transit depth but experience the greatest perturbation from a companion.

- 25,000 stars in *Kepler's* field-of-view are bright enough to be surveyed for a habitable-zone Saturn-like exoplanet hosting an Earth-mass exomoon. A Galactic plane survey with KCP should expand this survey to around 2 million stars within the detectable range.

- For the ideal, but realistic, case of a M2V dwarf at 10pc, KCP could detect a habitable-zone exomoon down to 0.2 $M_\oplus$.

- Saturn-Titan systems are not detectable with KCP but the results do suggest that large exomoons should be found in the coming years should they exist.

4. Characterizing Exomoons

4.1. Radius and Internal Structure

Once such a moon is detected through timing deviations, what more can be learnt of the satellite? The TTV and TDV effects provide the orbital period of the moon which may then be fitted through the data points to obtain an exomoon ephemeris. We also have multiple transit lightcurves of the planet-moon system. For a small moon, the signal will appear like a typical planetary signal and the moon signal will be at the same level as the noise. However, for each transit event we can subtract the planetary transit signal and then fold the residuals on the exomoon period. This will produce a composite exomoon transit lightcurve, which should beat down the noise at roughly the reciprocal of the square root of the number of transits observed.

With KCP, let us consider the 0.2 $M_\oplus$ habitable moon with an M2V host star. We expect the lunar radius to be $0.65 R_\oplus$ based upon a terrestrial planet model of Valencia et al. (2006) thus producing a transit depth of $\sim 140 \pm 17$ppm lasting for 50 minutes (using $b = 0.5$). However, with a period of 0.126 years, we would obtain around 28 transits over the *Kepler Mission* lifetime thus reducing the error from a 12% uncertainty to a 2% uncertainty (3.2 ppm). Obtaining the radius to this precision is critical in constraining the internal structure of the exomoon, i.e. distinguishing between icy and rocky (Valencia et al. 2006).

4.2. Exomoon Transmission Spectroscopy

In the same way, transmission spectroscopy could be potentially performed. A M2V star at 10pc may be apparent magnitude 9.5 in *Kepler*'s bandpass but is likely to be around 5.5 in K-band for infrared spectroscopy, i.e. approximately as bright as HD 189733. On this target, Knutson et al. (2007) obtained 8μm photometry with 0.4 second cadence and rms scatter 4.325 mmag with *Spitzer*. Over a 50 minute integration this reduces to 49.9 ppm. JWST will have a collecting area 44 times larger than *Spitzer* and thus we expect this same signal to be reduced to 7.5 ppm per transit, comparable to *Kepler* in the visible. The scale height of a planetary atmosphere is roughly $H = kT/g\mu$. For a terrestrial planet with mass 0.2 $M_\oplus$ and radius 0.65 $R_\oplus$, we have $H \simeq 20$km. For a strong molecular transition, the radius of the planet may appear to change by a few scale heights, say 3, which would change the transit depth by $\sim$ 4ppm. This should be detectable to 1-σ confidence by binning 4 transits together and to 3-σ by binning 32 transits.

4.3. Direct Imaging

Potentially, the most powerful method for characterizing an extrasolar moon would be a spectroscopic investigation through direct imaging. This would necessitate the ability to resolve the planet and moon contributions. Spatially resolving such signals would require multi-kilometric interferometers (Schneider et al. 2010) and is likely to be several decades away from being realized. However, it will be possible to make spectroscopic observations of the unresolved planet-moon pair with facilities like the Terrestrial Planet Finder Coronagraph (Ford et al. 2004) or the 42m European Extremely Large Telescope (Dierickx et al. 2004), both of which are designed with the required sensitivity to detect

Earth-sized planets. The combined light could then be resolved by subtracting the planet spectrum from the planet-moon spectrum during 'mutual events'; for example planet-moon eclipses or the disappearance of the moon in the shadow of the planet (Cabrera & Schneider 2007).

5. Conclusions

We have shown that exomoons, in particular habitable exomoons, are detectable by coupling transit timing variations (TTV) and transit duration variations (TDV) together down to 0.2 $M_\oplus$ for *Kepler*-class photometry. Around 25,000 stars could be surveyed within *Kepler*'s field-of-view for Earth-mass habitable exomoons. Indeed, we propose here that it may even be possible for gas giants of $\sim 10 M_J$ to harbour a super-Earth mass exomoon, or a 'super exomoon'. Subsequent characterization of exomoons could be achieved by using the planet-moon ephemeris derived from the transit timing signal. For the case of a system within 10pc and a 0.2 $M_\oplus$ habitable exomoon, we predict that molecular species could be found using transmission spectroscopy with JWST after the binning of $\sim$30 transit events.

References

Almenara, J. M. et al., 2009, A&A, 506, 337
Barnes, J. W. & O'Brien, D. P., 2002, ApJ, 575, 1087
Cabrera, J. & Schneider, J., 2007, A&A, 464, 1133
Canup, R. M. & Ward, W. R., 2006, Nat, 441, 834
Dierickx, P. et al. 2004, SPIE, 5489, 391
Domingos, R. C., Winter, O. C. & Yokoyama, T., 2006, MNRAS, 373, 1227
Ford, V., Lisman, P., Shaklan, S., White, M. & Hull, T., 2004, BAAS, 36, 792
Hartman, J. D., Bakos, G., Stanek, K. Z. & Noyes, R. W., 2004, AJ, 128, 1761
Holman, M. J. & Murray, N. W., 2005, Sci, 307, 1288
Kipping, D. M., 2009a, MNRAS, 396, 1797
Kipping, D. M., 2009b, MNRAS, 392, 181
Kipping, D. M., 2009c, in 'Molecules in the Atmospheres of Extrasolar Planets', ASP Conf. Series
Kipping, D. M., Fossey, S. J. & Campanella, G., 2009, MNRAS, 400, 398
Knutson, H. A. et al., 2007, Nat, 447, 183
Lewis, K. M., Sackett, P. D. & Mardling, R. A., 2008, ApJ, 685, L153
Pollacco, D. L. et al., 2006, PASP, 118, 1407
Sartoretti, P. & Schneider, J., 1999, A&AS, 134, 553
Schneider, J. et al., 2010, Astrobiology, in press
Valencia, D., O'Connell, R. J. & Sasselov, D., 2006, Icarus, 181, 545
Williams, D. M., Kasting, J. F. & Wade, R. A., 1997, Nat, 385, 234

Pathways Towards Habitable Planets
ASP Conference Series, Vol. 430, 2010
Vincent Coudé du Foresto, Dawn M. Gelino, and Ignasi Ribas, eds.

Pressure-temperature Phase Diagram of the Earth

E. G. Jones[1] and C. H. Lineweaver[1,2]

[1] Planetary Sciences Institute, Research School of Astronomy & Astrophysics, Australian National University, Mount Stromlo Observatory, Cotter Road, Weston Creek, Canberra, ACT, Australia

[2] Planetary Sciences Institute, Research School of Earth Sciences, Australian National University, Canberra, Australia

Abstract. Based on a pressure-temperature (P-T) phase diagram model of the Earth, Jones & Lineweaver (2010) described uninhabited terrestrial liquid water. Our model represents the atmosphere, surface, oceans and interior of the Earth - allowing the range of P-T conditions in terrestrial environments to be compared to the phase regime of liquid water. Here we present an overview and additional results from the Earth model on the location of the deepest liquid water on Earth and the maximum possible extent of the terrestrial biosphere. The intersection of liquid water and terrestrial phase space indicates that the deepest liquid water environments in the lithosphere occur at a depth of $\sim$ 75 km. 3.5% of the volume of the Earth is above 75 km depth. Considering the 3.5% of the volume of the Earth where liquid water exists, $\sim$ 12% of this volume is inhabited by life while the remaining $\sim$ 88% is uninhabited. This is distinct from the fraction of the volume of liquid water occupied by life. We find that at least 1% of the volume of liquid water on Earth is uninhabited. Better geothermal gradients in the Earth's crust and mantle will improve the precision and accuracy of these preliminary results.

1. Background and Motivation

Jones & Lineweaver (2010) (henceforth JL10) consider the question of whether there are regions of the Earth where liquid water is uninhabited. If such regions exist, this would imply that liquid water is not the sole control on life and that life is restricted from some liquid water environments by constraints of temperature, water activity, pressure, nutrients, or energy. This implies that a strategy of "following the water" is not sufficient in the search for life on other planets, and that subsurface modeling has a significant role to play in finding subsurface environments on other planets that are within the set of conditions that are hospitable for (at least) terrestrial life.

To clarify our results, consider the Venn diagram in Figure 1. The circle on the left represents the region of phase space where liquid water can exist. The circle on the right represents the region of phase space occupied by the Earth. The overlapping region is where there is water on Earth. The lower part of the overlap is inhabited water and the upper part is uninhabited water. In our model "Earth" and "Liquid Water" are represented by areas of pressure and temperature (P-T) in Figure 2. Note that in Figure 2, there are regions of phase space where there is Earth but no liquid water, and vice-versa. The P-T phase space of the Earth was constructed from the range of measured and

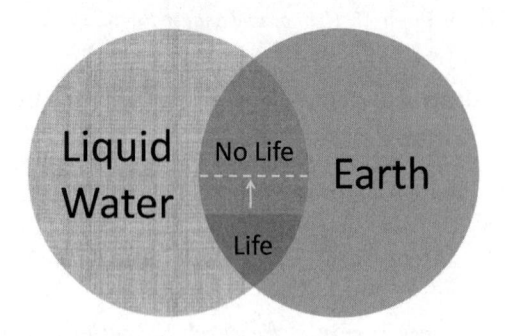

Figure 1. *Liquid Water, Earth, and Habitability.* We plot liquid water (left circle), all terrestrial environments (right circle), and divide the overlapping region into "Life" (inhabited terrestrial water) and "No Life" (uninhabited terrestrial water). Considering the intersection of Earth and liquid water in the phase space model (Figure 2), life was found in JL10 to occupy ∼ $\frac{1}{3}$ of that phase space area (lower part of overlap), while $\frac{2}{3}$ is uninhabited water (upper part of overlap; see Figure 5 JL10). If the high temperature limit for life is increased to 250°C (dashed line) then terrestrial life occupies ∼ $\frac{1}{2}$ of the overlap.

modeled geotherms in the subsurface and atmosphere. Densities were employed (see Table 2 of JL10) to determine the pressure gradients with depth. The range of conditions on the Earth's surface were also included. The liquid water phase space was constructed from the vapor, sublimation, and melting curves for pure water which were then modified via Raoult's equation to account for different salt content. Ocean salinity water, shown in Figure 2, corresponds to 3.5% salt by mass (Lide & Frederikse 1996). "Maximum liquid range" is our estimate of the broadest range of pressures and temperatures under which water can remain liquid due to increasing concentrations of solute and thin film effects at low temperatures (∼ lower water activity). The coldest liquid water on Earth is -89°C and has a triple point pressure of 3.2×10^{-7} bar (JL10).

In JL10 the P-T space occupied by terrestrial life was superimposed on Figure 2 and the area of intersection with liquid water was used to quantify what fraction of terrestrial water is uninhabited and to locate these environments. Life was found to be excluded from hot and deep regions of the Earth at temperatures above 122°C and to possible restrictions on pore space, nutrients, and energy. Additionally life was restricted from cold near surface regions in ice and permafrost where liquid water is available, but only as brines or thin films which have a low water activity. The limiting factors in this region of the P-T diagram are most likely low water activity ($a_w < 0.6$) which becomes an issue at temperatures below -20°C (Grant 2004). Finally all examples of life found at pressures less than 0.3 bar have been classified as dormant indicating that there may be a low pressure limit for active life, possibly due to some combination of low water activity, low levels of nutrients available at altitudes above ∼ 9 km, or low temperatures.

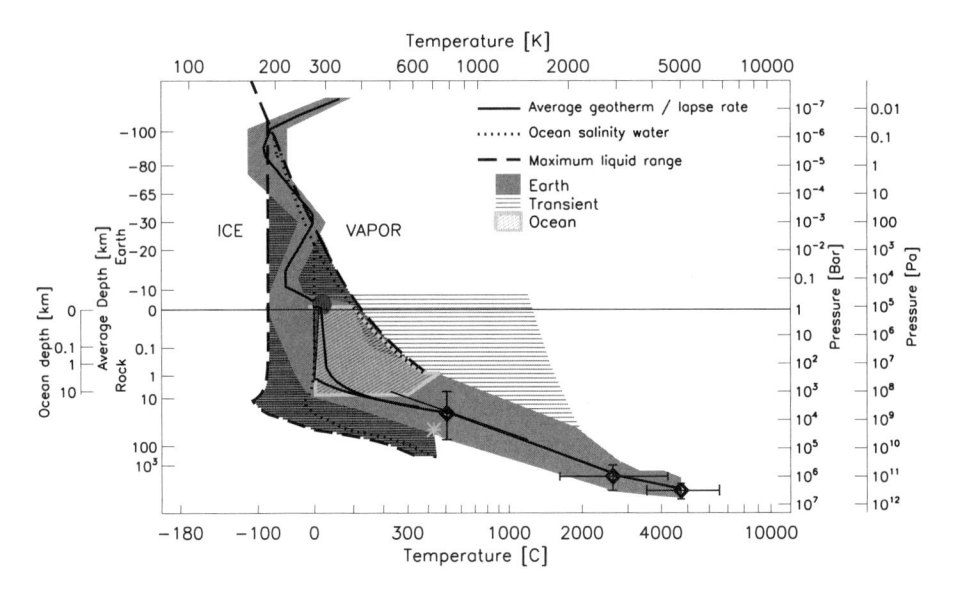

Figure 2. *Superposition of terrestrial environments on the P-T diagram of water.* The Earth's core, mantle, crust and atmosphere are shaded medium grey and are centered on the average geotherm (subsurface) and lapse rate (atmosphere). The ocean and crust geotherms are plotted separately and meet at $P = 2000$ bar, $T = 200°C$, corresponding to a depth of ~ 10 km. A transient region shows mantle that has intruded upwards (e.g., volcanism, geothermal vents). The horizontal thin line at 1 bar is the average sea level atmospheric pressure of the Earth. The parameters of our Earth model (e.g., core, mantle, continental crust, oceanic crust, geotherms and atmospheric lapse rates) are given in Appendix A of JL10. The dark vertical wedge identifies the majority of ocean water, while the light grey diagonally striped ocean area represents thermally heated water.

2. Results: Representing the Earth

Our model is shown in Figure 2 with details given in JL10. All known environments in the crust, oceans and surface lie within the pressure-temperature polygons.

2.1. Deepest Liquid Water

Our estimate of the deepest liquid water is shown by the light grey asterisk in Figure 2. This environment occurs in the crust at $T \sim 431°C$ and $P \sim 3 \times 10^4$ bar, corresponding to a depth of ~ 75 km. As this is the limit for liquid water on Earth, it represents the deepest possible extent of the terrestrial biosphere. This limit was obtained from the intersection of the phase space occupied by the Earth with the phase space of liquid water. The size of the area occupied by the Earth in the P-T space of Figure 5 is determined by a combination of uncertainty in geothermal gradients and variations (both spatial and temporal) in geotherms (Table A, JL10). Liquid water is stable at higher pressures and temperatures than this limit, but terrestrial environments at these pressures and temperatures do not exist. As an example, water exists as a liquid at 450°C and a few times 10^5 bar, however the geothermal and pressure gradients in the Earth's crust and mantle do not reach these conditions.

As the phase space of the Earth is modelled from our adopted estimates and uncertainties of pressure and temperature gradients with depth, it is possible that the boundaries of the Earth's crust and mantle will change as models improve. This may change our estimate of the deepest liquid water. We can assess how robust our estimate is by examining the range of geothermal gradients which are consistent with our phase space model. The mean geotherm is 25 K km^{-1} in the continental crust (Scheidegger 1976) and $\sim$ 35 K km^{-1} in oceanic crust. The crustal geotherms vary widely due to variations in heat flow and thermal conductivity of rock (with both parameters varying by over an order of magnitude) (Clauser & Huenges 1995), but are predominately between 5-70 Kkm^{-1} (Chapman & Pollack 1975). Figure 2 shows the average P-T value for the crust-mantle boundary as a diamond with error bars around it to reflect the variation and uncertainty in temperature, depth and pressure of this boundary. The shallowest Moho occurs at a depth of 10 km and at temperatures between 200-500°C (Mooney et al. 1998; Blackwell 1971; Huppert & Sparks 1988). Deep Moho beneath thick crust of 80 km occurs in the temperature range of $\sim$ 500°C (Hyndman et al. 2005; Priestley et al. 2008) to $\sim$ 1200°C (Jimènez-Munt et al. 2008). This range of crust-mantle boundary conditions requires geotherms between 6 - 49 K km^{-1}, which is consistent with the medium grey shaded "Earth" region in Figure 2. Unless the thickest regions of the crust are found to exist above mantle of cooler temperatures than those cited above, our estimate of the deepest terrestrial liquid water at 75 km is reasonable. Lithosphere above a depth of 75 km represents $\sim$ 3.5% of the volume of the Earth (see Figure 3).

We have not considered pore space in our model, however the available pore space and permeability may be a limiting factor for life at the deep ($\sim$5 km), hot end of the biosphere. As high temperatures are a limiting factor for life, life is restricted to the top $\sim$ 3-10 km of the subsurface. Pore space declines exponentially with increasing depth (Athy 1930) and crustal porosity is generally less than 5% at 10 km depth (Revelle et al. 1990). A related issue is the connectivity of the pore space which allows fluid to permeate though crustal rocks. In the deep crust ($>$ 10 km) the permeability is extremely low, on the order of 10^{-20} m^2 in unfaulted domains (Townend & Zoback 2000). Further work needs to be done to determine if this is a limiting factor for the deep hot biosphere (Gold 1999).

2.2. Interpreting the Amount of Uninhabited Water

JL10 concluded that the 88% of the volume of the Earth where liquid water exists is not known to host life. The uninhabited liquid water environments identified in the paper were predominantly at temperatures above $\sim$150°C and depths below $\sim$ 5 km in continental crust. This result was obtained by assuming that liquid water exists in all P-T environments on the Earth which intersect the dark striped liquid water region bounded by "Ocean salinity water" in Fig-

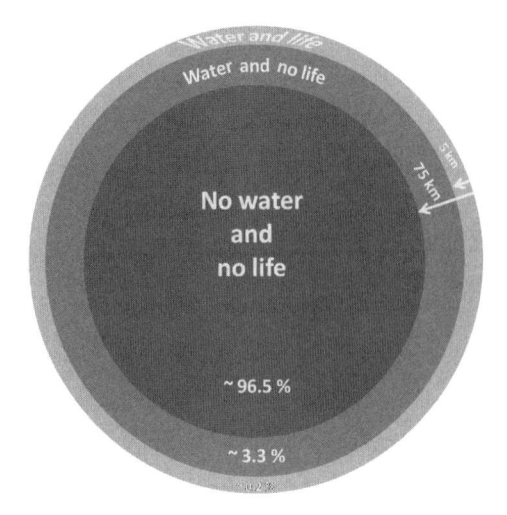

Pressure-temperature Phase Diagram of the Earth   149

Figure 3. *Earth's hydrosphere and biosphere.* In the context of the volume of Earth, we plot the regions shown in Figure 1: water and life (outer shell), water and no life (middle shell), and no water and no life (inner shell). The thickness of the known biosphere is ~ 5km which represents 0.2% of the volume of the Earth. Liquid water exists to a depth of ~ 75 km in the lithosphere which represents 3.5% of the Earth's volume. 3.3% of the volume of the Earth has uninhabited liquid water. In this calculation we have ignored the extent of the biosphere above the average surface (e.g., throughout the troposphere, on top of mountains).

ure 2. Expressed as a fraction of the volume of the Earth, this can be easily misinterpreted. Figure 4 is meant to clarify this point.

The volume of liquid water on Earth that is uninhabited can be estimated. The total volume of water on Earth is ~ 1.39×10^9 km^3 (Gleick et al. 2009; Rogers & Feiss 1998) of which ~ 1.36×10^9 km^3 exists as liquid water. There are a range of estimates for the total volume of subsurface liquid water, from 0.02% (Gleick et al. 2009) to a maximum of 0.6% of the Earth's total water budget (Lehr & Lehr 2000). We have chosen here to use the conservative lower estimate of 0.02%, corresponding to 2.34×10^7 km^3, for the total volume of groundwater within the crust. The volume of groundwater at temperatures below 122°C can be estimated using the average continental crust geotherm. From Rogers & Feiss (1998), 8.39×10^6 km^3 of groundwater is shallower than ~ 5km (leaving ~ 1.5×10^7 km^3 deeper groundwater). Using an average geotherm of 25 Kkm^{-1} gives ~5 km as the depth of the 122°C isotherm. Therefore by taking the ratio with the total volume of terrestrial liquid water indicates that the vast majority (~ 99%) of terrestrial liquid water is hospitable to life as it is within the range of temperature and pressure in which life has been found. Approximately 1.1% of terrestrial liquid water may not support life as it exists at temperatures above the current maximum known for active life (122°C, Takai et al. 2008). If this temperature limit is real this represents a significant volume of liquid water without life (~ 1.5×10^7 km^3), approximately 200 times the volume of the Caspian Sea (Peters et al. 2000). This estimate ignores the < 0.2 wt% of water

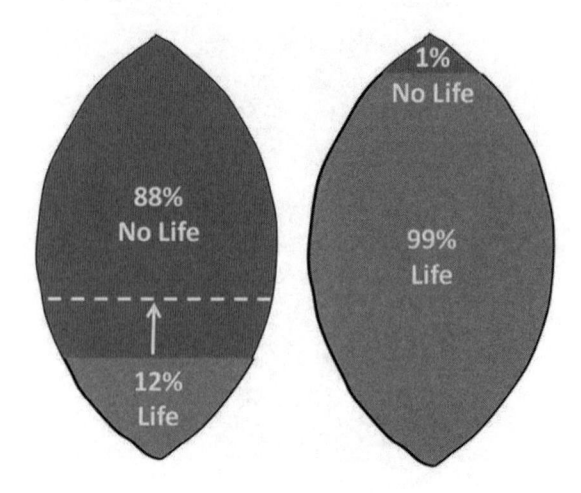

Figure 4. *Quantifying uninhabited water.* Considering the environments on Earth that have liquid water, we plot two ways of visualising what fraction of these environments are inhabited by life. *Left:*) Of the 3.5% of the volume of Earth where liquid water exists, 12% is inhabited, 88% is uninhabited. If the upper temperature limit for life is found to be 250°C (dashed line) then the inhabited fraction increases from 12% to 36% (JL10). *Right:* Of the volume of liquid water on Earth, $\sim$ 99% is inhabited. At least $\sim$ 1% of liquid water is uninhabited. The inhabited fraction of liquid water may increase as deep subsurface environments are searched for life. For example, the apparent depth limit to life (the deepest life is 5.3 km; Szewzyk & Szewzyk 1994) is a selection effect as we have found life as deep as we have looked.

in the mantle ($\sim 10^{-9}$ oceans) which exists as OH within hydrous minerals rather than free liquid H_2O (Ohtani 2005).

3. Conclusions

We have reviewed and clarified the results of JL10. We have quantified in P-T phase space the uninhabited hydrosphere of the Earth.

References

Athy, L. 1930, AAPG Bull. 14
Blackwell, D. 1971, in The Structure and Physical Properties of the Earth's Crust, ed. J. Heacock, Geophysical Monograph 14, 169
Chapman, D. & Pollack, H. 1975, Earth Planet. Sci. Lett. 28, 23
Clauser, R. & Huenges, E. 1995, in Rock Physics and Phase Relations: A Handbook of Physical Constants, ed. T. Ahrens, American Geophysical Union, 105
Gleick, P., Cooley, H., Morikawa, M., Morrison, J., & Palaniappan, M. 2009, The World's Water 2008-2009: the Bicentennial Report on Freshwater Resources (Washington, DC: Island Press), 6
Gold, T. 1999, The Deep Hot Biosphere (New York: Springer-Verlag)
Grant, W. 2004, Phil. Trans. R. Soc. Lond. B. 359, 1249
Huppert, H. & Sparks, R. 1988, J. Petrol. 29, 599

Hyndman, R., Currie, C., & Mazzotti, S. 2005, GSA Today 15, 4
Jimènez-Munt, I., Fernãndez,M., Vergès, J., & Platt, J. 2008, Earth Planet. Sci. Lett. 267, 276
Jones, E. & Lineweaver, C. 2010, Astrobiology (in press) (JL10)
Lehr, J. & Lehr, J. 2000, Standard Handbook of Environmental Science, Health and Technology (New York: McGraw-Hill Professional)
Lide, D. & Frederikse, H. 1996, CRC Handbook of Chemistry and Physics, 77th edn. (Boca Raton, FL.: CRC Press)
Mooney, W., Laske, G., & Masters, G. 1998, J. Geophys. Res. 103, 727
Ohtani, E. 2005, Elements, 1, 25
Peters, F., Kipfer, R., Achermann,D., Hofer, M., Aeschbach-Hertig, W., Beyerle, U., Imboden, D., Rozanski, K., & Frohlich,K. 2000, Deep Sea Res. Ocean. Res. 47, 621
Priestley, K., Jackson, J., McKenzie, D. 2008, Geophys. J. Int. 172, 345
Revelle, R., et al. 1990, Sea Level Change (Washington, DC: National Academies Press)
Rogers, J. & Feiss, P. 1998, People and the Earth: Basic Issues in the Sustainability of Resources and Environment (Cambridge, UK: Cambridge University Press), 126
Scheidegger, A. 1976, Foundations of Geophysics (Amsterdam, NY.: Elsevier Scientific)
Szewzyk, U. & Szewzyk, R. 1994, PNAS. 91, 1810
Takai, K., Nakamura, K., Toki, T., Tsunogai, U., Miyazaki, M., Miyazaki, J., Hirayama, H., Nakagawa, S., Nunoura, T., & Horikoshi, K. 2008, PNAS 105, 10949
Townend, J. & Zoback, M. 2000, Geology 28, 399

Pathways Towards Habitable Planets
ASP Conference Series, Vol. 430, 2010
Vincent Coudé du Foresto, Dawn M. Gelino, and Ignasi Ribas, eds.

Abiotic Buildup of Ozone

S. D. Domagal-Goldman and V. S. Meadows

Virtual Planetary Laboratory, University of Washington Astronomy Department, Seattle, WA, USA

Abstract. Two of the best biosignature gases for remote detection of life on extrasolar planets are oxygen (O_2) and its photochemical byproduct, ozone (O_3). The main reason for their prominence as biosignatures is that large abiotic fluxes of O_2 and O_3 are not considered sustainable on geological and astronomical timescales. We show here how buildup of O_3 can occur on planets orbiting M stars, even in the absence of the large biological fluxes. This is possible because the destruction of O_2 and O_3 is driven by UV photochemistry. This chemistry is much slower on planets around these stars, due to the smaller incident UV flux. Because the destruction of these gases is slower, O_3 can build up to detectable levels even if the O_3 source is small. We will present atmospheric profiles of these gases for planets around AD Leo (an M dwarf) as well as spectra that show the implications for missions such as Darwin and the Terrestrial Planet Finder (TPF).

1. Introduction

As we traverse "Pathways towards Habitable Planets" we should ponder our steps beyond these paths; specifically, we should consider the steps to search for life on these habitable worlds. A search for life on exoplanets will by necessity be remote and will occur via the identification of spectral biosignatures. These biosignatures are absorption features caused by the presence of atmospheric constituents that are predominantly or exclusively produced by life. Foremost amongst these gases is molecular oxygen (O_2) (Des Marais et al. 2002), a byproduct of oxygenic photosynthesis, the dominant primary producing metabolism on Earth. Oxygen has absorption features in the visible portion of the spectrum, and it leads to the photochemical buildup of ozone (O_3), which has absorption features in the infrared portion of the spectrum. O_2 would not accumulate in the atmosphere of the Earth without biological production, so O_2 and O_3 are considered to be good biosignature gases. Given the technical challenges life-surveying missions such as TPF or Darwin will face, it is tempting to suggest a narrowband mission that searches exclusively for O_2 or O_3. However, in this paper, we show that it is possible for O_3 to sustainably build up to detectable levels on planets without life, but that abiotic production can be identified via the presence or absence of other gases, such as O_2, methane (CH_4) and ethane (C_2H_6), if observations over a sufficiently large wavelength region are obtained.

1.1. Ozone Chemistry and Stellar Spectra

Below, we examine the effect of stellar spectral energy distribution on the buildup of atmospheric O_3 concentration. Following this discussion requires a rudimentary understanding of the processes that create and destroy O_3. Here, we briefly introduce the predominant photochemical O_3 sources and sinks:

$$O_2 + h\nu \xrightarrow{\lambda<180\text{nm}} O + O \quad (1)$$
$$O + O_2 \longrightarrow O_3 \quad (2)$$
$$O_3 + h\nu \xrightarrow{200\text{nm}<\lambda<300\text{nm}} O_2 + O \quad (3)$$
$$O + O_3 \longrightarrow O_2 + O_2 \quad (4)$$

Because reactions 2 and 3 are typically faster than reactions 1 and 4, O and O_3 are thought of as roughly interchangeable in the atmosphere, as a change in the concentration of one species will quickly lead to a corresponding change in the other. Photochemists often use the term "odd oxygen" to refer to the sum of the two species, and the relevance of reactions to ozone chemistry are measured by their ability to create or destroy odd oxygen. For example, reactions 2 and 3 are neutral in this regard, whereas reaction 1 creates two additional odd oxygen species and reaction 4 destroys two odd oxygen species.

2. Methods

We simulated the chemistry of exoplanet atmospheres with a 1-D (altitude) model (after Pavlov, Brown & Kasting 2001) designed for analyzing the atmosphere of the Archean Earth, a geologic eon that is marked by the presence of life and the absence of atmospheric O_2 (Holland 1984). There is no surface oxygen flux in our model; our boundary condition on both O_2 and O_3 is a constant deposition velocity that accounts for reactions with surface rocks and dissolution in surface water. The outputs from this model - vertical concentration profiles of various gases - were then used as inputs to a spectral mapping and radiative transfer model (SMART) that produces simulated spectra of the planet (Meadows & Crisp 1996). In both the photochemical and radiative transfer models, we used three stellar spectra: 1) the measured spectrum from the Sun (Segura et al. 2005); 2) the measured spectrum from AD Leo, a particularly active M dwarf (Segura et al. 2005); and 3) the modeled spectral flux from T3100K, a model star with a surface temperature of 3100K and no chromosphere (Segura et al. 2005). The latter star represents a low-UV, low-activity end-member case and is not meant to be representative of stars we expect to observe.

3. Results

3.1. Influence of the stellar spectra on atmospheric O_3

The stellar flux is predominantly controlled by two stellar properties: the temperature of the star determines the broadband blackbody radiation, and the amount of stellar activity (such as flares) results in the emission of high-energy photons. This latter radiation is in the proper wavelength range for O_2 photolysis (reaction 1), which leads to the production of two odd oxygen species. Planets that are orbiting stars with higher stellar activities will therefore have higher rates of O_2 photolysis, more odd oxygen, and more O_3. In addition to the

Chapman mechanism, there are also catalytic cycles that impact O_3 concentrations by destroying odd oxygen species. The most important of these involves OH:

$$H + O_2 \longrightarrow HO_2 \tag{5}$$
$$OH_2 + O \longrightarrow OH + O_2 \tag{6}$$
$$OH + O \longrightarrow H + O_2 \tag{7}$$

OH and HO_2 are part of a catalytic cycle that is the largest sink for odd oxygen in our standard "anoxic" planet orbiting the Sun. However, this cycle requires the presence of H, OH, and HO_2, three species whose greatest source is photolysis:

$$CO_2 + h\nu \longrightarrow CO + O \tag{8}$$
$$H_2O + h\nu \longrightarrow OH + O \tag{9}$$
$$H_2CO + h\nu \longrightarrow H_2 + CO \tag{10}$$
$$CH_4 + h\nu \longrightarrow CH_3 + H \tag{11}$$

In our simulations, reactions 8-10 are all much slower on planets around M dwarfs. This leads to lower concentrations of H, OH, and HO_2, a slower catalytic cycle, and higher odd oxygen concentrations. The decrease in H_2CO (reaction 10) also leads to smaller concentrations of CO, a major sink for O atoms (via $CO + O \to CO_2$). Thus, M dwarfs will have smaller photochemical sinks for O_2 and O_3. M dwarfs with high stellar activity will emit large amounts of Lyman-α radiation, providing photons for photodissociation of O_2 and production of odd oxygen (reaction 1). The high energy photons do not penetrate the lower atmosphere (where H_2O, H_2CO, and CH_4 exist) due shielding. The net result is an increase in O_3/O_2 ratio, further increasing the accumulation of O_3 in the atmosphere.

The effects of stellar radiation on O_3 chemistry can be seen in Figure 1, which shows vertical O_3 profiles from our simulations of an "anoxic" planet (one without oxygenic photosynthesis) in the habitable zone of three different stars: 1) the Sun, which represents our "standard" model; 2) AD Leo, a particularly active M dwarf with low blackbody radiation but high Lyman-α radiation; and 3) T3100K, a model M dwarf with no activity and little UV radiation from either source. The ground CH_4 and CO_2 mixing ratios in these simulations were 1000 ppm and 300 ppm, respectively. The modern-day O_3 profile is shown for comparison. Note the high amount of O_3 build up on the planet receiving AD Leo's stellar flux. This shows that planets whose atmospheres receive high Lyman-α radiation but otherwise low UV fluxes can contain O_3 concentrations approaching modern-day values, even in the absence of biological O_2 production. Unlike other abiotic sources of O_2 such as photodissociation of water during in a steam-filled, runaway greenhouse atmosphere (Schindler & Kasting 2000), this source of O_3 could be sustainable for long periods of time.

3.2. O_3 signal detectability

The key issue for the relevance of abiological O_3 production to future planet characterization missions is whether or not the O_3 will create a detectable signal. Figure 2 shows the spectrum from our model planet around AD Leo, along with

Abiotic Buildup of Ozone 155

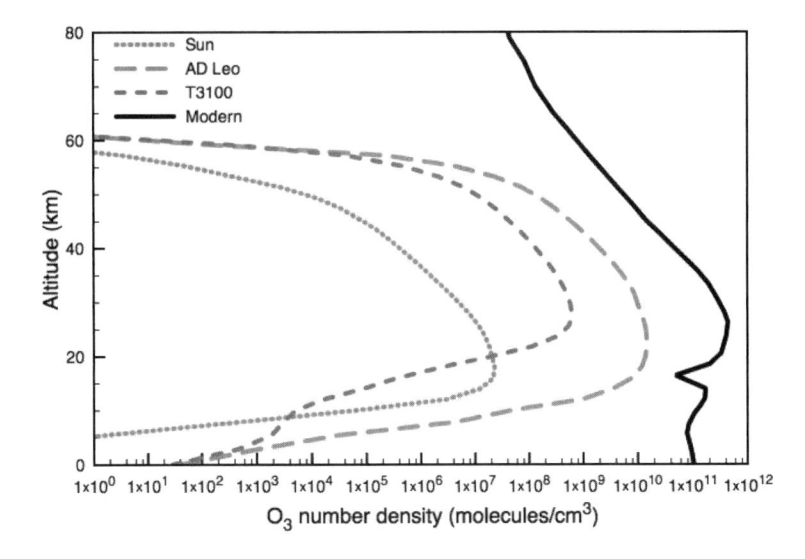

Figure 1. O_3 number density (molecules/cm^3) plotted as a function of altitude (km) for four planets. The solid line is for the modern-day Earth, and the three other lines are for anoxic planets in orbit around different star types. These planet around AD Leo can obtain O_3 number densities approaching that found in our atmosphere even without a biological O_2 flux.

lines representing the independent removal of O_3 and C_2H_6 from the spectral model. Note a divergence in the zero O_3 spectrum at 9.6 μm, the center of a major O_3 absorption band. This difference appears even at the low spectral resolution ($\lambda/\Delta\lambda = 20$) shown here. For the signal to be detectable at this resolution requires an interferometer with signal/noise (S/N) > 13.

3.3. O_3 source determination

Although O_3 buildup can occur in the absence of biological O_2 production, there are certain characteristics that should allow us to discriminate biological O_3 from abiological O_3. First, it appears that this chemistry is only possible in planets orbiting dim, active stars similar to AD Leo. Second, this signal was a strong function of other remotely detectable atmospheric constituents. As the CH_4/CO_2 ratio in an atmosphere climbs, CH_3 molecules that form during CH_4 photolysis become more likely to react with each other to form long-chain organic molecules that condense to form particles. These particles shield the lower atmosphere from incoming stellar radiation, thereby allowing O_3 concentrations to increase. Thus, the high-O_3 profiles shown in Figure 1 are the result of both the UV flux from the star and UV shielding in the atmosphere.

C_2H_6 is unlikely to reach detectable concentrations on planets with oxygenic photosynthesis. Thus, the simultaneous presence of C_2H_6 and O_3 may be a signal of abiological O_3. C_2H_6 would be detectable, as the absorption feature centered at 12μm caused by C_2H_6 (Figure 2) is much stronger than the one caused by

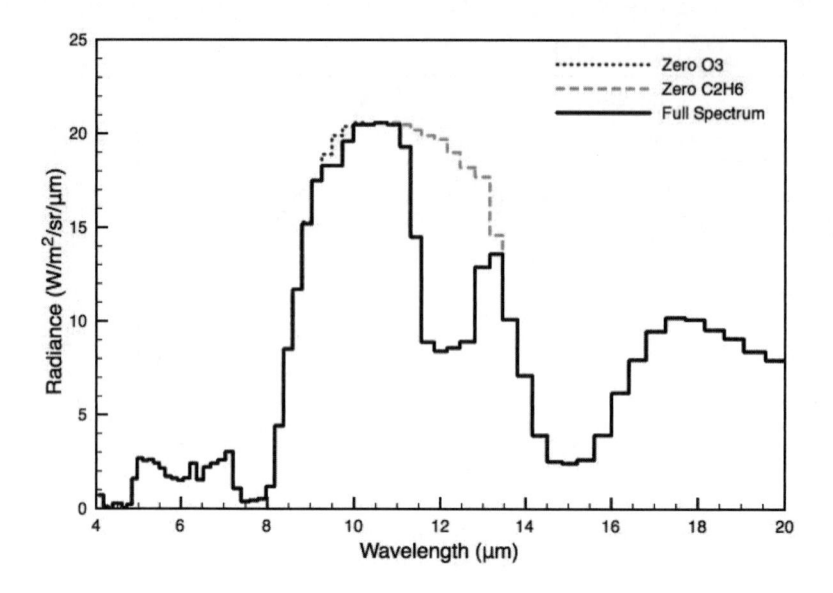

[Figure: radiance spectrum plot, x-axis Wavelength (μm) from 4 to 20, y-axis Radiance (W/m²/sr/μm) from 0 to 25, with three curves labeled "Zero O3" (dotted), "Zero C2H6" (dashed), and "Full Spectrum" (solid)]

Figure 2. Three spectra are shown here, at a resolution ($\lambda/\delta\lambda$) of 20. The full spectrum of our model planet around AD Leo is shown with a solid line; the same spectrum with O_3 and C_2H_6 features independently removed are respectively shown with dotted and dashed lines. The feature at 9.6μm disappears when O_3 is removed and the feature centered at 12μm disappears when C_2H_6 is removed.

O_3. Thus, any mission whose wavelength range extends past 12 μm and that has the requisite S/N to discover the "false O_3 signal" should also be able to detect C_2H_6. As long as the mission has adequate wavelength coverage we should be able to determine the source of the O_3. However, we must caution that we have not yet examined the potential of C_2H_6 to accumulate in oxidized atmospheres around M dwarfs. This buildup may be possible, and if it can occur it would eliminate this particular method of O_3 source determination.

It will also be possible to determine the O_3 source if the planet in question is examined using both TPF-I (interferometer) and TPF-C/O (coronagraph/occulter) mission concepts. If we have the capability to examine planets in the visible, we can search for a signal from the organic haze particles responsible for protecting O_3 from photolysis. Additionally, we will be able to search for O_2, which is not nearly as high in our "O_3-rich" models as it is on modern-day Earth. Our model atmospheres contain a detectable amount of O_3 but undetectable amounts of O_2, as the O_2 does not accumulate to high enough concentrations to create an absorption feature. Thus, planets that contain both O_3 and O_2 likely have a biological O_2 source.

4. Discussion

The simulations shown here have an atmosphere with 300 ppm CO_2 and a ground mixing ratio of 1000 ppm CH_4. We have not yet examined whether a planet with these characteristics could exist without life, as this relatively high CH_4/CO_2 ratio may only be possible in the presence of methanogens (microbes that produce CH_4). Alternatively, it is possible that intense serpentinzation could produce high CH_4/CO_2. Until we research this more fully with geophysical and ocean chemistry models, our conclusions should be considered preliminary.

An abiotic source of O_3 represents a potential "false positive" for life on exoplanets. However, this false positive should only pose a problem to missions that attempt to detect life exclusively through a search for O_3 absorption features. The presence of detectable O_3 concentrations on exoplanets without oxygenic photosynthesis requires an organic haze in the upper atmosphere of the planet. This haze will only be present if organic compounds such as C_2H_6 are also abundant in the atmosphere. Missions that include a wavelength range wide enough to detect the spectral features of these organic compounds will avoid misinterpreting the presence of O_3. Additionally, by examining planets in both the infrared and the visible, we will be able to look for the optical properties of the haze particles themselves, and search for O_2. If O_2 is absent, that would be indicative of a high O_3/O_2 ratio, which is one of the hallmarks of this abiological O_3 signal. The presence of O_2 would be indicative of a large surface flux of O_2 to the atmosphere and would be consistent with the presence of life on the planet. Our conclusions support a likely diversity of planetary environmental signatures and a pathway towards habitable planets that includes exoplanet characterization missions robust at both infrared and visible wavelength ranges.

Acknowledgments. We would like to thank the NASA Astrobiology Institute for supporting this research. We also acknowledge the insights of other VPL team members, in particular James F. Kasting and David Crisp.

References

Des Marais, D. J., Harwit, M. O., Jucks, K. W., Kasting, J. F., Lin, D. N. C., Lunine, J. I., Schneider, J., Seager, S., Traub, W. A. & Woolf, N. J. 2002, Astrobiology, 2, 153
Holland, H. D. 1984, The Chemical Evolution of the Atmosphere and Oceans, (Princeton University Press : Princeton, NJ)
Meadows, V. S. & Crisp, D. 1996, J. Geophys. Res., 101, 4595
Pavlov, A. A., Brown, L. L. & Kasting, J. F. 2001, JGR (Planets), 106, 23267
Schindler, T. L. & Kasting, J. F. 2000, Icarus, 145, 262
Segura, A., Kasting, J. F., Meadows, V. S., Cohen, M., Scalo, J., Crisp, D., Butler, R. A. H. & Tinetti, G. 2005, Astrobiology, 5, 706

Pathways Towards Habitable Planets
ASP Conference Series, Vol. 430, 2010
Vincent Coudé du Foresto, Dawn M. Gelino, and Ignasi Ribas, eds.

CoRoT-7b: The First Transiting Super-Earth Fully Characterized in Radius and Mass

D. Rouan,[1] A. Léger,[2] J. Schneider,[3] P. Barge,[4] M. Fridlund,[5]
B. Samuel,[2] D. Queloz,[6] C. Moutou,[4] F. Bouchy,[7] A. Hatzes,[8]
and all the CoRoT Exoplanet Science Team

Abstract. In February 2009, the transit detection by the *CoRoT* satellite of CoRoT-7b, a planet as small as R = 1.7 R$_{Earth}$ and closely orbiting a solar-type star, was announced at the *CoRoT* symposium in Paris. Here we present the arguments, largely based on follow-up observations from the ground, that leave practically no doubt on this detection and we give the last evaluation of the planet parameters. In addition, we show how the most recent results of radial velocity measurements with HARPS fully confirm the discovery and give now a firm evaluation of the mass of the planet, 4.8 M$_{Earth}$, and thus of its density which is equal to that of Earth, suggesting a similar composition dominated by silicates. Lastly, it is shown that a second planet, of only twice the mass of CoRoT-7b, is found on a slightly larger orbit, but still extremely close to the star. The discovery of the CoRoT-7 system is obviously an important milestone on the pathway to habitable planets.

1. Introduction

The first super-Earth of 5.5 M$_{Earth}$ radius was detected in 2006 thanks to gravitational lensing (Beaulieu et al. 2006). Three years later, there is a wealth of evidence that the short-period planet population features an important component of small objects in the so-called super-Earth range. At the time of the Barcelona meeting (September 2009), there were 32 planets with M < 0.1 M$_{Jup}$. The impressive results obtained by the HARPS team even indicates that 30%

---

[1]LESIA, UMR 8109 CNRS , Observatoire de Paris, UVSQ, Université Paris-Diderot, 5 place J. Janssen, Meudon, France

[2]Institut d'Astrophysique Spatiale, UMR 8617 CNRS, bat 121, Université Paris-Sud, Orsay, France

[3]LUTH, UMR 8102 CNRS, Observatoire de Paris-Meudon, 5 place J. Janssen, Meudon, France

[4]Laboratoire d'Astrophysique de Marseille, UMR 6110 CNRS, Technopôle de Marseille- Etoile, Marseille, France

[5]Research and Scientific Support Department, European Space Agency, ESTEC, Noordwijk, The Netherlands

[6]Observatoire de Genève, Université de Genève, 51 Ch. des Maillettes, Sauverny, Switzerland

[7]Observatoire de Haute Provence, USR 2207 CNRS, OAMP, St.Michel l'Observatoire, France

[8]Thüringer Landessternwarte Tautenburg, Sternwarte 5, Tautenburg, Germany

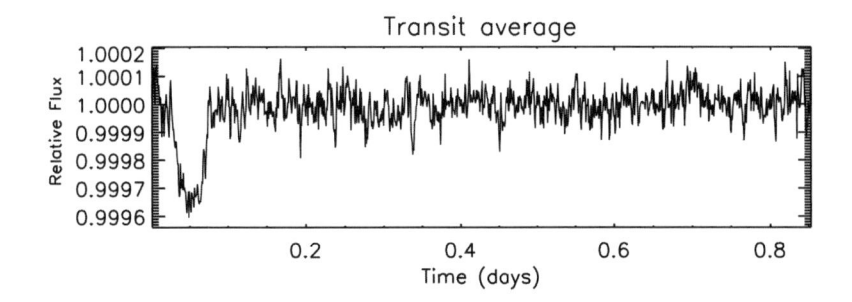

[Figure: Transit average light curve plot]

Figure 1. Mean value of the transit curves with a low pass filter preserving the transit shape.

of stars would harbour a planet of this category (Mayor & Udry 2008; Mayor et al. 2009; Bouchy et al. 2009).

The next step is to derive the radius of some members of this component of small objects, in order to access their density and thus their structure. This was one major goal of *CoRoT* at its launch and indeed it became clear, after the first *CoRoT* data analysis, that the achieved performances allowed it. CoRoT-7b is the first discovered case and hopefully not the last one.

2. The CoRoT-7b Detection

CoRoT-7 is a G9V star of $m_V = 11.7$, at a distance of 130 pc. It features strong activity, evidenced by emission features in the H and K Ca lines, as well as a significant variability that can be traced to be essentially caused by spots crossing the disk through rotation. A rotation period of 23 days is derived. The age, evaluated to be 1.2 Gyr, is consistent with this activity.

Despite the limitation in photometric accuracy resulting from this activity, the first clue of a transiting small planet was obtained after one month of observation with *CoRoT*. All together, there are 153 transit events that were recorded: they all can be seen individually when superimposed on a same period slot. The two most striking features of the transits are *a)* their extremely short period of 0.8536 days and *b)* the transit depth which is only of 0.033% ! Figure 1 shows the resulting average folded light curve.

The question of whether a small planet was indeed detected raised the issue of a possible false positive that can mimic such a transit of small amplitude: grazing eclipsing binary, eclipsing binary in a dwarf/giant system, eclipsing binary or transiting giant planet in a triple system or in the background or foreground.

Clearly an intensive campaign of follow-up observation from the ground was mandatory. It has been a huge effort of the whole team.

3. The Follow-up Observation

The observation campaign used various means:

- UVES and HARPS spectra

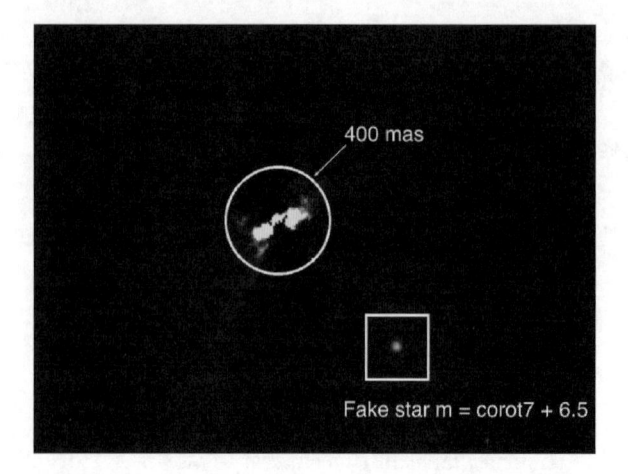

Figure 2. Central part of the NACO image, after substraction of rotated images and addition of a simulated star 6.5 magnitudes fainter than CoRoT-7 (within square). The circle of 400 mas radius defines the region in which such a star could still be confused with residual speckles.

The spectral type of the star CoRoT-7 was determined to be a K0V star and no blends were discovered, excluding a dwarf/giant binary system

- ON/OFF photometry of neighbors

  Using MEGACAM-CFHT and IAC-80, time-series CCD photometry was performed on CoRoT-7 and all stars within a $30''$ radius. For each one, we calculated the expected eclipse amplitude *if* this star was the source of the observed dips. No Δmag larger than those thresholds was found on any of the stars in the neighbourhood, excluding a BEB at $\theta > 4''$.

- Adaptive Optics + Angular Differential Image

  Thanks to NACO-VLT diffraction-limited imaging, combined with differential rotation, it was shown that no star could be found between $0.4''$ and $4''$, even 6.5 mag fainter than CoRoT-7, to explain the transit by a BEB. Figure 2 shows the adaptive optics image of the field after subtraction of rotated images (on the sky) and addition of a fake star 6.5 mag fainter than CoRoT-7 for comparison with residual speckles.

- IR spectra with CRIRES-VLT

  Because infrared spectroscopy is very sensitive to low mass stars (prominent CO lines) it was possible to put limits on the presence of a late K or M star earlier than M5V, in the vicinity of CoRoT-7. Such stars that would belong to a binary system, were excluded thanks to good S/N spectra taken with CRIRES.

- ROSAT all sky survey

  With a 0.9 days period, the components of a binary system would be so close that an intense X-ray emission would be produced. Using ROSAT maps we found no such emission indicating a very close X binary (e.g. similar to YY Gem) in the vicinity of CoRoT-7.

- *CoRoT* colors

  Can the source of the periodic dips in the light curve be a triple system with a Jupiter transiting a second star that would be fainter and redder than CoRoT-7? In that case, the information should translate on the colors since the transit signal would be essentially seen in red and would not show up in a blue channel. Three *CoRoT* light curves in Blue, Green, and Red are available for this object: they reveal that the transit is essentially achromatic, excluding this case.

- Radial Velocity with SOPHIE-OHP

  Two points on the radial velocity curve were obtained in Fall 2008 using the spectrograph SOPHIE at OHP: compared to the expected velocity curve in the case of a 10 and a 20 M_{Earth} planet, they clearly show that the latter case is excluded with a 95% confidence level. This means that there is no grazing Jupiter or white dwarf orbiting CoRoT-7 on a short orbit. On the other hand, those two cases are also excluded from simple timing considerations, given the duration and depth of the transit.

To summarize the results of this thorough follow-up study, we can say that all known cases of false positives were practically eliminated with a high level of confidence (risk of a background blend $< 8 \; 10^{-4}$). *The transit should be due to a super-Earth planet of radius 1.7 R_{Earth} on a very close orbit, with a fairly high probability.* This is the first time a planet has been characterized by its radius at such a small value. This result will appear in Léger, et al. (2009).

4. Radial Velocity Measurements with HARPS-VLT

As soon as CoRoT-7b became a serious candidate, a vigorous campaign of radial velocity measurements was started and conducted during several months using HARPS. One hundred and six measurements representing 70 h of observations were accumulated during 4 months. The main issue was clearly the strong variablity of CoRoT-7 that affects the cross correlation function. We used different criteria to assess those effects and to try to get rid of them. For instance, it was shown that the bissector and FWHM correlates with photometric variations. Also the Ca II H and K S index exhibits an obvious correlation with the rotation period of 23 days. Several types of analysis were conducted, and one of them, the Scargle analysis gave especially convincing results. After filtering of rotation (3 harmonics) a first peak shows up at 3.7 days. When substracted, the spectrum reveals a second peak at 0.85 days, i.e., the period found on *CoRoT* data. In addition the phase of this 0.85 days peak agrees with the phase of the transits. This is a clear confirmation of the reality of CoRoT-7b. The cherry on the cake is the detection of a second planet that is responsible for the peak at 3.7 days.

This planet is not detected on the light curve, indicating that the inclination of its orbit is large enough so that it does not produce transits.

CoRoT-7 is thus harbouring a unique planetary system with two super-Earths of 4.8 ± 0.8 and 8.4 ± 0.9 M_{Earth} respectively, on very close orbits of major axes 0.017 and 0.046 AU respectively. Those results will appear in Queloz et al. (2009). Table 1 summarizes the main characteristics of this system.

Table 1. Planetary Parameters

Parameter	CoRoT-7b Value	CoRoT-7b Uncertainty	CoRoT-7c Value	CoRoT-7c Uncertainty
Period (day)	0.853585	± 2.4 10^{-5}	3.698	± 3 10^{-3}
a (AU)	0.0172	± 2.9 10^{-4}	0.046	
R_{pl} / R_{Earth}	1.68	± 0.09		
M_{pl}/ M_{Earth}	4.8	± 0.8	8.4	± 0.9

5. The Nature of CoRoT-7b

CoRoT-7b can be considered the first solid evidence for a rocky planet. This opens the door to a lot of exciting physics (Léger et al. in prep). For instance :

- Tidal forces

   The planet must be phase-locked since the synchronisation time can be evaluated to be < 100 years, even if large uncertainties exist.

   The elongation under tidal forces must be limited to a typical value of 150 km which is not detectable but does not affect the radius derivation. On the other hand, the tidal heating can be extremely efficient if the eccentricity is different from zero, even with a value as small as 10^{-3}.

- Temperature

   Since there must be no efficient heat redistribution by the atmosphere or the ground, the dark side of the planet should be extremely cold, unless tidal heating is very efficient. We predict T ≈ 50 K, taking into account a geothermal origin. This is pretty cold, especially if we compare this temperature to that on the day side which should reach 1800 to 2600 K, depending on the albedo of the soil.

   Between 50 and 2000 K, there must be a temperate region (270-370 K) of 90 km width where, in principle, water would be liquid. However, in fact water would not survive, as it condenses on the cold hemisphere. Life is likely not possible on CoRoT-7b !

- Structure

   This is the first time that the density of a super-Earth can be measured, with a value of 5.6 g cm^{-3}, practically the Earth value. We conclude that CoRoT-7b must be a rocky planet, likely with a silicate mantle. This density is however compatible with 20% of water.

Figure 3. Artist's view of the possible appearance of CoRoT-7b, with realistic limits on the lava ocean and the solid crust as well as the frost hemisphere.

The probable origin of the planet is either a core of a Neptune or a rocky telluric planet that migrated.

With regards to the surface, it must be an ocean of liquid rock, for latitudes $> 40^o$ (with North taken towards the star), made of refractory oxides: CaO, Al_2O_3, with continents elsewhere on the dayside, while H_2O and CO_2 frost should prevail on the dark side. Figure 3 gives an idea of the likely appearance of CoRoT-7b.

- Atmosphere

  The escape of all volatiles should occur in a time scale of 0.1 to 1 Gyr, so that one cannot expect gas, other than a tenuous vapor of silicates, at a pressure of 0.05 to 5 10^{-5} bar.

In summary, CoRoT-7b is a planet of similar mass and size that makes it a cousin of our Earth, around a star which is a cousin of our Sun: we feel that this discovery represents a solid milestone on the pathway explored at this conference.

Acknowledgments. The authors wish to thank warmly all the teams who worked hard to make *CoRoT* and its follow-up such a successful venture.

References

Beaulieu, J.-P. et al. 2006, Nat, 439, 437
Bouchy, F. et al. 2009, A&A, 496, 527
Léger, A. et al. 2009, A&A, 506, 287
Mayor, M. & Udry, S. 2008, Physica Scripta Volume T, 130, 014010
Mayor, M. et al. 2009, A&A, 493, 693
Queloz, D. et al. 2009, A&A, 506, 303

Fabien Malbet and Pierre Léna discuss the proceedings during a coffee break.

Session V
Methods and Projects

Pathways towards Habitable Planets: JWST's Capabilities for Exoplanet Science

M. Clampin

NASA Goddard Space Flight Center, Greenbelt, MD, USA

Abstract. The James Webb Space Telescope (JWST) is a 6.5 meter cryogenic telescope scheduled to be launched in 2014. JWST has a complement of near-IR (1μm - 5μm) and mid-IR (5μm - 28μm) instruments that include modes suitable for detecting and characterizing exoplanets, and studying different stages of planetary system evolution. Direct imaging studies of young, self-luminous planets will be possible using coronagraphic imagers of the instruments. JWST's large collecting area will facilitate extensive characterization of the atmospheres of transiting exoplanets.

1. Introduction

The James Webb Space Telescope (JWST) is a large aperture space telescope designed to conduct imaging and spectroscopic observing programs over the wavelength range 1μm to 29μm (see Figure 1). JWST is scheduled to be launched to an L2 orbit aboard an Ariane 5 launcher in 2014. The Goddard Space Flight Center (GSFC) is the lead center for the JWST program, and manages the project for NASA. The prime contractor for JWST is Northrop Grumman Space Technology (NGST). JWST is a cooperative project with the European Space Agency (ESA) and the Canadian Space Agency (CSA).

The JWST observatory is designed to address four major science themes: first light and re-ionization; the assembly of galaxies; the birth of stars and protoplanetary systems; and the formation of planetary systems and origins of life (Gardner 2006). The formation of planetary systems and the origin of life theme seeks to determine the physical and chemical properties of planetary systems including our own, and investigate the potential for the origins of life in those systems. It is directly relevant to the objectives of the "Pathways Towards Habitable Planets" meeting reported in these proceedings and forms the basis of this paper.

JWST has a complement of four instruments (Figure 2), the Near Infrared Camera (NIRCam), the Near Infrared Spectrograph (NIRSpec), the Tunable Filter near-infrared camera (FGS-TF), the Mid Infrared Instrument (MIRI), and the Fine Guidance Sensor. The characteristics of the science instruments are summarized schematically in Table 1 to illustrate their major scientific objectives. The Near Infrared Camera (NIRCam) development team is led by the University of Arizona (Rieke et al. 2003). This instrument will be JWSTs primary imager in the wavelength range of 0.6μm to 5μm. Required by many of the core science goals, the instrument is particularly well suited to the task of detecting the first luminous sources that formed after the cosmological dark ages.

Figure 1. A schematic illustration of the James Webb Space Telescope.

It is primarily designed for deep, wide field imaging. It is also capable of coronagraphic observations, which will enable imagery of debris disks, and searches for massive giant planets around nearby stars. The NIRCam also fulfills the key role of wavefront sensor for the JWST telescope assembly. The multi-object Near Infrared Spectrograph (NIRSpec) is provided by the European Space Agency (Jakobsen et al. 2009). This instrument will serve as the principal multi-object spectrograph in the 0.6μm to 5μm wavelength range. Its ability to obtain simultaneous spectra of more than 100 objects in a 9-square arc-minute field of view at spectral resolutions of $\lambda/\Delta\lambda$ = 100, 1000, and 3000, enables high survey efficiency for a variety of compact sources including primordial galaxies. The Mid-Infrared Instrument (MIRI), provided by an international collaboration of agencies led by the Jet Propulsion Laboratory and the United Kingdom Advanced Technology Center, will provide broad-band imaging and integral field spectroscopy over the 5μm to 28μm spectrum (Wright et al. 2008). This instrument will study the creation of the first heavy elements and will reveal the evolutionary state of high redshift galaxies. It is uniquely capable of studying the very early stages of star and planet formation, in regions where all visible light is blocked by dust and most of the emission is radiated at mid-infrared wavelengths. The Tunable Filter Imager (TFI) enables extended objects at high redshift to be imaged in Lyman-α with diffraction limited angular resolution at $\lambda/\Delta\lambda$ = 100 (Doyon et al. 2008). This instrument is critical for emission line surveys of primordial galaxies and detailed morphological studies of galaxy nuclei and Galactic nebulae. TFI forms part of the Fine Guidance Sensor (FGS)

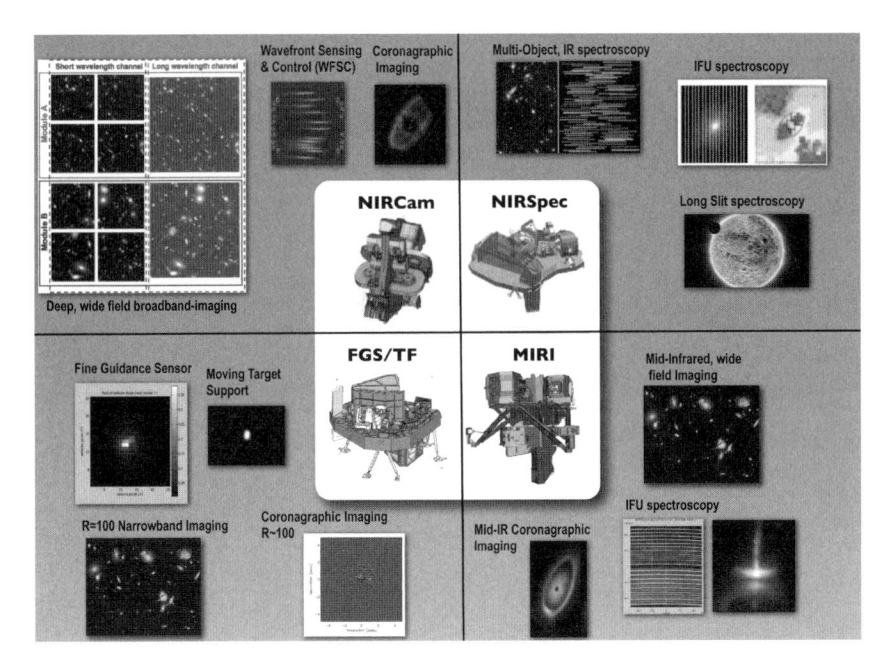

Figure 2. The JWST science instrument complement and their primary scientific capabilities.

instrument package, provided by the Canadian Space Agency. FGS will ensure the telescope can precisely point to a few milli-arcseconds.

2. High Contrast Imaging

JWST offers a range of coronagraphic capabilities designed to provide high contrast imaging to facilitate observations of exoplanets, and faint circumstellar structures such as debris disks. These capabilities are summarized in Table 1. JWST was not designed to conduct deep, wide-field surveys in the infrared, and so its architecture is not optimal for coronagraphic imaging. Its 18 segment mirror generates considerable diffraction structure in the telescopes point spread function compared to that of a monolithic mirror, and the design wavefront error of 150 nm at the NIRCam focal plane, is the limiting performance factor for the shorter wavelength coronagraphs. These are both important considerations when making comparisons to fully optimized visible and near-infrared coronagraphs designed for ground-based or space-based applications. However, JWST does offer low infrared backgrounds, by virtue of its orbit at L2 and the telescope's cryogenic thermal design. Furthermore, JWST's point spread function will be relatively stable over the 14-day periods between scheduled fine-phasing adjustments to the primary mirrors. Image stability will be primarily dominated by thermal drift in mirror alignments and mirror figures. The current prediction, reported during the Optical Telescope Element (OTE) critical design review, anticipates a maximum drift of ~31 nm over 14 days, calculated

Table 1. Overview of JWST's high contrast imaging capabilities. Predicted contrast ratios are presented for modes that have been modeled.

SI	Mode	λ (μm)	Contrast	radial Distance
NIRCam	Short λ Lyot Coronagraph	0.6 - 2.3		
NIRCam	Long λ Lyot Coronagraph	2.4 - 5.0	6×10^{-4}	$1''$
TFI	Multi-λ coronagraph	1.6 - 2.5	6×10^{-5}	$1''$
TFI	Multi-λ coronagraph	3.2 - 4.9	6×10^{-5}	$1''$
TFI	Non-redundant mask	1.6 - 2.5	10^{-4} - 10^{-5}	$\leq 0.55''$
TFI	Non-redundant mask	3.2 - 4.9	10^{-4} - 10^{-5}	$\leq 0.55''$
MIRI	Quadrant Phase Coronagraph	10.65		
MIRI	Quadrant Phase Coronagraph	11.4	10^{-4}	$1''$
MIRI	Quadrant Phase Coronagraph	15.5	4×10^{-5}	$1''$
MIRI	Lyot Coronagraph	23		

assuming the worst case thermal slew impulse. JWST's wavefront error stability combined with low infrared backgrounds offer unique advantages that make possible a range of observations that will complement other near-term ground-based approaches. JWST's discovery space for high contrast imaging in the thermal infrared is unique because of the low backgrounds, combined with excellent angular resolution offered by its 6.5 meter aperture.

NIRCam offers coronagraphic imaging in both its short and long wavelength channels via a set of five occulting masks (Krist et al. 2009). This mode is described in detail by Green et al. (2005) and Beichman et al. (2009). The masks comprise three circular and two wedge shaped masks. The masks are optimized for operating wavelengths of 2.1μm, 3.35μm, 4.3μm and 4.6μ. The predicted contrast at a wavelength of 4.6μm is 6×10^{-4} at an angular radius of $1''$. Optimum processing using the technique of roll-deconvolution or subtraction of a spectrally matched reference star might improve this figure by up to an order of magnitude (Krist 2007). Accurate predictions of the short wavelength channel's coronagraphic performance will depend upon the wavefront error of JWST's optical system. Makidon (2008) has shown that short wavelength performance especially might require careful scheduling to achieve optimum results.

The Tunable Filter Imager (TFI) features a similar coronagraph to NIRCam, however, the observing concept for the instrument is different since TFI is built around a tunable narrowband imager. TFI images in R$\sim$100 increments over the 1.6μm - 4.9μm bandpass. TFI coronagraphic observations will employ an analysis technique called Differential Speckle Imaging (Marois et al 2000) to achieve a contrast 10$\times$ greater than NIRCam. TFI also employs a non-redundant mask (NRM) which uses the technique of closure phase imaging (reference) to achieve a contrast in the region of 10^{-4}-10^{-5} out to a radius of $0.55''$ (Sivaramakrishnan et al. 2009).

The mid-infarared imager MIRI offers four self-contained coronagraphs, three of which are quadrant phase masks and the fourth a traditional Lyot corongraph. Quadrant phase masks work by breaking the image plane into four quadrants and applying a π phase-shift to two diagonally quadrants. This produces a deep null in the pupil plane which is then passed through a lyot stop and re-imaged. The quadrant phase mask images to $< \lambda/D$ although the contrast is significantly decreased at smaller inner working angles. Each MIRI coronagraph is designed for a specific wavelength, the three quadrant phase masks are assigned to wavelengths of 10.65μm, 11.4μm and 15.5μm, and the Lyot coronagraph to 23 μm.

The JWST coronagraph complement is optimized for near-IR and mid-IR high contrast imaging, making them excellent tools for the study of young, self-luminous planets. Recent modeling by Beichman et al. (2009) has shown that JWST should be able to detect 0.2 M_J planets as close as 50 AU. The TFI's non-redundant mask imager and MIRI provide sensitivity to comparable masses in the 10 to 20 AU and $\geq$40 Au, respectively. These capabilities will provide the tools to conduct surveys that could constrain planet formation models and planet properties. The low backgrounds offered by JWST also make surveys for planets around M stars especially attractive, as ground-based observations of these relatively faint systems will be limited by the sky background. JWST should be able to detect $\sim$2 M_J planets from at distances of a few AU and beyond.

3. Transiting Exoplanets

The study of transiting exoplanets has provided most of the key data to date on the properties of exoplanets, such as direct estimates of their mass and radius (e.g. Charbonneau 2007), and spectral diagnostics of their atmospheres (e.g. Swain et al. 2008). Observations of transiting exoplanets by the Hubble Space Telescope (HST) and Spitzer Space Telescope (SST) have both played lead roles in making demanding, high signal to noise observations of the light curves, and spectra of transiting exoplanets. The launch of JWST in 2014 will provide new capabilities for the characterization of transiting exoplanets via transit spectroscopy and high precision transit photometry. Spectroscopic characterization of transiting exoplanets demands extremely high signal to noise observations, which JWST will provide by virtue of its large 25 m^2 collecting area and low infrared backgrounds. JWST's orbit around L2 provides it with both excellent sky coverage and long dwell times on targets. JWST offers a broad suite of instrumentation which will provide capabilities for imaging spectroscopic observations of transiting systems. The most important instrument operating modes that address transit photometry and spectroscopy requirements are summarized in Table 2, with descriptions of their application to observations of transiting systems.

3.1. Imaging

Both NIRCam, TFI, and MIRI offer the opportunity to obtain high precision light curves, a capability that has served as the mainstay of Spitzer transit science. In combination with radial velocity measurements, high precision light curves yield exoplanet mass and radii. High precision light curves are also used to search for unseen companions via transit timing, search for exoplanet moons

Table 2. Overview of JWST's transit science instrument modes.

Instrument	λ (μm)	R ($\lambda/\Delta\lambda$)	Comments
NIRCam (Imaging)	0.6 - 2.3 2.4 - 5.0	4, 10 , 100 4, 10 , 100	High precision light curves of primary and secondary eclipses
NIRCam (Defocused Imaging)	0.6 - 2.3 2.4 - 5.0	4, 10 , 100 4, 10 , 100	High precision light curves for bright targets that need to be defocused to avoid rapid saturation of detectors
NIRCam (Spectroscopy)	2.4 - 5.0	1700	Transmission/emission spectroscopy spectroscopy of transiting planets
NIRSpec (Spectroscopy)	1.0 - 5.0	100, 1000 2700	Transmission/emission spectroscopy of transiting planets (1.6″×1.6″ slit)
TFI (Imaging)	1.6 - 2.6 3.2 - 4.9	100 100	High precision light curves of primary and secondary eclipses
MIRI (Imaging)	5 - 28	100 100	High precision light curves of secondary eclipses
MIRI (Spectroscopy)	5 - 11 5.9 - 7.7 7.4 - 11.8 11.4 - 18.2 17.5 - 28.8	100 3000 3000 3000 3000	Slitless spectroscopy of secondary eclipses Spectroscopy of secondary eclipses: suitable for specific spectral features e.g. CO_2 at 15 μm

and rings, and record reflectance and thermal phase variations across the duration of a system's light curve to study atmospheric dynamics. Each of the JWST cameras has the capability to use sub-arrays for detector readout, to increase the dynamic range for brighter targets. In addition to the standard imaging and spectroscopic modes, NIRCam has several instrument modes designed for phasing the primary mirror that also have specific application to the observation of transiting systems. NIRCam's wavefront sensing and control optics includes special lenses that have 4, 8, and 12 waves of defocus to facilitate phase retrieval measurements using instrument optics rather than the secondary mirror. A telescope as large as JWST will be limited to imaging observations of relatively faint stars when measuring photometric transits as the detectors pixels will saturate in even the shortest exposure times. However, the defocus lenses, combined with the use of sub-arrays allows NIRCam to image stars as bright as K$\sim$3 without saturating detector pixels in the minimum exposure time. NIRCam can employ this capability to obtain high-precision light curves of transiting terrestrial planets e.g. SNR 20-30 for a K=10 star in 6.5 hours (Beichman et al. 2009).

3.2. Spectroscopy

It is in the field of spectroscopic characterization that JWST has the capacity to make major contributions to exoplanet science. Impressive progress has been made with Spitzer (Swain et al. 2009) and HST (Charbonneau et al. 2002) in obtaining the first spectral diagnostics of exoplanets. Two techniques can

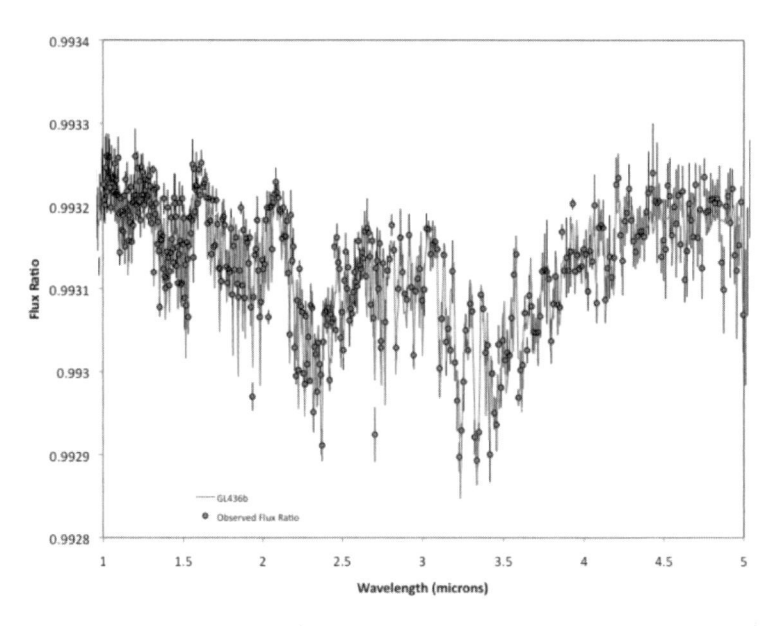

Figure 3. Simulated observation of Gj 436b, a hot neptune, with NIRSpec. The graph shows the simulated observation (filled circles with error bars), plotted over a model spectrum of GL 436B provided Sara Seager. The simulation includes instrumental and observatory pointing effects.

be used to probe transiting extrasolar planet atmospheres with JWST. The absorption spectrum of the planet can be measured by detecting the spectral signature imposed on stellar light transmitted through the planet's atmosphere during transit. The emission spectrum of the planet can also be measured during the secondary eclipse. Emission spectra produce potentially larger signals than transmission spectra at infrared wavelengths. However, features in transmission spectra will be present even in the extreme case when the atmospheric temperature profile of the exoplanet is isothermal - which would produce a featureless spectrum in emission. Gas giant planets will present many molecular features (H2, CO, H2O, CH4), strong atomic lines (Na, K), and a spectral shape (due to Rayleigh scattering) that leave distinct imprints on transmission spectra. High quality spectra also probe energy redistribution within the atmospheres. With its large collecting area, JWST will be able to conduct detailed comparative studies of gas giant atmospheres and their composition both in transmission and emission, including many of the transiting gas giants discovered by Kepler. JWST will be capable of R=100 to 3000 follow-up spectroscopy of gas giants expected to be found by ground and space-based surveys over the 0.7 μm to 10 μm wavelength range. For exoplanets with bright parent stars, it can deliver R 2700 spectra from the 1μm - 5 μm wavelength range, and for the first time provide high quality line diagnostics of these exoplanets. In the mid-IR it will be able to deliver R$\sim$100 spectra of gas giants in a single transit.

JWST's large collecting area makes it an obvious choice for characterization studies of intermediate and superearth-mass transiting planets. Transiting exoplanets around late-type stars are especially attractive (Charbonneau & Deming 2007) since the relatively small stellar radius yields transit depths that can enable low-resolution spectral characterization of some intermediate and superearth mass exoplanets, in emission and transmission. A recent example of acandidate system for JWST follow-up is GJ 436b, a hot Neptune, with a mass 0.072 MJ, and a period of 2.6 days, orbiting an M2.5V star. In Figure 3 we show a simulated NIRSpec observation of GJ 436, combining four transits to achieve a R-300 spectrum. The figure demonstrates that JWST will be able to make relatively high precision observations of intermediate mass transiting planets. The simulation includes the effects of detector pixel response functions, the JWST pointing budget, and expected detector flat field response.

Recent discoveries of Corot-7b (Leger et al. 2009) and GL1214 (Charbonneau et al. 2009) have increased interest in the question of whether JWST will be able to characterize superearths. Deming et al. (2009) have addressed this question in detail and find that depending upon the frequency of occurrence and the nature of their atmospheres, JWST can measure the temperature and identify molecular absorptions, such as water and CO_2, of superEarths orbiting lower main sequence stars. However, significant amounts of observing time will be required for such observations, together with suitable candidates for study. In the coming decade transiting planet surveys focused on late type stars with bright central stars are required to provide targets for observations of superearths with JWST. While JWST will not be able to address the question of the "earth analog" with a one-year period (Traub & Kalteneggar 2009), it will open new discovery space for superearths around late-type stars with short periods.

Acknowledgments. We wish to thank Chas Beichman for an advance copy of model results from the JWST Science Instrument Teams (Beichman et al. 2009), and Sara Seager for the GJ436b model spectrum.

References

Beichman, C. et al. 2009, PASP, 122, 162
Cavarroc, C. et al. 2008, SPIE 7010, 70100W
Charbonneau, D. et al. 2009, Nat, 462, 891
Charbonneau D. & Deming, L. 2007, arXiv0706.1047C
Charbonneau, D., Brown, T. M., Noyes, R. W. & Gilliland, R. L. 2002, ApJ, 568, 377
Deming, L. et al. 2009, PASP, 121, 952
Gardner, J. P. et al. 2006, SSR, 123, 485
Jakobsen, P. et al. 2010, BAAS 215, 41, 396
Krist, J. et al. 2009, SPIE 7440, 74400W
Krist et al. 2007, Conf. Proc. "In the Spirit of Bernard Lyot: The Direct Detection of Planets and Circumstellar Disks in the 21st Century", ed. P. Kalas, Univ. of California, Berkeley, CA, USA
Léger, A. et al. 2009, A&A, 506, 287
Makidon et al. 2008, SPIE 7010, 70100O
Marois, C., Doyon, R., Racine, R., & Nadeau, D. 2000, PASP, 112, 91
Rieke, M. J. et al. 2003, SPIE 4850, 478
Sivaramakrishnan, A. et al. 2009, SPIE 7440, 74000Y
Swain, M. R. et al. 2009, ApJ, 704, 1616
Swain, M. R. et al. 2008, Nat, 452, 329
Traub, W. A. & Kaltenegger, L. 2009, ApJ, 698, 519
Wright, G. S. et al. 2008, SPIE 7010, 70100T

Pathways Towards Habitable Planets
ASP Conference Series, Vol. 430, 2010
Vincent Coudé du Foresto, Dawn M. Gelino, and Ignasi Ribas, eds.

Radioastronomy and the Study of Exoplanets

P. Zarka
LESIA, Observatoire de Paris–CNRS–UPMC–Univ. Paris Diderot, Meudon, France

Abstract. The relatively high contrast between planetary and solar low-frequency radio emissions suggests that the low-frequency radio range may be well adapted to the direct detection and study of exoplanets. Study of solar system planetary radio emissions (auroral as well as satellite induced) show that their primary engine is the interaction of a plasma flow with an obstacle in the presence of a strong magnetic field (of the flow or of the obstacle). A "radio–magnetic" scaling law has been derived that relates the emitted radio power to the magnetic energy flux convected onto the obstacle. An RS CVn magnetic binary for which both quantities are available is found to be consistent with this scaling law. Extrapolating it to the case of exoplanets and their plasma interaction with their parent star, it is found that hot Jupiters may produce very intense radio emissions due to planetary magnetospheric interaction with a strong stellar wind, reconnection between planetary and stellar magnetic fields, or unipolar interaction between the planet and a magnetic star (or strongly magnetized regions of the stellar surface). Emitted radio power is expected in the hectodecameter range with intensity 10^3 to 10^6 times that of Jupiter (unless some unexpected "saturation" mechanism occurs). Corresponding flux densities should be detectable at the tens of parsecs range with modern radio arrays. Ongoing and future observational searches are briefly described, as well as the interests of direct radio detection, among which access to exoplanetary magnetic field measurements and comparative magnetospheric physics.

1. Remote Observation of Exoplanetary Magnetospheres?

Electromagnetic signatures of planet-star plasma interactions consist mainly of aurorae (UV, IR, optical) and associated radio emissions, as can be observed in our solar system. Contrary to the optical/UV (resp. IR) range, where the star/planet luminosity contrast is $\sim 10^9$ (resp. $\sim 10^6$), nonthermal low-frequency (LF) planetary radio emissions, such as Jupiter's, are nearly as intense as solar ones (see Figure 1 of Zarka 2007). Moreover, the radio window between wavelengths of a few mm and 30 m permits the use of large ground-based instruments. But LF radio observations also suffer strong limitations: galactic background luminosity, ionospheric perturbations, and intense natural and man–made radio interference (RFI). Assuming that the latter two are overcome using adaptive optics and RFI mitigation techniques, the ultimate limitation is imposed by the LF sky background, a bright extended source of temperature $T(K) \sim 1.15 \times 10^8/f^{2.5}$ with f in MHz, with spatial variations of amplitude × 2–3 across the sky. The detection limit is actually set by the statistical fluctuations of the sky background noise, which depends on the radiotelescope area A, observing frequency f, bandwidth b and integration time τ. With the largest available LF radiotelescopes ($A \sim 10^5$ m^2), it has been shown (Zarka 2010) that it would be possible to detect Jupiter's magnetospheric radio emissions up to no more than 0.01 pc (at $f \sim$10 MHz with $b \times \tau = 10^6 = 1$ MHz $\times$ 1 sec) to 2 pc (at $f \sim$100 MHz with $b \times \tau = 4 \times 10^{10} = 10$ MHz $\times$ 1 hour). Within such a short range, an exoplanetary LF radio search is hopeless, unless magnetospheric radio emissions much stronger than Jupiter's can be produced.

2. Planetary Radio Emissions Properties and Energy Sources in Planet-star Plasma Interactions

The most intense planetary auroral radio emissions are generated by the so-called cyclotron maser instability (CMI), which builds up on unstable keV electron populations in strongly magnetized and rarefied regions ($f_{pe} \ll f_{ce}$, with $f_{pe}(Hz) = 9 \times N_e^{1/2} (m^{-3})$ the plasma frequency and $f_{ce}(Hz) = 2.8 \times 10^6 B(G)$ the cyclotron frequency) such as planetary auroral regions. CMI directly converts up to $\geq$1% of the electron energy into coherent cyclotron waves (Treumann 2006; Zarka 1998). Intense radio emissions (with average powers from 10^6 W for Neptune to 10^{11} W for Jupiter) are produced at the local electron cyclotron frequency $f \sim f_{ce}$. The highest frequencies are thus emitted just above the ionosphere, and lower frequencies at increasing distances up to a few planetary radii. Overall bandwidth is broad ($\Delta f \sim f$), and covers a spectral range from $<$ 1 MHz to $\sim$ 40 MHz depending on the magnetic field intensity in the sources. Emission polarization is 100% circular or elliptical, consistent with extraordinary (X) magneto–ionic mode, and beaming is very anisotropic ($\Omega \ll 4\pi$) and mostly directed at large angle from the magnetic field in the source, causing modulation of the emission by the planetary rotation. Finer structure in bursts (from $\leq$1 sec to hours) is observed. Although the CMI is a well-developed quantitative theory, the emitted radio flux strongly depends on the shape of the unstable electron populations and thus on the acceleration processes at work. Those are not fully understood nor quantified, so that radio emission intensity is not (yet) predictable from basic theory and a few planet-star parameters.

However, a strong correlation exists between the energy input into the planet-star system and its output radio emission power. For magnetospheric auroral emissions, energy sources include the solar wind kinetic energy flux (Farrell et al. 1999) and Poynting (magnetic energy) flux on the magnetospheric cross-section (Zarka et al. 2001). For the auroral-like intense emissions induced by satellite-planet electrodynamic interactions (e.g., Io-Jupiter or Ganymede-Jupiter), the dominant energy input to the system is the Poynting flux of the magnetospheric magnetic field convected onto the satellite obstacle (Zarka 2007). This energy is dissipated either via the emission of Alfvén waves along magnetic field lines toward the central – planetary – body (in the case of Io, called "unipolar inductor"), or via magnetic reconnection of the planetary and satellite fields (in the case of Ganymede). Quite remarkably, the expression of the dissipated Poynting flux is similar in the solar wind-magnetosphere interaction and in both satellite-magnetosphere types of interaction:

$$P_d \sim (\varepsilon V B_\perp^2 / \mu_0) \, \pi R_{obs}^2 \qquad (1)$$

with R_{obs} the obstacle cross-section (magnetopause, exosphere or ionosphere), $B_\perp$ the component of the magnetic field convected by the incident flow perpendicular to the flow direction in the frame of the obstacle, and ε an efficiency factor of the order of 0.2 ± 0.1.

It is interesting to note that the Ganymede-Jupiter plasma interaction is a small-scale model of interacting magnetized binaries or magnetized star–planet systems, whereas the Io-Jupiter (unipolar) interaction may be a good small-scale model of magnetized star–unmagnetized planet systems. Possible optical signatures of such star-planet interactions have been obtained by Shkolnik et al. (2008) for the systems of HD 179949 and υ Andromedae.

3. Scaling Laws and Extrapolation to Hot Jupiters

Based on our experience of solar system radiophysics, intense CMI emissions are produced by flow–obstacle interactions when either (i) the obstacle, (ii) the flow, or (iii) both, are strongly magnetized, i.e. permit the existence of regions where $f_{pe} \ll f_{ce}$. This encompasses solar-wind–magnetosphere (i), Io-Jupiter (ii) and Ganymede-Jupiter (iii) interactions, but excludes cases such as the solar wind interaction with Mars or Venus.

When plotting the overall radio powers of all solar system CMI emissions (computed by integration of the received flux over the emission beam, frequency range, and time) against the dissipated magnetic power on the obstacle cross-section, a scaling law appears (Figure 1):

$$P_{Radio} \sim \eta \times P_d \qquad (2)$$

with slope ~ 1 and an empirical "efficiency" $\eta \sim 2 \times 10^{-3}$ (best fit for auroral radio emissions; satellite-Jupiter emissions are slightly above this line). This scaling law, where P_d is given by Equation (1) with ε set to 1, has been called "generalized radio–magnetic Bode's law" (Zarka et al. 2001; Zarka 2007). The correlation between the output radio power and the magnetic energy flux convected onto the obstacle holds over ≥ 4 orders of magnitude, with fluctuations of amplitude 2–3× of individual points relative to the best fit line.

Following Akasofu (1982) who concluded that the total energy output (or "consumption") of the Earth's magnetosphere, as defined by geomagnetic indices, correlates best with the reconnected magnetic power expressed by Equation (1) (rather than with e.g., the solar wind kinetic power), we draw the same conclusion for the emitted radio power. More recently, Jardine & Cameron (2008) computed the conditions under which reconnection of the stellar and planetary magnetic fields could provide the accelerated electrons necessary for the observed radio emissions, and laid the basis for a theoretical justification of Figure 1. We suggest that this scaling law characterizes a fundamental aspect of energy dissipation in a flow–obstacle system, in which the role of the magnetic field is a determinant for extracting part of the flow power and converting it to energetic particles, whatever the details of the extraction process (reconnection preceded or not by the draping of the magnetic field lines embedded in the flow around the obstacle, or Alfvén waves).

We can extrapolate the "generalized radio–magnetic Bode's law" defined by Equations (1) and (2) to the case of exoplanets. Out of >400 exoplanets (within

[Figure: Generalized radio–magnetic Bode's law plot with Radio power (W) on y-axis and Incident magnetic power (W) on x-axis]

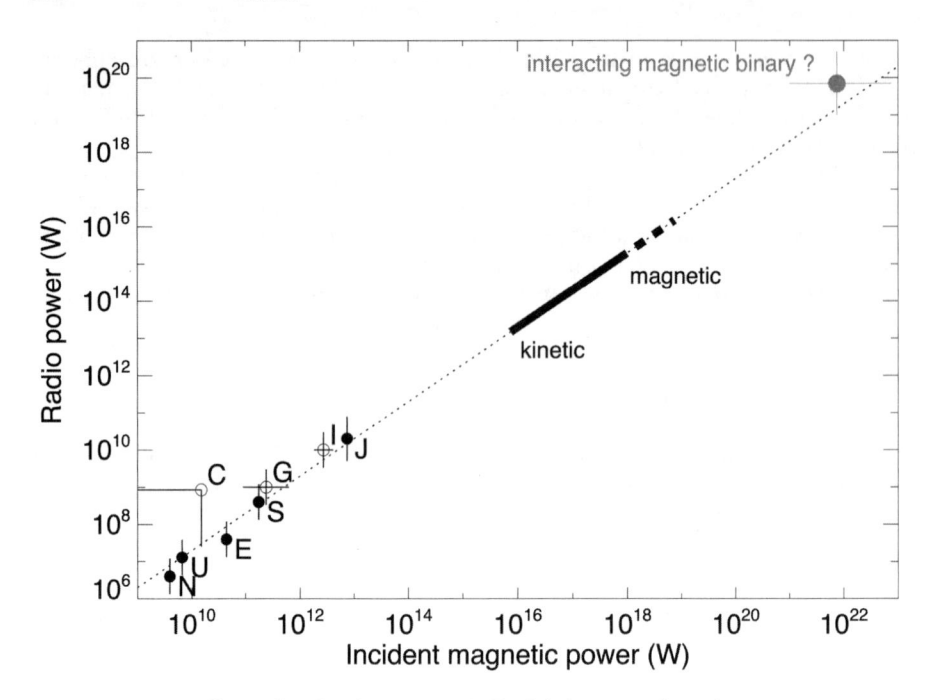

Figure 1. Generalized radio–magnetic Bode's law revealing the proportionality between output radio powers and input Poynting flux in flow–obstacle systems. Black dots refer to the solar wind interaction with the 5 magnetized radio-emitting planets (Earth, Jupiter, Saturn, Uranus and Neptune). Open dots refer to interaction of the satellites Io, Ganymede and Callisto with the Jovian magnetosphere (values for the latter are only upper limits). The dashed line has slope 1 and emphasizes the proportionality between ordinates and abscissae, with a coefficient 2×10^{-3}. The thick bar extrapolates to hot Jupiters the magnetospheric interaction (solid) and satellite-planet electrodynamic interactions (dashed). The grey dot illustrates the case of the RS CVn magnetic binary V711 τ discussed in the text.

> 300 systems) known at the time of this writing (cf. http://exoplanet.eu/), ~20% have a semi–major axis ≤ 0.05 AU (or ~10 R_S, with R_S the solar radius), and ~30% have a semi–major axis ≤ 0.1 AU. These planets are called "hot Jupiters" due to their strong irradiation. They also receive from the wind of their parent star a much larger energy flux than at the orbit of Jupiter (~5 AU). This energy flux depends on the stellar wind strength, and on the star's magnetic field carried away by the wind in the circumstellar medium. On the Sun, the large scale dipolar magnetic field has a surface intensity $1 - 1.5$ G. Magnetic spots and associated loops may reach 10^3 G on a few percent of the solar surface. Stellar surface magnetic fields of 10's-100's G are being discovered by optical spectropolarimetry (see e.g., Moutou et al. 2007), and values up to 10^3 G should not be uncommon (so-called magnetic stars).

Zarka et al. (2001) and Zarka (2007, 2010) showed that the input energy flux on an exoplanetary magnetosphere cross-section could be $10^3\times$ (kinetic) to

> $10^5 \times$ (magnetic) larger than for Jupiter. They noted that for hot Jupiters, which are likely to be spin-orbit synchronized due to strong tidal interactions, the planetary magnetic dipole field (empirically varying in P_{sid}^α, with P_{sid} the planet's sidereal rotation period and $-1 \leq \alpha \leq -1/2$ (see Farrell et al. 1999) may have decayed to a value much smaller than Jupiter's (∼4.2 G). In that case, the potential exoplanetary magnetospheric radio emission should decrease both in frequency and intensity. But if the central star is strongly magnetized, the star-planet system becomes a giant analog of the Io-Jupiter system, and their electrodynamic interaction may lead to radio emission $> 10^6 \times$ larger that the Io-Jupiter one. These large emitted powers will be possible only if no unexpected "saturation" mechanism limits the wave amplitude in the source.

These various predictions are sketched in Figure 1, together with an estimate of the emitted radio power and magnetic energy of interaction of the RS CVn magnetic binary V711 τ, adapted from Budding et al. (1998): radio detection at 8.4 GHz at a level ∼100 mJy from a range ∼35 pc implies an emitted radio power $\sim 7 \times 10^{19}$ W, while the dissipated power – assuming a field strength ∼10 Gauss, an inter-star stream velocity ∼300 km/s, and a characteristic size of interaction $\sim 10^8$m – is about $\sim 7 \times 10^{21}$ W. Although uncertainties on these estimates are large, the corresponding point falls remarkably close to the prediction of our generalized radio–magnetic Bode's law, suggesting that this law may remain valid up to 10 orders of magnitude above the solar system planets range. Further study of binary star cases is required to confirm this hypothesis.

With a conservative upscaling factor of 10^5 with respect to Jupiter's case, the radio emission of a hot Jupiter should be detectable above sky background fluctuations up to distances of 3 pc (f ∼10 MHz, $b \times \tau = 10^6$) to 700 pc (f ∼100 MHz, $b \times \tau = 4 \times 10^10$). This would make them accessible to large ground-based instruments such as UTR-2 (Ukraine), GMRT (India), LOFAR (The Netherlands) and SKA (tdb). Higher upscaling factors might be reached, either permanently due to stellar winds or magnetic fields much stronger than the Sun's e.g., in young stellar systems (Griessmeier et al. 2004), or sporadically due to stellar coronal mass ejections passing by the planet (Griessmeier et al. 2007) or radio scintillations (Farrell et al. 1999). Especially good candidates include τ Bootes, υ Andromedae, HD 189733, and HD 179949 (Griessmeier et al. 2007).

4. Observations

As 1 AU at 1 pc corresponds by definition to a separation of $1''$, source imagery of hot Jupiters, which would require resolutions $\leq 0.01''$, is not possible in the LF radio range. Ongoing radio searches thus intend to (i) detect a signal, and (ii) distinguish between a stellar or planetary origin (for which comparable intensities are expected). Planet identification will rely upon the measurement of circular or elliptical polarization (solar bursts are generally unpolarized), and upon occurrence, periodicity between a few hours and the planet's orbital period.

Detection is made difficult by the strong limitations suffered by LF radio observations (see above). Attempts have been made using:

- The VLA at 74 MHz (imagery of τ Bootes at several epochs between 1999 and 2003, with a sensitivity ∼200 mJy (Lazio & Farrell 2007, and references therein).

- The GMRT at 153, 244 and 614 MHz (imagery an tied array beam observations of τ Bootes, υ Andromedae, and HD 189733 at several epochs between 2005 and 2008, with a sensitivity down to ≤1 mJy (see the search for an anti-transit – radio occultation – of HD 189733b by Lecavelier des Etangs et al. (2009) and references therein).

- The UTR-2 array at 10−32 MHz (tied array beam observations – with simultaneous On & Off beams – of many exoplanetary targets at several epochs within 1997−2000 and 2006−2009, with a sensitivity ∼1 Jy (Ryabov et al. (2004); Zarka et al. in preparation).

Results of all these searches are still negative. Beyond the deep analysis of the UTR-2 campaign 2006−2009, performed with powerful digital receivers, the next attempts will be performed in 2010+ using LOFAR (cf. http://www.lofar.org) in the range 30−250 MHz, via imagery and tied array beam observations. On-line and offline digital processing, including RFI mitigation and correction for ionospheric perturbations, is expected to bring a sensitivity of ≤1 mJy, while the observation strategy will combine a systematic search of all exoplanetary candidates and nearby stars, as well as deeper targeted observations.

Beyond the detection of exoplanetary photons, the direct LF radio detection of exoplanets is expected to provide estimates of the planetary magnetic field (putting strong constraints on scaling laws and internal structure models) and of the planetary rotation period (confirming or not the spin–orbit locking), constraints on orbit inclination, as well as valuable inputs for extending comparative magnetospheric physics to a much broader range of parameters. Existence of an exoplanetary magnetosphere is also important to limit atmospheric erosion (Griessmeier et al. 2004).

References

Akasofu, S.-I. 1982, ARA&A, 20, 117
Budding, E., Slee, O. B., & Jones, K. 1998, PASA, 15, 183
Farrell, W. M., Desch, M. D., & Zarka, P. 1999, JGR, 104, 14025
Griessmeier, J.-M., Stadelmann, A., Penz, T., et al., 2004, A&A, 425, 753
Griessmeier, J.-M., Zarka, P., & Spreeuw, H. 2007, A&A, 475, 359
Jardine, M., & Cameron, A. C. 2008, A&A, 490, 843
Lazio, T. J. W., & Farrell, W. M. 2007, ApJ, 668, 1182
Lecavelier des Etangs, A., Sirothia, S. K., Gopal-Krishna, & Zarka, P. 2009, A&A, 500, L51
Moutou, C., Donati, J.-F., Savalle, R., et al. 2007, A&A, 473, 651
Ryabov, V. B., Zarka, P., & Ryabov, B. P. 2004, P&SS, 52, 1479
Shkolnik, E., et al. 2008, ApJ, 676, 628
Treumann, R. A. 2006, A&A Rev., 13, 229
Zarka, P. 1998, JGR, 103, 20159
Zarka, P. 2007, P&SS, 55, 598
Zarka, P. 2010, in Physics and Astrophysics of Planetary Systems, ed. T. Montmerle et al. (EDP Sciences), 41, 441
Zarka, P., Treumann, R. A., Ryabov, B. P., & Ryabov, V. B. 2001, Ap&SS, 277, 293

NAHUAL: A Next-Generation Near Infrared Spectrograph for the GTC

E. L. Martín,[1] E. Guenther,[2] C. del Burgo,[3,4] F. Rodler,[5] C. Álvarez,[5] C. Baffa,[6] V. J. S. Béjar,[5] J. A. Caballero,[7] R. Deshpande,[8] P. Esparza,[9] M. López Morales,[10] A. Moitinho,[11] D. Montes,[7] M. M. Montgomery,[8] E. Pallé,[5] R. Tata,[8] L. Valdivielso,[5] and M. R. Zapatero Osorio[1]

Abstract. The extension of the highly successful Doppler technique into the near-infrared (0.9-2.5 μm) is opening new parameter domains for extra-solar planet searches in terms of ages and masses of the primary stars. Here we present the current status of NAHUAL, a next generation near-infrared spectrograph for the 10.4-meter Gran Telescopio Canarias (GTC). Among its observing modes, it includes a hyper-stable high-resolution capability aimed at high-precision Doppler measurements of cool stars in the near-infrared. The sticky question of the best wavelength calibration for the near-infared is addressed in the last section of this paper.

1. Rationale

In less than two decades, more than 400 exoplanets and hundreds of brown dwarfs (BDs) and planetary mass objects (3-13 Jupiter mass free floating objects) have been identified. In addition, we have witnessed a revolution in the field of debris disk studies. Indeed extended surveys have identified about 300 stars that harbor debris dust, shedding light into the frequency of planetesimal formation (the building blocks of planets) and allowing the characterization of extra-solar planetesimal belts. Although a very interesting picture of this rich zoology is emerging, several important questions remain to be answered,

---

[1] Centro de Astrobiología (CAB-CSIC), Ctra. Ajalvir km. 4, Torrejón de Ardoz, Madrid, Spain

[2] Thuringer Landessternwarte, Tautenburg, Germany

[3] School of Cosmic Physics, Dublin Institute of Advance Studies, Ireland

[4] UNINOVA-CA3, Campus da Caparica, Caparica, Portugal

[5] Instituto de Astrofísica de Canarias, La Laguna, Spain

[6] Arcetri Observatory, Firenze, Italy

[7] Universidad Complutense de Madrid, Spain

[8] University of Central Florida, Physics Dept., Orlando, FL, USA

[9] Universidad de La Laguna, Facultad de Química Inorgánica, La Laguna, Spain

[10] Carnegie Institution, Washington D.C., USA

[11] Fac. de Ciencias, Universidade de Lisboa, Ed. C8, Campo Grande, Lisboa, Portugal

for example: What is the diversity of planetary systems, including long period planets (beyond the reach of radial velocity (RV) and transit surveys) and exotic systems? What is their dynamical evolution? What are the properties of exoplanets (orbits, masses and atmospheric properties)? What are the frequencies and properties of planetesimal belts and what are the implications for the frequency of terrestrial planet formation, for the habitability of the planetary systems and for water delivery into the terrestrial planet region? How common are Earth-like planets?

These, and other related questions, are of critical importance from an astrobiological point of view because they can help us place the evolution of the Solar System and the establishment of basic habitability conditions on at least one of its planets, the Earth, into perspective. Studies of extrasolar planets have revealed an astonishing diversity of masses, semi-major axes and eccentricities, from the short period hot Jupiters, to planets in very elongated orbits, to planetary systems with multiple Jupiter-mass planets, to the super Earth mass planets with orbital periods of a few days (Butler et al. 2006; Udry & Santos 2007). According to a recent report from the ExoPlanet Task Force (ExoPTF), organized jointly by the National Science Foundation and NASA (http://www.nsf.gov/mps/ast/exoptf.jsp), planet studies in the next decade must address three compelling questions: (1) What are the characteristics of Earth-mass/Earth-size planets in the habitable zone around nearby, bright stars? (2) What is the architecture of planetary systems? And (3), how do planets and planetary systems form? ExoPTF recommends that the near-term highest priority be given to improving RV precision and providing more telescope time to intensify RV studies with the goal of detecting (several) Earth mass planets around bright stars.

Very recently, a new population of low-mass planets in the Uranus to sub-Neptune mass range (5-21 Earth masses) with orbital periods of a few days, called super Earth mass planets, has been detected by the RV surveys (Mayor et al. 2009). The precision of these surveys has improved from 3-10 m/s before $\sim$2003 to $\sim$1 m/s afterwards. So far, a total of $\sim$20 such planets have been announced among $\sim$200 bright stars searched for super Earth mass planets, mostly using state-of-the-art high-resolution cross-dispersed echelle optical spectrographs such as HARPS at the ESO 3.6-m telescope and HIRES at the 10-m W. M. Keck Observatory. These discoveries usually require $\sim$30-100 observations with $\sim$1 m/s precision, since their velocity amplitudes are extremely small (a few m/s).

Although the sample of super Earths is still scarce, the observed properties show different trends with respect to the known giant planets. Unlike the giant planets, whose frequency scales as the square of the host-star metallicity (Gonzalez 2006), the super Earth frequency seems to weakly depend on the host-star metallicity. Also, the 3-day orbital period pile up observed for giant planets (Marcy et al. 2005) does not appear to be present for super Earths. These early results indicate that they may belong to a new planet population whose formation and evolution may be very different from that of the giant planets. For instance, they may be formed without accumulating a substantial amount of gaseous material, unlike the gas giant planets. In other words, they may be terrestrial rocky planets. Their lower masses may make them less capable of opening up gaps in protoplanetary disks, so they may have undergone a very different migration history than the gas giants, and some may have been formed in-situ.

Most (>80%) of the nearby stars and brown dwarfs are significantly cooler and less massive than the Sun. They have spectral types M, L, and T, corresponding to effective temperatures in their photospheres ranging from 3800 to 600K ("ultra-cool dwarfs"). They are very faint at optical wavelengths because the bulk of their luminosity is emitted at infrared wavelengths. Current strategies to search for exoplanets are focused in the optical wavelength domain, but it is very inefficient to observe ultra-cool dwarfs in the optical. As a consequence, exoplanet searches around ultra-cool dwarfs are limited to a small number of targets due to the large amount of telescope time needed to observe them.

A new generation of high-resolution near-infrared spectrographs capable of extending accurate Doppler measurements from the optical into the near-infrared (NIR) domain is currently under study (Martin et al. 2005; Oliva et al. 2004; Ramsey et al. 2008). To detect rocky planets in the habitable region around M dwarfs, the required RV precision is about 1 m/s, whereas it will take a precision of 10 cm/s to reach a similar science goal around G-type stars. GIANO for the Telescopio Nazionale Galileo in La Palma is the most advanced project (Oliva et al. 2004). It is a cross-dispersed echelle spectrograph providing a spectral resolution up to 46,000 and a very high spectral stability to measure RV with an accuracy of few m/s (better than 10 m/s). A reference spectrograph with very-high accuracy in the visible range is HARPS at the ESO 3.6-m telescope in La Silla, having a measured short-term stability of 0.2 m/s during one night, and a long-term stability of ∼1 m/s, these values being the present limit of standard technologies. Despite of the fact that HARPS only has one highly specialized observing mode, it is one of the most demanded instruments at ESO.

Infrared RV monitoring has been carried out for intrinsically red objects such as pre-main sequence stars, red giants, very low-mass stars and brown dwarfs. For example, Huélamo et al. (2008) used VLT/CRIRES to monitor TW Hya for 6 nights and reported an rms dispersion of 35 m/s, which did not confirm the presence of a short-period giant planet around this young star. CRIRES has also been used to follow a red giant for 5 hours with a RV accuracy of 20 m/s (Seifahrt & Kaüfl 2008) and some nearby late-M dwarfs during 6 months with an accuracy of 5 m/s (Bean et al. 2010). IRTF/CSHELL has been used to obtain RV measurements for a sample of young stars and reached a precision of 100 m/s (Prato et al. 2008). Gemini-South/PHOENIX was utilized during 5 nights to monitor a sample of L dwarfs and obtain rms dispersion of 300 m/s (Blake et al. 2007). Using Keck/NIRSPEC, Martin et al. (2006) obtained a dispersion of 110 m/s for the dM8 vB10 over 1 hour, and a dispersion of 360 m/s for the bdM9.5 LP944-20 over 6 nights. The same instrumental setup was used by Zapatero Osorio et al. (2007) to infer an rms dispersion of 500 m/s for a T4.5 dwarf over several days; and by the same team (Zapatero Osorio et al. 2009) to report a dispersion of 300 m/s for vB10 over seven years. The latter paper hinted at RV variability in VB10 in accordance with astrometric observations that indicated the presence of a giant planet around this star (Pravdo & Shaklan 2009), and make use of PHOENIX models that have provided good fits to T dwarf NIRSPEC spectra (del Burgo et al. 2009). However, a revisit of that data using deconvolution techniques has not confirmed the RV variability of VB10 and does not support the presence of a giant planet around it (F. Rodler et al., in preparation).

All of the previous work has made use of spectrographs that are not stabilized and do not have specific tools for high-precision RV calibrations. It has been demostrated by those authors that calibrations using telluric absorption lines and/or ThAr emission arcs can provide an accuracy down to 20 m/s and perhaps even 10 m/s. To improve this level of precision, the use of gas cells has been proposed (Martin et al. 2005; Mahadevan & Ge 2009) and is starting to be implemented (Ramsey et al. 2008; Bean et al. 2010). However, there are many issues involved in the future development of a new generation of high-precision near-infrared spectrographs that can yield accurate RV measurements that are comparable to the state-of-the-art in the optical. In the next section we will mention of some those issues and in the rest of the paper we will deal with the two main wavelength calibration options.

2. Current Status of NAHUAL

In September 2008, the fifth NAHUAL science meeting[1] was held in the dream Island of Fuerteventura, and the following optimal spectrograph for accurate NIR RV work was defined:

- Hyper-stabilization in pressure and temperature. No moving parts. The main problem with this requirement is to stabilize the spectrograph when operating at cryogenic temperatures.

- High quality detector with cosmetics, read-out noise and dark current comparable with CCDs.

- Accurate and simultaneous wavelength calibration. Probably more than one method should be used in the near-infrared. In the optical regime, two different wavelength reference systems are used: ThAr and an absorption cell. More recently the laser-comb method (Murphy et al. 2007) has been investigated in the laboratory but it has not been fully implemented yet for RV surveys. Moreover, it has been proven to work only in a narrow wavelength range.

- High enough spectral resolution to resolve the spectral lines; simulations indicated that a resolving power of at least 60,000 would be adequate for late-M dwarfs, and broad spectral coverage to maximize the amount of Doppler information in each spectrum.

- Homogeneous illumination of the spectrograph entrance. With HARPS image scrambling is used for this purpose. The image scrambler contains one optical system to transform the light at the exit of the optical fibre into parallel light, and another system that the re-injects the light into a second fibre. At the exit of the second fibre there is another optical system that changes the f-ratio to the one of the spectrograph.

---

[1]http://www.ucm.es/info/Astrof/nahual/index.html

In January 2010, a one year contract was signed between the NAHUAL team and Grantecan. The latter institution has provided funding for a one year conceptual design study. This concept will include the high-resolution hyper-stable mode, which will be fiber-fed. It will also include a mid-resolution multi-object mode (resolving power between 10,000 and 20,000). The cost envelope for the whole instrument is estimated at 6 million euros and the GTC would be accepting it by 2016.

3. What Wavelength Calibration in the Near-Infrared?

The main difference between the two currently used main wavelength calibration modes is that the spectral lines of the gas absorption cell are superimposed onto the spectrum of the star, whereas in the ThAr-method there are two beams: one containing the spectrum of the star, the other the calibration lamp. The ThAr-method works only if the internal variation of the shift between the two spectra is smaller than the desired accuracy of the RV-measurements. The mechanical requirements and the optical design of both systems are thus quite different.

In all cases the RV-accuracy that can be reached in the IR also depends on the quality of the IR detectors and the optical design of the spectrograph. Experiments have shown that a regular pattern on detector (like even-odd effect) has strong effects on accuracy. The optical design has to be made in such a way that the astrometric distortions can be calibrated with as high as possible a number of references lines.

In order to minimise the shift of the two spectra in the case of the ThAr-method, the light of the star and the light of the calibration lamp are fed into the spectrograph via fibres. In order to avoid that tracking errors of the telescope induce shifts of the spectra, instruments like HARPS use double image scramblers in addition to fibres. We will discuss these devices in more detail below.

A second disadvantage is that in principle two fibres are required: one for the starlight, a second for the ThAr lamp. Thus, the orders have to be sufficiently well separated, which requires a large dispersion of the cross-disperser, and the format of the detector has to be wide enough.

A third important issue is the number of spectral lines per wavelength interval. RV-measurements are essentially determinations of the position of lines on the detector. This means that it is possible to measure the "astrometric" distortion on the detector with sufficient accuracy if the number of lines available is sufficiently high. The atlas of Kerber et al. (2008) lists about 1500 ThAr-lines in the near-infrared domain. This number sets a strong constraint on the optical design. It is required to calibrate the distortions to an accuracy that the wavelength calibration has an error of the order of 1 m/s or less.

A fourth issue are aging effects of the ThAr-lamps. It is well known that the relative intensities of the ThAr-lines change with time. In the case of blended lines, this results in wavelength shifts. It is thus required to calibrate the lamp that is being used against some wavelength reference that is known to be constant. In the case of HARPS several lamps are used that are calibrated against each other. The key point is that the Argon lines are less variable than the Thorium lines. Another issue is the cross-talk between the fibres. In the case of

HARPS, the stability is so good that the simultaneous wavelength calibration is actually not used during the night; ThAr-spectra are only taken at the beginning and at the end of the night. In this way the problem of the cross-talk of the fibres is avoided. In the infrared the cross-talk is certainly smaller.

Summary of pros and cons of the ThAr-method:

- Pro: 1 m/s, even 0.3 m/s, stability demonstrated with HARPS.
- Con: Needs at least two fibres (star and calibration, separately).
- Con: Needs image scramblers.
- Con: Age effect of lamps.
- Con: Needs very high stability because calibration and star-light are in different fibres.
- Con: Sufficient number of lines is required in the wavelength range of operation.

For the ThAr-method, we require an optical system that works efficiently in the Y, J, and H bands. Estimating the efficiency of the scramblers is very difficult. In the case of the optical spectrograph SOFIE, there are two modes: one with scrambler and one without. The difference in efficiency between these modes is a factor of 2.40±0.02. Since both modes have a 3 arcsec entrance aperture, we may estimate the efficiency of the scrambler to be 42%. A similar figure is also estimated for HARPS. However, in order to obtain a better estimate of the efficiency of an image-scrambler in the IR, a test model is required. More details can be found in Queloz et al. (1999).

For the absorption cell method, in the optical regime, an iodine cell is used. Iodine produces a very dense grid of lines in the wavelength regime between 500 and about 620 nm. For higher temperatures of the cell, the lines extend further into the red but even with a cell temperature of 200 K, the lines only extend up to about 900 nm. The RV shift is obtained by modelling the observed spectrum of the star together with the superimposed forest of absorption lines. Very high-resolution spectra of the iodine cell and of the star are convolved with the PSF of the spectrograph. The latter determined iteratively during the process. In the model, the RV of the star and the instrumental shift are free parameters that are determined when the observed spectrum is modelled. In the optical regime, the very high-resolution spectrum of the star is usually obtained by de-convolving a spectrum of that star taken without the cell. Because the instrumental shift is obtained in the same beam as the starlight, the constraints on mechanical stability are much lower. Since the iodine cells are unsuitable in the infrared, it is required to use a different gas. Laboratory development of gas cells by our team will be presented elsewhere (L. Valdivielso et al., in preparation).

The pros and cons of the absorption cell method are summarized below:

- Pro: Only one fibre needed.
- Pro: No strict need for image scramblers, although they might be helpful.
- Pro: No known aging effect.

- Pro: No need of very high mechanical stability.
- Con: Gases, which produce a dense grid of lines, are needed.
- Con: Superposition of lines makes it difficult to use the spectrum for other purposes than RV.
- Con: Throughput loss because of cell continuum absorption.
- Con: Precision better than 1 m/s not demonstrated.

To summarize, our current understanding of this problem indicates that the ThAr method could work in the Y and J-bands, while the cell absorption method could be preferable in the H and K-bands because of the insufficient density of ThAr lines at those longer wavelengths. These studies are of interest not only for NAHUAL but also for other similar projects, such as CARMENES (Quirrenbach et al. 2009). Hence, we are considering the possibility of setting up a testbed for comparison of different calibration methods in the near future. Simulations of the RV precision indicate that for late-M dwarfs (e.g., Rodler et al. (2010)) there is significant Doppler information in all the near-infrared bands, and thus calibration methods for all of them should be pursued.

Acknowledgments. This work has been supported by the Spanish Ministerio de Eduación y Ciencia through grant AYA2007-67458 and by the Portuguese FCT grant PTDC/CTE-AST/68444/2006.

References

Bean, J. L., et al. 2010, ApJL, 711, 1, L19
Blake, C. H., et al. 2007, ApJ, 666, 1198
Butler, R. P. et al. 2006, ApJ, 646, 505
del Burgo, C., et al. 2009, A&A, 501, 1059
Huélamo, N. et al. 2008, A&A, 489, L9
Gonzalez, G. 2006 PASP, 118, 849, 1494
Kerber, F., Nave, G., & Sansonetti, C. J. 2008, ApJS, 178, 374
Mahadevan, S. & Ge, J. 2009, ApJ, 692, 1590
Marcy, G. et al. 2005, ApJ, 619, 570
Martín, E. L. et al. 2005, AN, 326, 1015
Martín, E. L. et al. 2006, ApJ, 644, L75
Mayor, M. et al. 2009, A&A, 493, 639
Murphy, M. T. et al. 2007, MNRAS, 380, 839
Oliva, E., Origlia, L., Maiolino, R., & Gennari, S. 2004, in Ground-based Instrumentation for Astronomy SPIE, 5492, 1274
Prato, L. et al. 2008, ApJ, 687, L103
Pravdo, S. H. & Shaklan, S. B. 2009, ApJ, 700, 623
Queloz, D., Casse, M. & Mayor, M. 1999, ASC, 185, 13
Quirrenbach, A. et al. 2009, arXiv:0912.0561
Ramsey, L. W. et al. 2008, PASP, 120, 887
Rodler, F. et al. 2010, this volume
Seifahrt, A. & Kaüfl, H. U. 2008, A&A, 491, 929
Udry, S. & Santos, ARA&A, 45, 397
Zapatero Osorio, M. R. et al. 2007, ApJ, 666, 1205
Zapatero Osorio, M. R. et al. 2009, A&A, 505, L5

Pathways Towards Habitable Planets
ASP Conference Series, Vol. 430, 2010
Vincent Coudé du Foresto, Dawn M. Gelino, and Ignasi Ribas, eds.

Infrared Detection and Characterization of Debris Disks, Exozodiacal Dust, and Exoplanets: The FKSI Mission Concept

W. C. Danchi,[1] R. K. Barry,[1] B. Lopez,[2] S. A. Rinehart,[1] O. Absil,[5] J.-C. Augereau,[3] H. Beust,[3] X. Bonfils,[3] P. Bordé,[4] D. Defrère,[5] P. Kern,[3] P. R. Lawson,[6] A. Léger,[4] J.-L. Monin,[3] D. Mourard,[2] M. Ollivier,[4] R. Petrov,[2] W. A. Traub,[6] S. C. Unwin,[6] and F. Vakili[2]

Abstract. The Fourier-Kelvin Stellar Interferometer (FKSI) is a mission concept for a nulling interferometer for the near-to-mid-infrared spectral region. FKSI is conceived as a mid-sized strategic or Probe class mission. FKSI has been endorsed by the Exoplanet Community Forum 2008 as such a mission and has been costed to be within the expected budget. The current design of FKSI is a two-element nulling interferometer. The two telescopes, separated by 12.5m, are precisely pointed (by small steering mirrors) on the target star. The two path lengths are accurately controlled to be the same to within a few nanometers. A phase shifter/beam combiner (Mach-Zehnder interferometer) produces an output beam consisting of the nulled sum of the target planet's light and the host star's light. When properly oriented, the starlight is nulled by a factor of 10^{-4}, and the planet light is undiminished. Accurate modeling of the signal is used to subtract the residual starlight, permitting the detection of planets much fainter than the host star. The current version of FKSI with 0.5-m apertures and waveband 3-8 μm has the following main capabilities: (1) detect exozodiacal emission levels to that of our own solar system (Solar System Zodi) around nearby F, G, and K stars; (2) characterize spectroscopically the atmospheres of a large number of known non-transiting planets; (3) survey and characterize nearby stars for planets down to 2 $R_\oplus$ from just inside the habitable zone and inward. An enhanced version of FKSI with 1-m apertures separated by 20 m and cooled to 40 K, with science waveband 5-15 μm, allows for the detection and characterization of 2 $R_\oplus$ super-Earths and smaller planets in the habitable zone around stars within about 30 pc.

---

[1] NASA Goddard Space Flight Center, Greenbelt, MD, USA

[2] Observatoire de la Côte d'Azur (OCA), Boulevard de l'Observatoire, Nice, France

[3] Laboratoire d'Astrophysique de Grenoble (LAOG), Observatoire de Grenoble, 414, Rue de la Piscine, Domaine Universitaire, Saint-Martin d'Héres, France

[4] Institut d'Astrophysique Spatiale (IAS), Centre universitaire dOrsay, Orsay, France

[5] Université de Liége, Liége, Belgium

[6] Jet Propulsion Laboratory, California Institute of Technology, Pasadena, CA, USA

1. Introduction

A great deal of progress has been made in the study of exoplanetary systems in the last decade, with more than 400 planets discovered. Some key questions of immediate importance are: (1) What are the atmospheres of non-transiting extrasolar planets composed of, and what are the physical conditions in their atmospheres? (2) How many 2-$R_\oplus$ planets are in our solar neighborhood and what are their atmospheres composed of? (3) What is the amount of exozodiacal dust around nearby stars? Is it comparable to that of our own solar system? (4) What is the spatial distribution of debris material around stars like our sun, how is it sculpted by the presence of planets, and how does this material form and evolve? The thermal infrared spectral region has distinct advantages compared to the visible region because the star-to-planet contrast ratio in the habitable zone is about 3 orders of magnitude less than at visible wavelengths, and the thermal emission from warm (room temperature) material emits most strongly in the infrared.

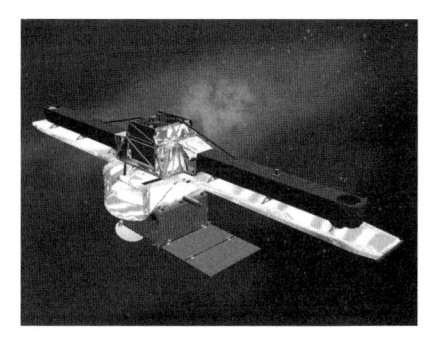

Figure 1. Artist's conception of current FKSI system as in operation at L2.

The Fourier-Kelvin Stellar Interferometer (FKSI) is a structurally connected infrared space interferometer with 0.5 m diameter telescopes on a 12.5-m baseline, and is passively cooled to 60 K. FKSI operates in the thermal infrared from 3-8 μm in a nulling (or starlight suppressing) mode for the detection and characterization of faint material around relatively bright stars such as exoplanets, debris disks, emission levels of extrasolar zodiacal dust disks. Figure 1 displays an artist's conception of the spacecraft in operation at L2. FKSI will have the highest angular resolution of any infrared space instrument ever made with its nominal resolution of 40 mas at a 5 μm center wavelength. This resolution exceeds that of Spitzer by a factor of 38 and JWST by a factor of 5. The FKSI mission is conceived as a Probe class or mid-sized strategic mission that utilizes technology advances from flagship projects like JWST, SIM, Spitzer, and the technology programs of TPF-I/Darwin.

2. History and Current Status

The FKSI mission concept has been under development for a number of years at NASA's Goddard Space Flight Center (Danchi et al. 2003; Danchi & Lopez 2007) and it has a budget that fits into the strategic mission or Probe-class category with a lifecycle cost of around $600-800 million. During the last few years technology development funded by NASA and ESA for TPF-I, Darwin, and JWST have retired most of the major risks. Most of the key technologies are at a Technical Readiness Level (TRL) of 6 or greater, which means that, if funded, FKSI could enter into Phase A within the next two years (Danchi et al. 2008). The FKSI mission is designed to answer major scientific questions on the pathway to the discovery and characterization of Earth-twins around nearby F, G, and K main sequence stars.

The FKSI mission is realistic, cost-effective, and low risk. First, it has been studied extensively since 2002 and its baseline design is at an advanced state of development. Consequently a detailed mass chart exists, which has been used in connection with both grass roots and parametric cost estimates. Second, it is cost effective because many of its components can be adapted from or simply copied from those of JWST. These include sun shades, cryocoolers, detectors, mirrors, and precision cryogenic structures. Indeed these components are at a TRL of 6 since JWST passed its Technical Non-Advocate Review (TNAR) and is in Phase C/D. Third, the only significant components that are below TRL 6 are optical fibers used for wavefront cleanup and the nuller instrument itself both of which require cryogenic testing. These final technology activities can be accomplished during Phase A.

The FKSI mission has broad support in the US and European communities and has been endorsed by the Exoplanet Community Forum as a Probe class or mid-sized strategic mission by the US community. In France, FKSI has been included in the plans by the French space agency (CNES) as a mission of opportunity. It has also been included in plans by the European Space Agency (ESA) in a similar context.

3. Performance Estimates for the Current FKSI Design

One of the most vexing questions in the study of exoplanetary stellar systems is the location and amount of emission of warm dust in the habitable zone of these stars, analogous to the zodiacal light in our own solar system. This emission around nearby stars is called exozodi emission and is measured in units relative to that of our own solar system (defined as one solar system zodi, or SSZ). Current consensus in the exoplanet community is that a small (non-flagship) space-based mission is required to measure exozodi levels to that required to properly scope and specify a flagship mission (see for example the panel summary on this topic in this volume by Absil et al. 2010). Figure 2 is a comparison of the ability of ground- and space-based concepts to measure this emission for nearby solar type stars. This figure demonstrates that FKSI is the only instrument capable of measuring exozodi levels down to one SSZ in the habitable zone, and given the relatively short integration times required, within a few months of operation FKSI can measure the zodi levels for all TPF and Darwin target stars of interest.

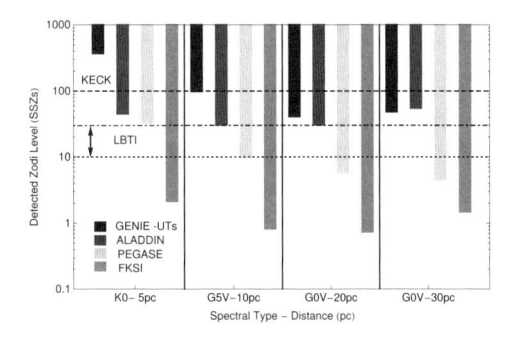

Figure 2. Comparison of exozodiacal detection limits for two ground-based concepts (GENIE with UTs on the VLTI and ALADDIN), and two space mission concepts (PEGASE and FKSI). Four different test cases were studied and displayed from left to right in the figure: a K0 star at 5 pc, a G5V star at 10 pc, a G0V star at at 20 pc, and a G0V star at 30 pc. The three vertical lines separate the four cases (adapted from Defrère et al. 2008).

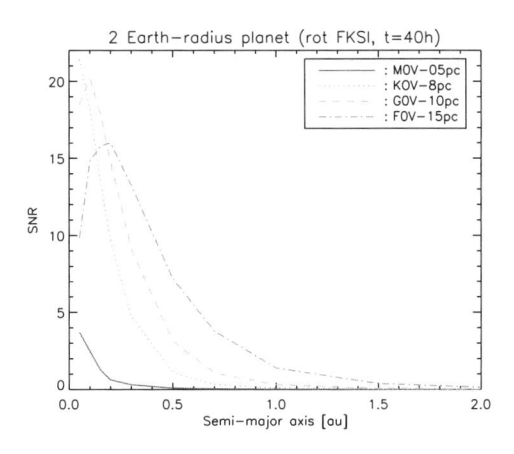

Figure 3. Simulations for 2 $R_\oplus$ super-Earth detection show that FKSI can detect many rocky planets around nearby F, G, K, and M stars Defrère (private communication).

Another major scientific area is characterizing the atmospheres of exoplanets discovered through radial velocity and other techniques. Currently only a small fraction of exoplanet atmospheres can be studied spectroscopically using transit methods, and these exoplanets are largely the ones with short periods that are very close to their stars. Not being limited to transiting planets, FKSI will greatly extend the size of the sample, which will have an enormous impact on models of planet formation and evolution. A third area is the search for rocky planets of size 2-$R_\oplus$ super-Earths or smaller in the habitable zone of nearby stars. Figure 3 displays a simulation of the FKSI's capability for super-Earth detection.

4. Performance Estimates for an Enhanced Design

As part of our submission to the Astro2010 Request for Information for the Program Prioritization Panel (PPP) on Electromagnetic Observations from Space (EOS), we began to look at an enhanced version of FKSI, with parameters more ideally suited for exoplanet characterization, namely moving the wavelength band from 3-8 μm to 5-15 μm, thus being centered on 10 μm, near the emission peak of the Earth's spectrum.

Figure 4 displays the discovery space of FKSI for exoplanets as a function of semi-major axis and compares it to other missions. The enhanced version of the FKSI system, with its approximate discovery space marked by the dark dashed lines, has a greatly enhanced phase space for detection compared to the current design, and is the only mission within the projected Probe cost cap that is capable of detecting Earth-twins in the habitable zone of nearby stars. Figure 5 displays an engineering model of the enhanced design of FKSI, which has been shown to be technically feasible and within the projected cost envelope of the Probe class mission category.

We used the TPF performance simulator of Dubovitsky & Lay (2004) and computed the number of 1-$R_\oplus$ and 2-$R_\oplus$ sized planets we could detect using standard assumptions that were used for similar studies of the flagship TPF Interferometer mission, as a function of aperture size from 1-m to 2-m diameter, with a 40 K telescope temperature, 20-m separation, and $\pm$ 45deg field-of-regard. The key result from these simulations is that even with a 1-m aperture diameter, we were able to detect and characterize up to four 1-$R_\oplus$ Earth-twins around very nearby G and K stars, and detect 34 2-$R_\oplus$ super-Earths and characterize 16 of them. Figure 6 displays a summary comparison of key design features of the current design compared to those of the enhanced design, as well as the assumptions, and the key results of the simulation. Figure 7 displays the results of a more complex numerical simulation of the performance of FKSI for super-

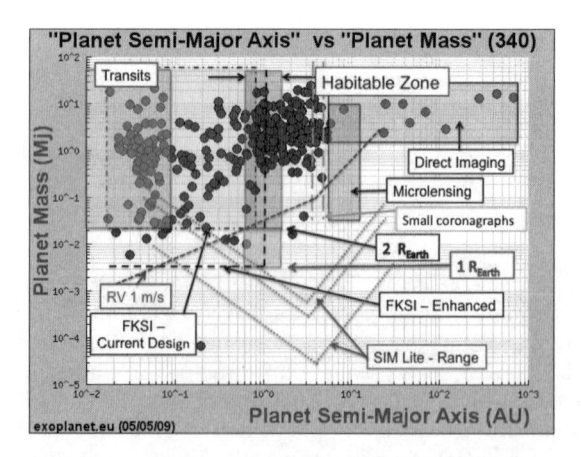

Figure 4. Discovery space for exoplanets for FKSI and other mission concepts and techniques.

FKSI Mission Concept 193

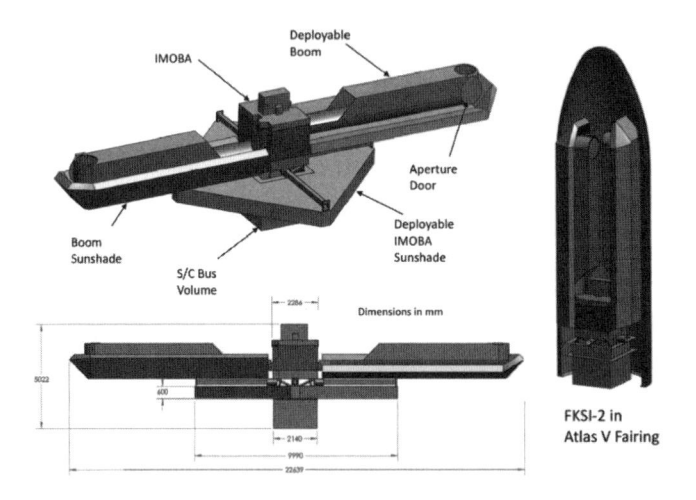

Figure 5. Engineering model of the enhanced FKSI design, called FKSI-2.

Enhancements to Current Design Allow Detection of Earth-like Planets

Current design			Enhanced design		
Telescope diameter	0.5 m		Telescope diameter	from 1 to 2 m	
Baseline	12.5 m		Baseline	20 m	
Wavelength range	from 3 to 8 µm		Wavelength	from 5 to 15 µm	
Telescope temperature	down to 60 K		Telescope temperature	down to 40 K	
Field of regard / Sun shade	+/- 20 °		Field of regard / Sun shade	> +/- 45 °	

Basic Assumptions:
- SNR = 5 for detection
- SNR = 10 for spectroscopy (R = 20 at 10 µm)
- 3 visits
- < 2 years total
- < 7 days total per star
- T_{earth} = 288 K
- Earth albedo = 0.3
- Inclination angle of planet orbit = 45°
- Sunshade FOR = +/- 45°

Reference: "TPF Interferometer Performance Model," Dubovitsky & Lay 2004.

Enhanced design, Tel = 1 m					
R_{Planet}	Total	N_F	N_G	N_K	N_{Spec}
1 R_{Earth}	4	0	1	3	4
2 R_{Earth}	34	6	16	12	16
Tel = 1.5 m					
R_{Planet}	Total	N_F	N_G	N_K	N_{Spec}
1 R_{Earth}	15	0	7	8	4
2 R_{Earth}	95	35	48	12	27
Tel = 2.0 m					
R_{Planet}	Total	N_F	N_G	N_K	N_{Spec}
1 R_{Earth}	29	3	14	12	12
2 R_{Earth}	138	65	61	12	43

Figure 6. Simulations of FKSI performance with 1-2 m class telescopes at 40K and a 20-m baseline demonstrate that many 2 $R_\oplus$ super-Earths and a few Earth-twins can be discovered and characterized within 30 pc of the Sun.

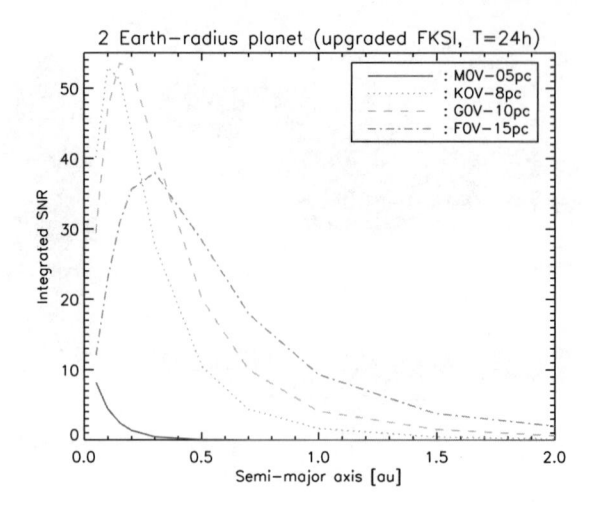

Figure 7. Simulations for 2 $R_\oplus$ super-Earth detection for the enhanced FKSI (1-m apertures, 40 K telescope temperature, 20-m separation) show that FKSI can detect many rocky planets *in the habitable zone* around nearby F, G, K, and M stars (Defrère private communication).

Earth detection (see Defrère et al. (2008) for more details of the simulation), which is consistent with the results presented in Figure 6.

References

Absil, O., et al., 2010, in this volume
Danchi, W. C., et al., 2003, ApJ, 597, L57
Danchi, W. C., & Lopez, B. 2007, Compte Rendu Physique, 8, 396
Danchi, W. C., et al. 2008, SPIE, 7013, 70132Q
Defrère, D., private communication
Defrère, D., et al., 2008, A&A, 490, 435
Dubovitsky, S. & Lay, O., 2004, SPIE, 5491, 284

Pathways Towards Habitable Planets
ASP Conference Series, Vol. 430, 2010
Vincent Coudé du Foresto, Dawn M. Gelino, and Ignasi Ribas, eds.

Japanese Space Activity on Exoplanets and SPICA

T. Nakagawa

Institute of Space and Astronautical Science, JAXA, 3-1-1 Yoshinodai, Sagamihara, Kanagawa, Japan

Abstract. We review Japanese activities on space science missions for the study of stellar and planetary formation with special emphasis on the SPICA (Space Infrared Telescope for Cosmology and Astrophysics) mission and its impact on the study of exoplanets. With a 3m-class telesope (3.5 m in the baseline design) cooled to 6K, SPICA is optimized for mid- and far-infrared astronomy. One of the focal plane instrumets proposed for SPICA is a stellar coronagraph for the study of exoplanets, protoplanetary and debris disks. SPICA's telescope has a monolithic structure and is expected to have smooth point spread function, which should be a great advantage of coronagraphy. SPICA is an international mission, and European participation in SPICA is approved as a candidate for future missions under the framework of ESA Cosmic Vision. The target launch date is 2018.

1. Space Science Programs in Japan

In the first section, we present an overview of current and future space astrophysics programs in Japan and their relations with the study of stellar and planetary formation.

1.1. Infrared Astrophysics

AKARI AKARI (ASTRO-F) is the first Japanese satellite dedicated to infrared astronomy, and was launched in 2006 (Murakami et al. 2007). AKARI has a 68.5 cm cooled telescope, together with two focal-plane instruments, which survey the sky in six wavelength bands from the mid- to far-infrared. The instruments also have the capability for imaging and spectroscopy in the wavelength range 2 – 180 μm in the pointed observation mode. The in-orbit cryogen lifetime was one and a half years. The All-Sky Survey covered more than 90 percent of the whole sky with higher spatial resolution and wider wavelength coverage than that of the previous IRAS all-sky survey. The pointed observations are used for deep surveys of selected sky areas and systematic observations of important astronomical targets. AKARI ran out of its stored liquid helium in 2007, and is now in the "warm mission" phase, in which AKARI is dedicated to near-infrared observations with mechanical cryocoolers.

AKARI has been making systematic observations of star-forming regions both in the All-Sky Survey and pointed observation modes. The AKARI All-Sky Survey (Figure 1) showed a class of new debris disk sources. AKARI made systematic observations debris disk sources also in pointed observations mode.

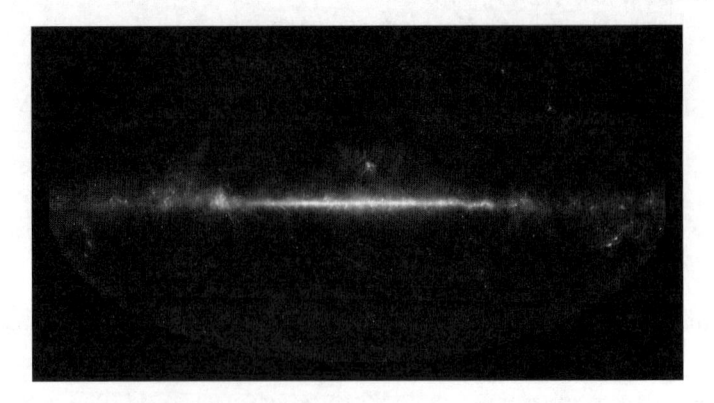

Figure 1. AKARI all-sky image at 9 μm.

AKARI also observed various sources in our solar system, which are expected to served as references for the study of exoplanets.

SPICA Following the highly successful AKARI mission, the next-generation mission SPICA (Space Infrared Telescope for Cosmology and Astrophysics; Figure 2) mission (Nakagawa 2008) is now being studied as one of JAXA's "pre-projects" in Japan. SPICA is an astronomical mission with a cryogenically cooled 3m-class (3.5m in the baseline design) telescope optimized for mid- and far-infrared astronomy. To reduce the mass of the whole mission, SPICA carries no cryogen. SPICA will be launched at ambient temperature and cooled down on orbit by on-board mechanical coolers with an efficient radiative cooling system. This combination enables a 3m-class cooled (< 6 K) telescope in space.

European participation in SPICA has been studied as one of candidates for future European missions under the frame work of the "ESA Cosmic Vision 2015-2025". The target year of the SPICA launch is 2018.

Because of its high spatial resolution and unprecedented sensitivity in the mid- to far-infrared, SPICA can address a number of key problems in current astrophysics, including characterization of exoplanets through direct spectroscopic observations. Expected contributions by SPICA on the study of exoplanets will be discussed in detail in a separate section.

JTPF Japanese Terrestrial Planet Finder (JTPF) (Tamura et al. 2010) is the Japanese activity aimed at directly detecting terrestrial planets in other solar systems. Currently, two concepts are being studied: one is an optical high contract telescope with coronagraphic capability and the other is an infrared interferometric mission. JTPF is proposed to be conducted under international collaboration.

1.2. High-Energy Astrophysics

Suzaku Suzaku (ASTRO-EII) is the fifth Japanese X-ray astronomy mission and was launched in 2005 (Mitsuda et al. 2007). Suzaku has been performing various kinds of observational studies for a wide variety of X-ray sources, over

Figure 2. SPICA in orbit.

a wider energy range (from 0.3 to 600 keV) than ever before achieved. Suzaku has been observing X-ray emission from various sources including protostars and star-forming regions.

MAXI Monitor of All-sky X-ray Image (MAXI) (Matsuoka et al. 2009) is the first astrophysical payload which is mounted on the Japanese Experiment Module (JEM) Exposed Facility of International Space Station (ISS). MAXI started its observation in 2009. MAXI is designed to monitor the variability of a large number of X-ray sources.

ASTRO-H ASTRO-H (Takahashi et al. 2008) is the next-generation X-ray observatory mission aiming at comprehensive understanding of high-energy non-thermal phenomena by means of extra-fine spectroscopy in X-ray band below 10 keV and imaging spectroscopy in the 10-80 keV band. ASTRO-H is expected to make both high-resolution spectroscopy and wide-band imaging observations of protostars and T-Tauri stars. ASTRO-H is proposed to be launched in 2013.

1.3. Radio Astrophysics

ASTRO-G Following the success of the first space VLBI mission "HALCA", JAXA is now working on the ASTRO-G (VSOP-2) mission (Saito et al. 2009). ASTRO-G, together with ground-based VLBI arrays, will be able to image the innermost parts of both protostars and AGN accretions disks. ASTRO-G will have a deployable 9-m off-axis paraboloidal antenna operating at 8, 22, and 43 GHz. The 22 and 43 GHz bands will use cooled receivers for good sensitivity. With an apogee height of 25,000 km, 43 GHz observations will have an angular resolution of 38 μ-arcseconds.

2. SPICA and Exoplanets

2.1. Exoplanet Characterization in the Infrared

Among the missions mentioned above, SPICA has the largest impact on the study of exoplanets. In this section, we briefly describe the SPICA's capability in the study of exoplanets' atmosphere. Please see SPICA Study Team Collaboration 2009 for more details.

The mid-infrared region (from ~ 3 to 30 μm) is very important in the study of planetary atmospheres as it spans both the peak of thermal emission from the majority of exoplanets so far discovered, and is particularly rich in molecular features that can uniquely identify the composition of planetary atmospheres and trace the fingerprints of primitive biological activity. SPICA has unique capability to characterize the atmospheres of exoplanets through the application of infrared spectroscopy in two ways: direct spectroscopy and transit spectroscopy.

2.2. Direct Imaging and Spectroscopy of Exoplanets

SPICA Coronagraph Instrument (SCI) is a unique instrument proposed for SPICA as one of focal plane instruments. Enya et al. (2010) discuss its detailed design. Here we concentrate on its scientific capability.

The most important capability of SIC is that it can undertake "direct" spectroscopy of exoplanets in the critical MIR domain (using a grism/prism providing R $\sim 20 - 200$) in addition to imaging. This unique capability is enabled by (1) a simple and clean PSF of the SPICA telescope because of its monolithic mirror and (2) SPICA's actively cooled telescope (< 6 K), which is optimised for mid- and far-infrared astronomy.

This spectroscopic capability in a continuous wavelength domain rich in chemical signatures ($\sim 3.5 - 27$ μm) represents a unique science possibility of SPICA.

Key population of outer and young exoplanets can not be studied through transit experiments but will be observed spectroscopically only by SCI. In fact, in the next decade, SCI will be the only instrument available to characterize outer and cool exoplanets through direct spectroscopy in the MIR, which will play a key role in constraining the temperature and atmospheric composition of exoplanets.

SCI utilizes a simple binary mask type approach for its coronagraphy (Enya et al. 2010) and SCI is expected to achieve a contrast of 10^{-6} at the equivalent to ~ 9 AU ($\sim$ Saturn's orbit) at 5 μm for a star at 10 pc (IWA ~ 1").

At this wavelength we probe the younger end of the planet age range (~ 100 Myr to 1 Gyr, see Figure 3). SCI is expected to observe ~ 200 targets and produce an "spectral atlas" of several outer exoplanets, therefore completing the discovery and characterization space of other telescopes and methods. According to their expected flux at long wavelengths ($\lambda > 5\mu$m), the main targets for SCI spectroscopy will be $\sim 2 - 3 M_{\rm Jup}$ planets around solar-type stars and lower mass planets, $\sim 1 M_{\rm Jup}$, around M dwarfs. An integration time of about one hour is required for each exoplanet.

SCI will also perform imaging surveys integrating a few minutes per target (both young M stars: ~ 300 Myr, < 50 pc; and FGK stars: ~ 1 Gyr, < 25 pc).

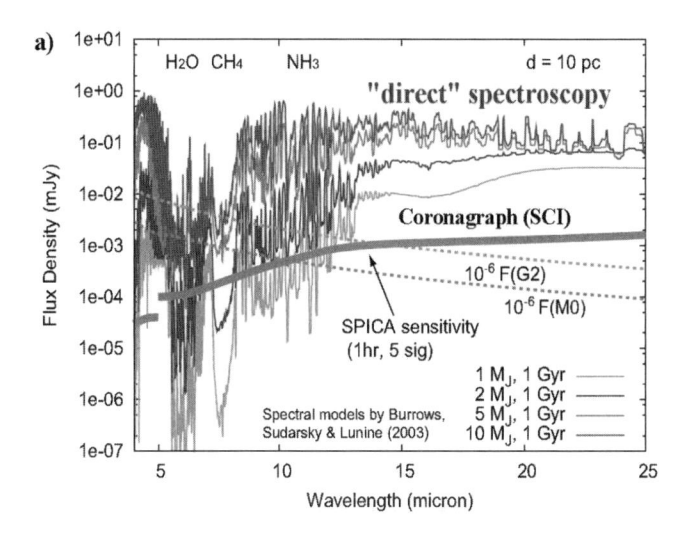

Figure 3. Simulated spectra for a range of exoplanet masses at an age of 1 Gyr and a distance of 10pc. The sensitivity of SCI (SPICA Coronagraph Instrument) is also shown.

The most significant atmospheric features expected in the exoplanet spectra can be summarized as follows:

1. The $\sim 4-5\mu$m "emission bump" due to an opacity window in exoplanets and cooler objects with temperatures between 100 and 1000 K

2. Molecular vibration bands of H_2O ($\sim 6-8\mu$m), CH_4 (~ 7.7 μm), O_3 (~ 9.6 μm), silicate (~ 10 μm), NH_3 (~ 10.7 μm), CO_2 (~ 15 μm) and many other trace species such as hydrocarbons. Among them, O_3 is regarded as one of the most important bio-markers. The relative abundance of all these species could be compared among different exoplanets and with solar system planets. Note that giant planets like Saturn in the solar system show strong NH_3, PH_3 and H_2O features around ~ 5 μm, which is within SCI's spectral coverage. SPICA will also have access to the PAHs band features which are excellent tracers of UV-radiation fields.

3. He-H_2 and H_2-H_2 collision induced absorption band features as tracer of the He/H relative abundance.

4. Features to distinguish cool brown dwarfs from "real" exoplanets such as those from deuterated molecular species (e.g., CH_3D at $\sim 8.6\mu$m) and non-equilibrium species (e.g., PH_3 at ~ 8.9 and ~ 10.1 μm). Were the spectral coverage to be extended down to 3.5 μm then H_3^+ (~ 4 μm), which is the key to understand the cooling processes in the upper atmosphere, would also be accessible.

2.3. Transit Photometry and Spectroscopy of Exoplanets

The SPICA MIRACLE, MIRMES and MIRHES instruments will cover the mid infrared range with low, medium, and high spectral resolution (up to $R \sim 30000$). Following the successful observations of hot-Jupiters by Spitzer, these instruments will be used to study primary and secondary transits of exoplanets orbiting close to the star. SPICA's spectrometers will be used to perform (1) multi-wavelength transit photometry of hot Jupiters routinely, and (2) transmission and occultation spectroscopy of an appropriate sample of bright exoplanets.

Hot-Jupiters, in particular, are bright and their detailed atmospheric spectroscopy will allow us to study exo-atmospheres in great detail. This will help to develop the techniques and models that will be needed to interpret the spectra of habitable exo-Earths in the future. We estimate that the 24 μm thermal emission of gas giant planets similar to HD 209458b could be extracted from secondary transit observations around stars as far as $\sim$ 150 pc (a few hundred star targets) as the contrast requirement is relatively modest ($\sim 0.1\%$).

Moreover, super-Earths could be observed by SPICA. The location of the habitable zone (i.e., the distance from the host star to a planet where liquid water could exist) scales with $\sim L_{star}^{1/2}$, and thus M stars have habitable zones much closer ($\sim 0.02 - 0.5$ AU) than Sun-like stars. Hence super-Earths around M stars would be suitable for transit observations. A super-Earth (2-3 R_{earth}; $T_p \sim 300$ K) orbiting in the habitable zone around a cool M8 star (T $\sim$2500K) will produce an intrinsic flux of $\sim 25\mu$Jy at MIR wavelengths, roughly a few times higher than the projected 1σ-1min photometric sensitivity of SPICA's MIR instruments. These photometric flux levels are accessible for SPICA secondary transit studies with a contrast of $\sim 0.1\%$.

SPICA mid-infrared spectra of transiting exoplanets will provide a unique opportunity to determine the properties of their atmospheres through detailed analysis of band profiles, e.g., to extract molecular abundances and isotopic ratios accurately.

In summary, SPICA will be an important milestone in exoplanet research, both in science achievements (direct spectroscopy of outer exoplanets for the first time in the MIR, atmospheric characterization of transiting exoplanets at long wavelengths) and in the required technological developments for future longer-term missions (e.g., coronagraphic techniques, cryogenic systems).

References

Enya, K., et al. 2010, this volume
Matsuoka, M. et al. 2009, PASJ, 61, 999
Mitsuda, K. et al. 2007, PASJ, 59, S1
Murakami, H. et al. 2007, PASJ, 59, S369
Nakagawa, T. 2008, SPIE, 7010, 70100H
Saito, H. et al. 2009, ASPC, 402, 25
SPICA Study Team Collaboration 2009, SPICA Assessment Study Report for ESA Cosmic Vision 2015-2025 Plan, arXiv:1001.0709
Takahashi, T. et al. 2008, SPIE, 7011, 70110O
Tamura, M. et al. 2010, this volume

SOFIA: On the Pathway toward Habitable Worlds

R. D. Gehrz,[1] D. Angerhausen,[2] E. E. Becklin,[3] M. A. Greenhouse,[4] S. Horner,[5] A. Krabbe,[2] M. R. Swain,[6] and E. T. Young[3]

Abstract. The U.S./German Stratospheric Observatory for Infrared Astronomy (SOFIA), a 2.5-meter infrared airborne telescope in a Boeing 747-SP, will conduct 0.3 - 1,600 μm photometric, spectroscopic, and imaging observations from altitudes as high as 45,000 ft., where the average atmospheric transmission is greater than 80 percent. SOFIA's first light cameras and spectrometers, as well as future generations of instruments, will enable SOFIA to make unique contributions to the characterization of the physical properties of proto-planetary disks around young stellar objects and of the atmospheres of exoplanets that transit their parent stars. We describe several types of experiments that are being contemplated.

1. Overview

The Stratospheric Observatory for Infrared (IR) Astronomy (SOFIA; figure 1), a joint project of NASA and the German Aerospace Center (DLR), is a 2.5-m

figure 1. SOFIA checkout flight in May 2007 (left) and nighttime flight line operations tests at Palmdale, CA in March, 2008 (right). NASA photos.

---

[1] Department of Astronomy, School of Physics and Astronomy, University of Minnesota, Minneapolis, MN, USA

[2] Deutsches SOFIA Institut (DSI), University of Stuttgart, Stuttgart, Germany

[3] Universities Space Research Assoc., NASA Ames Research Center, Moffett Field, CA, USA

[4] NASA Goddard Space Flight Center, Greenbelt, MD, USA

[5] Lockheed Martin, Sunnyvale, CA, USA

[6] JPL/Caltech, Pasadena, CA, USA

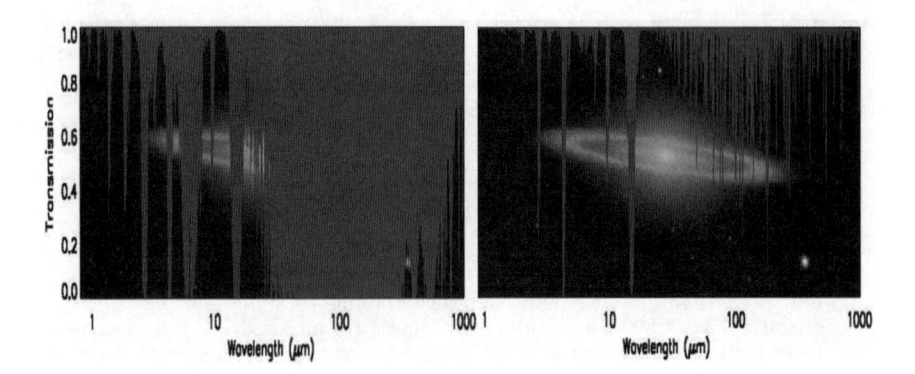

Figure 2. Left: Atmospheric transmission versus wavelength on a good night at Mauna Kea, HI (13,800 ft. MSL) Right: Typical atmospheric transmission at an altitude of 45,000 ft. from SOFIA. Courtesey of NASA/JPL-Caltech.

telescope in a Boeing 747SP aircraft designed to make sensitive 0.3 μm to 1.6mm measurements of astronomical objects. SOFIA will fly as high as 45,000 feet (13.72 km), above 99.8% of the obscuring atmospheric H_2O vapor where the atmospheric transmission is $\geq$ 80% transmission over its entire operating range (Figure 2). The home base for the SOFIA aircraft (*Clipper Lindbergh*) is NASA's Dryden Aircraft Operations Facility in Palmdale, CA. SOFIA will also operate from other bases worldwide to enable observations at any declination and to facilitate timely observations of transient events such as extrasolar planet transits. First science flights will commence in 2010, with regular science operations following in 2012 and continuing until the mid-2030s. Detailed descriptions of the specific scientific and operational advantages of the SOFIA observatory, its development and deployment schedule, and opportunities for participation by observers and instrument developers are described more fully by Gehrz et al. (2009), and Becklin and Gehrz (2009).

2. The SOFIA Observatory and Telescope

The SOFIA Observatory and telescope characteristics are summarized in Figure 3. The telescope is a bent Cassegrain with a 2.7-m (2.5-m effective aperture) parabolic primary mirror and a 0.35-m diameter hyperbolic secondary mirror. The scientific instruments (SIs) operate at a f/19.6 Nasmyth infrared focus fed by a 45° dichroic tertiary mirror that transmits the optical light. A flat behind the dichroic feeds the visible light into an optical focal plane guiding camera system, the Focal Plane Imager (FPI), at a second Nasmyth focus. Two other independent imaging and guiding cameras, attached to the front ring of the telescope, are available: the Wide Field Imager (WFI) and the Fine Field Imager (FFI). The dichroic tertiary can be replaced by a fully reflecting tertiary for applications requiring maximum throughput at the short wavelengths. The secondary mirror can be chopped with amplitudes up to $\pm$ 5' at frequencies between 1 and 20 Hz.

SOFIA: On the Pathway toward Habitable Worlds 203

Nominal Operational Wavelength	0.3 to 1600 μm	Diffraction Limited Wavelengths	≥ 15 μm
Primary Mirror Diameter	2.7 meters	Optical Configuration	Bent Cassegrain with chopping secondary mirror and flat folding tertiary
System Clear Aperture Diameter	2.5 meters	Chopper Frequencies	1 to 20 Hz for 2-point square wave chop
Nominal System f-ratio	19.6	Maximum Chop Throw on Sky	+/- 5 arcmin (unvignetted)
Primary Mirror f-ratio	1.28	Pointing Stability	= 1.0" rms at first light = 0.2" rms in operations
Telescope's Unvignetted Elevation Range	20 to 60 degrees	Pointing Accuracy	= 0.5" with on-axis focal plane tracking
Unvignetted Field-of-View Diameter	8 arcmin	Total Emissivity of Telescope (Goal)	15% at 10 μm with dichroic tertiary 10% at 10 μm with aluminized tertiary
Image Quality of Telescope Optics at 0.6 μm	1.5 arcsec on-axis (80% encircled energy)	Recovery Air Temperature in Cavity (and Optics Temperature)	= 240 K
Diffraction Limited Image Size	0.1" • λ μm FWHM		

Figure 3. SOFIA system characteristics.

3. SOFIA Instrumentation and Sensitivity

SOFIA's eight first generation SIs (Figure 4) include high speed photometers, imaging cameras, and spectrographs capable of resolving both broad features due to dust and large molecules, and molecular and atomic gas lines at km/s resolution. Three US facility class SIs (FLITECAM, FORCAST, and HAWC)

SOFIA Instrument	Description	Built by / PI	λ range (μm) spec res (λ/Δλ)	Field of View Array Size	Available
FORCAST	Faint Object InfraRed CAmera for the SOFIA Telescope Facility Instrument - Mid IR Camera and Grism Spectrometer	Cornell T. Herter	5 - 40 R ~ 200	3.2' x 3.2' 256 x 256 Si:As, Si:Sb	2010
GREAT	German Receiver for Astronomy at Terahertz Frequencies PI Instrument - Heterodyne Spectrometer	MPIfR, KOSMA DLR-PR, MPS R. Güsten	60 - 200 R = 10⁶ - 10⁸	Diffraction Limited Single pixel heterodyne	2010
FIFI LS	Field Imaging Far-Infrared Line Spectrometer PI Instrument w/ facility-like capabilities - Imaging Grating Spectrometer	MPE, Garching A. Poglitsch	42 - 210 R = 1000 - 3750	30"x30" (Blue) 60"x60" (Red) 2 - 16x5x5 Ge:Ga	2010
HIPO	High-speed Imaging Photometer for Occultation Special PI Instrument	Lowell Obs. E. Dunham	.3 - 1.1	5.6' x 5.6' 1024x1024 CCD	2012
FLITECAM	First Light Infrared Test Experiment CAMera Facility Instrument - Near IR Test Camera and Grism Spectrometer	UCLA I. McLean	1 - 5 R~2000	8.2' x 8.2' 1024x1024 InSb	2012
CASIMIR	CAltech Submillimeter Interstellar Medium Investigations Receiver PI Instrument - Heterodyne Spectrometer	Caltech J. Zmuidzinas	200 - 600 R = 3x10⁴ - 6x10⁶	Diffraction Limited Single pixel heterodyne	2012
HAWC	High-resolution Airborne Wideband Camera Facility Instrument - Far Infrared Bolometer Camera	Univ of Chicago D. Harper	50 - 240	Diffraction Limited 12x32 Bolometer	2013
EXES	Echelon-Cross-Echelle Spectrograph PI Instrument - Echelon Spectrometer	UT/UC Davis NASA Ames M. Richter	5 - 28 R = 10⁵, 10⁴, or 3000	5" to 90" slit 1024x1024 Si:As	2013

Figure 4. SOFIA's first light instrument complex.

will be maintained, operated, and their pipeline data archived by the science staff of the SOFIA Science and Mission Operations Center (SSMOC) for the general science community. Three US principal investigator (PI) class SIs (CASIMIR, EXES, and HIPO) and two German PI class SIs (FIFI LS, GREAT), will be maintained and operated by the PI teams at their home institutions, with general investigators using the SIs in collaboration with the PI team. FIFI LS will also be available to US observers as a facility-like SI under a special arrangement with the PI team. SOFIA can rapidly incorporate instrumentation upgrades in response to new technological developments. It is expected that there will be a new instrument or instrument upgrade per year for the lifetime of the program. The low ambient flight altitude temperature and 2.5-meter aperture of the SOFIA telescope will enable it to match the sensitivity of the Infrared Space Observatory (ISO) at high spectral resolution (Figure 5). SOFIA's spectrometers will be able to detect important atomic and ionic fine structure lines as well as ro-vibrational transitions of many simple molecules such as H_2O, CH_4, and C_2H_2 that exist in protoplanetary disks and exoplanet atmospheres.

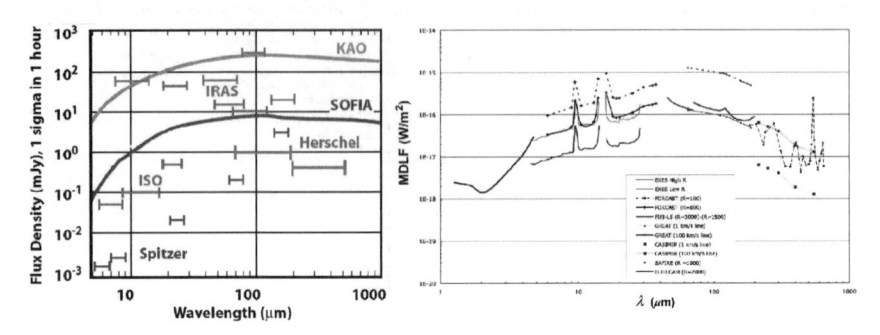

Figure 5. Photometric sensitivity in mJy, 1σ in 1 hour (left) and line sensitivities with spectrometers in Wm^{-2}, 10σ in 900 seconds.

4. SOFIA Studies of Exoplanetary Systems: Protoplanetary Disks

A detailed understanding of the origin of the solar system requires studies of circumstellar disks around young stellar objects (YSOs) where planets may be forming. SOFIA's spectrometers will elucidate the kinematics, composition, and evolution of these disks. For example, water (H_2O), the dominant reservoir of oxygen under nebular conditions, is important in the formation and early evolution of planetary systems, and water ice condensation should dominate the mass budget of newly-formed planetesimals. Therefore, studies that can reveal the origin and distribution of liquid and frozen water in proto-planetary disks are crucial to our understanding of both the abundance of water on terrestrial planets in the inner habitable zones around stars and the formation of comet nuclei and giant planets in the outer reaches of the disks. The resolved line profiles of atomic and molecular species in combination with Kepler's law will yield the distribution of the water and other biogenic molecules in the emitting layers of the disk (see Figure 6). Observations of 2.0 to 2.4 μm water emission lines from very hot gas close to the protostar in have already revealed the power of such molecular line studies. The 6 μm region is sensitive to the warm gas in the terrestrial planet zone and near the outer zones where ices form. This region is yet to be fully explored but is accessible to SOFIA. The strength of the lines will provide direct measurements of the temperature and column density of water in these disks. The technique can be applied to other molecules such as CH_4, and C_2H_2.

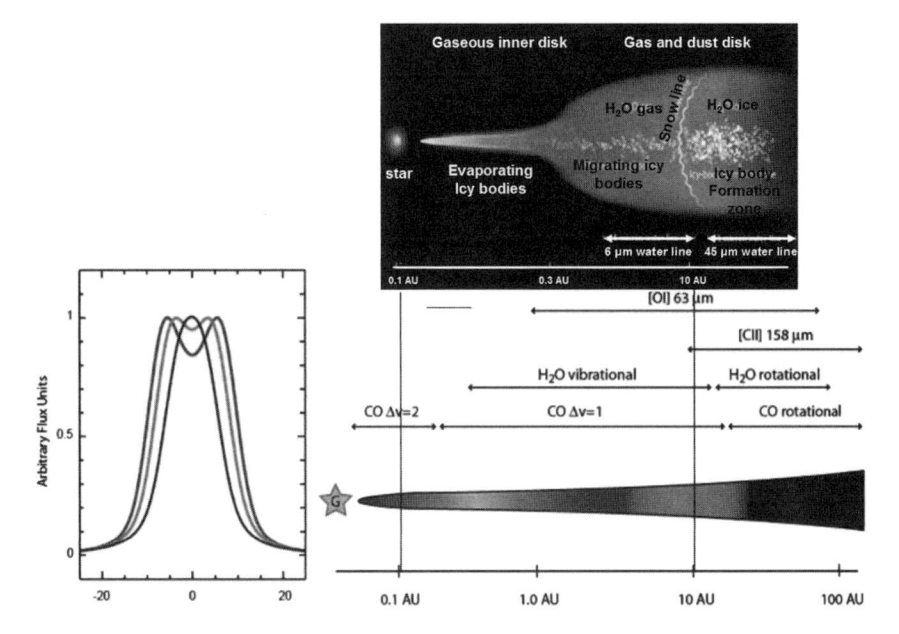

Figure 6. Spatial distribution of the three phases of water (upper right) and models of velocity-resolved lines in a disk (lower left). Doppler separation of the peaks gives the location within the disk of the chemical causing the line emission (lower right) Based on models by van Boekel (2007) and Kamp & Dullemond (2004).

5. SOFIA Studies of Exoplanetary Systems: Transits of Exoplanets

SOFIA flies above the scintillating component of the atmosphere so that several of SOFIA's first generation instruments have the potential to make high signal to noise observations of the transits of exoplanets of their parent stars. Extrapolation of groundbased models to SOFIA's service altitude suggests that a 15 minute integration will yield a scintillation noise floor of 4×10^{-5} (Sandell et al. 2003). HIPO is designed to provide simultaneous high-speed, time- resolved imaging photometry at two optical wavelengths. FLITECAM will provide moderate resolution grism spectroscopy 2000) from 1 to 5.2 μm. HIPO and FLITECAM can be mounted simultaneously to enable data acquisition at two optical and one near-IR wavelength on the same object in the sky. An additional advantage of SOFIA is its ability to fly to the optimal geographical observing location for exoplanet transits to observe an entire event continuously with a very stable platform. Spectroscopic observations of molecular emission and absorption can be used to probe the chemical composition and physical structure of the atmosphere of an exoplanet as it transits or is eclipsed by its parent star. In fact, such observations are emerging as the most powerful tool available for characterizing exoplanet atmospheres. The detection of biogenic molecules such as H_2O, CO_2, CH_4, and C_2H_2 in exoplanet atmospheres, an important goal for understanding the origins and evolution of habitable planetary systems, has al-

206   *Gehrz et al.*

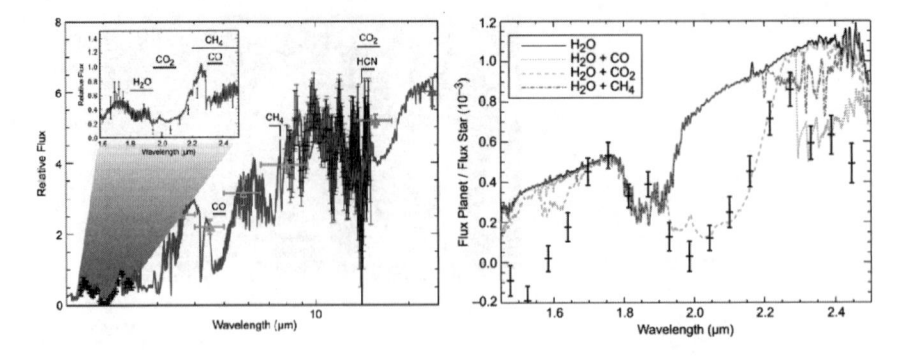

Figure 7. Synthetic spectra of hypothetical exoplanet atmospheres with significant abundances of CH_4, H_2O, and CO_2. These atmospheres produce significant absorption and emission features at wavelengths obscured from the ground but available to SOFIA. Data are for the hot-Jupiter HD 189733b. Figures from Swain et al. 2009 (see the captions and references therein for details). Right hand panel by kind permission of M. R. Swain et al. (2009).

ready been accomplished for a hot, transiting hot-Jupiter exoplanet with both HST and the Spitzer Space Telescopes (Figure 7). These molecules are believed to be the progenitors and building blocks of life, and the detection of their presence in an exoplanetary atmosphere is an important step towards the eventual detection of these same molecules in planets that might be habitable. Detection of these molecules in both transits (which probe the terminator at relatively high pressure depths in the atmosphere, e.g. 10^{-2} to 10^{-4} bar) and eclipses (which probe the day-side photospheric emission deep in the atmosphere at pressure-depths from 0.01 to 1 bar) in the near-IR has allowed some degree of knowledge about the diurnal atmospheric conditions. These observations have shown detectable spectroscopic variability that should be explored further to evaluate diurnal and seasonal effects. Our expectation is that the spectral resolution and sensitivity of SOFIA's instruments will enable us to answer the detailed questions about exoplanets that are only possible with molecular spectroscopy.

Acknowledgments. We thank the SOFIA team for their tireless effort. R. D. G. was supported by USRA, NASA, and the US Air Force.

References

Becklin, E. E., & Gehrz, R. D. 2009, SPIE, 7453, 745302-01
Gehrz, R. D., et al. 2009, AdSpR, 44, 413
Kamp, I., Dullemond, C.P. 2004, ApJ, 615, 991
Sandell, G., Becklin, E. E., & Dunham, E. W. 2003, in ASP Conf. Ser. 294 Scientific Frontiers in Research on Extrasolar Planets, eds. D. Deming & S. Seager, (San Francisco: ASP) 591
Swain, M. R. et al. 2009a, Astro2010: The Astronomy & Astrophysics Decadal Survey, Science White Paper no. 291
Swain, M. R. et al. 2009b, ApJ, 704, 1616
van Boekel, R. 2007, Nat, 447, 535

Probing the Extrasolar Planet Diversity with SEE-COAST

P. Baudoz,[1] A. Boccaletti,[1] J. Schneider,[2] R. Galicher,[2] and G. Tinetti[3]

[1] *Observatoire de Paris/LESIA UMR 8109, 5 Place Jules Janssen, Meudon, France*

[2] *Observatoire de Paris/LUTH, 5 Place Jules Janssen, Meudon, France*

[3] *University College London, Gower Street, London, UK*

Abstract. SEE-COAST is a space mission concept designed for the study of physical and chemical characteristics of exoplanets. The science driver of SEE-COAST is the exploration of the planetary diversity, which has been unveiled in the past 15 years. Both spectral and polarimetric properties of the parent starlight reflected by the planets are analyzed to better constraint the atmosphere of planets. The SEE-COAST concept relies on a series of high contrast imaging techniques like coronagraphy, differential imaging and possibly wavefront control. The science objectives as well as the wavelength range (0.4-1.25 microns) are complementary to the future near-infrared ground-based coronagraphic instruments and to the exoplanet imaging capability of the JWST. We propose SEE-COAST as a relatively secure step to prepare the more ambitious characterization mission of Earth-like planets. The strategy of the mission is presented and the instrumental concept is briefly introduced.

1. Introduction

Following the discovery of a giant planet orbiting the solar-type star 51 Peg (Mayor & Queloz 1995), planet hunters have detected more than 400 exo-worlds (http://exoplanet.eu). While radial velocity surveys led to the discovery of the majority of the known planets, transit search surveys are starting to produce a large number of detections using both ground-based telescopes and space-based instruments (CoRoT and Kepler). Other techniques like microlensing or pulsar timing are further helping to increase the number and diversity of known exoplanets. Finally some planetary mass objects have already been imaged in some favorable conditions from the ground with current 8-m class telescopes equipped with AO facilities and from space using Hubble Space Telescope. These measurements are mostly for very young systems with large separations (>100 AU) and small mass ratios (Lagrange et al. 2009; Marois et al. 2008; Kalas et al. 2008). Even if the number of planets identified using direct detection is small (11 planets), several instruments dedicated to direct detection are being developed and should increase this number. In a few years (2010-2011), the same ground-based telescopes will benefit from extreme AO systems with advanced coronagraphic devices and differential imaging techniques, like SPHERE at the VLT (Beuzit et al. 2008) and GPI on Gemini (Macintosh et al. 2008). These instruments will be specialized for the search and spectral-polarimetric characterization of young and/or massive planets in the solar neighborhood achieving contrast of 10^6 to 10^7. In some particular cases Jupiter mass planets could become detectable if physically close to their very nearby host stars (0.5-1 AU) to increase the reflected light contribution. On the longer term (2018-2020), high contrast imaging instruments like EPICS (Kasper et al. 2008) on the Extremely Large Telescopes (ELT) are foreseen to reach higher contrast ($10^8 - 10^9$). We expect spectral characterization of giant gas planets to be feasible in the near IR and polarimetric characterization in the visible. Ideally, a few Super Earths would become detectable and possibly roughly characterized. METIS, the mid-IR instrument studied for the European ELT is also a possible contributor to planet hunting (Pantin et al. 2008).

As for space, by 2015 JWST will probably reach the realm of mature massive planets (1-5 Gyr) and will be able to characterize some of them on 5 to 10 AU orbits and in the 2-15 μm spectral range (Boccaletti et al. 2005; Baudoz et al. 2006a; Green et al. 2005). On the longer term, the strategy for direct detection projects from space has been revisited lately and a need for a clear roadmap has been expressed worldwide. This exercise has been done twice in the US, with the Exoplanet Task Force and the Exoplanet Forum. Similarly in Europe, two teams have been appointed (one by ESA, the ExoPlanet Roadmap Advisory Team and one by the community, the Blue Dots Team) both with the same objective of tracing a roadmap towards the ultimate goal: spectroscopy of terrestrial planets in the habitable zone. Several techniques are being proposed in this context. While an ambitious space mission is necessary to perform deep characterization of planetary atmospheres, the exact technical solution is not yet clearly defined. Indirect detections are needed to provide the proportion of planets around stars, the distribution of their mass, orbit, and multiplicity. It is also mostly accepted that the spectral characterization of Earth-like planets may require previous mission(s) to identify the targets or to better define the stellar environment (exo-zodiacal dust). These missions have to be considered in the roadmap and phased accordingly to the final characterization mission(s), which is (are) the final ambitious goal of this roadmap.

2. The Science Case: A Huge Variety of Planets

The detection of the first exoplanetary systems has shown a diversity of mass and orbital distributions and has indicated that the solar system is not representative of planetary systems in general. Globally, these discoveries include more than 30 multi-planetary systems and have brought to light the existence of planets with a huge variety of characteristics, opening unexpected questions about planet formation processes. Planet searches have revealed planets with orbital periods as short as 0.85 days (Léger et al. 2009), or as long as hundreds of years (Kalas et al. 2008; Marois et al. 2008). Some of the planets are on eccentric orbits more typical of some comets in the solar system. While the most recently discovered planet has a diameter less than twice the Earth diameter (Léger et al. 2009), some planets have more than 15 times the mass of Jupiter. According to the planet formation theories which were used 15 years ago, a lot of these objects were not supposed to exist. Diversity and open-mindedness must thus be the keywords for future exoplanet exploration.

Importantly, a consensus is now emerging that habitable planets are not restricted to Earth-sized bodies: super-Earths with masses up to tens of Earth's mass may well be habitable. Super-Earths are defined as planets for which terrestrial concepts apply: ocean/continents, planet tectonics, volcanism, habitability. As described in the previous section, current and committed projects from ground and space are not capable of studying these super-Earths, except for the ELTs which presumably will obtain some rough characterization in the near-IR. We suspect that these worlds are numerous and many detections are still to come from radial velocity (RV) surveys. Therefore we consider that a mission optimized to study this class of objects is at the moment a valuable program rather than simply an identification mission for finding Earth-mass planets. In the framework of exoplanet exploration, a space telescope operated in the visible would be very much complementary to what is foreseen on ground-based projects. Therefore, this approach has the advantage to provide scientific data to characterize planets and explore their diversity without the need of a previous mission to find these targets. This was precisely the purpose of the SEE-COAST (Super Earth Explorer - Coronagraphic Off-Axis Space Telescope) proposal to the ESA Cosmic Vision in 2007. Several mission concepts for direct detection were proposed at that time but none were selected. A new call for proposals will be issued in a few years and in the meantime we intend to progress with the concept definition of SEE-COAST.

3. SEE-COAST Science Drivers

The core science program of SEE-COAST is to explore this diversity of planets especially focusing on super-Earths and mature Jovian planets. These classes of objects represent an intermediate topic between the capabilities of ground-based instruments and future long-term missions like DARWIN. Therefore, a mission like SEE-COAST could be operated in parallel to ELTs on the ground with the objective to contribute to the characterization of the lowest mass gaseous planets and the most massive telluric ones. Setting such an ambitious goal requires not only spectroscopic characterization but also measuring the polarization of light reflected by the planet. The observation strategy is not a survey mode to search for or just detect a large number of planets, but an extensive characterization mode of a limited number of planets previously detected by RV and by astrometry (PRIMA/ESPRI at the VLT and Gaia). We believe that one snapshot is not sufficient to understand how planets work, and therefore emphasize observing time variations. A precise and continuous time variation measurement is unique to space-based telescopes and is especially interesting for clouds and planet rotation measurements.

The SEE-COAST program focuses especially on the following physical parameters :

- *Mass* : The mass is derived from RV and astrometric measurements coupled with several observations along the orbit to constrain the inclination angle of the planet orbit.

- *Radius* : The radius case is more delicate if it is not derived from the few cases of transits in front of nearby stars. Otherwise the estimation of the radius will rely on a comparison between the atmospheric model and measured spectrum.

- *Atmosphere* : The visible and near-infrared spectrum is rich and provides spectral features of water, oxygen, methane, and CO_2 (at 1.25 µm). We plan to use both spectral and polarimetric information to better constrain the planetary characteristics.

- *Surface* : In the case of super-Earths, the albedos of oceans and/or continents could be measured, as well as variation during the planet rotation.

4. Mission Concept

Based on astrophysical requirements, we can derive the technical aspects of the SEE-COAST project. The philosophy is a simple, compact telescope and spacecraft to reduce cost and development time, the complexity and requirements for achieving high contrast being relayed to the focal instrument. A 1.5-2 m class off-axis telescope is needed to reach a reasonable number of targets (giant planets and close-by super-Earths) provided that high contrast can be obtained at 2 or 3 λ/D between 0.4 and 1.25 µm. For the focal instrument, spectroscopy and polarimetry are necessary. It is probably necessary to separate the visible and near-IR parts into two channels for technical issues (detectors, coatings, dimensioning, etc). Here, spectroscopy refers to spectral-imaging to maximize detection capabilities in a reasonable field of view and with a modest spectral resolution of 40 to 80. For that, the heritage of SPHERE and EPICS will be a major advantage for the technical aspects and instrumental modeling.

In the previous SEE-COAST study we identified three main critical aspects or sub-systems, namely: (1) the wavefront errors (WFE), (2) the system to suppress the starlight (coronagraph), and (3) the system to perform calibration and hence further improvement of the contrast. WFE requirements can be obtained with good optical quality at the primary mirror but the implementation of a deformable mirror (DM) may be mandatory. The implementation of a DM in a space telescope has already been studied in the context of TPF-C (Trauger & Traub 2007) and SPICA (Enya et al. 2008). While the two solutions are different, wavefront correction for space telescope is becoming feasible and should be considered in our approach for assessment.

The coronagraphic device should provide high contrast ($10^6 - 10^7$), high throughput (>50%) and smooth chromaticity. In that context, several concepts exist, many are being prototyped across the world, and some have already demonstrated capabilities that are compatible with our goal. Among them, the Annular Groove Phase Mask (Mawet et al. 2005), the Multi-Stage 4 Quadrant Phase Mask (Baudoz et al. 2008), and the Phase-Induced Amplitude Apodization Coronagraph (Guyon et al. 2005) are quite promising. The contrast requirement should be compatible with the flux ratio of the planets we are looking at, $\approx 10^8 - 10^{10}$. Such performance has been met already in laboratory demonstrations (Trauger & Traub 2007; Baudoz et al. 2008).

In addition to wavefront quality that ensures a low speckled halo, improvement of rejection is feasible by the use of differential imaging, e.g., by using spectral, polarimetric, or coherence characteristics or possibly a combination of

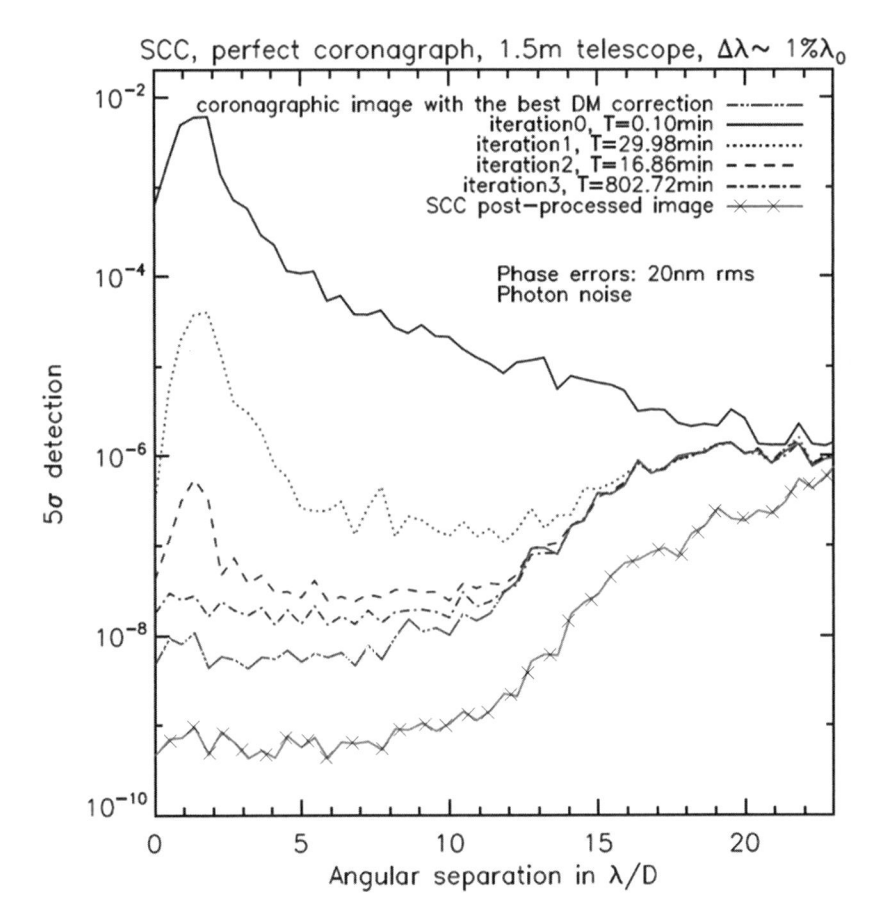

Probing the Extrasolar Planet Diversity with SEE-COAST 211

Figure 1. Self-Coherent Camera (SCC) performance for SEE-COAST: 5 σ detection as a function of angular separation in λ/D calculated with the same hypotheses than in Galicher et al. (2008) except for the telescope diameter (1.5 m instead of 8m). The second lowest curve shows the lowest reachable detection after a perfect correction using a 32x32 actuator DM. This level is rapidly reached thanks to the Self-Coherent Camera (SCC) ability to directly measure wavefront defects in the final focal plane. The SCC further helps to improve the detection level by calibrating the residual speckles (lowest curve).

them. A significant gain is to be expected although a thorough system design compatible with such techniques is required. A technical solution that combines a new promising concept of wavefront measurement and correction using a DM and calibration of the residual speckles is the Self-Coherent Camera (Baudoz et al. 2006b; Galicher et al. 2008, 2010), for which the performance on a 1.5 m telescope is illustrated in Figure 1.

Preliminary performance of SEE-COAST was addressed with simulations for the Cosmic Vision proposal and more details of the capability of the SEE-COAST project can be found in Schneider et al. (2009).

5. Conclusion

To explore planetary diversity we propose a coronagraphic mission with a modest size telescope (1.5 to 2m), that can accomplish high contrast imaging by coupling low wavefront errors, coronagraphy, and calibration of the post-coronagraphic residual speckles. This mission, called SEE-COAST, is proposed to measure the starlight which is reflected by exoplanets, from the far blue to the near infrared, with a spectral resolution of 40 to 80. The spectral distribution of the light and its degree and direction of linear polarization will be recorded to characterize the physical-chemical properties of giant planets and some super Earths. SEE-COAST is an interesting precursor mission before more ambitious space missions like large visible coronagraphic (4-8 m class) missions, large infrared interferometers, or external occulters. In addition, the characterization in the visible spectral range, only feasible from space, is complementary to the near-infrared and mid-infrared instruments that are foreseen in the near future on ground-based telescopes.

References

Baudoz, P., Boccaletti, A., Riaud, P., et al. 2006a, PASP, 118, 765
Baudoz, P., Boccaletti, A., Baudrand, J., & Rouan, D. 2006b, in IAU Colloq. 200 Direct Imaging of Exoplanets: Science and Techniques, ed. C. Aime & F. Vakili, (Cambridge: Cambridge University Press), 553
Baudoz, P., Galicher, R., Baudrand, J., & Boccaletti, A. 2008, SPIE, 7015, 70156C
Beuzit, J.-L., Felt, M., Dohlen, K. et al. 2008, SPIE, 7014, 701418
Boccaletti, A., Baudoz, P., Baudrand, J. et al. 2005, Adv. Space Res., 36, 1099
Enya, K., Abe, L., Haze, K. et al. 2008, SPIE, 7010, 70102Z
Galicher, R., Baudoz, P., & Rousset, G. 2008, A&A, 488, 9
Galicher, R., Baudoz, P., Rousset, G., Totems, J., & Mas, M., 2010, A&A, 509, A31
Green, J., Beichman, C., Basinger, S. et al. 2005, SPIE, 5905, 58050L
Guyon, O., Pluzhnik, E., Galicher, R. et al. 2005, ApJ, 622, 744
Kalas, P., Graham, J., Chiang, E. et al. 2008, Sci322, 1345
Kasper, M., Beuzit, J.-L., & Verinaud, C. 2008, SPIE, 7015, 70151S
Léger, A., Rouan, D., Schneider, J. et al. 2009, A&A, 506, 287
Lagrange, A.-M., Gratadour, D., Chauvin, G. et al. 2009, A&A, 493, L21
Macintosh, B., Graham, J., Palmer, D. et al. 2008, SPIE, 7015, 701518
Marois, C., Macintosh, B., Barman, T. et al. 2008, Sci, 322,1348
Mawet, D., Riaud, P., Absil, O., & Surdej, J. 2005, ApJ, 633, 1191
Mayor, M., & Queloz, D., 1995, Nat, 378, 355
Pantin, E., Siebenmorgen, R., Cavarroc, C., & Sterzik, M. 2008, SPIE, 7014, 70142D
Schneider, J., Boccaletti, A., Mawet, D. et al. 2009, Exp. Astronomy, 23, 357
Trauger, J. & Traub, W. 2007, Nat, 446, 771

Pathways Towards Habitable Planets
ASP Conference Series, Vol. 430, 2010
Vincent Coudé du Foresto, Dawn M. Gelino, and Ignasi Ribas, eds.

The SIM Lite Mission

M. Shao
California Institute of Technology, Jet Propulsion Laboratory, Pasadena, CA, USA

1. Introduction: The Changing Landscape of Exoplanet Science

There are over 400 exoplanets that have been discovered to date. Most of them were detected by radial velocity (RV) down to a size of 4.5 M_{Earth} in orbits of a few days. In space, the Kepler mission is in orbit and operating successfully. For the future, Kepler's discoveries will be coupled with the James Webb Space Telescope (JWST) to obtain spectra of hundreds of Jovian-like planets. Also there is considerable technology progress for direct imaging from space as well as proposals for all-sky transit missions. On the ground, extreme adaptive optics (AO) coronagraphs on 8-10m telescopes are expected to come on-line in 2010/2011 and will yield spectra of self luminous Jupiters. In the far future, perhaps, spectra of giant planets in reflected light will be obtained with giant ground-based telescopes (TMT, ELT).

The next major (space) advance in exoplanet research is the discovery of terrestrial planets in the habitable zone. The Space Interferometry MIssion (SIM Lite) plays a pivotal and unique role. The SIM Lite (Figure 1) mission is ready to be started in the coming decade. All the technology needed for an astrometric mission is complete. The SIM Lite mission is capable of the detection of planets and the estimation of planetary mass. After discovery with SIM Lite, follow-on large space coronagraphs and/or interferometers will look for biosignatures via spectroscopic characterization of these terrestrial Earth-clones.

Figure 1. Artist's impression of SIM Lite.

To be accepted as a valid discovery of a earth-like, terrestrial planet, the claimed detection must satisfy three criteria: the probability of a false alarm must be low (<< 10%); the location in the habitable zone MUST be made on the basis of a measured orbit and not a guess or estimate; and finally, the mass estimate MUST be a true measurement and not a guess or estimate from other properties such as photometric colors. SIM Lite discoveries of terrestrial planets in the habitable zones of nearby solar-type stars will meet these criteria.

2. Detection of Exoplanets with SIM Lite

We will examine the SIM Lite mission and describe its capabilities and its potential for detecting planets and measuring masses. SIM Lite is unique in its ability to estimate mass apart from a planet's color and brightness. Since SIM Lite determines masses and distances directly, SIM Lite has the ability to calibrate planetary masses, as well as the parent star's luminosity and age relationships.

SIM Lite is an optical Michelson interferometer with a 6-meter baseline. Fringe detection with a CCD allows measurement of wavefront delays, leading to an angular precision of 1.0 μas in a single measurement and a systematic error floor below 0.1 μas for longer integration times as verified with the SIM optical test bed at JPL. In about 20 minutes SIM Lite can measure angular positions of stars to within 1.0 μas for stars as faint as 10th mag (in V) relative to 10th mag reference stars within approximately 1 degree.

The planet detection thresholds for SIM Lite follow directly. As a benchmark, an Earth-mass planet orbiting 1 AU from a solar mass star at 6 pc induces an angular wobble with a radius of 0.5 μas. SIM Lite can detect this wobble since its demonstrated systematic error floor is below 0.1 μas. With N measurements spanning a few orbits, the detection improves as the square root of N because the signal's periodicity is coherent and inconsistent with random or systematic errors.

SIM Lite detects the presence of a planet by measuring the periodic reflex motion of the star as the planet orbits the star. Ideally, the motion is observed for at least one orbital period. If the orbital period is longer than the duration of observations, planet detectability is degraded due to confusion with proper motion. After a series of observations has been made, we may compute the periodogram power with the detection threshold set so that the false alarm probability is less than 1%.

The discovery of exoplanets with SIM Lite consists of three major survey components. The first is a habitable planet survey focusing on the nearest ~60 stars. By "habitable" we mean planets between 0.3 and 10 times the mass of the Earth, orbiting in the habitable zone, which ranges from 0.7 to 1.4 AU from the star, scaled appropriately for stellar luminosity. The second component is a broad planet survey of around 2100 stars which will have approximately 4 times lower astrometric accuracy but will still be able to find terrestrial mass planets in five year orbits around low mass stars. The third component is a survey focusing on Jupiter/Saturn mass planets orbiting very young stars, some of which are young enough that the planet formation and migration processes may still be underway.

3. Exoplanet Discovery Space

Figure 2 shows the exoplanet discovery space for current radial velocity (RV) capabilities, for ESA's GAIA mission, and for SIM Lite. One obtains a high detection sensitivity by concentrating 40% of the observing time over a subset of stars. While radial velocity (RV) detection is more sensitive to close-in orbits, astrometry is more sensitive for more distant orbits (Figure 2). Around stars more luminous than the Sun, an Earth in the habitable zone (HZ) can be detected by SIM Lite in a large volume of space in the solar neighborhood.

SIM Lite could detect Earth-mass planets orbiting at mid-habitable zone around every one of the most favorable (∼60) stars in the solar neighborhood. Many of the terrestrial planets found by SIM Lite can be followed up by ambitious direct imaging observations made with telescopes on the ground and in space. Moreover, SIM's direct measurement of the astrometric wobble yields the mass of the planet with no $sin(i)$ ambiguity. Equally important, the full three-dimensional orbit is derived from SIM Lite astrometry, yielding the orbital eccentricity of single planets and the coplanarity of multiple planets in a system. Coplanarity is often assumed, but without any observational evidence. SIM Lite measures coplanarity unequivocally.

SIM Lite can also help to establish how planetary systems form and evolve. For instance, astrometry can find gas giants within 1-5 AU of parent stars, and make mass measurements which are critical to evolutionary models. These measurements help to answer the questions; what fraction of young stars have gas-giant planets? Do gas-giant planets form at the "water-condensation" line?

By contrast, ground-based adaptive optics (AO) imaging can find distant planets (> 10 AU), but are unable to determine either their mass or detect any inner planets. Effects such as stellar variability, spectral jitter and rotation

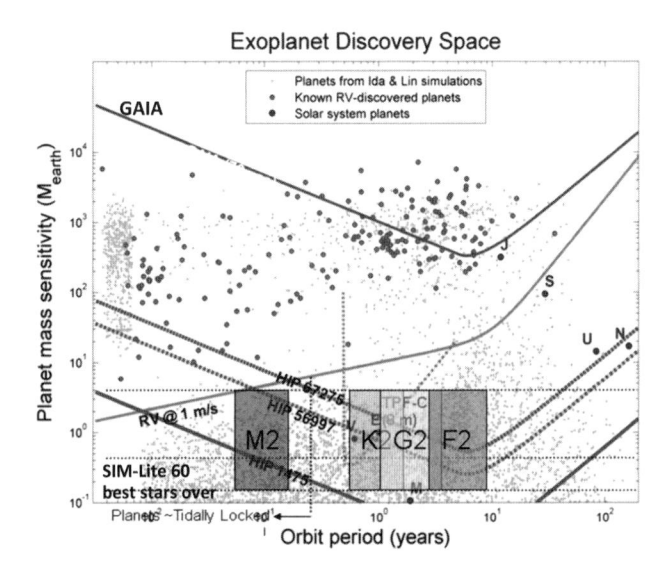

Figure 2. Exoplanet discovery space compared to other space missions.

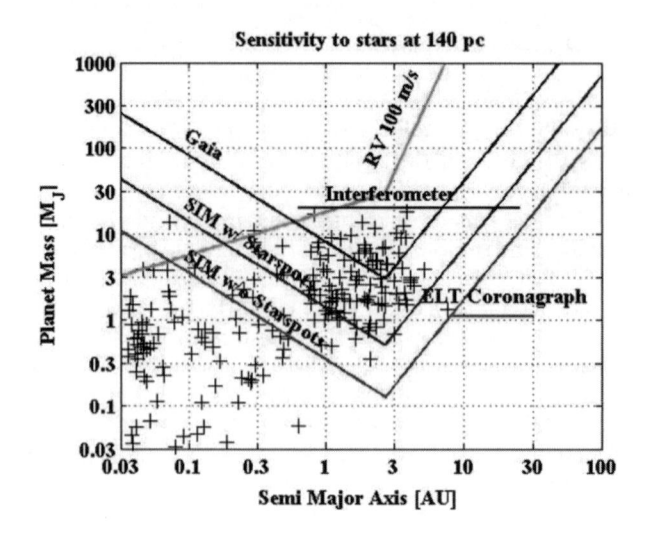

[Figure: Sensitivity to stars at 140 pc — plot of Planet Mass [M_J] vs Semi Major Axis [AU], showing regions for Gaia, SIM w/ Starspots, Interferometer, RV 100 m/s, and ELT Coronagraph]

Figure 3. SIM sensitivity to planets orbiting young stars at distances of 140 pc. For comparison, the crosses denote planets detected via RV measurements around mature stars.

in excess of 100 m/s preclude planet detection either by radial velocity or by transits.

SIM Lite has the ability to probe a broad planet mass range around young stars (Figure 3). SIM Lite surveys critical mass and age range to determine:

1. Gas giants around youngest stars (1-10 Myr at 140 pc)
2. Gas and icy giants around closest stars (>10 Myr at < 50 pc)
3. Large rocky planets around nearby, young, low mass stars

By contrast GAIA is capable of finding Jupiters around the closest young stars (d < 50 pc and ages > 10 Myr). SIM Lite is sensitive to stellar variation such as those caused by star spots. However, for stars with photometric variability $\Delta V < 0.05$ mag, the astrometric variability at 140 pc will be < 2 μas and will not prevent the detection of gas giant planets.

The ability of SIM Lite to detect planets has been demonstrated in a recent double blind study (Shao et al. 2009). Four independent teams were given simulated data sets of 48 planetary systems including a total of 95 planets and approximately 300 asteroids. There were 48 detectable planets when observed over the 5 year mission life time at a signal to noise ratio of 5.8. As shown in Figure 4, planets with masses as low at 1 M_{Earth} were detected by the teams. More importantly, all habitable zone planets and all Earth-sized planets were detected with no false positives.

4. SIM Lite Technology Status

The SIM Lite technology development program was completed to Technical Readiness Level (TRL) 6 in June 2005. It has been extensively reviewed and

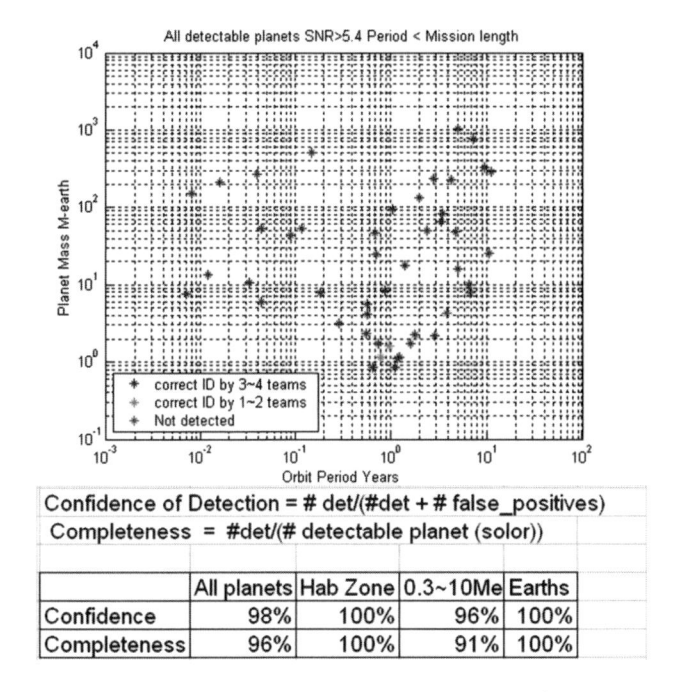

Figure 4. Major results of SIM Lite double blind study (Shao et al. 2009).

was formally signed off by NASA HQ in March 2006. The technology program has demonstrated the extreme performance (1.0 μas narrow-angle single-measurement and 4.0 μas wide-angle mission accuracy) needed for exoplanet detection with substantial margins. Instrumental errors in the SIM Lite testbed (chopped) integrate down as $1/\sqrt{T}$ to a systematic error floor of less than 100 nano-arcsec (Figure 5).

5. Summary

SIM Lite is a much reduced cost version of SIM that retains the potential to detect Earth clones around 60 of the nearest stars at a total cost around $1.1B. It has the capability to detect a 1 M_{Earth} planet orbiting at 1AU around a solar twin. SIM Lite will survey the locations of the nearest potentially habitable planets. A successful execution of SIM Lite will guide the way for future planet-finding missions. SIM Lite also has a strong astrophysics program, (dark matter, stellar physics, compact objects). The technology for SIM Lite is ready. Flight designs and models exist for many flight components that must operate at picometer levels. SIM Lite has completed virtually all of its formulation phase (A/B) work, including completion of technology development as described above. This work has resulted in a mature design with well understood cost and schedule of $940M to launch and $170M to operate. The budget has also been verified by NASA external independent reviews.

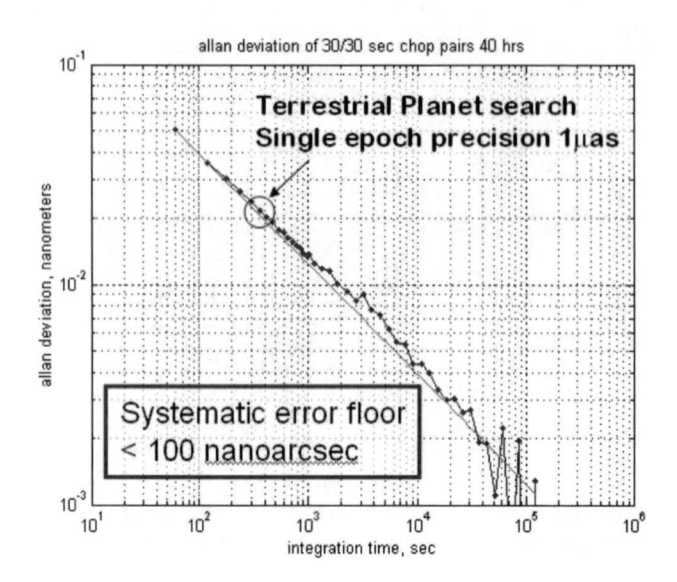

[Figure: Allan deviation plot showing "Terrestrial Planet search Single epoch precision 1μas" and "Systematic error floor < 100 nanoarcsec", x-axis: integration time, sec (10¹ to 10⁶), y-axis: allan deviation, nanometers]

Figure 5.   Results of instrumental errors in SIM test bed (Shao et al. 2009).

The past two Decadal Surveys in Astronomy and Astrophysics recommended the completion of a space-based interferometry mission, for its unique ability to detect and characterize nearby rocky planets (Bahcall et al. 1991; McKee & Taylor 2001), as well as contributions to a broad range of problems in astrophysics. Numerous committees of the National Research Council as well as NASA Roadmaps have similarly highlighted SIM Lite as the one technology that offers detection and characterization of rocky planets around nearby stars and which is technically ready. This mission has been submitted for evaluation to the Astro2010, the Astronomy and Astrophysics Decadal Survey conducted by the US National Academy of Sciences.

Acknowledgments.   The assistance of B. Martin Levine is acknowledged in the preparation of this paper. This work was performed at the Jet Propulsion Laboratory, California Institute of Technology, under contract to the National Aeronautics and Space Administration.

References

Bahcall, J.N. 1991, The Decade of Discovery in Astronomy and Astrophysics, (Washington DC: National Academies Press) ISBN 0-0309-04381-6
McKee, C. & Taylor, J. 2001, Astronomy and Astrophysics in the New Millennium (Washington DC: National Academies Press) ISBN309-0731-2
Shao, M. et. al. 2009, Astrometric Detection of Earthlike Planets, Science White Paper 67 submitted to the Astro2010 review committee, http://sites.nationalacademies.org/BPA/BPA_050603

Pathways Towards Habitable Planets
ASP Conference Series, Vol. 430, 2010
Vincent Coudé du Foresto, Dawn M. Gelino, and Ignasi Ribas, eds.

Darwin in the Context of Cosmic Vision 2015-2025

R. Liseau
Radio and Space Science, Chalmers University of Technology, Göteborg, Sweden

Abstract. The present status of the Darwin mission will be briefly reviewed, with particular focus on various developments since 2007. Of special interest is the readiness level (TRL) of critical mission technologies. While Darwin has essentially been put in limbo in Europe, continued research by the TPF-I team in the USA has demonstrated the high level of maturity which has recently been achieved for the critical technologies. This should encourage us to prepare for the next Cosmic Visions call by ESA. This call is expected to be issued in 2010/2011, reasonably well-timed with the upcoming US Decadal Survey. We argue that the SIM Lite mission would be an important milestone on the road toward Darwin-TPF and that ESA should join NASA in this endeavor.

1. Introduction

By the time of the conference *Darwin and Astronomy - The Infrared Space Interferometer* held in Stockholm in 1999, many of us were convinced that we would launch the Darwin mission in 2009. The timing would have been excellent, not the least to properly celebrate the 200th birthday of Charles Darwin and the 150th anniversary of his *On the Origin of Species by Means of Natural Selection, or the Preservation of Favoured Races in the Struggle for Life*. But also because a thorough study together with space industry (Alcatel, France) had revealed that, although a number of technical issues remained to be resolved, there were no show-stoppers. Our colleagues in the USA arrived at a similar conclusion for their mission concept Terrestrial Planet Finder (TPF).

1.1. Brief Historical Flashback

In 1994, Alain Léger and collaborators responded to a call for mission proposals within the Horizon 2000+ programme of the European Space Agency (ESA) with a concept for the detection of rocky planets around stars in the solar neighbourhood and for the detection of signs of life elsewhere. The basic ideas had previously been laid out by Bracewell (1978), Owen (1980) and e.g. Woolf & Angel (1990), and described in more detail by Léger et al. (1996). The proposal was highly successful and the mission was selected for further study.

The initial design of the interferometer involved telescopes on a long boom, but Robin Lawrence, the visionary ESA engineer who was assigned to lead the technical study, abandoned the concept of a connected structure, because he realized that the inevitable vibrations would not be sufficiently damped. Consequently, he proposed the idea of the free-flying flotilla of telescopes on individual

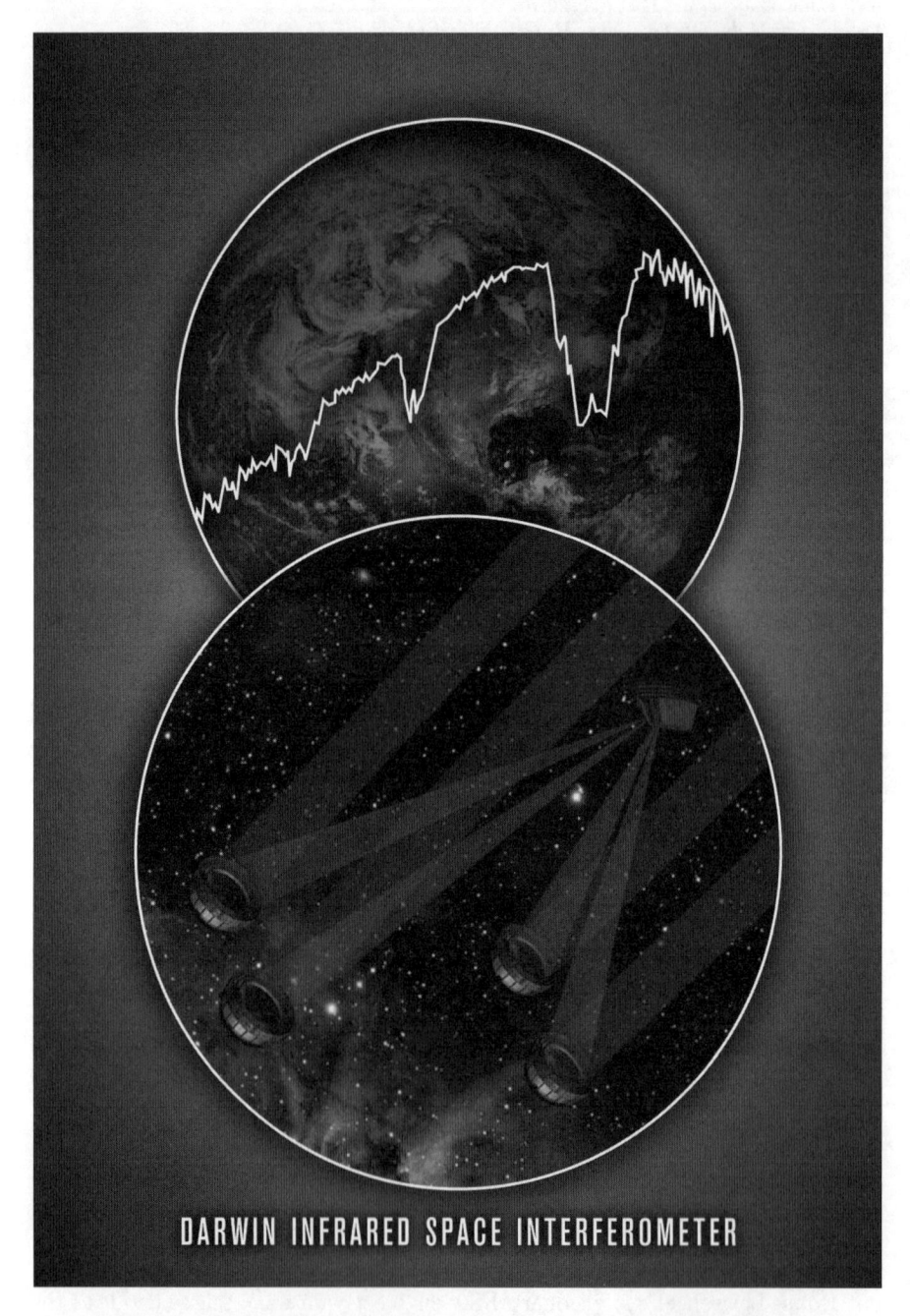

Figure 1. Bottom: an artist's impression of the X-array Darwin-TPF in space (Th. Jikander 2007). Top: an image of the Earth with its mid-infrared spectrum at low resolution superposed. Such or similar data, when seen from a distance, would enable us to infer that this planet likely has developed organic life forms.

Darwin and Cosmic Vision 2015-2025   221

spacecraft (Figure 1) which remains the basic design concept of the Darwin mission and which was further developed by Anders Karlsson.
For the optical design, J.M. Mariotti played an important and decisive role. He realized that detecting extrasolar rocky planets should be possible, if the feeble signals were modulated internally. Together with B. Mennesson, he had demonstrated, by means of numerical simulations, that the Darwin Nulling Interferometer was indeed capable of recovering Venus, Earth, and Mars around the Sun at 10 pc distance. The invaluable contributions to the Darwin project by J.M. Mariotti and R. Laurance have been summarized in the obituary by Léger (2000).

2. ESA's Cosmic Vision 2015-2025 Programme

In 2007, ESA issued a call for mission proposals for the newly created Cosmic Vision 2015-2025 programme (CV1525). This entailed four themes, the first of which was: *What are the conditions for planetary formation and the emergence of life?* Thus, in 2007, a proposal for an *X-array chopped Nulling Interferometer for the Infrared* was submitted to ESA with A. Léger as PI (see, e.g., Cockel et al. 2009) and 51 co-investigators. In addition, this document was endorsed by 650 supporting scientists from 20 countries. Needless to say, the scientific interest in this theme is enormous. Unfortunately, Darwin was not selected for further study, but both the AWG and SSAC made "very strong recommendations to initiate the necessary technology development for Darwin."

3. Recent Developments

3.1. Technology Readiness Level

The response by ESA to these recommendations has been only partially exhaustive. On September 3, 2009, the head of ESA's Advanced Technologies Section communicated "We have 3 technology development activities (TDAs) ongoing, which have relation to the former Darwin: (1) Integrated optics, (2+3) Single mode waveguide (two parallel contracts). All these TDAs are roughly at midterm and are progressing well." A more detailed plan will be presented in 2010 by the *Exo-Planet Roadmap Advisory Team* appointed by ESA in 2008 (see ibid., contribution by Artie Hatzes).

The NASA scheme of Technology Readiness Levels (TRLs) benchmarks various degrees of maturity, with TRL 1 (basic principles observed and reported) being the lowest. The technology development programs and their nomenclature of ESA are somewhat different. However, for a mission to enter its Definition Phase (Phase B), critical technologies must have reached a technological maturity corresponding to TRL 5 (component and/or breadboard validation in relevant environment). At the start of the Implementation Phase (C/D), TRL 5 to 6 (system/subsystem model or prototype demonstration in a relevant environment - ground or space) is required.

The proposed NASA mission Terrestial Planet Finder-Interferometer (TPF-I) is an identical twin to Darwin, since the adopted mission concepts have been developed in collaboration between the European and US teams. Luckily, scien-

Table 1. Technology Readiness Level of TPF-I (31 March 2009)[a]

Critical Technology	Readiness Level	Comment
Broadband Starlight Suppression	TRL 4	→ TRL 5[b]
mid-IR Spatial Filtering	TRL 4	→ TRL 5[b]
Adaptive Nulling	TRL 4	→ TRL 5[b]
Optical Delay Lines	TRL 6	OK
Cryocoolers	TRL 6	OK
Thermal Shields	TRL 6	OK
Detector Technology	TRL 6	OK
Precision Formation Flying	TRL 4	Formation Control Testbed, JPL
Precision Formation Flying	TRL 8/9	tbd, Prisma (2009/10)[c]
Precision Formation Flying	TRL 8/9	tbd, Proba-3 (2013)[d]

Notes to the Table:
[a] First part of Table based on Lawson et al. (2009).
[b] TRL 5 requires cryotesting.
[c] Swedish led two-spacecraft mission (SSC/SNSB), in collaboration with German DLR and French CNES (launch from Yasny, Russia).
[d] Two-spacecraft mission of ESA.

tists and engineers in the US have continued their efforts to validate critical technologies for Darwin-TPF. The present status, including expected (near-term) future results, are presented in Table 1, from which it is evident that the future of the technology readiness for Darwin-TPF is looking extremely bright, provided the remaining issues become resolved with active research. These include high-precision formation flying, the feasibility of which will soon be demonstrated in space, i.e., in the 2009/10 time frame[1] (see Table 1).

3.2. New ESA Call for CV1525

ESA intends to issue a new call for Cosmic Vision concepts in 2010/11. It is imperative that the scientific community supporting the elements of Theme 1 (*1. What are the conditions for planetary formation and the emergence of life?*) make its voice heard and respond to this call. Thus the Darwin-TPF/BlueDot teams should initiate interest among **scientists in a wide variety of disciplines**. Among fellow astronomers, a commonly felt reluctance against interferometry should be addressed and surmounted. The assistance offered by the ALMA Regional Centres for scientific support (ARC nodes) will most certainly be of value.

The power of **public outreach** should not be underestimated. This should be pursued in Europe much more vigorously than has been done so far. We have to make politicians and other potential donors aware of the fact that answering

---

[1] http://www.ssc.se/?id=7611

one of the most intriguing questions of humanity is within the realm of known technology - but simply lacks the necessary financial resources. In this context one notes that a European organization similar in size to ESA (18 member states), viz. CERN (20 member states), was able to raise the funds to look for answers to other important questions. The costs for this project, i.e., the Large Hadron Collider[2] (LHC) in Geneva, Switzerland, exceed 3.7 billion Euro, i.e., approaching the budget of a decade of ESA-science.

The **science case** for Darwin-TPF is already excellent and needs only minor adjustments. For instance, doubts regarding the *general* validity of life on Earth have frequently been raised. Such misunderstandings are readily obviated by less focusing on the Earth *per se*, but more clearly state the fact that our planet serves as the only good example we have so far (Figure 1). Darwin-TPF will provide us with a more general view. To do this properly, much higher spectral resolution than that depicted in the figure should eventually be aimed for.

4. Toward Darwin-TPF with SIM-Lite

The close collaboration with colleagues in the USA is already well established and has led to fruitful synergies. It would be very natural to join their proposed intermediate step toward Darwin-TPF, i.e., the Space Interferometry Mission Lite (SIM-Lite) to measure with extremely high accuracy the astrometric shifts of the centre of gravity due to the presence of orbiting planets. This method is complementary to the highly successful radial velocity technique and eliminates uncertainties regarding the planet masses due to inclination. The detection of Earth-like rocky planets appears feasible and will determine their masses and precise location at any one time (see contributions from Traub and Shao, this volume). This will pave the road for the future investigation of the physics and chemistry of these planets by means of spectrophotometric observations. Knowing precisely beforehand what targets to go for will enable Darwin-TPF to seriously and efficiently search for life outside our solar system. The development of interferometry for SIM-Lite would reduce technical risk for a future Darwin-TPF.

It is by no means excluded that, in 2010, the US National Academy's Decadal Survey will endorse an astrometry mission. In that case, the implementation of SIM-Lite may become realized and an ESA partnership with NASA would be an obvious and desirable option. Most importantly, this step would indeed do justice to the recognized importance of Theme 1 of the Cosmic Vision programme: *What are the conditions for planetary formation and the emergence of life?*

5. Conclusions

The general mission concept of Darwin-TPF matches the description of Theme 1 of the Cosmic Vision 2015-2025 perfectly. Unfortunately, a proposal submitted to ESA was not selected for future study. Here, we have addressed the

---

[2]http://en.wikipedia.org/wiki/Large_Hadron_Collider

Technology Readiness Level of various components of Darwin-TPF and argue that Europe should join the USA in efforts to complete the technology needed for a common Darwin-TPF mission. An ESA partnership in the US mission SIM-Lite would be a valuable component of a long-term collaboration to find and characterize other habitable worlds.

Acknowledgments. The author would like to thank the organizers for the kind invitation and the highly interesting conference in such a fabulous environment.

References

Bracewell, R.N., 1978, Nat, 274, 780
Cockell, C.S., Herbst, T., Léger, A., Kaltenegger, L., Absil, O., Beichman, C., Benz, C., and 40 coauthors, 2009, ExA, 23, 435
Lawson, P.R. et al., 2009, TPF-I Astro 2010 Technology Whitepaper & Addendum: *Technology for a Mid-IR Flagship Mission to Characterize Earth-like Exoplanets*
Léger, A., Mariotti, J.M., Mennesson, B., Ollivier, M., Puget, J.L., Rouan, D., & Schneider, J., 1996, Icarus, 123, 249
Léger, A., 2000, Proc. Conf. *Darwin and Astronomy - The Infrared Space Interferometer*, Stockholm, Sweden, 17-19 November 1999, ESA SP-451, 5
Owen, T., 1980, Strat. Search Life/Universe, Montreal, Canada, ASSL, 83, 177
Woolf, N. & Angel, R., 1990, in AIP Conf. Proc., 207, Astrophysics from the Moon, (New York: AIP)95

Pathways Towards Habitable Planets
ASP Conference Series, Vol. 430, 2010
Vincent Coudé du Foresto, Dawn M. Gelino, and Ignasi Ribas, eds.

MOA-II Microlensing Exoplanet Survey

T. Sumi
Solar-Terrestrial Environment Laboratory, Nagoya University, Nagoya, Japan

Abstract. We summarize the results of the exoplanet search via gravitational microlensing. The microlensing technique is unique and complementary to other methods because its sensitivity is very different from others. Microlensing is sensitive to low-mass planets down to Earth mass at relatively wide separation (1-6 AU) around dim K, M-dwarfs. Planets more distant than the snow line with large semi-major axes can be detected instantaneously without waiting for one orbital period to elapse. Most significantly, since it does not rely on the light from the primary, distant planetary systems up to the Galactic bulge at 8 kpc can be probed, which helps us to understand the Galactic distribution of planetary systems. We present the current status of the observations focusing on the survey by Microlensing Observation in Astrophysics (MOA) collaboration and summarize the features of the detected planetary systems.

1. Introduction

Since the first discovery of exoplanets orbiting around a main sequence star in 1995 (Mayor & Queloz 1995; Marcy et al. 2005), more than 300 exoplanets have been detected via the radial velocity (RV) method (Mayor et al. 2004), more than 50 exoplanets via the planetary transit method (Udalski et al. 2004; Konacki et al. 2005) and some by astrometry and direct imaging. Furthermore nine exoplanets have been discovered by gravitational microlensing (Udalski et al. 2005; Beaulieu et al. 2006; Gould et al. 2006; Dong et al. 2008; Gaudi et al. 2008; Bennett et al. 2008; Dong et al. 2009; Janczak et al. 2009).

Mao & Paczyński (1991) first proposed the method of exoplanet searches via microlensing. If a massive lens object, like a star, perfectly aligns along the line of sight toward the background source star, the light from the source is bent by the gravitational field of the lens object and the source brightness is magnified by up to several thousands. The timescales of magnifications are typically for a few weeks between a few days and a year. If there is a planet around the lens star, the gravity of a planet and the host star make small caustics which cause small deviations in standard single lens microlensing light curves (Paczyński 1986) as shown in Figure 1.

Compared to other techniques, microlensing has a sensitivity to low-mass planets down to Earth-mass in relatively wide orbits of 1-6 AU. Figure 2 shows the known exoplanets as a function of mass vs. semi-major axis, along with the predicted sensitivity curves for various methods. As you can see, microlensing is the only method that can detect low-mass planets outside of 1 AU and so it is complementary with other methods. Because the microlensing does not rely on the light from the host star (lensing star), it is sensitive to faint host stars like

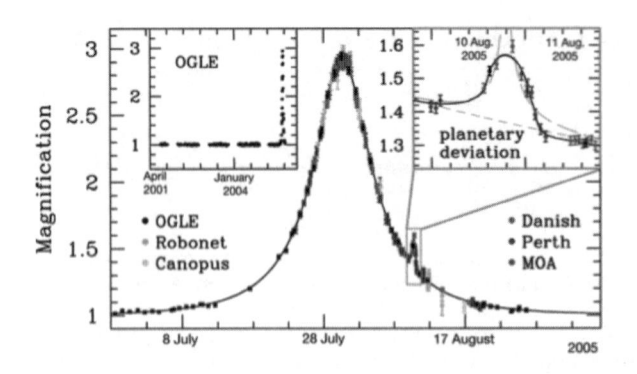

Figure 1. Light curve of 5.5 Earth mass planetary event OGLE-2005-BLG-390. The different shades indicate the data from different telescopes. The solid and dashed lines represent models with and without a planet, respectively.

M-dwarfs and even brown dwarfs. Since it is sensitive to the system at distances of several kpc away from the Sun, we can study the Galactic distribution of exoplanetary systems. Furthermore, numerical simulations predict that the orbit of the multiple-planet system becomes unstable and some of them will be rejected from the gravity field of the host star. Such planets unbounded by host stars are called free floating planets. Only microlensing can detect such free floating planets as a short timescale single lens event.

1.1. Observation

The event rate of microlensing is very low, about one in a million stars per month. Furthermore the probability of finding a planet among them is only a few percent. So we have to monitor ∼100 million stars in the Galactic bulge every day for years to detect the microlensing events by a star whose typical timescale is a few weeks. On the other hand, the timescale of the planetary signal is very short, a few days for Jupiter-mass planets and a few hours for Earth-mass planets. Thus, high cadence observations are needed to characterize the planetary signal accurately. To fulfill such a requirement, several groups around the world are collaborating in a global microlensing observation network. The two survey groups, MOA[1] and the Optical Gravitational Lens Experiment (OGLE[2]) are conducting a long-term continuous wide area survey by aiming wide field of view (FOV) telescopes toward the Galactic bulge to find the microlensing events by stars and issue the alerts all around the world. Then, the follow-up groups, Microlensing Follow Up Network (μFUN), Probing Lensing Anomalies NETwork (PLANET), RoboNet-II and MiNDSTEp carry out high

---

[1] https://it019909.massey.ac.nz/moa/

[2] http://ogle.astrouw.edu.pl/

Exoplanet Discovery Potential

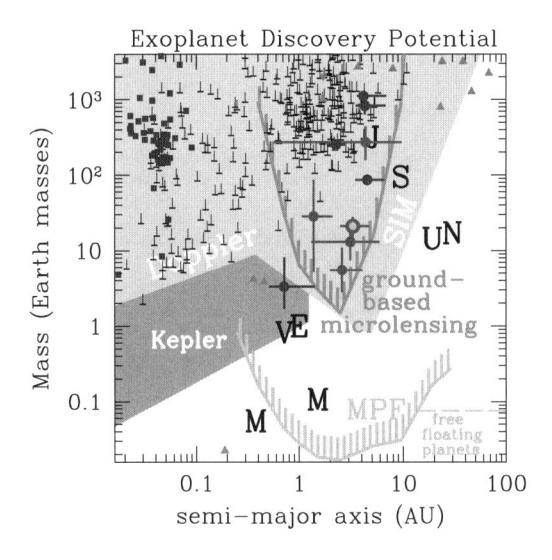

Figure 2. Known exoplanets as a function of mass vs. semi-major axis, along with the predicted sensitivity curves for various methods. The dark and light circles with error bars indicate the microlensing planets with direct mass measurements and mass estimated by Bayesian analysis, respectively. The squares represent the planets first detected via transit. The black bars with upward-pointing error bars (indicating $1\,\sigma\,\sin i$ uncertainty) are the radial velocity planet detections. The triangles indicate the planets found by timing (including the pulsar planets) and by direct detection, respectively. The shaded regions indicate the expected sensitivity limits of the radial velocity *Kepler* and *SIM* space missions. The upper and lower curves indicate the predicted lower sensitivity limits for a ground-based and space-based microlensing planet search program, respectively. The Solar System's planets are indicated with black letters.

cadence 24-hour follow-up observations of each alerted event to find a planetary signal, using a world-wide network of telescopes. Since the source can be magnified by more than several hundreds or thousands of times, even small, ∼25 cm, amateur telescopes are taking important roles in the follow-up observations.

MOA started the microlensing survey in 1995 at Mt. John observatory in New Zealand, using the 61-cm telescope. In 2005, we constructed the new 1.8-meter telescope, equipped with an 8k×10k-pixel CCD camera with very wide FOV of 2.2 deg.² (MOA-II). This is the largest dedicated telescope for microlensing survey so far. Thanks to the wide FOV, we can carry out the high-cadence (once an hour for 45 deg.², and covering 4 deg.² every 10 minutes) survey, by which we can find the beginning of the short planetary deviation and issue an alert to prompt the follow-up observations. Observed images are reduced by difference image analysis (DIA) (Alard & Lupton 1998; Alard 2000) in real time. If the planetary signal is detected, we sometimes take high-resolution imaging observations by HST or adaptive optics (AO) on the Very Large Telescope (VLT) and Keck telescopes. Using these high-resolution images, we can separate any stars that are blended in the ground-based images from the lens and source star.

Using this information, we can constrain the mass of the host stars, which is very important because we can usually extract only the mass ratio of the planets and host stars from the light curve modeling.

2. Results

In 2003, the first exoplanet was discovered using microlensing (Bond et al. 2004). In 2005, we found a 5.5 Earth-mass planet (Figure 1), which was the smallest planet at that time (Beaulieu et al. 2006), and a 3.5 Jupiter-mass (M_J) planet around the M-dwarf, which was the most massive M-dwarf planetary companion (Dong et al. 2008). Forming such a massive planet around a small M-dwarf is challenging under the core accretion scenario for the reasons discussed below.

As of 2005, two Jupiter-mass and two Neptune-mass exoplanets had been found. Taking into account the fact that the detection efficiency of Jupiter-mass planets is about three times higher than for Neptune-mass planets, these discoveries indicate that such Neptune-mass planets are more common than gas giants around the M-dwarfs, with a frequency of ∼40% (≥16% at 90% confidence) (Gould et al. 2006). This result is consistent with the theoretical simulation based on the core accretion scenario (Ida & Lin 2004), in which the lighter the host star, the lighter the protoplanetary disk, making lighter planets more common. This theoretical prediction was first confirmed by microlensing observations. So far eight planetary systems (nine exoplanets) (Udalski et al. 2005; Beaulieu et al. 2006; Gould et al. 2006; Dong et al. 2008; Gaudi et al. 2008; Bennett et al. 2008; Dong et al. 2009; Janczak et al. 2009) have been found by microlensing, and this trend seems to be continuing.

Furthermore, in 2006, the first system with multiple planets via microlensing was discovered in the event OGLE-2006-BLG-109 (Gaudi et al. 2008). Based on detailed analysis, we found that the lensed system has two planets with 0.71 M_J and 0.27 M_J gas giants, with separations of 2.3 AU and 4.6 AU, respectively, around 0.5 $M_\odot$ M-giant. This system is about a half analog of our Sun/Jupiter/Saturn. More interestingly, the effective temperature of these planets estimated from the brightness of the host and their distances, are only ∼ 30% cooler than our Jupiter/Saturn. It means that their orbits are beyond the snow line of this system (Ida & Lin 2004; Laughlin, Bodenheimer & Adams 2004; Kennedy, Kenyon & Bromley 2006), where the core accretion theory predicts that the most massive planets should form because water can condense there. In 25 planetary systems detected so far by the radial velocity method, none of them has multiple giant planets beyond their snow line. To explain them, the theories predict that they were actually formed beyond the snow line, then migrated inward. The OGLE-2006-BLG-109 observation was the first case that such giant planets actually exist beyond the snow line besides our Solar System, which supports the core accretion scenario. Furthermore, this model predicts that giants can grow larger in regions closer to the star, which is also consistent with these discovered planets, as well as our Jupiter/Saturn.

In 2007, a planetary system with the smallest host star (which could be a very late M-dwarf or a brown dwarf) was found in the event MOA-2007-BLG-192 (Bennett et al. 2008). After detailed analysis of the MOA and OGLE light curves, we found that the $3^{+3}_{-1} M_\oplus$ planet orbiting at 0.6 AU around the host star with the mass of $0.06^{+0.03}_{-0.02} M_\oplus$. This was the first such case where the planet was found without follow-up observations, thanks to the high cadence survey observations by MOA-II. VLT/NACO AO images were taken during the event immediately. No flux from the lens was detected in the AO image, which is consistent with the result that the lens is a late M-dwarf or brown dwarf. Whether the lens is an M-dwarf or brown dwarf can be determined by *HST* observation about five years later. Although protoplanetary disks have been found around brown dwarfs, this system may be the first detection of a planet around a brown dwarf.

To see the Galactic distribution of planetary systems, we plotted the host mass of seven detected systems as a function of the distance to the system, D_L, from the Sun towards the Galactic center (GC) at 8kpc in Figure 3. This shows that they are distributed uniformly along the way to the GC, even though ~70% of microlensing events are caused by bulge stars. This indicates that planetary systems may be rare in the bulge, although we should check to see if there is any observational bias before drawing a definite conclusion.

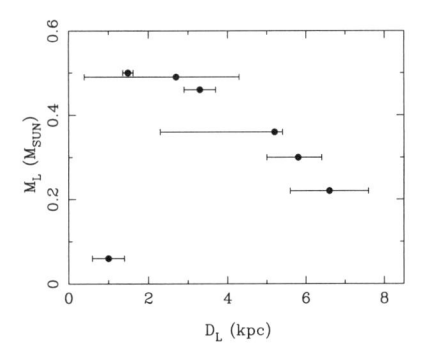

Figure 3. Host mass vs. distance to detected planetary systems.

Thanks to our high cadence survey observation, we found dozens of single lens events with very short timescale of fewer than 2 days in 2006-2007 as shown in Figure 4. Such short timescale events are most likely caused by free floating planets. We are currently estimating how many such free floating planets exist.

3. Summary

Microlensing finds different types of exoplanetary systems than those discovered with other techniques, which means that microlensing is complementary to other methods. It is very important to understand the formation mechanism of exoplanets, and so far, the results support the core accretion model. The Kepler satellite was successfully launched last March and uses the transit method and is sensitive to lower-mass, down to the super Earth, planets. These techniques, along with microlensing, are expected to find Earth-mass planets within a few years. The Microlensing Planet Finder (MPF) space mission for searching for exoplanets by microlensing, is proposed by NASA to launch in 2014 or later. It is expected to find 150 Earth-mass planets, assuming each of its target stars

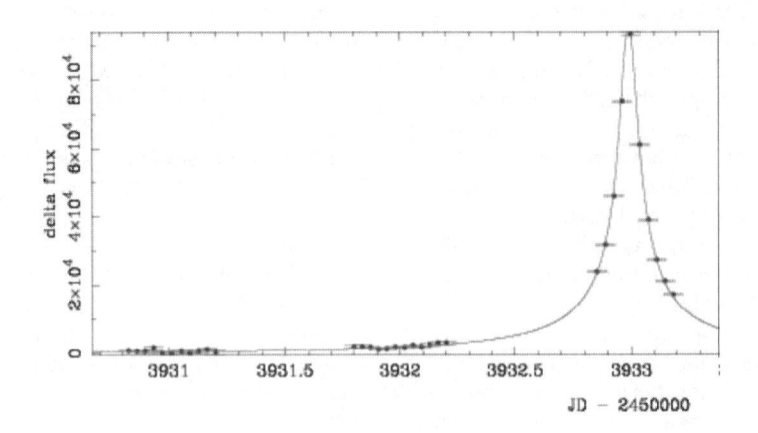

Figure 4. The light curve of the free floating planet candidates.

have one Earth-mass planet. Thus, in the near future we may know how many "Earths" exist in the universe.

Acknowledgments. This research was supported by MEXT Japan, Grant-in-Aid for Young Scientists (B), 18749004 and Grant-in-Aid for Scientific Research on Priority Areas, "Development of Extra-solar Planetary Science", 19015005.

References

Alard C. 2000, A&AS, 144, 363
Alard, C., & Lupton, R.H. 1998, ApJ, 503, 325
Beaulieu, J.-P., et al. 2006, Nat, 439, 437
Bennett, D. P., et al. 2008, ApJ, 684, 663
Bond, I. A., et al. 2004, ApJ, 606, L155
Dong, S. et al. 2008, ArXiv e-prints, 804, arXiv0804.1354v2
Dong, S. et al. 2009, ApJ, 695, 970
Gaudi, B. S., et al. 2008, Sci, 319, 927
Gould, A., et al. 2006, ApJ, 644, L37
Ida, S., & Lin, D.N.C. 2004, ApJ, 616, 567
Janczak, J., et al. 2009, arXiv0908.0529v1
Kennedy, G. M., Kenyon, S. J., & Bromley, B. C. 2006, ApJ, 650, L139
Konacki, M., et al. 2005, ApJ, 624, 372
Laughlin, G., Bodenheimer, P., & Adams, F.C. 2004, ApJ, 612, L73
Mao, S., & Paczyński, B. 1991, ApJ, 374, L37
Marcy, G. W., et al. 2005, ApJ, 619, 570
Mayor, M., et al. 2004, A&A, 415, 391
Mayor, M., & Queloz, D. 1995, Nat, 378, 355
Paczyński, B. 1986, ApJ, 304,1
Udalski, A., et al. 2004 AcA, 54, 313
Udalski, A., et al. 2005 ApJ, 628, L109

Pathways Towards Habitable Planets
ASP Conference Series, Vol. 430, 2010
Vincent Coudé du Foresto, Dawn M. Gelino, and Ignasi Ribas, eds.

Direct Detection of Giant Extrasolar Planets with SPHERE on the VLT

J.-L. Beuzit,[1] A. Boccaletti,[7] M. Feldt,[2] K. Dohlen,[3] D. Mouillet,[1]
P. Puget,[1] F. Wildi,[4] L. Abe,[5] J. Antichi,[6] A. Baruffolo,[6] P. Baudoz,[7]
M. Carbillet,[5] J. Charton,[1] R. Claudi,[6] S. Desidera,[6] M. Downing,[8]
C. Fabron,[3] P. Feautrier,[1] E. Fedrigo,[8] T. Fusco,[9] J.-L. Gach,[3] E. Giro, [6]
R. Gratton,[6] T. Henning,[2] N. Hubin,[8] F. Joos,[11] M. Kasper,[8]
A.-M. Lagrange,[1] M. Langlois,[3] R. Lenzen,[2] C. Moutou,[3] A. Pavlov,[2]
C. Petit,[9] J. Pragt,[10] P. Rabou,[1] F. Rigal,[10] S. Rochat,[1] R. Roelfsema,[10]
G. Rousset,[7] M. Saisse,[3] H.-M. Schmid,[11] E. Stadler,[1] C. Thalmann,[2]
M. Turatto,[6] S. Udry,[4] F. Vakili,[5] A. Vigan,[3] and R. Waters[12]

Abstract. Direct detection and spectral characterization of extra-solar planets is one of the most exciting but also one of the most challenging areas in modern astronomy. The challenge consists in the very large contrast between the host star and the planet, larger than 12.5 magnitudes at very small angular separations, typically inside the seeing halo. The whole design of a "Planet Finder" instrument is therefore optimized towards reaching the highest contrast in a limited field of view and at short distances from the central star. Both evolved and young planetary systems can be detected, respectively through their reflected light and through the intrinsic planet emission. We present the science objectives, conceptual design, and expected performance of the SPHERE instrument.

---

[1]LAOG, CNRS/Université J. Fourier, B.P. 53, Grenoble, France

[2]Max Planck Institute for Astronomie, Konigsthul 17, Heidelberg, Germany

[3]LAM, CNRS/Université de Provence, B.P. 8, Marseille, France

[4]Observatoire de Genève, 51 ch. des Maillettes, Sauverny, Switzerland

[5]Laboratoire H. Fizeau, UNS/CNRS/OCA, Nice, France

[6]Osservatorio Astronomico di Padova, INAF, Vicolo dellOsservatorio 5, Padova, Italy

[7]LESIA, CNRS/Observatoire de Paris, 5 place J. Janssen, Meudon, France

[8]European Southern Observatory, Karl-Schwarzschild-Strasse 2, Garching, Germany

[9]ONERA, B.P. 72, Chatillon, France

[10]NOVA/ASTRON, Oude Hoogeveensedijk 4, Dwingeloo, The Netherlands

[11]Institute of Astronomy, ETH, Zurich, Switzerland

[12]Universitat van Amsterdam, Kruislaan 403, Amsterdam, The Netherlands

1. Introduction

The prime objective of the Spectro-Polarimetric High-contrast Exoplanet Research (SPHERE) instrument for the VLT is the discovery and study of new extra-solar giant planets orbiting nearby stars by direct imaging of their circumstellar environment. The challenge consists in the very large contrast between the host star and the planet, larger than 12.5 magnitudes (10^5 in flux ratio), at very small angular separations, typically inside the seeing halo. The whole design of SPHERE is therefore optimized towards reaching the highest contrast in a limited field of view and at short distances from the central star. Both evolved and young planetary systems will be detected, respectively through their reflected light (by visible differential polarimetry) and through the intrinsic planet emission (using IR differential imaging and integral field spectroscopy). Both components of the near-infrared arm of SPHERE will provide complementary detection capabilities and characterization potential, in terms of field of view, contrast, and spectral domain.

SPHERE is built by a consortium of eleven institutes of five European countries, together with ESO. The consortium includes the following entities: Laboratoire d'Astrophysique de l'Observatoire de Grenoble (LAOG; P.I. institute, Grenoble, France), Max Planck Institute for Astronomy (MPIA; Co-P.I. institute, Heidelberg, Germany), Laboratory of Astrophysics (LAM; Marseille, France), Laboratoire d'tudes spatiales et d'instrumentation en astrophysique (LESIA; Paris, France), Fizeau (Nice, France), Osservatorio Astronomico di Padova/Istituto Nazionale di Astrofisica (INAF; Padova, Italy), Observatoire de Genève (Geneva, Switzerland), Eidgenssische Technische Hochschule Zrich (ETHZ; Zurich, Switzerland), University of Amsterdam (Amsterdam, The Netherlands), Netherlands Institute for Radio Astronomy (NOVA/ASTRON; Dwingeloo, The Netherlands), ONERA (Chatillon, France). The Final Design Review took place end of 2008 and SPHERE in now in its integration phase, with a delivery foreseen by middle of 2011 and a start of science observations in the Spring of 2012.

2. Science Objectives

The primary goal of extra-solar planet science in the next decade will be a better understanding of the mechanisms of formation and evolution of planetary systems. The fundamental observational parameter is the frequency of planets as a function of mass and separation. Theoretical models of planet formation predict that the peak of formation of giant planets is found close to the snowline, thanks to the availability of a larger amount of condensate in the proto-planetary disk. In outer regions, the longer timescales involved should make planet formation a less efficient process. Migration mechanisms and long term orbit instabilities will alter the original distribution. Determination of the frequency of giant planets in wide orbits (further than 5-10 AU) will allow testing basic aspects of the planet formation models.

While radial velocity spectroscopy remains the best technique currently available to study the inner side of the planet distribution with semi-major axis smaller than 5 AU, high-resolution, high contrast imaging like that provided by

SPHERE is expected to be the most efficient technique to discover planets in the outer regions of planetary systems. With its enhanced capabilities (a gain of two orders of magnitudes in contrast with respect to existing instruments) and a list of potential targets including several hundred stars, SPHERE will provide a clear view of the frequency of giant planets in wide orbits. With the number of expected detections (several tens), the level of the large separation wing of the distribution with semi-major axis can probably be estimated with an accuracy of about 20-30%, good enough for a first statistical discussion of the properties of planetary systems. Beside frequency, it would also be interesting to derive the distributions of planets parameters such as mass, semi-major axis and eccentricities.

The main scientific goal of SPHERE is the description of the properties of young planets in the expected peak region of gas giant formation and in the outer regions of the systems. Imaging of planets already detected by radial velocity and/or astrometry would additionally represent a major breakthrough thanks to the availability of dynamical constraints (or even full orbit determinations) on the planet masses and on the orbital elements. Therefore, these objects will represent the ideal benchmarks for the calibration of models for sub-stellar objects. Furthermore, a direct imager like SPHERE provides the only way of obtaining spectral characteristics for outer planets. Finally, the SPHERE differential polarimetric channel (ZIMPOL) might allow detecting a few planets shining by reflecting stellar light. Such an instrument will then provide invaluable information with which to hone models in preparation for the Darwin and ELT era. SPHERE will be then highly complementary to current and contemporaneous studies of extrasolar planets.

An instrument with the capabilities to achieve these goals will also be able to make great advances in related areas of study such as brown dwarfs, star and planet formation (e.g. via imaging of disks), solar system objects (asteroids), etc. These domains will nicely enrich the scientific impact of the instrument.

2.1. Target Classes

These science objectives fully justify a large effort in an extended observational survey of several hundred nights concentrating on selected classes of targets: *Nearby young associations* (10-100 Myr, 30-100 pc) offering the best chance of detecting low mass planets, since they will have brighter sub-stellar companions; *Young active F-K dwarfs* of the solar neighbourhood (ages less than 1 Gyr, d < 50 pc); *Nearest stars* (all ages within 20 pc of the Sun) for probing the smallest orbits and also giving the only opportunities for detecting planets by directly reflected light; *Stars with known planets*, especially any that exhibit long-term residuals in their radial velocity curves, indicating the possible presence of a more distant planet (F-G-K stars within 50-100 pc); *Young early type stars*; *Planet candidates from astrometric surveys*.

2.2. High Level Requirements

The key high level requirements derived from the science analysis and driving the design of the instrument are summarized below.

- High contrast capability to detect giant planets 15 magnitudes fainter than their host star at $0.5''$ (for host stars with J < 6).

- Access to very small angular separations, 0.1″ to 3″ from the host star.
- Optimal performance for targets up to visible magnitude ∼ 9, for building a large enough target list (a few hundred targets).
- Access to an extended spectral domain at low resolution, for the characterization of the detected objects, at a resolving power ∼ 30.
- Sensitivity to extended sources down to ∼ 17 magnitudes per arcsecond^2 at less than 0.2″ from the host star.

3. Instrument Concept

SPHERE is divided into four subsystems: the Common Path and Infrastructure (CPI) and the three science channels, a differential imaging camera (IRDIS, InfraRed Dual Imaging Spectrograph), an Integral Field Spectrograph (IFS) and a visible imaging polarimeter (ZIMPOL, Zurich Imaging Polarimeter). The Common Path includes pupil stabilizing fore optics (tip-tilt and rotation), calibration units, the SAXO extreme adaptive optics system, and the NIR and visible coronagraphic devices. ZIMPOL shares the visible channel with the wavefront sensor through a beamsplitter, which can be a grey (80% to ZIMPOL) beamsplitter, a dichroic beamsplitter, or a mirror (no ZIMPOL observations). IRDIS is the main science channel and does imaging over a square field of 11″ in one or two simultaneous spectral bands or two orthogonal polarizations and low and medium resolution long slit spectroscopy. The IFS, working from 0.95 to 1.65 microns, provides low spectral resolution (R ∼ 30) over a limited, 1.8″ x 1.8″, field of view. For the main survey mode, expected to be used for a very large fraction of the observations, a photon sharing scheme has been agreed upon between IRDIS and IFS, allowing IFS to exploit the NIR range up to the J band, leaving the H-band, deemed optimal for the differential mode, for IRDIS. This multiplexing approach optimizes the observational efficiency. A short description of the SPHERE main sub-systems is given below, a more detailed description of the instrument can be found in Beuzit et al. (2008) and Wildi et al. (2009).

3.1. Common Path and Infrastructure

The Common Path and Infrastructure is a large bench mounted on a triplet of active dampers, to which each science instrument will dock as a whole. When in operation on the Nasmyth platform A of the VLT UT3, SPHERE will be entirely enclosed in a thermal/dust cover. Besides classical optical components, the common path embeds various innovative components which have been the subject to a specific R&D effort in the project: toroidal mirrors manufactured by spherical polishing of pre-stressed substrates, achromatic 4-quadrant coronagraphs, apodized Lyot coronagraphs and a variable spatial filter.

3.2. Extreme Adaptive Optics

Three different loops and one off line calibration compose the SAXO extreme adaptive optics system (Petit et al., 2008). The *main AO loop* corrects for atmospheric, telescope and common path defects. Its main impact is the increase of detection signal to noise ratio through the reduction of the smooth PSF halo due to turbulence effects. The *differential tip-tilt loop* ensures a fine centering of the beam on the coronagraphic mask (correction of differential tip-tilt between the visible and IR channels). It will therefore ensure an optimal performance of the coronagraphic devices. The *pupil tip-tilt loop* corrects for pupil shift (telescope and instrument). It will ensure that uncorrected instrumental aberrations effects in the focal plane will always be located at the same position and thus will be canceled out by a clever post-processing procedure. Finally, *non common path aberrations* will be measured with phase diversity, and their pre-compensation will lead to the reduction of persistent speckle.

SAXO includes a 41 × 41 actuator high-order deformable mirror from CILAS with a maximum stroke > ± 3.5 microns, and a 40 × 40 lenslet visible Shack-Hartmann wavefront sensor, based on the dedicated 240 × 240 pixel electron multiplying CCD220 from EEV achieving a temporal sampling frequency of 1.2 kHz with a read-out-noise < 1 electron and a 1.4 excess photon noise factor. The wavefront sensor is equipped with a focal plane spatial filter for aliasing control. At the heart of the AO system is the ESO standard real-time computer platform called SPARTA providing a global AO loop delay < 1 ms. SPARTA allows to control the 3 system loops while also providing turbulent parameters and system performance estimation as well all the relevant data for an optimized PSF reconstruction and a clever signal extraction from scientific data.

3.3. Coronagraphy

Efficient coronagraphy is a key for reaching the SPHERE science goals (Boccaletti 2008). Its action is two-fold: reduce the intensity of the stellar peak by a factor of at least 100 and eliminate the diffraction features due to the pupil edges. The base-line coronagraph suite will include an achromatic four-quadrant phase mask coronagraph based on precision mounting of four half-wave plates, and both a classical Lyot coronagraph and an apodized Lyot coronagraph. Other options include the classical four-quadrant phase mask, which is now very well mastered and tested in the laboratory and on the sky with VLT/NACO. More explorative devices are also being investigated with options like a broad-band versions of the four-quadrant phase mask based on zero-order gratings or its circularly symmetric version, the annular groove phase mask. Stellar coronagraphy is a quickly evolving research field and it is important to leave the instrument open for future evolutions by allowing exchangeable masks both in the coronagraphic focus and in its entrance and exit pupil planes.

3.4. Infra-Red Dual-beam Imaging and Spectroscopy

The IRDIS science module (Dohlen et al. 2008) covers a spectral range from 0.95-2.32 microns with an image scale of 12.25 mas consistent with Nyquist sampling at 950 nm. The FOV is $11'' \times 12.5''$, both for direct and dual imaging. Dual band imaging is the main mode of IRDIS, providing images in two neighboring spectral channels with < 10 nm rms differential aberrations. Two parallel images are projected onto the same 2k × 2k detector with 18-micron square pixels, of which they occupy about half the available area. A series of filter couples is defined corresponding to different spectral features in modeled exoplanet spectra. The classical imaging mode allows high-resolution coronagraphic imaging of the circumstellar environment through broad-, medium-, and narrow-band filters throughout the NIR bands including Ks. In addition to these modes, long-slit spectroscopy at resolving powers of 50 and 500 is provided, as well as a dual polarimetric imaging mode. A pupil-imaging mode for system diagnosis is also implemented.

3.5. Integral Field Spectroscopy (IFS)

IFS are very versatile instruments, well adapted for spectroscopic differential imaging as needed for detection of planets around nearby stars (Claudi et al. 2008). The main advantage of IFS is that differential aberrations can be kept at a very low level; this is true in particular for lenslet-based systems, where the optical paths of light of different wavelength within the IFS itself can be extremely close to each other. Additionally, IFS provide wide flexibility in the selection of the wavelength channels for differential imaging, and the possibility to perform spectral subtraction, which in principle allows recovering full information on the planet spectra, and not simply the residual of channel subtraction, as in classical differential imagers. The main drawback of IFS is that they require a large number of detector pixels, resulting in a limitation in the field of view, which is more severe for lenslet-based systems. Classical differential imagers and IFS are then clearly complementary in their properties, and an instrument where both these science modules are available may be extremely powerful for planet search.

3.6. Imaging Polarimeter

ZIMPOL is located behind the SPHERE visible coronagraph. Among its main specifications are a bandwidth of 600-900 nm and an instantaneous field of view of $3''$ x $3''$ with access to a total field of view of $8''$ in diameter by an internal field selector (Thalmann et al. 2008). The ZIMPOL optical train contains a common optical path that is split with the aid of a polarizing beamsplitter in two optical arms, each with its own detector. The common path contains common components for both arms like calibration components, filters, a rotatable half wave plate and a ferroelectric liquid crystal polarization modulator. The two arms have the ability to measure simultaneously the two complementary polarization states in the same or in distinct filters. The images on both ZIMPOL detectors are Nyquist sampled at 600 nm. The basic ZIMPOL principle for high-precision polarization measurements includes a fast polarization modulator with a modulation frequency in the kHz range, combined with an imaging photometer that demodulates the intensity signal in synchronism with the polarization modulation. The polarization modulator and the associated polarizer convert the degree-of-polarization signal into a fractional modulation of the intensity signal, which is measured in a demodulating detector system by a differential intensity measurement between the two modulator states. Each active pixel measures both the high and low states of the intensity modulation and dividing the differential signal by the average signal eliminates essentially all gain changes, notably changes of atmospheric transparency or electronic gain drifts.

4. Expected Performances

We illustrate here the expected performance for the IRDIS dual-band imaging (DBI) mode only. See Claudi et al. (2008), Desidera et al. (2008), Dohlen et al. (2008) and Thalmann et al. (2008) for detailed simulations of the expected performances of the the various SPHERE observing modes.

Numerical simulations of the IRDIS performance in DBI mode have been made using our end-to-end model, considering a 1-hour integration time achieved by imagery at both sides of the H-band methane absorption edge. High contrast imaging has to deal at first order, with two components: a speckled halo which is averaging over time and a static speckle pattern originating from quasi-static aberrations evolving occurring with a much longer lifetime than atmospheric residuals. Because the DBI mode is performing simultaneous differential imaging, performances are mostly limited by the quasi-static aberrations upstream of the coronagraph and by the spectral separation between DBI filters. Figure 1 shows plots of different companions intensity compared with radial variance profiles in the processed images for several models of planets in various conditions. At 1.6 microns and for bright sources not limited by photon noise, it is expected to reach a 5σ contrast of $\sim 2\ 10^{-5}$ at its inner working angle of $0.1''$, and $\sim 10^{-7}$ at $0.6''$.

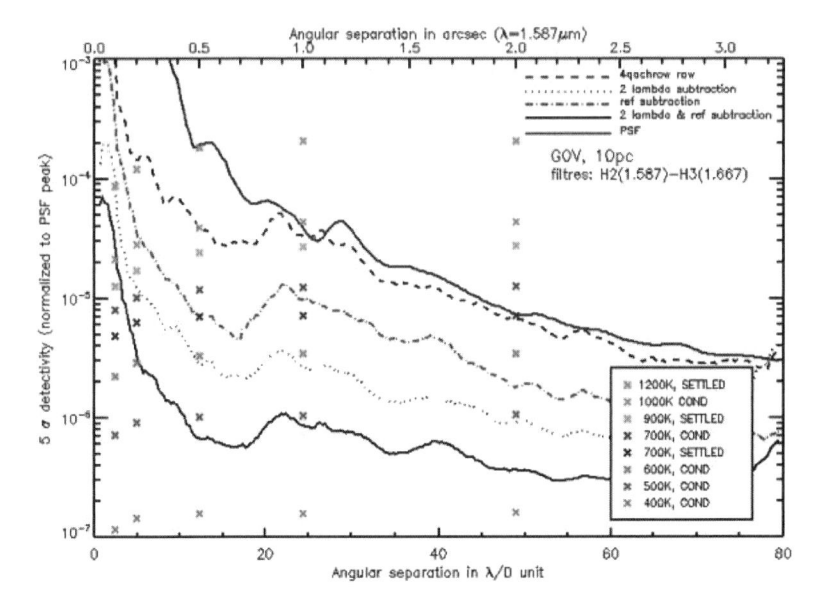

Figure 1. Estimated detectivity at 5σ in 1 hour for various models of planets, for a G0V star at 10pc observed in dual band imaging (H2H3). COND refers to models for a condensed atmosphere free of dust, SETTLED refers to atmospheres with rainout of refractory material.

5. Observing Strategy

Three main observing modes have been defined in order to draw the maximum benefit of the unique instrumental capabilities of SPHERE.

The *NIR survey mode* is the main observing mode which will be used for a large fraction of the observing time. It combines IRDIS dual imaging in H-band with imaging spectroscopy using the IFS in the Y-J bands. This configuration benefits simultaneously from the optimal capacities of both dual imaging over a large field (out to $\sim 5''$ radius) and spectral imaging in the inner region (out to at least $0.8''$ radius). In particular, it reduces the number of false alarms and confirms potential detections obtained in one channel by data from the other channel. This will be a definitive advantage in cases of detections very close to the limits of the system.

The *NIR characterization mode*, in which IRDIS is used alone in its various modes, will allow observations with a wider field of view in all bands from Y to short-K, either in dual imaging or dual polarimetry, or in classical imaging using a variety of broad- and narrow-band filters. This will be especially interesting in order to obtain complementary information on already detected and relatively bright targets (follow-up and/or characterization). Spectroscopic characterization at low or medium resolution will be possible in long-slit mode. Additional science cases will also benefit from these observing modes (disks, brown dwarfs, etc.).

The *visible search and characterization mode*, will benefit from ZIMPOL polarimetric capacities to provide unique performance in reflected light very close to the star, down to the level required for the first direct detection in the visible of old close-in planets, even if on a relatively small number of targets. ZIMPOL also provides classical imaging in the visible, offering unique high-Strehl performance.

References

Beuzit, J.-L. et al. 2008, SPIE 7014, 42.
Boccaletti, A. et al. 2008, SPIE 7015, 46
Claudi, R. et al. 2008, SPIE 7014, 118
Desidera, S. et al. 2008, SPIE 7014, 127
Dohlen, K. et al. 2008, SPIE 7014, 126
Petit, C. et al. 2008, SPIE 7015, 65
Thalmann, C. et al. 2008, SPIE 7014, 120
Wildi, F. et al. 2009, SPIE 7440, 24

Pathways Towards Habitable Planets
ASP Conference Series, Vol. 430, 2010
Vincent Coudé du Foresto, Dawn M. Gelino, and Ignasi Ribas, eds.

Resolved Imaging of Extra-Solar Photosynthesis Patches with a "Laser Driven Hypertelescope Flotilla"

A. Labeyrie,[1] H. Le Coroller,[2] S. Residori,[3] U. Bortolozzo,[3] J.-P. Huignard,[4] and P. Riaud[5]

[1] *Collège de France, 11 place Marcelin Berthelot, Paris, France*

[2] *Observatoire Astronomique Marseille, Provence, France*

[3] *Institut Non Linéaire de Nice, Université de Nice-Sophia Antipolis, CNRS, 1361 route des Lucioles, Valbonne, France*

[4] *Thales Research & Technology, 1 Avenue A. Fresnel, Palaiseau, France*

[5] *60 rue des Bergers, Paris, France*

Abstract. Formation-flying arrays of many apertures in space, in the form of a "hypertelescope" imaging interferometer, can produce direct images of habitable planets. Designs proposed (Labeyrie et al. 2009) to NASA and ESA, however, require several actuators and sensors per spaceship to accurately control the formation flight, as is the case for other proposed interferometer flotillas. The ensuing complexity and cost has led these agencies to postpone the development of all such flotillas, in spite of their breakthrough resolution capability . The theory of hypertelescope imaging shows that more sub-apertures of smaller size produce more science for a given collecting area and array size. This suggests sub-apertures as small as 3 to 10 cm, in the form of laser-trapped mirrors. The mirrors are trapped axially in interference standing waves formed by a pair of counter-propagating laser beams, and have a deviating prismatic edge for transverse trapping. The flotilla of miniature satellites is fully passive, yet controlled with sub-wavelength accuracy, and can be deployed from a small delivery package. Following numerical simulations of the dynamic behaviour, some of us (SR & UB) began a laboratory experiment with a mirror suspended in a vacuum. Further testing aboard the International Space Station is considered in a second step before designing a full system with a kilometric size. Much larger sizes are possible in theory, toward a 100-1000 km Exo-Earth Imager capable of resolving colored patches of photosynthetic activity on habitable planets.

1. Introduction

The Exo-Earth Imager concept previously described by Labeyrie (1999) involves a "hypertelescope" formation flight of many sub-apertures forming a highly diluted aperture at the scale of 100 km. It can, in principle, produce direct images of habitable planets, showing resolved details of their morphology, including colored patches of photosynthetic activity.

Hypertelescopes are a new form of imaging interferometers which use many apertures and a densified pupil to produce direct snapshot images. The sensitivity is greatly increased with respect to the optical forms of aperture synthesis which were previously considered. The gain arises from the more dense wavefront sampling achieved by the coherent combination of optical vibrations received from all apertures simultaneously, rather than the incoherent combination of fringe data received at different times from pairs or triplets, etc., of apertures, forming different baselines. On angularly small sources such as an exo-Earth or other habitable planet, most photons collected by a hypertelescope are effectively directed toward each of the image resels corresponding to a point of the source, as is the case in conventional telescope imaging. Imaging simulations have shown that a 150 km hypertelescope having 150 apertures of 3 m can produce resolved images of an Earth-like planet at 3 pc containing 30 x 30 resolved elements (resels) within a 3 hour exposure.

Here we discuss the possibility of building hypertelescopes in even more diluted form, using a flotilla of much smaller mirrors which are driven by laser-trapping effects. This theoretical possibility was briefly mentioned in Labeyrie et al. (2009), and we now describe the concept in more detail.

2. Hypertelescopes and the Sensitivity Gain with Many Small Apertures

Following the introduction of the hypertelescope principle (Labeyrie 1996; Labeyrie et al. 2006), its imaging properties have been studied in some detail (Lardière et al. 2007; Labeyrie et al. 2008; Aime 2008; Labeyrie et al. 2009; Patru et al. 2009), and miniature versions were demonstrated (Pedretti et al. 2000; Gillet et al. 2003). A prototype is under construction and preliminary testing at Haute-Provence observatory and has an 18m maximal aperture, with engineered hypertelescope optics including a fixed spherical primary array, a Mertz corrector of spherical aberration, a pupil densifier, and a white laser metrology system for sensing the array geometry with micron accuracy (Le Coroller et al. 2004). The design concept is an optical and diluted version of the Arecibo radiotelescope. A deep glacial valley site has been selected in the Aragonese Pyrenees for testing the feasibility of a larger version, with the aperture size reaching 200 m.

It has been shown (Labeyrie et al. 2009) that the science yield of hypertelescopes increases greatly when the size of their elementary mirrors is reduced, at a given total array size and collecting area. This results from the enriched sampling of the incoming optical wavefront, give a wider "direct imaging field" and a better dynamic range. There is, however, a minimal usable size for the mirror elements, defined by their diffractive lobe, the spread of which should not exceed the size of the collecting focal optics. This minimal mirror size is on the order of a few centimeters for the 100-1000 km flotilla considered here. Infra-red observers have asked whether small mirrors would not degrade the star/sky contrast, as happens with a single telescope. The answer, calculated in Labeyrie (2007), is that the contrast does not depend on the subaperture size, at constant collecting area, nor on the amount of pupil densification, which however improves the signal/(photon noise) ratio (Labeyrie et al. 2008). This also likely applies to the case of a planet's image which is contaminated by light from its parent star when the pair is not resolved by the sub-apertures. The coronographic attenuation of the star's light may then require some clustering of the small mirrors to form groups spanning a few meters, across a giant diluted aperture spanning tens of kilometers.

3. Laser-trapped Hypertelescope Flotilla

Following the early suggestion (Labeyrie 1979) to fabricate large mirrors in space by trapping nanoparticles with a laser, the concept was studied theoretically by Guillon (2000), who also trapped oil micro-droplets in the laboratory. As discussed in Labeyrie et al. (2005) and in the report of a NASA/NIAC study (McCormack et al. 2006), the feasibility of large mirrors depends on issues such as the in-situ production of suitable nanoparticles, the loading of the laser trap, and the damping mechanisms. Preliminary calculations of the trapping stability, in the presence of random impacts by thermal photons from the cold universe, have suggested that broad particles, shaped like thin rigid plates or membranes, whether rigid or elastically deformable, are better than weakly bound and inelastic sheets of sub-wavelength nanoparticles. As may happen in soap bubbles, damped visco-elastic deformation tends to smooth, through spatio-temporal averaging, the local bumpiness induced by the photon impacts. These are mostly infrared ,originating from radiative thermal exchange rather than from the laser. Instead, a malleable membrane, such as a sheet of weakly-bound sub-wavelength nanoparticles, tends to become increasingly bumpy under such disturbances, thus requiring more laser power to restore its shape. Pending more detailed modelling to find optimal materials and plate sizes, rigid or visco-elastic plates spanning a few centimeters are considered here.

As sketched in Figure 1, their axial trapping involves a pair of monochromatic and coherent laser beams which propagate in nearly opposite directions.

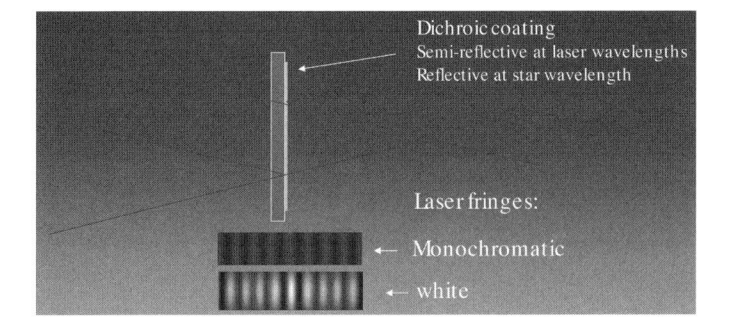

Figure 1. Principle of laser-trapped mirror: a pair of laser beams propagate in nearly opposite directions, producing an interference pattern in the form of standing waves. The mirror is semi-transparent at the laser wavelength. Because the intensity of both emerging beams is modulated by the interference of their transmitted and reflected components, the radiation pressure which they exert on the mirror is sensitive to its axial position and reverses at the internodes. The mirror remains trapped in one of the minima of the standing wave pattern. A tunable laser, with color shifted repeatedly from red to blue, attracts the mirror toward the central fringe, where the optical path difference is null.

They produce an interference pattern in the form of standing waves, a periodic stratification of the light intensity. A thin semi-reflective plate having zero absorption, oriented parallel to the standing wave and having a low initial velocity, tends to become trapped axially within one of the periods. This results from the reversing radiation pressure exerted on the plate if it moves axially. Indeed, such motion causes alternating extinctions of the light emerging on both sides of the plate. The extinctions themselves result from the destructive interference of both contributing light waves, the reflected and the transmitted. The former has a phase which is sensitive to the plate position, while the latter has a constant phase. In the absence of light absorption, the destructed emerging light on one side necessarily implies a reinforced light intensity emerging on the other side, produced by a constructive interference. The varying balance of emerging light on both sides of the plate applies to it a varying radiation pressure (Mansuripur 2004), the direction of which reverses if the plate moves by a quarter of the laser wavelength. Hence the trapping situation, at nodes which are spaced half a wavelength apart if the incident beams are exactly counter-propagating.

For constraining the plate to a single well-defined trapping node, rather than a periodic stratification of adjacent traps, polychromatic laser light is needed. A white laser produces a white central fringe at the location where the optical path difference between both beams is zero, which defines a privileged trapping position. It is of interest to use instead a tunable laser, with wavelength repeatedly swept from red to blue. This shrinks the fringe stratification toward the central fringe, thus attracting the mirror plate toward it, and all the way if the blue-sweeping action is repeated cyclically. With the small plates considered,

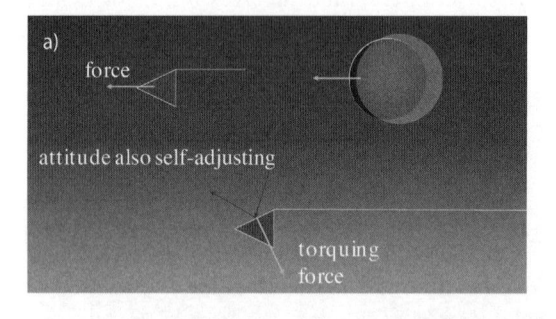

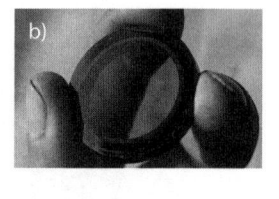

Figure 2. a) Transverse trapping is also achievable, if for example the mirror's edge is an annular deviating prism, shaped somewhat like the edge of a biconvex lens. The radial radiation force then exerted by a pair of laser beams having the same diameter as the mirror tends to correct the centering errors. The coarse attitude of the mirror can also be stabilized by laser light reflected on faces of the prism, which generates a restoring torque. Fine corrections of attitude are achieved by the standing wave. b) Example of commercial "pellicle beam splitter", potentially usable as a trapped mirror if its rigid frame, accurately polished flat before stretching the membrane, is made of glass with a prismatic cross-section. The nitro-cellulose membrane, moderately stable in space, is potentially replaceable by materials such as silicon nitride or CVD diamond. A non-absorbing dielectric coating is needed for 50% reflection and transmission at the laser wavelength, combined with maximal reflection in the science band.

and the large curvature radius of the giant diluted mirror, their shape can be made flat while still matching the curved locus within Rayleigh's 1/8th wave tolerance.

In practice, the plate must also be trapped in the transverse directions, with a position tolerance relaxed to millimeters rather than tens of nanometers. This is also achievable with radiation pressure from the same pair of laser beams, as sketched in Figure 2 and commonly observed in "laser tweezer" devices used to manipulate small objects such as bacteria, etc. The attitude of a plate can be finely controlled by the standing wave once it has been coarsely adjusted, which can be done by the reflected laser light on the plate's conical edge, as shown also in Figure 2.

In the absence of damping, oscillations tend to arise in all trapping modes. These can be attenuated with a visco-elastic flexing membrane serving as the mirror plate, or with micro-mechanical dampers if the plate is rigid, using for example a small mass attached to a visco-elastic flexing whisker. Active damping can also be considered, by moving the standing waves as needed to attenuate the oscillations, but this would require sensors and actuators which may be difficult to operate in systems using thousands of plates. Some simple modelling of these dynamic effects has been initiated, pending more detailed simulations, and verifications in the laboratory to optimize the plate design.

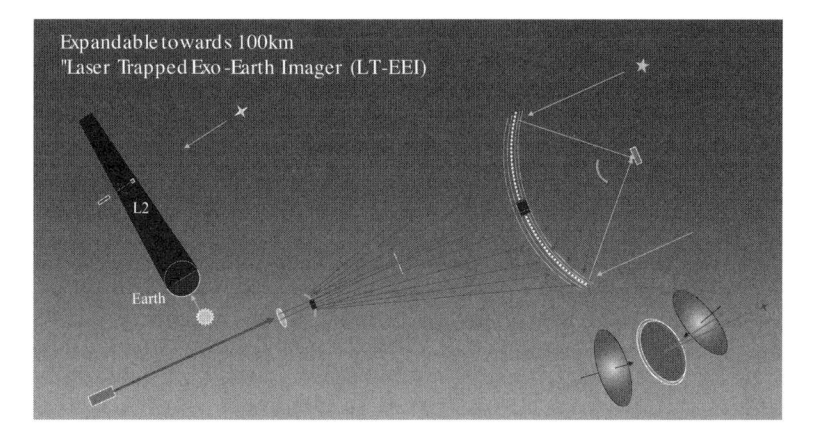

Figure 3. General arrangement of a large diluted mirror using small trapped elements. A single laser, illuminating a beam splitter and a pair of small diverging mirrors, provides a broad standing wave, the shape of which can be made paraboloidal by appropriately shaping the diverging mirrors. It can trap many small mirrors, which focus a star's image on a camera at the focal point. With adequate phasing, expected to be provided passively by the laser trapping, a bright and sharp interference peak formed in the stellar spread function allows direct imaging with high angular resolution. In accordance with hypertelescope theory, a pupil densifier micro-optical component (not shown) is needed near the camera for using efficiently the highly diluted aperture. It intensifies the image a million times or more. For smooth gravity and partial shading of sunlight by the Earth, a particularly interesting location is the L2 Lagrange point of the Sun-Earth system. Then, the laser can be located some distance away in full sunlight for powering its solar cells.

The general arrangement of a large diluted mirror using small trapped elements according to the above description is shown in Figure 3. A single laser illuminates all plates with as many pairs of beams, in such a way that a single giant mirror locus be assigned for all plates, then behaving as small pieces of the giant mirror. Since narrow laser beams, just covering each plate, are needed for the transverse trapping but also for economy of laser power, a micro-lens array is inserted near the laser source for "fanning" the beam, in such a way as to direct separate beams onto each mirror plate.

Although passively trapped, the flotilla's trapping involves no electronic feedback loops, it behaves as an adaptive dilute mirror since the standing wave serves both as a wavefront sensor and an actuator array. Indeed, it locally applies a restoring force related to the wavefront error, sensed in the form of the unbalance between the pair of emerging beams, itself responding to the plate's position error.

In theory a small flotilla can be expanded for a higher angular resolution just by angularly spreading the laser beams. The laser power needed is not increased, at constant mirror count, unless the trapping stability becomes limited by the gravity gradient (very small at the Lagrange L2 point considered) or the need for fast pointing.

4. Feasibility of Laser Cooling for Infrared Work

Laser cooling has been very successfully applied to atoms having resonant spectral lines. If it proves applicable to the macroscopic plates considered here, the cooling effect could be of interest for infrared observing and may suppress the need for liquid helium, heretofore used for cooling infrared space telescopes. The thermal jitter of atoms within a trapped plate causes some global jitter of the plate, the energy of which is drained by the laser trapping, with its damper or active damping system. The trapped mirror, therefore, comprises a potential well plus a dissipative mechanism. If the damping is large enough, the global energy of the system is expected to decrease in time, thus cooling down the mirror to temperatures which may be expected to reach a few Kelvin. Similar experiments have been performed in optical cavity configurations, where a stable optical trap for a macroscopic mirror was demonstrated with mirror cooling to an effective temperature of 0.8 K (Corbitt et al. 2007).

5. Deployment

The very weak radiation pressure, amounting to $2P/c = 60$ picoNewtons for a mirror receiving $P = 10$ milliWatts of illuminating laser light, requires a low level of disturbances such as the solar illumination, thermal photons absorbed and emitted, micro-meteorites, electrostatic effects, etc. The solar illumination can be minimized by pointing the instrument at right angle from the Sun, and also by using locations such as Lagrange's L2 point, in the Earth's shadow. Its partial darkness and smooth gravity are of interest for minimal laser power requirement. At least two macro-satellites are needed (Figure 4) to carry respectively: 1- the laser, with its solar cells and also possibly the main beam splitter with one of the beam launchers; 2 - the focal camera and the second beam launcher. With a

Figure 4. Flotilla with only two macro-satellites. One carries a blue-swept tunable laser and a beam splitter with a phasing actuator to stabilize the central fringe at the flotilla. The second carries a diverging laser mirror, and the stellar camera with its pupil densifier and other optics. The two satellites can be spaced 6000 km apart, the first being outside of the Earth's shadow for powering the laser with solar cells, and the second close to the mirror flotilla at L2. Stars are pointed by moving the laser satellite with conventional thrusters or possibly a solar sail.

blue-sweeping laser, the main beam-splitter can be carried by a phasing actuator to stabilize the position of the central fringe where the flotilla is trapped. The error signal may be provided by the laser light emerging from both sides of the central mirror in the flotilla, and returning to the main beam splitter where it produces Michelson fringes with zero path difference.

Among the advantages of optical apertures built from many small mirrors is the low bulk and mass, resulting from the fact that such mirrors are very thin, even perhaps down to microns if frameless membranes prove usable with their flexure controlled by the standing wave. With a more conservative mirror thickness, of the order of 0.1 to 3 mm for adequate rigidity, the collecting area equivalent to the 6.5 m JWST can be delivered in a suitcase containing for example 40,000 coin-sized mirrors. The deployment of all mirrors is expected to be achievable by the laser rlight rays themselves, if the divergence among the pair of fan beams is driven driven to increase after the initial trapping of all mirrors as a compact flotilla.

6. Numerical and Laboratory Simulations

It is of interest to verify the calculated behaviour of small trapped mirror plates before attempting to design a space hypertelescope. Steps are taken toward a laboratory experiment using a beam-splitter suspended in a vacuum from a wire. As depicted in Figure 5a, the experiment uses a "garden-gate" pendulum, according to the classical design which accounts for the high sensitivity of horizontal seismometers. The mass, which in our case is a dielectric beam splitter (BS), moves in a nearly horizontal plane around a nearly vertical axis. The restoring force is $g\sin\alpha$, where α is the angle between the vertical and the rotation axis, and the pendulum natural frequency $\sqrt{g\sin\alpha/L}$, where L is the wire length, can become extremely small for small α. In seismographic systems, displacements as small as 1 nm, with frequencies as low as 0.05 Hz, were previously measured (see, e.g., Bormann 2000).

246   Labeyrie et al.

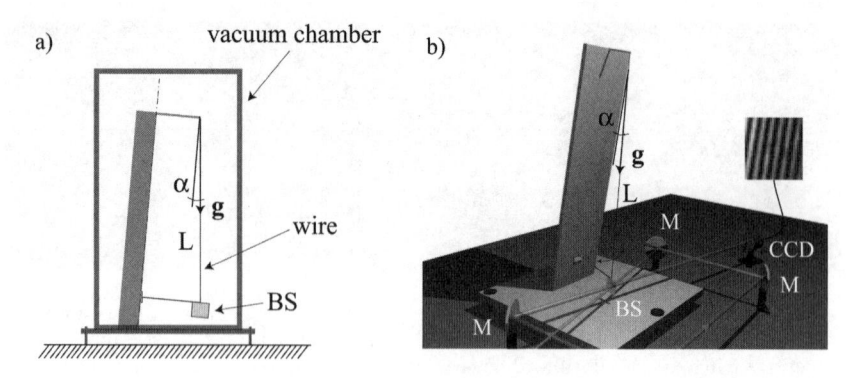

Figure 5. Laboratory testing with a suspended mirror: a) schematic diagram of a "garden gate" pendulum; the vacuum chamber is needed for the radiation pressure effect to become dominant with respect to the Brownian motion of the environmental gas particles. b) Setup for the interferometer with a trapped mirror BS. The inset shows a typical interference pattern which was recorded by the CCD camera during the motion of the mirror BS.

The principle scheme of the "garden-gate" pendulum inserted in an interferometer was tested by some of us (S.R., U.B.) in the laboratory. The setup is shown in Figure 5b. As a mirror we have used a dielectric beam splitter, diameter 7 mm, weight 0.44 g. The light source is an Ar^+ laser at $\lambda = 514$ nm, beam diameter 6 mm. The wire length is $L = 0.83$ m and α can be finely tuned by changing the tilt angle of the rotation axis of the pendulum. For $\alpha = 3°$ the natural frequency is 0.1252 Hz, corresponding to a period of 8 s.

It is worth noting that in order to obtain the same sensitivity with a standard pendulum a much longer wire, $L = 15.9$ m, should have been employed. The laser is separated in two optical beams of equal power, that are made to interfere on the BS, having a coefficient 0.5 of transmission and reflection. The tilt angle of the BS is adjusted in such a way that the transmission of one beam and the reflection of the opposite beam are colinear, thus producing a standing wave along the direction of the mirror axis. The position of the trapped beam splitter will be controlled by moving the standing wave, which is achieved with micrometric actuators carrying one of the mirrors.

After validating the setup, we have equipped the system with a vacuum chamber, where the whole pendulum was inserted. Preliminary tests have shown the necessity to reach vacuum levels on the order of 10^{-6} mbar, at least. Indeed, vacuum is needed for the radiation pressure effect to become dominant with respect to the Brownian motion of the environmental gas particles. An ion vacuum pump is used to reach ultra high vacuum condition with no moving parts, thus avoiding vibrations that would disturb the interferometric setup.

Finally, for accurately sensing the plate motion, slow light techniques are considered. Some of us (S.R, U.B) have recently shown that group velocities as small as 0.2 mm/s can be reached by performing two-wave mixing experiments in a liquid crystal cell (Residori et al. 2008). Correspondingly, a light pulse "sees" a very large group index, which can be exploited to greatly increase the sensitivity of certain types of interferometers, thus, allowing displacement measurements with subnanometer accuracy. Optical phase modulation methods can also be employed for detecting the mirror displacements with ultra-high accuracy (Bortolozzo et al. 2009).

Following such laboratory work it will be of interest to pursue the testing in the micro-gravity environment of the International Space Station (ISS) or other satellite. The vacuum racks arranged for experiments in the Columbus module of the ISS are of particular interest for such testing.

7. Conclusions and Future Work

Unlike the concept of a giant monolithic "laser trapped mirror", previously considered and subjected to preliminary investigations with nanoparticles, the adoption of small solid mirrors trapped as a large formation flight appears capable of reaching much higher angular resolution since the available laser power is fanned across a much larger diluted aperture, to be exploited according to the hypertelescope principle which itself favors many small apertures rather than a few large ones. The behaviour of the small trapped mirrors is much easier to model than for nano-particle mirrors, and it can be tested in the laboratory. If successful, further testing can be planned in the International Space Station for a robust design leading to an actual free-flying hypertelescope. The high angular resolution of such "Exo-Earth Imager" instruments thus buildable if the flotilla size can reach tens of kilometers, will be arc-microseconds at visible wavelengths. This is enough to resolve habitable planets and attempt to detect life signatures in the form of colored photosynthetic patches. Ways of achieving the required coronagraphic cleaning must be explored in more detail. Extra-galactic sources such as active galactic nuclei and remote galaxies will also be directly observable.

References

Aime, C. 2008, A&A, 483, 361
Bormann, P. 2002, New Manual of Seismological Observatory Practice, (Potsdam: GeoForchungs Zentrum)
Bortolozzo, U., Residori, S., & Huignard, J. P. 2009, Opt. Lett. 34, 2006
Corbitt, T., et al. 2007, Phys.Rev.Lett, 98, 150802
Gillet, S., et al. 2003, A&A, 400, 393
Guillon, M. 2000, Ph.D Thesis
Labeyrie, A. 1979, A&A, 77, L1
Labeyrie, A. 1996, A&AS, 118, 517
Labeyrie, A. 1999, in ASP Conf. Ser. 194, Working on the Fringe: Optical and IR Interferometry from Ground and Space, ed S. Unwin & R. Stachnik (San Francisco: ASP), 350
Labeyrie, A., Fournier, J. M. R., & Stachnik, R. V. 2004, SPIE, 5514, 365
Labeyrie, A., Guillon, M., & Fournier, J. -M. 2005, SPIE, 5899, 307
Labeyrie, A. Lipson, S. G., & Nisenson, P. 2006, An Introduction to Optical Stellar Interferometry (Cambridge: Cambridge University Press)
Labeyrie, A. 2007, C. R. Physique, 8, 426
Labeyrie, A., Le Coroller, H., & Dejonghe, J. 2008, SPIE, 7013, 70133J
Labeyrie, A., et al. 2009, Exp. Ast., 23, 463
Lardière, O., Martinache, F., & Patru, F. 2007, MNRAS, 375, 977
Le Coroller, H., Dejonghe, J., Arpesella, C., et al. 2004, A&A, 426, 721

McCormack, E. F., et al. 2006, Laser-Trapped Mirrors in Space, NASA/NIAC report, http://www.niac.usra.edu/files/library/meetings/annual/oct06/1202McCormack.pdf
Mansuripur, M., 2004, Optics Express, 12, 5375
Patru, F., Tarmoul, N., Mourard, D. & Lardiere, O. 2009, MNRAS, 395, 2363
Pedretti, E., Labeyrie, A., Arnold, L., et al. 2000, A&AS, 147, 285
Residori, S., Bortolozzo, U., & Huignard, J. P. 2008, Phys.Rev.Lett, 100, 203603

Pathways Towards Habitable Planets
ASP Conference Series, Vol. 430, 2010
Vincent Coudé du Foresto, Dawn M. Gelino, and Ignasi Ribas, eds.

Astrometric-Radial-Velocity and Coronagraph-Imaging Double-Blind Studies

W. A. Traub

Jet Propulsion Laboratory, California Institute of Technology, M/S 301-451, 4800 Oak Grove Dr., Pasadena, USA

Abstract. I describe two double-blind studies regarding the detection of Earth-size exoplanets, as follows. (1) Can we successfully detect Earth-like planets around nearby stars, in multi-planet systems, with the combination of a space astrometric mission such as SIM Lite plus ground-based radial-velocity (RV) observations? This study is now complete, and the answer to the question is yes, Earth-like planets have been detected by multiple teams, in numerical simulations with realistic noise and other factors, subject to specific simplifying assumptions, with accuracies that are close to the theoretically-expected values. (2) Can Earth-like planets be detected and characterized with imaging coronagraphs, in the presence of realistic instrumental noise and astrophysical confusion from exozodiacal dust disks, and with the possible help of information from SIM Lite? This study is about to begin.

1. Introduction

Two major space missions that have been proposed for exoplanets are the Space Interferometer Mission (SIM Lite) and the Terrestrial Planet Finder Coronagraph or Occulter (TPF-C or TPF-O). SIM Lite would detect exoplanets by astrometry, i.e., measuring the reflex motion of a parent star in the plane of the sky. TPF would characterize exoplanets by suppressing the light of the parent star and imaging the surrounding planet system in selected spectral bands. Both missions would be designed to measure planets as small in mass or radius, respectively, as the Earth. Both missions would target roughly 100 stars in the solar neighborhood.

Before committing to any major mission, it is important to carry out numerical simulations of performance. The studies described here address key aspects of the performance of SIM Lite and TPF.

For SIM Lite we addressed this question: Can SIM Lite, in conjunction with RV observations, detect an Earth-mass planet in the habitable zone (HZ) of its star, in systems that also contain giant planets?

For TPF we addressed this question: Can an imaging coronagraph detect and characterize an Earth-like planet in the HZ of its star, in systems that also contain a zodiacal light disk and other stars?

In both cases we make the simulation as realistic as possible, within bounds. Specifically, we do include expected instrument performance, realistic noise, realistic observing scenarios, and actual nearby stars. However we ignore effects that we expect can be modeled in a noise-free fashion, including for example relativistic effects, aberration of light, background clutter, and *n-body* effects that produce orbits different from a simple superposition of Keplerian orbits.

Our simulations are designed to be double-blind in the sense that the input parameters of the simulations are hidden from the people who distribute the simulated data sets as well as from the people who analyze the data. We do this to ensure that the analyzers see data as if it had been obtained at a telescope, with no hint as to what to expect.

2. The Astrometric-RV Double-Blind Study

The astrometric signature of an Earth at 1 AU around a Sun at 10 pc is about 0.3 μas, and the RV signature is about 0.1 m/s. Both values are the amplitude (half of peak to valley) of an edge-on system; for a face-on system the astrometric signature increases in the sense of going from one dimension to two, and the RV signal decreases proportional to the sine of the inclination.

For this study we take the root-mean-square (RMS) measurement noise to be 1.4 μas for astrometry, and 1.0 m/s for RV. The astrometric noise is derived from the difference of two 1.0 μas observations on the target and reference star, respectively, and these in turn are based on the photon noise of typical targets and integration times, plus expected instrument noise based on lab performance of component parts. The RV noise is based on the experience of observers who expect that most stars will have an intrinsic RV jitter from photospheric motions that limits the measurement accuracy.

For this study we assume that all other astrophysical effects can be perfectly removed from the data, and therefore do not contribute to the noise. This assumption can and does generate controversy owing to the fact that these effects have not yet been fully studied and cannot therefore be dismissed, but it is a comfortable assumption for others including some who are engaged in analyzing real RV data.

As examples of effects that we ignore we list relativistic effects of the planets, *n-body* interactions among the planets, relativistic effects of the observer, aberration of starlight, deflection of light by Jupiter, parallax and proper motion of the reference stars, and instrumental drifts.

We set a criterion that the false-alarm probability (FAP) of detecting a planet should be less than 1% at the end of the mission. This requires that the signal-to-noise ratio (SNR) of the measured stellar motion at the end of the mission should be 5.8 or larger. Thus to achieve SNR = 5.8 for the astrometric observations by themselves requires $(5.8 \times 1.4/0.3)^2$ or 740 observations. For RV we need 3400 observations.

Both cases assume that the noise can be beat down proportional to the square root of the number of measurements, which has been shown to be true for the astrometric case in the lab, and is simply assumed here for the RV case since the nature of stellar oscillation noise is not yet well understood.

The number of astrometric observations is within the scope of the SIM Lite mission, and is less than the current limit for the number of observations ($\leq$ 1600) that follow the square-root relation.

The number of RV measurements would require one observation per night for nearly 20 years, assuming that the star is available for half of the year. This is not realistic, and in addition it ignores the possibility that 1/f noise may limit the effective number of observations.

For these reasons, the simulation assumes that SIM and RV would work together, with a 5-year SIM mission providing the accuracy needed for Earth-HZ detection, and a 15-year RV data set providing information on the long-term, Jupiter-like orbits that cannot be fully observed during the SIM mission. In fact, this is one of the outstanding results of this simulation, to show that the partial orbit of a star owing to a Jupiter can be successfully removed with the RV data. Without this help from RV, there is vast potential for confusing a partial 12-year orbit with the incorrect combination of a proper motion plus a shorter few-year orbit.

The double-blind exercise engaged five teams of modelers who generated over 750 model planetary systems, one team who generated over 155 sets of simulated astrometric and RV data, and five teams of analyzers who extracted planetary mass and orbit estimates from the data. Details are provided in a brief preliminary report (Traub et al. 2010a) and in a comprehensive final report (Traub et al. 2010b).

The overall results of this double-blind exercise can be summarized in two statistics, as follows.

The first statistic is the completeness of detection of planets, which is defined as the percentage of correct detections compared to the number which should have been expected. Here "correct" means that the planet must have an estimated SNR of at least 5.8, and a period of less than 15 years, and that its estimated mass and period must be equal to the input values to within 3 times the Cramer-Rao bound for that planet considered in isolation. As a preliminary result, we find that the overall completeness is about 89%, with the value for terrestrial planets in the HZ being slightly better, 94%. We believe that this is an excellent result, which could only improve as the experience of the analyzing teams improves.

The second statistic is the reliability of detection of planets, which is defined as the percentage of correct detections compared to total corrections. The design goal is a reliability of $1 - FAP$ or 99%. Again, as a preliminary result, we find that the overall reliability of detecting all planets is about 96%, with the reliability of terrestrial planets in the HZ being even better, 100%. We believe that this result, too, is excellent.

Overall this exercise has shown that, within the limits of the assumptions and approximations made, the combination of a SIM Lite mission, working at its design accuracy, with a 10-year longer record of RV observations, working at the nominally expected accuracy, can successfully detect terrestrial mass planets in their HZ, even in the presence of more massive giant planets whose signatures substantially exceed those of terrestrial planets.

3. The Coronagraph-Imaging Double-Blind Study

An outstanding issue for coronagraphs is the question of how well can a planet be detected in an image that also contains the star's zodi cloud, other planets, speckles from incomplete suppression of scattered light from the star, photon noise, and read noise? We plan to address this issue using the same general technique as above, using teams of competitively selected analysts. We sketch here some considerations regarding the input parameters for this exercise.

The star, as seen in the focal plane of a coronagraph, can be modeled as a highly suppressed (but not zero) diffraction-limited source. The shape and brightness of the point spread function (PSF) of the star will therefore be a function of the pupil shape and coronagraph rejection capability.

The image of the zodi cloud will be a function of the inclination, optical thickness, and radial and azimuthal density variation of the face-on disk. The latter three parameters will be derived from theoretical model disks, convolved with the PSF.

Other planets in the system will be taken from the model systems in the SIM-RV exercise, with appropriate guesses for the albedos as a function of planet mass and semi-major axis.

The speckle pattern for an internal coronagraph will be simulated from estimates of the power spectral density of residual wavefront errors in an internal coronagraph, transformed to the focal plane, and modified by the expected ability of the coronagraph to suppress such speckles in the range between the inner and outer working angles. These working angle limits will depend on the design of the coronagraph, the diameter of the primary mirror, wavelength, and number of elements in the deformable mirror.

The speckle pattern for an external coronagraph will depend on the intrinsic suppression of a perfect occulter as well as the power spectral density of the errors in the shape of the occulter, including positioning errors. These will be estimated from diffraction theory and expectations for manufacturing accuracy, deployment accuracy, and stability.

Photon noise as well as read and cosmic ray noise will be added according to theory and practice, respectively, with parameters being the target brightness, integration time, bandwidth, and type of detector.

The scenes to be imaged may or may not contain planets. The scenes also may be either single snapshots or a sequence of images over a period sufficient to allow a planet to move into or out of the observable range of separation angles. We may also supply simulated information from a SIM-RV exercise, as above, thereby providing estimates of mass and orbit, but with the proviso that any orbit prediction from a SIM-RV mission will necessarily have uncertainties, such that although a given planet may be known to be present with high confidence, its predicted position may not be accurate to within a substantial fraction of an orbit, owing to expected errors of estimating its orbital parameters.

We expect that this exercise may take over a year to complete.

Acknowledgments. Part of the research described in this paper was carried out at the Jet Propulsion Laboratory, California Institute of Technology, under a contract with the National Aeronautics and Space Administration. (c)2009. All rights reserved.

References

Traub, W.A.,et al. 2010a, Detectability of Earth-Like Planets in Multi-Planet Systems: Preliminary Report, PASP, submitted

Traub, W.A., et al. 2010b, Detectability of Terrestrial Planets in Multi-Planet Systems, Using Astrometry and Radial Velocity, in preparation

Pathways Towards Habitable Planets
ASP Conference Series, Vol. 430, 2010
Vincent Coudé du Foresto, Dawn M. Gelino, and Ignasi Ribas, eds.

Gaia and the Astrometry of Giant Planets

M. G. Lattanzi and A. Sozzetti

INAF - Ossservatorio Astronomico di Torino, Pino Torinese, Italy

Abstract. The scope of this contribution is twofold. First, it describes the potential of the global astrometry mission Gaia for detecting and measuring planetary systems based on detailed double-blind mode simulations and on the most recent predictions of the satellite's astrometric payload performances (launch is foreseen for late Summer 2012). Then, the identified capabilities are put in context by highlighting the contribution that the Gaia exoplanet discoveries will be able to bring to the science of extrasolar planets in the next decade.

1. Gaia in a Nutshell

The Gaia mission is the new global, all-sky, astrometric initiative of the European Space Agency with a possible launch in late Summer 2012. A Soyuz-Fregat launcher will take the Gaia module to a transfer orbit, which in one month will allow the satellite to reach its operational environment in a Lissajous orbit at Sun-Earth L2, 1.5 million kilometers away from Earth. During its 5 years of operational lifetime (with the possible extension of an extra year), Gaia will monitor all point sources in the visual magnitude range 6 − 20 mag, a huge database of some $\sim 10^9$ stars, a few million galaxies, half a million quasars, and a few hundred thousand asteroids.

As for the observing strategy, Gaia's mode of operation has adopted the principles successfully experimented with the Hipparcos mission (ESA 1997). In particular, it will continuously scan the sky implying that all detected objects, irrespective of their magnitudes, are observed for the same amount of time during each field-of-view crossing, with mission-end observing time mainly depending on ecliptic latitude (Lindegren 2010).

In this way, it is anticipated that Gaia will determine the five basic astrometric parameters (two positional coordinates, two proper motion components, and the parallax) for all objects, with end-of-mission (sky-averaged) precision between 7-25 μas (microarcsec) down to the Gaia magnitude $G^1 = 15$ mag and a few hundred μas at $G = 20$ mag, depending on color. Objects redder than $V - I = 0.75$ are expected to have better astrometry, while that of extreme blue targets is estimated to degrade by a factor of two.

A combination of an ambitious science case, wishing to address breakthrough problems in Milky Way astronomy, and lessons learned from the Hipparcos experience made European astronomers realize that spectrophotometry is essential for modern astrometry. Therefore Gaia's astrometry is complemented

---
[1]close to Johnson's *R*.

by on-board spectrophotometry and also, for objects brighter than $G = 17$), with radial velocity information. These data have the precision necessary to quantify the early formation, and subsequent dynamical, chemical, and star formation evolution of our Galaxy. The broad range of crucial issues in astrophysics that will be addressed by the wealth of the Gaia data is summarized by e.g., Perryman et al. (2001). One of the relevant areas on which the Gaia observations will have great impact is the astrophysics of planetary systems (e.g., Casertano et al. 2008), in particular when seen as a complement to other techniques for planet detection and characterization (e.g., Sozzetti 2009).

2. Project Organization

The main partners behind the Gaia project are: i) the European Space Agency (ESA), which has the overall project responsibility for funding and procurement of the satellite, launch, and operations. Of interest is the fact that in this case satellite procurement includes the payload and its scientific instruments, unlike ESA's other science missions for which scientific instruments are usually PI-lead and funded (or, at least, co–funded) by participating national space agencies; ii) EADS Astrium, which was selected in 2006 as the prime industrial contractor to design and build the satellite according to the scientific and technical requirements formulated to fulfill the mission science case as approved at time of selection (ESA 2000); and iii) the Gaia Data Processing and Analysis Consortium (DPAC), charged with the design, implementation, and operation of a complete software system for the scientific processing of the satellite data, resulting in a Gaia Catalogue a few years after the end of the operational (observation) phase.

DPAC was formed in 2006 in response to an Announcement of Opportunity issued by ESA. The Consortium currently lists nearly 400 individual members in more than 20 countries, including a team at the European Space Astronomy Center near Madrid. Six data processing centers participate in the activities of the consortium, which is organized in eight coordination units. Each coordination unit is responsible for the development of one part of the software like, e.g., *simulations*, *core processing* (global astrometry), *photometry*, and *non-single stars*, the unit devoted to the processing of astrometrically "noisy" stars, which will include potential planetary systems. Most of the financial support is provided by ESA (for the team at ESAC) and by the various national space agencies through a legally binding long–term funding agreement, a real first for ESA-run missions.

There will be no proprietary periods for the scientific exploitation of the data. The final Gaia catalogue will be produced and immediately delivered to the worldwide astronomical community as soon as ESA and DPAC agree that the processed data have reached the targeted (science) quality. This catalogue is expected to be ready three years after the end of operations. Finally, intermediate releases, of some provisional results, are planned after a few years of observations.

More information can be found in Lindegren (2010), while other organizational details and the latest news on payload and satellite developments are available on the Gaia web pages at http://www.rssd.esa.int/gaia/.

3. Gaia and Extrasolar Planets

3.1. What will Gaia see?

As explained above, Gaia's mode of operation[2] is such that there cannot be any optimization to the case of extrasolar planets. The fundamental requirement, i.e. to have sufficient astrometric accuracy at magnitudes brighter than $V = 13$, was established at time of the science case definition.

Since little can be done with the photometric and spectroscopic capabilities aboard the satellite, and these cannot compete with present and planned ground-based facilities for very high precision radial-velocity measurements (Pepe & Lovis 2008) and space-borne observatories devoted to ultra-high precision transit photometry (e.g., Sozzetti et al. 2010), the potential contribution of Gaia to exoplanets science must be gauged purely in terms of its astrometric capabilities.

3.2. The Gaia double-blind tests campaign

A number of authors have tackled the problem of evaluating the sensitivity of the astrometric technique required to detect extrasolar planets and reliably measure their orbital elements and masses (Sozzetti 2005, and references therein). Those works mostly relied on simplifying assumptions with regard to (*a*) the error models to be applied to the data (e.g., simple Gaussian distributions, perfect knowledge of the instruments) and (*b*) the analysis procedures to be adopted for orbit reconstruction (mostly ignoring the problem of identifying adequate configurations of starting values from scratch). The two most recent exercises on this subject (Casertano et al. 2008; Traub et al. 2009) have revisited earlier findings using a more realistic double-blind protocol. In this particular case, several teams of "solvers" handled simulated datasets of stars with and without planets and independently defined detection tests, with levels of statistical significance of their choice, and orbital fitting algorithms, using any local, global, or hybrid solution method that they judged was best. The solvers were provided no information on the actual presence of planets around a given target.

In the large-scale, double-blind test (DBT) campaign carried out to estimate the potential of Gaia for detecting and measuring planetary systems, Casertano et al. (2008) showed that (*a*) planets with $\alpha \simeq 6\sigma$ (where σ is the single-measurement error) and orbital periods shorter than the nominal 5-year mission lifetime could be accurately modeled, and (*b*) for favorable configurations of two planet systems with well-separated periods (both planets with $P \leq 4$ yr and $\alpha/\sigma \geq 10$, redundancy over a factor of 2 in the number of observations) it would be possible to carry out meaningful coplanarity tests, with typical uncertainties on the mutual inclination angle of ≤ 10 degrees. Subtle differences as well as significant discrepancies were found in the orbital solutions carried out by different solvers. This constitutes further evidence that the convergence of non-linear fitting procedures and the quality of orbital solutions (particularly for multiple systems and for systems with small astrometric signals) can be significantly affected by the starting guesses for the parameters in the orbital

---

[2] A magnitude–limited, or better, S/N threshold-limited survey, uneven coverage, including time sampling and scanning geometry, depending on ecliptic latitude.

256 Lattanzi and Sozzetti

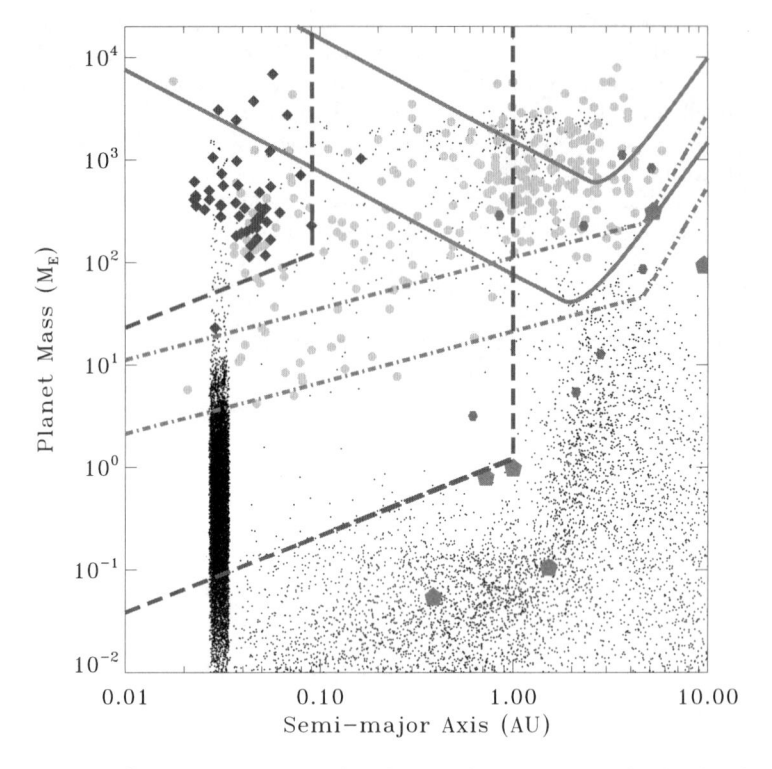

Figure 1. Gaia discovery space for planets of given mass and orbital radius compared to the present-day sensitivity of other indirect detection methods, namely Doppler spectroscopy and transit photometry. Detectability curves are defined on the basis of a 3-σ criterion for signal detection. The upper and lower solid curves are for Gaia astrometry with $\sigma_A = 10$ μas, assuming a 1-$M_\odot$ G dwarf primary at 200 pc and a 0.4-$M_\odot$ M dwarf at 25 pc, respectively. The nominal survey duration is set to 5 yr. The radial velocity curves (dashed-dotted lines) assume $\sigma_{RV} = 3$ m s^{-1} (upper curve) and $\sigma_{RV} = 1$ m s^{-1} (lower curve), $M_\star = 1 M_\odot$, and 10-yr survey duration. For visible-light transit photometry (long-dashed curves), the assumption are $\sigma_V = 5 \times 10^{-3}$ mag (upper curve) and $\sigma_V = 1 \times 10^{-5}$ mag (lower curve), $S/N = 9$, $M_\star = 1$ $M_\odot$, $R_\star = 1$ $R_\odot$, uniform and dense (>> 1000 datapoints) sampling. The light-grey circles indicate the inventory of Doppler-detected exoplanets as of December 2008. Transiting systems are shown as dark-grey filled diamonds, while the grey hexagons are planets detected by microlensing. Solar System planets are also shown as large grey pentagons. The small black crosses represent a theoretical distribution of masses and final orbital semi-major axes from Ida & Lin (2008).

Gaia and the Astrometry of Giant Planets

Δd (pc)	N_*	Δa (AU)	ΔM_p (M_J)	N_d	N_m	Case	Number of Systems
0-50	1×10^4	1.0 - 4.0	1.0 - 13.0	1400	700	Detection	~ 1000
50-100	5×10^4	1.0 - 4.0	1.5 - 13.0	2500	1750	Orbits and masses to better than 15-20% accuracy	$\sim 400 - 500$
100-150	1×10^5	1.5 - 3.8	2.0 - 13.0	2600	1300	Successfull coplanarity tests	~ 150
150-200	3×10^5	1.4 - 3.4	3.0 - 13.0	2150	1050		

Figure 2. Left: Number of giant planets that could be detected (N_d) and measured (N_m) by Gaia, as a function of increasing distance (see Casertano et al. 2008 for details). Starcounts are obtained using models of stellar population synthesis to V≤13 (Bienaymé et al. 1987), while the Tabachnik & Tremaine (2002) model for estimating planet frequency as a function of mass and orbital period is used. Right: Number of planetary systems that Gaia could potentially detect, measure, and for which coplanarity tests could be carried out successfully.

fits, by using different statistical indicators of the solution quality, and by varied levels of significance of the latter.

Overall, the authors concluded that Gaia could discover and measure massive giant planets ($M_p \geq$2–3 M_J) with 1 < a < 4 AU orbiting solar-type stars as far as the nearest star-forming regions, as well as explore the domain of Saturn-mass planets with similar orbital semi-major axes around late-type stars within 30–40 pc (see Figure 1). These results can be used to infer the number of planets of a given mass and orbital separation that can be detected and measured by Gaia, using Galaxy models and the current knowledge of exoplanet frequencies. By inspection of the tables in Figure 2, one then finds that Gaia's main strength will be its ability to accurately measure orbits and masses for thousands of giant planets[3], and to perform coplanarity measurements for a few hundred multiple systems with favorable configurations.

4. The Gaia Legacy

Gaia's main contribution to exoplanet science will be its unbiased census of planetary systems orbiting hundreds of thousands nearby (d < 200 pc), relatively bright ($V \leq 13$) stars across all spectral types, screened with constant astrometric sensitivity. Gaia data have the potential to:

a) *Significantly refine our understanding of the statistical properties of extrasolar planets*: for example, the predicted database of several thousand extrasolar planets with well-measured properties will allow users to test

---

[3]To avoid saturation, Gaia will reject stars brighter than R=6; the effect will be a drop of as much as 30% in the number of targets screened for planets in the 0–50 pc bin of Fig. 2, but of only a few percent when compared to $\sim$ 500,000 useful targets out to 200 pc.

the fine structure of giant planet parameters distributions and frequencies, and to investigate their possible changes as a function of stellar mass, metallicity, and age with unprecedented resolution;

b) *Help crucially test theoretical models of gas giant planet formation and migration*: for example, specific predictions on formation time-scales and the role of varying metal content in the protoplanetary disk will be probed with unprecedented statistics due to the thousands of metal-poor stars and hundreds of young stars screened for giant planets out to a few AUs ;

c) *Achieve key improvements in our comprehension of important aspects of the formation and dynamical evolution of multiple-planet systems*: for example, the measurement of orbital parameters for hundreds of multiple-planet systems, including meaningful coplanarity tests will allow discrimination between various proposed mechanisms for dynamical interaction;

d) *Aid in the understanding of direct detections of giant extrasolar planets*: for example, actual mass estimates and full orbital geometry determination for suitable systems will inform direct imaging surveys about the epoch and location of maximum brightness, in order to estimate optimal visibility, and will help in the modeling and interpretation of giant planets' phase functions and light curves;

e) *Provide important supplementary data for optimal target selection for future observatories aiming at the direct detection and spectral characterization of habitable terrestrial planets*: for example, all F-G-K-M stars within the useful volume (~ 25 pc) will be screened for Jupiter- and Saturn-sized planets out to several AUs, and these data will help probe the long-term dynamical stability of their Habitable Zones, where terrestrial planets may have formed, and may be found.

5. Conclusions

Gaia will collect the largest compilation of astrometric orbits of giant planets (in many cases signposts of more interesting systems!), unbiased across all spectral types up to $d \simeq 200$ pc. In combination with present-day and future extrasolar planet search programs, these data will allow Gaia to make important contributions to several aspects of planetary systems astrophysics including formation theories and dynamical evolution.

Acknowledgments. Financial support from the Italian Space Agency (ASI) through contract I/037/08/0 (Gaia Mission - The Italian Participation in DPAC) is gratefully acknowledged.

References

Bienaymé, O., Robin, A. C., & Crézé, M. 1987, A&A, 180, 94
Casertano, S., Lattanzi, M. G., Sozzetti, A., et al. 2008, A&A, 482, 699
ESA 1997, *The Hypparcos and Tycho Catalogues*, ESA SP-1200
ESA 2000, *Gaia: Composition, Formation, and Evolution of the Galaxy*, ESA-SCI(2000)4
Ida, S., & Lin, D.N.C. 2008, ApJ, 685, 584
Lindegren, L. 2010, Proc. IAU Symp. 261, 296
Pepe, F., & Lovis, C. 2008, Physica Scripta, 130, 014007
Perryman, M. A. C., et al. 2001, A&A, 369, 339
Sozzetti, A. 2005, PASP, 117, 1021
Sozzetti, A. 2009, EAS Publication Series, in press (arXiv:0902.2063)
Sozzetti, A., et al. 2010, to appear in ASP Conf. Ser. (arXiv:0912.0887)
Tabachnik, S., & Tremaine, S. 2002, MNRAS, 335, 151
Traub, W. A., et al. 2009, EAS Publication Series, in press (arXiv:0904.0822)

Pathways Towards Habitable Planets
ASP Conference Series, Vol. 430, 2010
Vincent Coudé du Foresto, Dawn M. Gelino, and Ignasi Ribas, eds.

PLATO : PLAnetary Transits and Oscillations of Stars - The Exoplanetary System Explorer

C. Catala,[1] T. Arentoft,[2] M. Fridlund,[3] R. Lindberg,[3]
J. M. Mas-Hesse,[4] G. Micela,[5] D. Pollacco,[6] E. Poretti,[7] H. Rauer,[8]
I. Roxburgh,[9] A. Stankov,[3] and S. Udry[10]

Abstract. PLATO's objective is to characterize exoplanets and their host stars in the solar neighbourhood. While it builds on the heritage from CoRoT and *Kepler*, the major breakthrough will come from its strong focus on bright targets ($m_V \leq 11$). The PLATO targets will also include a large number of very bright ($m_V \leq 8$) and nearby stars. The prime science goals of PLATO are: (i) the detection and characterization of exoplanetary systems of all kinds, including both the planets and their host stars, reaching down to small, terrestrial planets in the habitable zone; (ii) the identification of suitable targets for future, more detailed characterization, including a spectroscopic search for bio-markers in nearby habitable exoplanets. These ambitious goals will be reached by ultra-high precision, long (few years), uninterrupted photometric monitoring in the visible of very large samples of bright stars, which can only be done from space. The resulting high quality light curves will be used on the one hand to detect planetary transits, as well as to measure their characteristics, and on the other hand to provide a seismic analysis of the host stars of the detected planets, from which precise measurements of their radii, masses, and ages will be derived. The PLATO space-based data will be complemented by ground-based follow-up observations, in particular very precise radial velocity monitoring, which will be used to confirm the planetary nature of the detected events and to measure the planet masses. The full set of parameters of exoplanetary systems will thus be measured, including all characteristics of the host stars and the orbits, radii, masses, and ages of the planets, allowing us to derive planet mean densities, and estimate their temperature and radiation environment. Finally, the knowledge of the age of the exoplanetary systems will allow us to put them in an evolutionary perspective.

[1]Observatoire de Paris, LESIA, 5 place Jules Janssen, Meudon, France

[2]Aarhus University, Denmark

[3]ESTEC, ESA, Noordwijk, The Netherlands

[4]Centro de Astrobiologia (CSIC-INTA), Madrid, Spain

[5]INAF - Osservatorio Astronomico di Palermo, Italy

[6]Queens University, Belfast, UK

[7]INAF - Osservatorio Astronomico Brera, Merate, Italy

[8]DLR, Berlin, Germany

[9]Queen Mary College, University of London, UK

[10]Observatoire de Genève, Switzerland

1. Main Science Goals

PLATO will address the question of the existence, distribution, evolutionary state, and characteristics of exoplanets in the solar neighbourhood. Answers to these questions are essential to understand how planetary systems, including our own, are formed and evolve, and also as a first and necessary step to understand whether life can exist elsewhere in the Universe, and locate potential sites for life. Since the discovery of the first exoplanet in 1995, this field has seen a remarkable development, with about 400 exoplanets known as of the end of October 2009. Most of these objects are giant planets in close-in orbits, but continuous progress in the precision of radial velocity observations is now enabling the detection of "super-Earths", with masses just a few times that of the Earth.

The field of exoplanet search has been recently boosted by the launch of the CoRoT satellite in Dec. 06, followed by that of *Kepler* in March 09. The discovery of CoRoT-7b, the very first small telluric, rocky planet with measured radius and mass, and therefore with a known density, has opened up a new era, in which the CoRoT extended mission and *Kepler* will now play a major role.

Both CoRoT and *Kepler* target faint stars, up to $m_V = 15$ and beyond, which makes their ground-based follow-up difficult, in particular in radial velocity monitoring. As a consequence, ground confirmation and mass measurements are restricted to the largest of the CoRoT and *Kepler* planets, which severely impacts the scientific return of these missions. While we can today with CoRoT, and soon *Kepler*, detect the passage of a planet the size of our own world, it is impossible to confirm the presence of any such object found by either spacecraft. Moreover, even in cases where radial velocities can be measured to the required precision to confirm the planetary nature of the detected event and measure the planet-to-star mass ratio, our knowledge of the faint host stars is still too poor to allow us to derive estimates of the planet radii, masses and ages to a sufficient accuracy to significantly constrain their structure and state of evolution.

The goal of PLATO is to alleviate these difficulties by focusing on bright stars, 3 to 4 magnitudes brighter than CoRoT and *Kepler*, and also by including in its target list a large sample of very bright ($m_V \leq 8$) and nearby stars. This will bring three decisive advantages: (i) the ground-based follow-up observations will be highly facilitated, and the required precision will be reached to confirm small, terrestrial planets in the habitable zone and to measure their masses; (ii) the host stars of the detected planets will be studied in detail, in particular via seismic analysis using the PLATO data themselves; seismic analysis, i.e. the measurement of stellar oscillations, will be used to probe the internal structure of these stars, and determine their radii, masses, and age in a precise and reliable way; (iii) the detection of exoplanets orbiting very bright and nearby stars will provide the best targets for subsequent detailed follow-up observations, both from space (e.g., JWST) and from the ground (e.g., E-ELT), including in particular spectroscopy of their surfaces and atmospheres, in the search for bio-markers.

2. Observation Strategy

The main PLATO science product will be a very large sample of ultra-high precision stellar light curves, obtained on very long time intervals (up to 3 years) with very high duty cycle ($\geq$95%). The requirement is to obtain a photometric precision better than 2.7×10^{-5} in 1 hr for more than 20,000 cool dwarfs/subgiants brighter than $m_V \approx 11$, and 8×10^{-5} in 1 hr for 250,000 stars down to $m_V \approx 13$-14.

In order to reach this goal, PLATO will monitor two successive very wide fields, for 2 to 3 years. These two long monitoring sequences will be followed by a one- or two-year step&stare phase, during which a number of fields will be monitored for several months each. This step&stare phase will bring flexibility to the mission, allowing for instance to survey a very large fraction of the whole sky (up to 40% depending on the selected concept), as well as to re-visit particularly interesting targets identified during the long monitoring phases.

The spacecraft is intended to be launched in 2017 on a Soyuz-Fregat rocket for injection into an orbit around the L2 Lagrangian point for a nominal lifetime of 6 years, which is compatible with the observation strategy outlined above.

3. Payload

PLATO was subject to three independent studies, two by industrial contractors and one by the PLATO PayLoad Consortium (PPLC). All three studies have been running in parallel and were all completed at the end of summer 2009.

The three resulting concepts differ significantly from one another, but have in common very wide fields of view, and overall large collecting areas. Wide fields-of-view are required to obtain large samples of bright stars, while large collecting areas are necessary to reach the desired photometric precision. In all three concepts, this is achieved by using a collection of small, wide-field telescopes, each with its own CCD-based focal plane. The light and centroid curves from each telescope unit are transmitted to the ground, where they are co-added to reach the desired precision. All three designs are compliant with the science requirements, and feasible at no technological risk.

We describe here only the PPLC concept, based on a set of 42 refractive telescopes. The telescopes are grouped in 4 sections, each section having its line of sight offset from the next one by one-half of the field of view. This overlapping line-of-sight arrangement is depicted in Figure 1 and results in an overall instrument field of view of about 1,800 deg^2, each star being observed either by 10, 20, or 40 telescopes, with an effective collecting area of 0.12, 0.24, or 0.48 m^2, depending on its position in the field. The surveyed area after two long pointings will thus be 3,600 deg^2, while it will go up to 18,000 deg^2 (i.e. more than 40% of the whole sky) when a two-year step&stare phase is added during which 4 fields per year are monitored for 3 months each.

The telescopes include fully refractive optics with 6 lenses, with a 12cm pupil. Each one is equipped with its own focal plane made of four $3,584^2 \times 18\mu m$ pixel CCDs, covering a square field of view of about 28°. Two out of 42 telescopes are devoted to very bright stars with m_V between 4 and 8, and have their CCDs operated in frame transfer mode, while the remaining forty telescopes observe

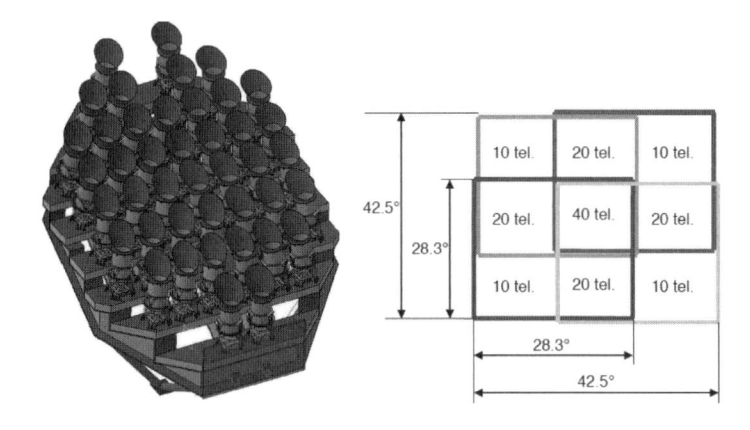

Figure 1. The PLATO payload in the PPLC concept. *Left:* global view; *Right:* overlapping line-of-sight concept.

stars fainter than m_V=8, with their CCDs working in full frame mode. The total dynamic range of the full PLATO instrument is 4 $\leq m_V \leq$ 13.

4. Performances

A full end-to-end simulator was used to estimate the expected level of noise. It takes into account all sources of noise, including photon noise, confusion by neighbouring sources, readout noise, satellite jitter, etc., by including all characteristics of the observed field (star positions and magnitudes, zodiacal light, etc) and of the instrument (optical PSF, detector characteristics, etc).

Figure 2 presents these results for a representative fraction of the PLATO field, and shows that photon noise level is approached closely at magnitudes brighter than m_V=10, and that non photonic noise remains well below photon noise at least down to m_V=11.5. A level of noise of 2.7×10^{-5} in 1 hr is reached down to m_V=11, while a noise of 8×10^{-5} in 1 hr is obtained around m_V=12.5. At fainter magnitudes, the noise becomes significantly higher than pure photon noise, due to the contribution of contaminating sources.

Table 1 shows the expected numbers of dwarfs/subgiants later than F5, observable with PLATO and *Kepler* at various levels of noise, and down to various magnitudes. The three PLATO concepts are considered in this table, concept C corresponding to the PPLC design presented here. The gain compared to *Kepler* is clearly shown in this table, indicating in particular that PLATO will observe as many stars down to 8th magnitude as *Kepler* down to 11th.

The comparison with *Kepler* is further illustrated by Figure 3, showing the expected total number of detectable planets as a function of mass and orbital semi-major axis, for both missions. The numbers indicated are those of the expected detectable planets by both missions, while the sizes of the coloured regions show the respective discovery potential of both missions (blue: PLATO and green: *Kepler*). There is no underlying planet formation model, instead we

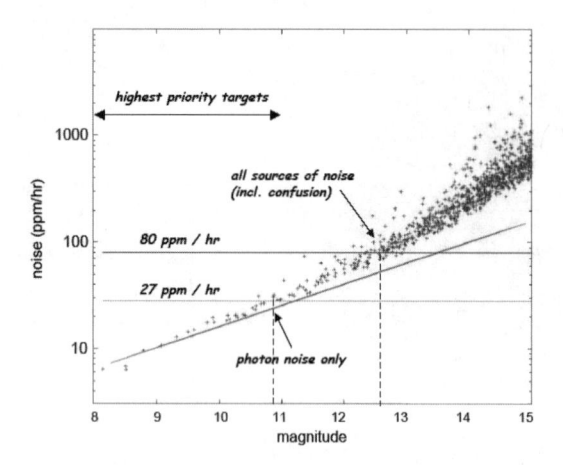

Figure 2. Noise level measured in the PLATO end-to-end simulation for stars in a representative subset of the PLATO field, in the PPLC concept.

design surveyed area	PLATO concept A 3600 deg²		PLATO concept B 1250 deg²		PLATO concept C 3600 deg²		Kepler 100 deg²	
noise level (10^{-5}/hr)	# stars	mag lim	# stars	mag lim	# stars	mag lim	# stars	mag lim
2.7	24,000	10.4	21,000	11.1	23,000	9.8-11.1	1,300	11.2
8.0	374,000	12.7	257,000	13.5	316,000	11.8-12.9	25,000	13.6
magnitude	# stars		# stars		# stars		# stars	
6					90		0	
8	1,350		675		1,350		30	
9	3,800		1,320		3,800		100	
10	13,500		4,700		13,500		370	
11	48,300		16,800		48,300		1,300	

Table 1. Numbers of targets from PLATO and *Kepler*, at various photometric noise levels and various magnitude limits. All star numbers refer to cool dwarfs and subgiants only.

simply assume that each star has one (and one only) planet in the parameter range considered. The planet is considered detectable if it can be seen in transit by the satellite AND confirmed by follow-up radial velocity measurements with a reasonable amount of telescope time.

These results take into account all sources of noise for the radial velocity follow-up (oscillation, granulation, activity level), limiting the required observing time to reasonable values. Future developments in our understanding of the interplay between activity level and induced radial-velocity jitter might help correct for the spurious effect and further improve our characterization ability.

As can be seen in Figure 3, we expect a vastly increased number of planet detections with PLATO , in particular in the lower mass range, where *Kepler* is not expected to produce confirmable planets. The main reason is related to the much larger field of view of PLATO , and consequently the brightness of the candidates. The main conclusion is that, although *Kepler* will bring considerable

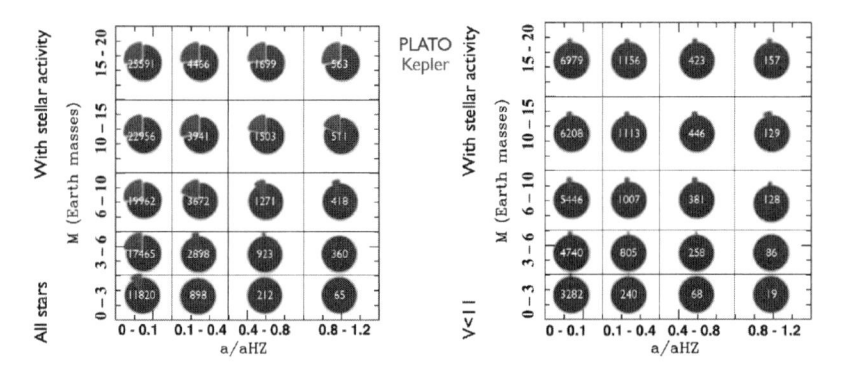

PLATO: PLAnetary Transits and Oscillations of Stars 265

Figure 3. Estimated total numbers of detected transiting planets, which can be confirmed by ground-based radial velocity observations, for PLATO (in the PPLC concept) and *Kepler*. *Left:* for all stars; *Right:* for stars with mV ≤11 only (see text for details).

progress, it is unlikely that it can detect unambiguously small planets in faraway orbits, mainly due to the faintness and the limited number of its targets. Only PLATO with its extended surveyed area and its main focus on bright cool dwarfs, will allow us to reach real Earth analogue systems and extend the search for exoplanets to small terrestrial planets in the habitable zone of their stars.

Thanks to its focus on bright stars, PLATO will enable the study of stellar light reflected on the planet surface, both for transiting and non transiting systems. For transiting planets, secondary transits will also be detected. Stellar reflected light and secondary transits will in particular enable the measurement of the planet albedo. In addition, astrometric measurements will be used to detect giant planets around the brightest and most nearby stars.

The stellar reflected light analysis and astrometric measurements will constitute a powerful tool for identifying exoplanetary systems around nearby stars, out to distances of 15 – 20 pc, and therefore can help select targets for further follow-up observations aiming at characterising their atmospheres, for instance in spectroscopy, interferometry and coronography.

Pathways Towards Habitable Planets
ASP Conference Series, Vol. 430, 2010
Vincent Coudé du Foresto, Dawn M. Gelino, and Ignasi Ribas, eds.

EUCLID: Dark Universe Probe and Microlensing Planet Hunter

J. P. Beaulieu,[1,2,3] D. P. Bennett,[4,3] V. Batista,[1,3] A. Cassan,[1,3]
D. Kubas,[1,3] P. Fouqué,[5,3] E. Kerrins,[6] S. Mao,[6] J. Miralda-Escudé,[7]
J. Wambsganss,[8] B. S. Gaudi,[9] A. Gould,[9,3] and S. Dong[10]

Abstract. There is a remarkable synergy between requirements for dark energy probes by cosmic shear measurements and planet hunting by microlensing. Employing weak and strong gravitational lensing to trace and detect the distribution of matter on cosmic and galactic scales, but as well as to the very small scales of exoplanets is a unique meeting point from cosmology to exoplanets. It will use gravity as the tool to explore the full range of masses not accessible by any other means. EUCLID is a 1.2 m telescope with optical and IR wide field imagers and slitless spectroscopy, proposed to ESA Cosmic Vision to probe for dark energy, baryonic acoustic oscillation, galaxy evolution, and an exoplanet hunt via microlensing. A three-month microlensing program will already efficiently probe for planets down to the mass of Mars at the snow line, for free floating terrestrial or gaseous planets and habitable super-Earth. A 12+ month survey would give a census on habitable Earth planets around solar like stars. This is the perfect complement to the statistics that will be provided by the KEPLER satellite, and these missions combined will provide a full census of extrasolar planets from hot, warm, habitable, frozen to free floating.

1. Introduction

In the last fifteen years, astronomers have found over 400 exoplanets (Schneider 2009), including some in systems that resemble our very own solar system (Gaudi et al. 2008). These discoveries have already challenged and revolutionized our theories of planet formation and dynamical evolution. Several different

---

[1]Institut d'Astrophysique de Paris, Paris, France

[2]Department of Physics and Astronomy, University College London, London, UK

[3]HOLMES collaboration

[4]University of Notre Dame, Department of Physics, Notre Dame IN, USA

[5]Observatoire Midi-Pyrénées, UMR 5572, Toulouse, France

[6]Jordrell Bank Center for Astrophysics, Univ. of Manchester, Manchester, UK

[7]ICREA/ICC-IEEC, Univ of Barcelona, Barcelona, Spain

[8]Astronomisches Rechen-Institut, Zentrum fur Astronomie, Heidelberg, Germany

[9]Department of Astronomy, Ohio State University, Columbus OH, USA

[10]Institute for Advanced Studies, School of Natural Sciences, Princeton NJ, USA

methods have been used to discover exoplanets, including radial velocity, stellar transits, and gravitational microlensing. Exoplanet detection via gravitational microlensing is a relatively new method (Mao & Paczynski 1991; Gould & Loeb 1992; Wambsganss 1997) and is based on Einstein's theory of general relativity. So far nine exoplanets have been published with this method. While this number is relatively modest compared with that discovered by the radial velocity method, microlensing probes a part of the parameter space (host separation vs. planet mass) not accessible in the medium term to other methods (see Figure 1).

The mass distribution of microlensing exoplanets has already revealed that cold super-Earths (at or beyond the snow line and with a mass of around 5 to $15 M_\oplus$) appear to be common (Beaulieu et al. 2006; Gould et al. 2006, 2007; Kubas et al. 2008). Microlensing is currently capable of detecting cool planets of super-Earth mass from the ground and, with a network of wide-field telescopes strategically located around the world, could detect planets with mass as low as the Earth. Old, free-floating planets can also be detected; a significant population of such planets are expected to be ejected during the formation of planetary systems (Juric & Tremaine 2008). Microlensing is roughly uniformly sensitive to planets orbiting all types of stars, as well as white dwarfs, neutron stars, and black holes, while other method are most sensitive to FGK dwarfs and are now extending to M dwarfs. It is therefore an independent and complementary detection method for aiding a comprehensive understanding of the planet formation process. Ground-based microlensing mostly probes exoplanets outside the snow line, where the favoured core-accretion theory of planet formation predicts a larger number of low-mass exoplanets (Ida & Lin 2005). The statistics provided by microlensing will enable a critical test of the core-accretion model.

Exoplanets probed by microlensing are much further away than those probed with other methods. They provide an interesting comparison sample with nearby exoplanets, and allow us to study the extrasolar population throughout the galaxy. In particular, the host stars with exoplanets appear to have higher metallicity (Fischer & Valenti 2005). Since the metallicity is on average higher as one goes towards the galactic centre, the abundance of exoplanets may well be somewhat higher in microlensing surveys.

2. Basic Microlensing Principles

The physical basis of microlensing is the deflection of light rays by a massive body. A distant source star is temporarily magnified by the gravitational potential of an intervening star (the lens) passing near the line of sight, with an impact parameter smaller than the Einstein ring radius R_E, a quantity which depends on the mass of the lens, and the geometry of the alignment. For a source star in the Bulge, with a 0.3 $M_\odot$ lens, $R_E \sim 2$ AU, the angular Einstein ring radius is ~ 1 mas, and the time to transit R_E is typically 20-30 days, but can be in the range 5-100 days. The lensing magnification is determined by the degree of alignment of the lens and source stars. The closer the alignment the higher the magnification.

268   Beaulieu et al.

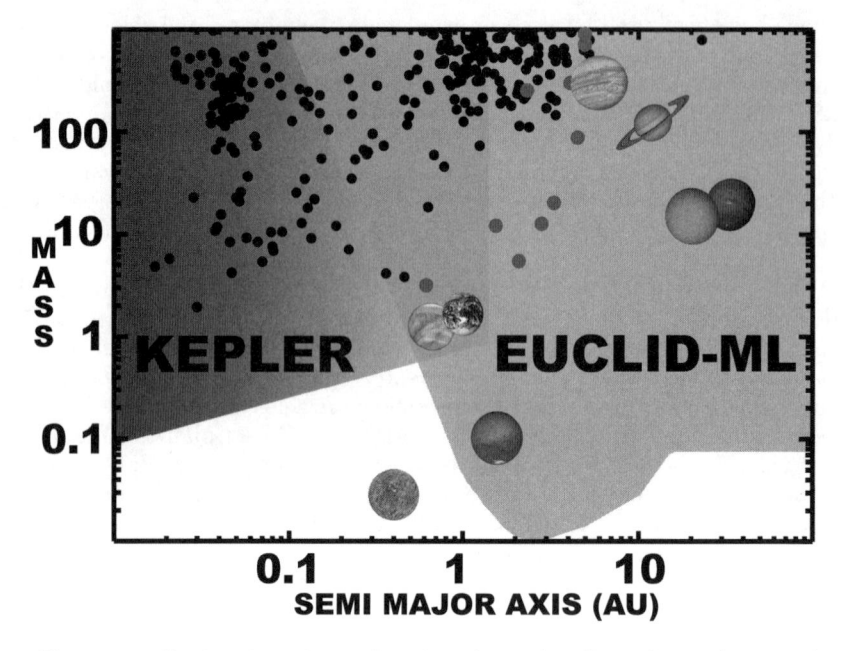

Figure 1. Semi major axis as a function of mass for all exoplanets discovered as of September 2009 (microlensing planets are plotted as red dots) and the planets from our solar system. We also plot the sensitivity of Kepler and of space based microlensing observations.

A planetary companion to the lens star will induce a perturbation to the microlensing light curve with a duration that scales with the square root of the planet's mass, lasting typically a few hours (for an Earth) to a few days (for a Jupiter). Hence, planets can be discovered by dense photometric sampling of ongoing microlensing events (Mao & Paczynski 1991; Gould & Loeb 1992). The limiting mass for the microlensing method occurs when the planetary Einstein radius becomes smaller than the projected radius of the source star (Bennett & Rhie 1996). The $\sim 5.5 M_\oplus$ planet detected by Beaulieu et al. (2006) is near this limit for a giant source star, but most microlensing events have G or K-dwarf source stars with radii that are at least 10 times smaller than this. High angular enough resolution to resolve dwarf sources of the Galactic Bulge (≤ 0.5 arcsec) will open the sensitivity below a few Earth masses (Figure 2).

The inverse problem, finding the properties of the lensing system (planet/star mass ratio, star-planet projected separation) from an observed light curve, is a complex non-linear one within a wide parameter space. In general, model distributions for the spatial mass density of the Milky Way, the velocities of potential lens and source stars, and a mass function of the lens stars are required in order to derive probability distributions for the masses of the planet and the lens star, their distance, as well as the orbital radius and period of the planet by means of Bayesian analysis. With complementary high angular resolution observations, currently done either by HST or with adaptive optics, it is possible to

EUCLID 269

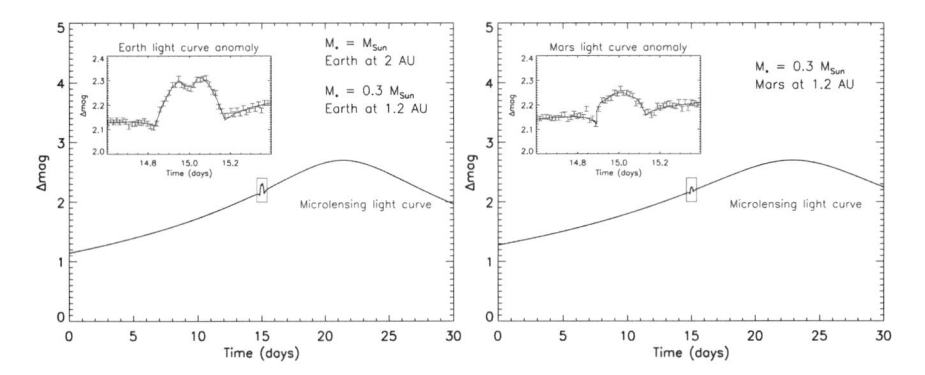

Figure 2. The two figures illustrate the detection capability of the microlensing technique in the very low-mass exoplanet regime. Here, the source star and the lens (the planet host star) are both located in the Galactic Bulge. The sampling interval is twenty minutes, and the photometric precision is one percent. The planetary signal on the left figure is expected from an Earthmass planet at 2 AU around a solar star, or from an Earth at 1.2 AU but orbiting an $0.3\,M_\odot$ M-dwarf star. A planet of the mass of Mars ($0.1M_\oplus$) at 1.2 AU can also be detected around such a low-mass host star (right figure). These are typical examples of low mass telluric planets to be detected by EUCLID.

further constrain system parameters, and determine masses to 10% by directly constraining the light coming from the lens and measuring the lens and sources relative proper motion (Bennett et al. 2007a,b; Dong et al. 2009). A space-based microlensing survey can provide the planet mass, projected separation from the host, host star mass, and its distance from the observer for most events using this method.

Different papers have presented the future strategies in the near, medium and long term, with the ultimate goal of achieving a full census of Earth-like planets with either a dedicated space mission (Microlensing Planet Finder, MPF) or advocating for synergy between dark energy probes and microlensing. There is a general consensus in the microlensing community about these milestones, and this consensus has been endorsed by the US ExoPlanet Task Force (ExoPTF). White papers submitted to the ExoPTF (Bennett et al. 2007a; Gould et al. 2007), the exoplanet forum (Gaudi et al. 2009a), the JDEM request for information, ESA-EPRAT (ExoPlanetary Roadmap Advisory Team) (Beaulieu et al. 2008) and Astro2010 PSF (Bennett et al. 2009, Gaudi et al., 2009b), and the Pathways conference in Barcelona.

3. A Program Onboard EUCLID to Hunt for Planets

Space based microlensing observations
The ideal satellite is a 1m class space telescope with a focal plane of 0.5 square degree or more in the visible or in the near-infrared. The Microlensing Planet Finder is an example of such a mission (Bennett et al. 2007a), which has been proposed to NASA's Discovery program, and endorsed by the ExoPTF. Despite

the fact that the designs were completely independent, there is a remarkable similarity between the requirements for the missions, one aimed at probing dark energy via cosmic shear (Refregier et al. 2010) and the other, a microlensing planet hunting mission (Beaulieu et al. 2008; Bennett et al. 2007b, 2009). EUCLID is a proposed mission to measure parameters of dark energy using weak gravitational lensing and baryonic acoustic oscillation, test the general relativity and the cold dark matter paradigm for structure formation submitted to the ESA COSMIC VISION program. It is a 1.2m Korsch telescope with a 0.48 square degree imager in a broad optical band consisting of R+I+Z (0.1 arcsec per pixel) and in the Y, J, H band (0.3 arcsec per pixel). Microlensing benefits from the strong requirement from cosmic shear on the imaging channel, and does not add any constraint to the design of EUCLID.

Observing strategy

We will monitor 2 square degree of the area with highest optical depth to microlensing from the Galactic Bulge with a sampling rate once every twenty minutes. Observations will be conducted in the optical and NIR channel.

Angular resolution is the key to extend sensitivity below few earth masses

Microlensing relies upon the high density of source and lens stars towards the Galactic Bulge to generate the stellar alignments that are needed to generate microlensing events, but this high star density also means that the Galactic Bulge main sequence source stars are not generally resolved in ground-based images. This means that the precise photometry needed to detect planets of $\leq 1M_\oplus$ is not possible from the ground unless the magnification due to the stellar lens is moderately high. This, in turn, implies that ground-based microlensing is only sensitive to terrestrial planets located close to the Einstein ring (at ∼2-3 AU). The full sensitivity to terrestrial planets in all orbits, from $0.5AU$ to free floating, comes only from a space-based survey (see Figure 1). In Figure 2 we give examples of simulated detections of an Earth and a Mars-mass planet.

Microlensing from space yields precise star and planet parameters

The high angular resolution and stable point-spread-functions available from space enable a space-based microlensing survey to detect most of the planetary host stars. When combined with the microlensing light curve data, this allows a precise determination of the planet and star properties for most events (Bennett et al. 2007a).

Probing a parameter space out of reach of any other technique

The Exoplanet Task Force (ExoPTF) recently released a report (Lunine et al. 2008) that evaluated all of the current and proposed methods to find and study exoplanets, and they expressed strong support for space-based microlensing stating that: *Space-based microlensing is the optimal approach to providing a true statistical census of planetary systems in the galaxy, over a range of likely semi-major axes, and can likely be conducted with a Discovery-class mission.* It can also be accomplished as a program on board the EUCLID M class mission, with a Dark Energy probe as the primary objective.

A EUCLID microlensing survey provides a census of extrasolar planets that is complete (in a statistical sense) down to $0.1M_\oplus$ at orbital separations ≥ 0.5 AU. When combined with the results of the Kepler mission (and ground-based radial velocity surveys) EUCLID will give a comprehensive picture of all types of extrasolar planets with masses down to well below an Earth mass. This fundamental exoplanet census data is needed to gain a comprehensive understanding of the processes of planet formation and migration, both of which are key to understanding the requirements for habitable planets and the development of life on extrasolar planets.

A subset of the science goals can be accomplished with an enhanced ground-based microlensing program (Gaudi et al. 2009a,b) , which would be sensitive to Earth-mass planets in the vicinity of the snow-line. But such a survey would have its sensitivity to Earth-like planets limited to a narrow range of semi-major axes, so it would not provide the complete picture of the frequency of exoplanets down to $0.1 M_\oplus$ that a space-based microlensing survey would provide. Furthermore, a ground-based survey would not be able to detect the planetary host stars for most of the events, and therefore will not provide systematic data on the variation of exoplanet properties as a function of host star type that a space-based survey will provide.

Duration of the program

One of the remarkable features of the EUCLID microlensing program is its linear sensitivity to allocated time and area of the focal plane. The minimal time allocation of three months will already give important statistics on planets at the snow line, down to the mass of mars, and of free floating planets. Habitable super-Earths will also be probed. Longer observing times (12 months of Galactic Bulge observing) would lead to sensitivity to study true analogues of habitable Earth-mass planets orbiting solar-like stars.

Acknowledgments. We acknowledge the support of HOLMES ANR-06-BLAN-0416.

References

Beaulieu, J. P., et al., 2006, Nat, 439, 437
Beaulieu, J. P., et al., 2008, ESA EPRAT, arXiv:0808.0005v1
Bennett, D. P. & Rhie, S. H., 1996, ApJ, 472, 660
Bennett, D. P. & Anderson, J. & Gaudi, B. S., 2007a, ApJ, 660, 781
Bennett, D.P., et al., 2007b, arXiv:0704.0454v1
Bennett D.P., et al., 2009, arXiv:0902.3000v1
Dong S., et al., 2009, ApJ, 695, 970
Fischer, D. A. & Valenti, J., 2005, ApJ, 622, 1102
Gaudi B.S. et al. 2008, Science 319, 927
Gaudi B.S. et al. 2009a, in "Exoplanet Community Report on Microlensing"
Gaudi B.S., et al., 2009b, arXiv:0903.0880v1
Gould A., & Loeb A., 1992, ApJ, 396, 104
Gould A., et al., 2006, ApJ, 644, L37
Gould A., Gaudi B.S., Bennett D.P., 2007, arXiv:0704.0767v1
Ida S. and Lin D.N.C., 2005, ApJ, 626, 1045
Juric M. & Tremaine S., 2008, ApJ, 686, 603
Kubas D., et al., 2008, A&A, 483, 317
Lunine J., et al., 2008, in "Exoplanet Task Force Report", arXiv:0808.2754v2
Mao S. and Paczynski B., 1991, ApJ, 374, L37
Refregier A., et al. 2010, Euclid Imaging Consortium Science Book, arXiv:1001.0061v1
Schneider J., 2009, http://exoplanet.eu/
Wambsganss J., 1997, MNRAS, 284, 172

Pathways Towards Habitable Planets
ASP Conference Series, Vol. 430, 2010
Vincent Coudé du Foresto, Dawn M. Gelino, and Ignasi Ribas, eds.

The Habitable Zone Planet Finder Project: A Proposed High Resolution NIR Spectrograph for the Hobby Eberly Telescope (HET) to Discover Low Mass Exoplanets around M Stars

S. Mahadevan,[1,2] L. Ramsey,[1,2] S. Redman,[1] S. Zonak,[1] J. Wright,[1,2] A. Wolszczan,[1,2] M. Endl,[3] and B. Zhao[4]

[1]*Department of Astronomy & Astrophysics, The Pennsylvania State University, University Park, PA, USA*

[2]*Center for Exoplanets and Habitable Worlds, The Pennsylvania State University, University Park, PA, USA*

[3]*McDonald Observatory, University of Texas, Austin, TX, USA*

[4]*Department of Astronomy, University of Florida, Gainesville, FL, USA*

Abstract. Radial velocity precision in the NIR is now approaching the level necessary to detect exoplanets around mid-late M stars that are very faint in the optical and emit most of their flux in the NIR. The Penn State Pathfinder prototype instrument has already demonstrated 7-10 ms^{-1} precision on sunlight, and similar precision has been reported at the Pathways conference using CRIRES and an ammonia gas-cell. We discuss the science goals that motivate a stable cross-dispersed, high-resolution NIR spectrograph on a large telescope, as well as the path leading from the Pathfinder prototype to one such possible instrument - the fiber-fed Habitable Zone Planet Finder (HZPF) on the Hobby Eberly Telescope (HET). We also discuss wavelength calibration issues specific to the NIR, and our ongoing exploration with Pathfinder to mitigate these issues.

1. Scientific Goal: Low Mass Planets Around Mid-late M Dwarfs

With the discovery of over 400 extrasolar planets, considerable interest is now focused on finding and characterizing terrestrial-mass planets in habitable zones around their host stars. Such planets are extremely difficult to detect around F, G, and K stars, requiring either very high radial velocity precision ($<<1$ ms^{-1}) or space-based photometry to detect a transit. Along with ongoing radial velocity (RV) programs, transit efforts such as the MEarth project are also now beginning to focus on M dwarfs (Irwin et al. 2009) as their lower luminosity (compared to the Sun) shifts the habitable zone (HZ), a region around a star where liquid surface water may exist on a planet, much closer to the star. The lower stellar mass of the M dwarfs, as well as the short orbital periods of HZ planets, increases the Doppler wobble caused by a terrestrial-mass planet. Planets around early M stars are already detectable with the current radial velocity precision obtained with current optical high-resolution echelle spectrographs. A number of the more massive M dwarfs are being surveyed in the optical with existing precision radial velocity instruments. A total of 17 exoplanets in 11 planetary systems have, to date, been discovered around M stars, including the low mass planetary system around GJ581 (Mayor et al. 2009) that hosts one of the lowest mass exoplanet known, GJ581e, as well as GJ581d which is close to the habitable zone. These observations suggest that, while hot Jupiters may be rare (Endl et al. 2006), lower mass planets do exist around M stars and may be rather common.

Climate simulations of planets in the HZ around M stars (Joshi et al. 1997) show that tidal locking does not necessarily lead to atmospheric collapse, and a surface pressure of 1-2 bars allows liquid water to exist. The habitability of terrestrial planets around M stars has been explored by Tarter et al. 2007 and Scalo et al. (2007), who find that, despite their flares and low flux, M dwarfs are good candidates for hosting habitable planets. Most of the M stars in current optical radial velocity surveys are typically earlier in spectral type than M4 because late-type M stars emit most of their flux in the 0.9-1.8μm wavelength region, the near-infrared (NIR) Y (0.9-1.1 μm), J (1.1-1.4 μm) and H (1.45-1.8 μm) bands. Existing precision radial velocity optical spectrographs with simultaneous Th-Ar calibration currently cover only the 0.38-0.68μm range, and ones using an I_2 cell are limited to 0.5-0.62μm for extracting precision velocities with PSF modeling, thus observing late-type M stars strains the capabilities of current precision radial velocity instruments, even on the largest telescopes. However, it is the low mass late-type M stars, which are the least luminous, where the velocity amplitude of a terrestrial planet in the habitable zone is highest, making them very desirable targets. Unlike early type stars (A, B, early F), M stars do have sharp absorption lines and molecular features, the positions of which encode significant radial velocity information, making them suitable for exoplanet searches. Since the flux distribution from M stars peaks sharply in the NIR (Pavlenko et al. 2006), a stable high-resolution NIR spectrograph capable of delivering high radial velocity precision can observe several hundred of the nearest M dwarfs to examine their planet population.

Using the Penn State brassboard Pathfinder instrument (Ramsey et al. 2008) we have already demonstrated 7-10 ms^{-1} precision on sunlight. Lessons learned from our ongoing Pathfinder efforts will enable us to build such an instrument, capable of providing 3 ms^{-1} or better long-term radial velocity precision. Our goal is to ultimately build such a facility-class instrument for the 9m Hobby Eberly Telescope (HET), called the Habitable Zone Planet Finder (HZPF). The NIR spectral region is rich in atmospheric absorption lines and OH nightglow lines, requiring a high spectral resolution to avoid these features, as well as to achieve good wavelength calibration. The HZPF will deliver a nominal resolution of R>50,000 to help meet these requirements.

1.1. Targets for a Survey with the HET

To ensure the feasibility of our scientific goals we have already begun to generate target lists for M stars in the mass range 0.08-0.4 M$_\odot$, spanning a factor of 5 in mass and extending the parameter space of discovery to complement ongoing radial velocity surveys around 0.4-1.5 M$_\odot$ stars. The majority of our targets will be within 20 pc. As demonstrated by the RECONS team (www.recons.org) 68% of the 354 objects currently known to be within 10pc of the Sun are M stars, with the majority being low mass M stars suitable for a NIR radial velocity survey. As part of an ongoing effort to measure the rotational velocity of possible targets for a high precision RV survey we have acquired observations of 56 main sequence stars of spectral type M4-M7 with the red orders of the High Resolution Spectrograph (HRS) on the HET (Jenkins et al. 2009). When combining our work with known *vsini* measurements from the literature we find 138 M4-M9 stars known to have rotational velocities less than 10 km s^{-1}. Since the radial velocity information content available in the M star spectra degrades by a factor of 3.5 as $v$sin$i$ rises 2 km s^{-1} to 10 km s^{-1} (Bouchy et al. 2001), slower rotators are better targets for a radial velocity survey. The target selection will however also include some faster rotators to mitigate selection effects, especially at M7 and later spectral types since slow rotating stars are relatively uncommon in those regimes. We are continuing to refine our target sample with an ongoing survey at the HET and hope to have a well defined target sample of 300 M4-M9 stars over the next few years.

2. The HZPF Design Concept

A stable NIR spectrograph is an essential tool to discover terrestrial mass planets around these relatively unexplored stars. In the 0.85-1.7μm region that where M dwarfs have significant radial velocity information content there are no molecular gas cells (like I_2) that have a dense set of sharp absorption lines over a significant bandwidth that one can use, though ammonia gas can be used in part of the K band (see Dreizler et al. these proceedings). We conclude that our requirements are most readily achieved with an NIR fiber-fed instrument that has a second calibration fiber that uses an emission lamp like Th-Ar to track instrument drifts. Most existing NIR high resolution spectroscopic instruments either have very small simultaneous wavelength coverage, or are not fiber-fed, or cannot cover the information rich Y band. The HZPF on the HET will be a cooled fiber-fed instrument on a large telescope capable of high-resolution, high stability, and large simultaneous wavelength coverage. The proposed UPF for the UKIRT 4m telescope and our HPZF instrument derive their heritage from the Gemini commissioned precision radial velocity instrument (PRVS) instrument study, in which Penn State was a partner. Our HZPF optical and system design draws heavily on the lessons learned in the PRVS design study. We have explored various design options that meet the science requirements and the baseline approach we adopt is the well proven quasi-Littrow white pupil design with a monolithic off-axis parabolic collimator, a 200 mm by 800 mm 31.6 g/mm replicated mosaic grating from Newport RGL blazed at 75 degrees, a ZnSe grism cross disperser with a 150 l/mm 5.4 deg blaze grating diamond turned on it, and an f/2.3 refractive camera made with standard Schott and Ohara glasses. The white pupil design enables good control of scattered light, and the grism cross-disperser enables a compact configuration. Significant progress has recently been made by Kuzmenko et al. 2008 in fly cutting grooves on ZnSe using ultra-precise diamond lathe; and we are currently working closely with II-VI Infrared on R&D to develop and test a prototype to lower this risk, or to identify it and switch to our low risk backup option (a 150 l/mm reflection grating from Newport) very early in the design process. The beam size of 150mm enables all the refractive optics to be of reasonable size, while still enabling a 0.5 arc second slit to be sampled by 3 pixels on the H2RG (3.6 pixels on the CCD) to yield a resolution of R ~50,000. Figure 1 shows the quantum efficiency of a deep-depletion CCD and the H2RG and also the order format and the split over the CCD and H2RG although this may be revised as we continue our trade studies. The simultaneous use of a CCD enables the instrument to be tested while warm, as well as providing an independent source of RVs that are not affected by any of the possible systematics of NIR arrays. The monolithic collimator mirror, Echelle grating, fold mirror, and cross disperser grating will all be manufactured on Zerodur substrates to reduce temperature sensitivity, and will all be gold coated to increase efficiency. The well-understood white pupil design enables good control of the scattered light, as well as minimizing the size of the cross-disperser and camera apertures. RMS spot sizes from our preliminary optical design are all within 1 pixel for the H2RG detector and within 1.5 pixels for the CCD.

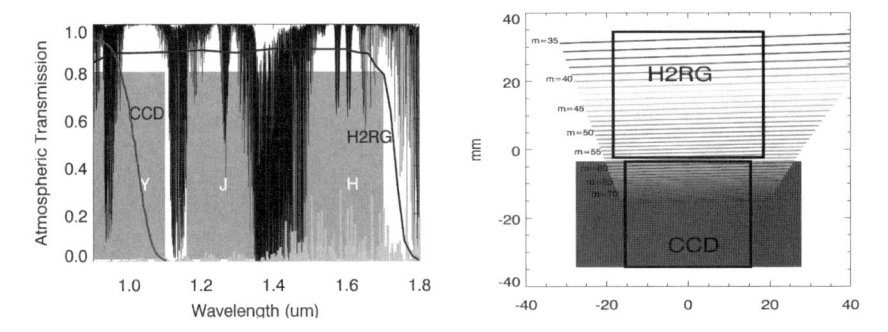

Figure 1. *Left:* The NIR Y, J and H Bands, atmospheric transmission and OH emission, and QEs of CCD and H2RG. *Right:* The spectral order format of the HZPF design.

Successful implementation of the calibration technique we plan to utilize requires high thermal stability in the spectrograph. HARPS has pointed the way to achieving high stability by placing the instrument in a vacuum in a stable thermal environment. HZPF will also be in a vacuum vessel, but has the additional complication of needing to be cooled so that thermal radiation is not a significant contributor to the system background. As part of the PRVS detailed study, we determined that the spectrograph components and enclosing structure need to be at < 200 K. This temperature will yields a thermal background per pixel in the H2RG a factor of two lower than the requirement. Thus, HZPF will be structurally more similar to cryogenic infrared instruments than optical instruments. Our baseline design approach consists of a vacuum chamber which is supported with optical bench style vibration damping supports. All the spectrograph optical components plus the detector are mounted on an optical support structure constructed of aluminum hollow rectangular beams-weldments or light-weighted aluminum plates so as to have the necessary stiffness. The optical support structure will be supported at three points by flexures from the bottom of a radiation shield which is itself supported by insulated flexures from the bottom of the vacuum chamber. The top of the radiation shield and vacuum chamber will be removable for access to the optical support structure via pneumatic lifts. Figure 2 shows the design concept for the spectrograph, support structure, radiation shield, and vacuum vessel. The optical support structure and radiation shield will be maintained at ~195 K by using resistive heating elements on the optical support structure and a vibration isolated CT1050 closed-cycle cooler. Using a Lakeshore 340 temperature controller we expect temperature stability to better than 0.03 K. Active temperature stabilization of HZPF is critical for reducing the instrument drift and achieving high precision radial velocities. Small temperature gradients between different parts of the instrument are acceptable as long as the gradients are stable. A second closed-cycle cooler will be used for the detector. Fiber inputs to the vacuum chamber are baselined as optical feed-throughs. By the time of a possible HZPF commissioning the Hobby Eberly Telescope Dark Energy Experiment (HETDEX) instruments will be operational and the HZPF provides a perfect bright time complement to the dark-time HETDEX project.

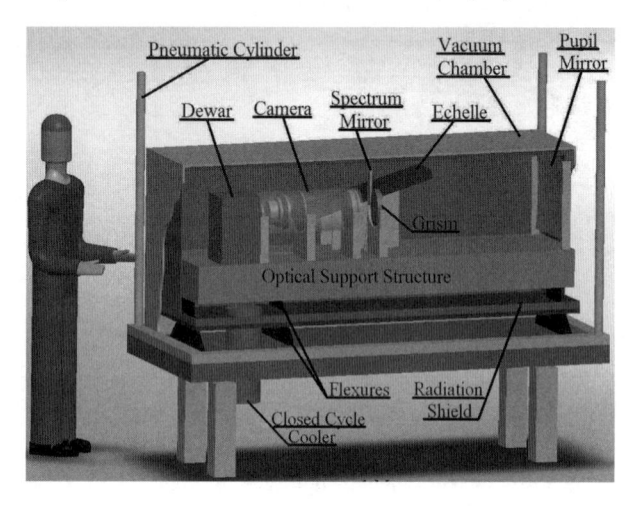

Figure 2. Cutout of the HZPF instrument design.

3. Calibration Challenges in the NIR

There are no suitable gas cells like iodine covering the Y, J, and H band in the NIR, but the fiber-fed design of the Pathfinder allows a second calibration fiber to be used to simultaneously track the instrument drift during an object exposure. The major requirements on the calibration source are that there be sufficient number of bright stable emission lines in each echelle order to achieve $< 1 ms^{-1}$ by tracking drifts of the spatial response function. Inexpensive hollow cathode lamps with stable heavy elements, like thorium (^{232}Th), have sharp emission lines, making them suitable for such applications. Such lamps also have a filler gas, usually argon (Ar) or neon (Ne). Th-Ar emission lines span the UV-NIR regions, making them a very useful and convenient source of wavelength calibration, and are in use with the fiber-fed HARPS spectrograph to achieve radial velocity precisions better than 1 ms^{-1} (Mayor et al. 2009) and discover super-Earth planets. In the NIR however the Ar lines are very bright, completely dominating the flux output of the lamp (Kerber et al. 2008). There are many suitable thorium lines for use in wavelength calibration, but unlike the situation in the optical, the flux output of these lines can be smaller by a factor of 102 -104 compared to the bright argon lines. The argon lines are unsuitable for use for precision radial velocity measurements (Lovis & Pepe 2008). With the Pathfinder longer integrations are therefore needed to have sufficient intensity in the Th lines to allow wavelength calibration and the Argon lines also cause significant scattered light within the instrument. Our tests with Pathfinder show that lamps with neon as a filler gas have fewer bright lines in the NIR than Argon, causes less scattered light and pixel saturation, making Th-Ne more attractive. For the H2RG IR array we will enable sub-array readout to ensure that the brightest Ne/Ar and Th lines can be read out quickly, allowing the fainter Th lines to gather sufficient signal during the exposure time. However not even letting bright lines enter the instrument will lead to a decrease in the scattered light. Our lab experiments with uranium lamps show that U-Ar has a large number of emission lines in the Y band, (210 lines in the same region that has 140 Th-Ar lines), making it an attractive alternative or supplement to thorium. Uranium (^{238}U) fulfills many of the requirements needed of an element in a hollow cathode lamp for wavelength calibration - it is a heavy element (heavier than thorium), has zero nuclear spin, and a long half-life. Its only disadvantage is the presence of $\sim 0.7\%$ ^{235}U in all naturally occurring uranium, which can lead to additional lines with isotope shifts – we continue to explore the suitability of such lamps with lab tests, and test thorium-neon and uranium-neon lamps with Pathfinder.

Acknowledgments. We acknowledge the contributions HZPF has inherited from the design study of the PRVS team. We thank Hugh Jones for input and comments on the HZPF design effort. The Center for Exoplanets and Habitable Worlds is supported by the Pennsylvania State University, the Eberly College of Science, and the Pennsylvania Space Grant Consortium.

References

Bouchy, F., Pepe, F., & Queloz, D. 2001, A&A, 374, 733
Endl, M., et al., 2006, ApJ, 649, 436
Irwin, J., et al. 2009, ApJ, 701, 1436
Jenkins, J. S., et al., 2009, ApJ, 704, 975
Joshi, M. M., Haberle, R. M., & Reynolds, R. T. 1997, Icarus, 129, 450
Kerber, F., Nave, G., & Sansonetti, C. J. 2008, ApJS, 178, 374
Kuzmenko, P. J., Little, S. L., Ikeda, Y., & Kobayashi, N. 2008, Proc. SPIE, 7018, 70184Q
Lovis, C., & Pepe, F. 2007, A&A, 468, 1115
Mayor, M., et al. 2009, A&A, 507, 487
Pavlenko, Y. V. et al., 2006, A&A, 447, 709
Ramsey, L. W., Barnes, J., Redman, S. L., Jones, H. R. A., Wolszczan, A., Bongiorno, S., Engel, L., & Jenkins, J. 2008, PASP, 120, 887
Scalo, J., et al. 2007, Astrobiology, 7, 85
Tarter, J. C., et al. 2007, Astrobiology, 7, 30

Pathways Towards Habitable Planets
ASP Conference Series, Vol. 430, 2010
Vincent Coudé du Foresto, Dawn M. Gelino, and Ignasi Ribas, eds.

The Fresnel Interferometric Imager

L. Koechlin,[1] J.-P. Rivet,[2] R. Gili,[2] P. Deba,[1] T. Raksasataya,[1] and D. Serre[3]

[1] *Laboratoire d'Astrophysique de Toulouse Tarbes, Université Paul Sabatier, CNRS, 14 avenue Edouard Belin, Toulouse, France*

[2] *Observatoire de la Côte d'Azur, Boulevard de l'Observatoire, B.P. 4229, Nice, France*

[3] *Leiden Observatory, Leiden University, P.O. Box 9513, Leiden, the Netherlands*

Abstract. We present a new "pathway" in the form of an innovative space-based telescope: the Fresnel Imager. It is a concept of two spacecrafts flying in formation, one satellite holding a diffraction array acting as entrance pupil and providing a very high wavefront quality in the visible and UV domains, the other one holding the focal instrumentation and detectors. The distance between spacecrafts would vary between 1 and 100 km, depending on the specifications, and the aperture size would be 3 to 100 meters. We present the validation prototypes realized, the spectral domains that can be explored, and the astrophysical targets.

1. Introduction

Fresnel arrays are an alternative to mirrors for large apertures in space, allowing lightweight and position-tolerant instruments. Diffractive optics based on solid transparent foils have been proposed for space telescopes by Chesnokov (1993), Hyde (1999) and Massonnet (2003). Metal Fresnel zone plates have also been proposed (Baez 1961), as well as "photon sieves" for X-ray, UV and visible focusing (Kipp et al. 2001). The principle of Fresnel Imagers is presented in detail in Koechlin et al. (2005) and Serre et al. (2009): a thin opaque foil forming the main aperture, the "Fresnel Array", is punched which numerous and specially designed subapertures which cover close to 50% of the surface. Those apertures lead to a constructive interference of the incident light, and are shaped and positioned to optimize the point spread function (PSF) while keeping the mechanical cohesion of the foil. Recently built Fresnel arrays achieve up to 7.5% light transmission efficiency (fraction of light that ends up in the central lobe of the PSF).

As light has travelled all the way to prime focus without encountering an optical surface, a very high quality wavefront is generated, which provides high resolution and high dynamic range images. On compact objects, the field covers usually 1000 Airy radii, and high dynamic range is present in all the field except on four spikes. On extended objects (larger than 50 Airy radii), high resolution remains, but the high dynamic range does not.

Figure 1. The 116 Fresnel zones array of testbed generation 1: CuBe foil, 75 μm thick.

There are drawbacks, e.g. long focal lengths and chromatism. Long focal lengths require formation flying of two spacecraft separated by a few kilometers for the smallest arrays (3.6 m), to hundreds kilometers for the very large arrays (100m diameter). In order to correct axial chromatism inherent to diffractive focusing, the receiver spacecraft placed at the focal plane of one of the observed wavelengths, features a reduced-size optical train, the beam cross-section being an order of magnitude smaller than the main aperture. Small diffractive optics are placed in a pupil plane (Schupmann 1899; Faklis & Morris 1989; Serre et al. 2009) and yield $\Delta\lambda/\lambda = 20\%$ bandpass for up to six scientific channels that can be set anywhere between 100 nm and 25 μm wavelengths.

During the last few years, we have worked on testbeds, and on science cases. In the second section of this paper we present the progress made on testbeds, in the third section we present astrophysical targets and strategies of development.

2. Prototypes

Different prototypes have been gradually developed or currently are, aiming to validate different aspects of the concept: global validation of the optical principle, angular resolution and dynamic range assessment, on-sky validation, and validation in the UV domain.

2.1. Generation I and II Testbeds

Generation I prototype, built in 2006, has validated wavefront quality, broad band imaging, and 10^{-6} dynamic range. It had a 8 cm apodized square aperture (Nisenson & Papaliolios 2001) made of 116 Fresnel zones carved in a metal sheet, and was working in the visible wavelengths domain. The drawing of the cutting can be seen in Figure 1. A detailed presentation can be found in Serre et al. (2009).

Figure 2. Orthocircular Fresnel array in testbed Generation 2, carved in 50 μm thick opaque matal sheet. Only the first 12 Fresnel zones are shown here. The Fresnel rings are held in place by bars every n Fresnel zones. n is adjusted for optimal dynamic range. The actual array features 348 Fresnel zones from center to limb (696 from center to corner.) The bars are not equidistant.

Built in 2009, the Generation II prototype is for validation on sky objects. It features a 20 cm apodized square aperture with 696 Fresnel zones and 18-m focal length at $\lambda = 800$ nm. Its optimized primary array yields a higher dynamic range and almost double throughput than Gen I: the orthogonal bars holding opaque rings have been adjusted for a tradeoff between mechanical rigidity (i.e. wavefront quality) and dynamic range (Figure 2). Other setups using different bars positioning laws, or de-centered ringed support bars that focus their diffracted light outside the field, resulted in good optical results but insufficient rigidity and have thus been discarded.

The two modules of the Generation II prototype have been set up at Nice observatory in "piggyback" of the 17.89 meter focal length Grand Refracteur. The global orientation of the Fresnel Imager to the sky targets is adjusted by the mount of the refractor, and the relative orientation of the focal module to point the array center is adjusted by two additional degrees of liberty: to and away from the main optical axis (Figure 3). Although the tip-tilt correction and nominal cameras were not present in summer 2009, we could get the first light with a "Generation I.5" testbed and the preliminary images are presented in Figures 4 and 5.

A nominal configuration should be ready by January 2010, and we plan to observe targets with high dynamic ranges such as Mars disk and satellites, Sirius AB, and other targets.

2.2. Generation III Testbed and Studies for a Space-based Fresnel Imager

As we propose UV domain science cases at high angular resolution and high dynamic range, the technologies involved in the Fresnel Imager should be validated in the specific UV domain. A laboratory testbed is currently at the conception stage, and we plan to publish the results in the next two years.

The Fresnel Interferometric Imager 281

Figure 3.   Nice refractor with Gen 1.5 prototype set in parallel.

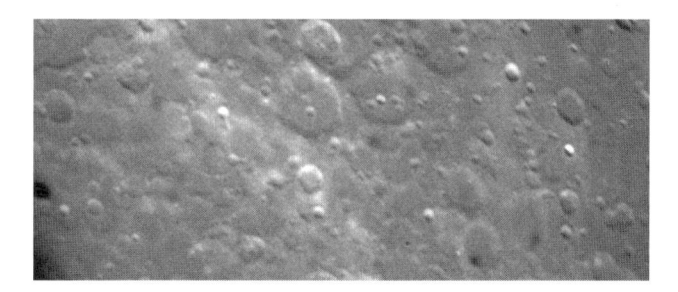

Figure 4.   520"x210" field on the moon, taken October 7th at 02:18 U.T. North is on the right. The white diagonal stripe is a ray from Tycho.

Figure 5.   Binary star STF 2726 (52 Cyg) of magnitudes 4.2 / 9.5, separation 6 arc seconds. The brighter star is over-exposed.

A "Phase zero" study for a space-based project has been carried out by Centre National d'Etudes Spatiales, and ESA is currently studying larger membrane telescopes using binary Fresnel arrays.

3. Space Applications

In parallel to the concept and technology validations, we are setting up a group to investigate the science cases that are within reach of Fresnel imagers. In the frame of this work, a workshop was been held in Nice on September 23-25, 2009, in which were presented various opportunities.

3.1. Rationale for Space Mission Based on a New Technology

The visible and IR domains should be explored with large telescopes in the 2020-2030s, and 10-15 meter class Fresnel arrays would be required to be competitive in terms of angular resolution and luminosity, and benefit from their space-based status. On the other hand, a Fresnel Imager would be competitive only if the cost is low enough to overcome the barrier associated with a technology never before tested in space.

This dilemma could be solved by choosing the UV domain, the wavefront quality of a Fresnel array not being wavelength dependent. A deployable 3–4 meter Fresnel array will provide a 7 mas, diffraction limited angular resolution at a $\lambda/50$ wavefront quality. For the "after focus" instrumentation, the UV requires high wavefront quality field optics and diffractive chromatic correction: this what will be tested in our third generation testbed, in collaboration with the instrumentation groups at Network for Ultra Violet Astronomy (NUVA).

3.2. Examples of Scientific Objectives

The mission targets and instrument performances will depend on many parameters, and the best tradeoff is not defined yet, as the diffraction optics technology progresses but the astrophysical problems also evolve: some that are now very hot, may be solved by the time a mission can be launched.

However, if we provide unmatched performances in a relatively unexplored optical domain, like the UV, we play safer. For example in the domain of telluric exoplanet direct imaging, studies in the visible or IR would require a 10-meter size Fresnel array, whereas in the UV domain Fresnel arrays will be competitive even at small sizes (3-meter). Other research areas are also investigated such as solar system objects (for example multiple asteroids and Kuiper belt), stellar disks and environments (with the NUVA consortium, we are working on the preparation of a space mission on stellar disks and young planetary systems: the "Disk Evolution Watcher"), planetary systems formation, galaxies, and extragalactic targets.

4. Conclusion

Since 2004 when we started the concept, we have been testing and improving Fresnel arrays for space imaging. The results from ground-based testbeds are promising: we have validated the angular resolution, the dynamic range and the spectral bandwidth of Fresnel imagers. For the time being, we have done this on apertures of 20 cm or less, and in the visible or close IR domains. We are working on closely interdependent fronts: optical conception, space orbits and guiding, and astrophysical science projects. We investigate science cases in the IR, visible and UV domains for which the Fresnel Imager would bring significant improvements, and have found a "niche" for a first space mission in the visible and UV.

Acknowledgments. The studies on which this paper is based are financed by Université de Toulouse, Centre National de la Recherche Scientifique, Centre National d'Etudes Spatiales, Thales Alenia Space, the "Sciences et Technologies pour l'Aéronautique et l'Espace" (STAE) foundation, and Observatoire de la côte d'azur.

References

Baez, A. 1961, Journal of the Optical Society of America, 51, 405
Chesnokov, Y. M. 1993, Space Bulletin, 1, 18
Faklis, D. & Morris, G-M., 1989, Optical Engineering, 28, 592
Hyde, R. A. 1999, Appl.Optics, 38, 4198
Kipp, L. et al., Nat, 414, 184
Koechlin, L., Serre, D., & Duchon, P. 2005, A&A, 443, 709
Massonnet, D. 2003, C. N. E. S. patent: un nouveau type de tlescope spatial, Ref. 03.13403
Nisenson, P. & Papaliolios, C. 2001, ApJ, 548, L201
Schupmann, L. 1899, Die medial-fernrohre: eine neue konstruktion für grosse astronomische instrumente, Teubner B G
Serre, D., Deba, P., & Koechlin, L. 2009, Appl.Optics, 48, 2811

SPICA Coronagraph

K. Enya,[1] T. Kotani,[1] T. Nakagawa,[1] H. Kataza,[1] K. Komatsu,[1]
H. Uchida,[1] K. Haze,[1,2] S. Higuchi,[1,3] T. Miyata,[4] S. Sako,[4]
T. Nakamura,[4] T. Yamashita,[5] N. Narita,[5] M. Tamura,[5] J. Nishikawa,[5]
H. Hayano,[5] S. Oya,[5] E. Kokubo,[5] Y. Itoh,[6] M. Fukagawa,[7] H. Shibai,[7]
M. Honda,[8] N. Baba,[9] N. Murakami,[9] M. Takami,[10] T. Matsuo,[11,5]
S. Ida,[12] L. Abe,[13] O. Guyon,[14] M. Venet,[15] T. Yamamuro,[16]
and P. Bierden[17]

Abstract. We present the SPICA Coronagraph Instrument for the direct imaging and spectroscopy of exoplanets. The SPICA mission gives us a unique opportunity for high-contrast observations because of the large telescope aperture, the simple pupil shape, and the capability for infrared observations from space. The primary goal of this coronagraph is the direct detection and spec-

---

[1] Inst. of Space and Astronautical Science/Japan Aerospace Exploration Agency, 3-1-1 Yoshinodai, Sagamihara, Kanagawa, Japan

[2] Graduate University for Advanced Studies, 3-1-1 Yoshinodai, Sagamihara, Kanagawa, Japan

[3] Department of Physics, Graduate School of Science, The University of Tokyo 7-3-1 Hongo, Bunkyo-ku, Tokyo, Japan

[4] Inst. of Astronomy, School of Science, Univ. of Tokyo, 2-21-1 Osawa, Mitaka, Tokyo, Japan

[5] National Astronomical Observatory of Japan, 2-21-1 Osawa, Mitaka, Tokyo, Japan

[6] Graduate School of Science, Kobe University, 1-1 Rokkodai, Nada, Kobe, Hyogo, Japan

[7] Dept. of Earth and Space Science, Graduate School of Science, Osaka University, 1-1 Machikaneyama, Toyonaka, Osaka, Japan

[8] Dept. of Information Science, Kanagawa Univ., 2946 Tsuchiya, Hiratsuka, Kanagawa, Japan

[9] Division of Applied Physics, Hokkaido University, Sapporo, Japan

[10] Inst. of Astronomy and Astrophysics, Academia Sinica, Taipei, Taiwan, R.O.C.

[11] Jet Propulsion Laboratory, Pasadena, CA, USA

[12] Dept. of Earth and Planetary Sciences, Tokyo Institute of Technology, 2-12-1 Ookayama, Meguro-ku, Tokyo, Japan

[13] Laboratoire A.H. Fizeau, Université de Nice Sophia-Antipolis, CNRS, Parc Valrose, Nice, France

[14] Subaru Telescope, National Astronomical Observatory of Japan, 650 North A'ohoku Place, Hilo, USA

[15] Observatoire Astronomique de Marseille-Provence, Pôle de l'Etoile Site de Château-Gombert 38, rue Frédéric Joliot-Curie, Marseille, France

[16] Optcraft Corp., 3-16-8-101, Higashihashimoto, Sagamihara, Kanagawa, Japan

[17] Boston Micromachines Corp., 30 Spinelli Place, Cambridge, MA, USA

troscopy of Jovian exoplanets. The specifications, performance, and the design of the instrument are shown. The main wavelengths and the contrast required for the observations are 3.5 – 27μm, and 10^{-6}, respectively. We also show the progress of the development of key technology to realize this instrument. The non-coronagraphic mode of this instrument is potentially useful for characterization of inner planets via observation of planetary transit and Color Differential Astrometry(CDA). We expect the SPICA coronagraph will provide drastic progress for understanding various planetary systems by it's unique capability, and will be a fruitful precursor for a future mission targeting terrestrial planets.

1. Introduction

We consider the systematic characterization of exoplanets one of the most important issues for space science in the near future. Since the first report by Mayor & Queloz 1995, more than 400 extra-solar planets (hereafter exoplanets) have been discovered, with the radial velocity method finding the most planets. Indirect detection by measuring the influence of a planet on the parent star usually does not provide the atmospheric spectral features of the planet.

Recently by using monitoring observations of planetary transits, atmospheric spectral features of some exoplanets were observed. Tinetti et al. (2007) and Swain, Vasisht & Tinetti (2008) reported the detection of H_2O and CH_4 features in an exo-planet HD189733b observed by the Spitzer Space Telescope and the Hubble Space Telescope. Though this method is quite interesting, it should be noted that the targets are biased to planets which are very close to the parent star, i.e., "hot Jupiters". The direct observation of exoplanets spatially resolved from the parent star is a clear way to approach the characterization of exoplanets. Recently, the first direct detection of an exo-planet was finally reported by Marois et al. (2008) and Kalas et al. (2008). However, these observations are limited in detection and spectroscopy is not achieved.

Considering these situations, we are developing the SPICA coronagraph instrument (SCI) with spectroscopic mode, in order to carry out the systematic characterization of exoplanets by direct imaging and spectroscopy.

2. SCI and its Performance

The specifications of SCI are shown in Table 1. An overview of SCI is shown in Figure 1. SCI has the capability to provide high contrast (10^{-6}) point spread function (PSF) and continuous wavelength coverage of spectroscopy, while the non-coronagraphic mode is also prepared to use SCI thanks to the mask changer. The core wavelength region is 3.5 – 27 micron, and equipped wavelength resolution is $\sim$20 and $\sim$200.

Figure 2 shows the expected spectrum of exoplanets with various masses and ages (Burrows, Sudarsky, & Hubeny 2004), derived by simulation with observation performance (Fukagawa, Itoh, & Enya 2009). SCI suppresses the parent starlight, and its spectroscopy mode reveals essentially important spectral features in the mid-infrared, CH_4, H_2O, CO_2, CO, NH_3. Jovian exoplanets around 1) nearby stars(<10pc) and young and 2) young and modestly old stars (<1Gyr

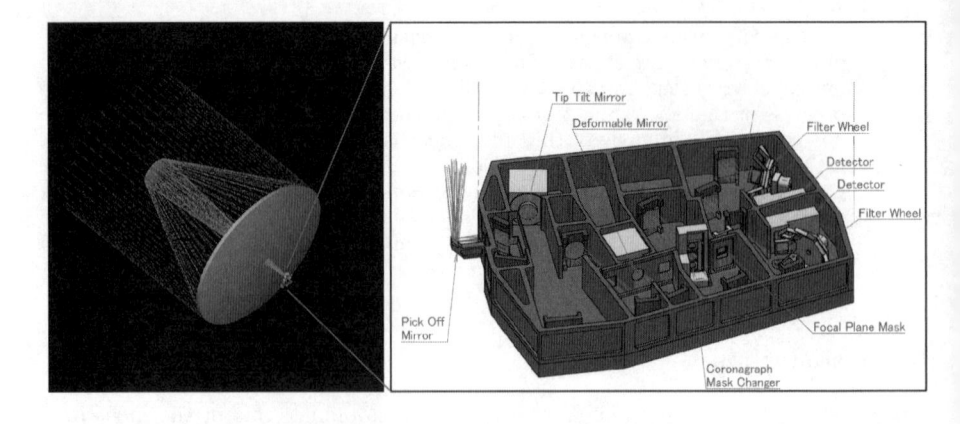

Figure 1. Overview of the design of SCI at the focal plane of the telescope.

old) are suitable targets of SCI. The former are suitable for detailed spatially resolved observations, and the later are suitable to understand the history of the planetary system formation. SCI is expected to make an atlas of various spectra of exoplanets as the result of survey observations of ∼100s targets.

We are planning to use SCI in the non-coronagraphic mode for the characterization of planets by the transit method (Yamashita, Narita, & Enya 2009). Coronagraphic observation by SCI covers "outer planets", e.g., planets ∼10AU or more from the parent star. So coronagraphic observations and observations of the planetary transits are complementary. The advantage of SCI as a transit monitor instrument is its capability of simultaneous observation with two detectors (InSb and Si:As for short and long wavelength channel, respectively), the best pointing stability of the SPICA instruments (0.03 arcsec realized by internal tip-tilt mirror), and use of a deformable mirror for defocusing to avoid satura-

Table 1. Specifications of SCI

Wavelength	3.5 – 27 μm
	(In 1 – 3.5 μm, SCI has sensitivity
	though high contrast is not guaranteed).
Observation mode	Coronagrahpic/Non-coronagraphic,
	Imaging/Spectroscopy
Coronagraph	Binary-shaped pupil mask
Contrast	10^{-6}
Inner working angle	3.3 λ/D
Outer working angle	16 λ/D
Detector	Si:As 1k×1k (long wavelength channel)
	InSb 1k×1k (short wavelength channel)
Field of View	1.0 × 1.0 arcmin^2
Spectral resolution	R = 20, 200

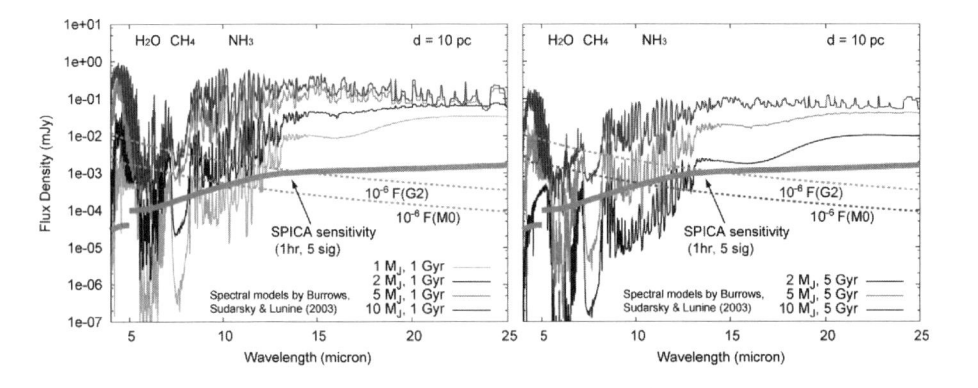

Figure 2. Left and right panels show the detection limit of SPICA and the spectrum of Jovian exoplanets with various mass and age derived by Burrows, Sudarsky, & Hubeny 2004. The suppressed light of the parent star is also shown by the dashed curves. Solid gray curves present the sensitivity limit of imaging with the SPICA telescope.

tion. Color Differential Astrometry (CDA) is also considered for observation of non-transiting inner planets (Abe et al. 2009).

Using the data set of spectra of various exoplanets obtained with SCI, we expect that our understanding for the whole planetary system will be improved drastically. Finally we should not discount the possibility of trying to detect the more challenging bio-marker, O_3, with SCI.

3. Development of Key Technologies

This section introduces the progress of the development of key technology to realize SCI. More detail is shown in Enya et al. (2009) and its references.

Laboratory Demonstration of Coronagraph

The SPICA coronagraph has to work in the mid-infrared wavelength region in a cryogenic environment. The coronagraph should be robust against telescope pointing errors caused by vibration of the mechanical cryo-cooler system and the altitude control system of the satellite. Achromatism was also assumed to be a beneficial property to realize spectroscopy mode. After considerations for these points, we selected a coronagraph implemented by a binary-shaped pupil mask as the primary candidate to be studied because of its physical properties and its feasibility. The purpose of the first step of our coronagraphic experiment is the demonstration of the coronographic principle. So we started experiments in the air using substrate-based binary-shaped pupil masks and visible laser as a light source. The electron beam drawing method was used to manufacture pupil masks on glass substrates. Taking into account the convenience of fabrication,

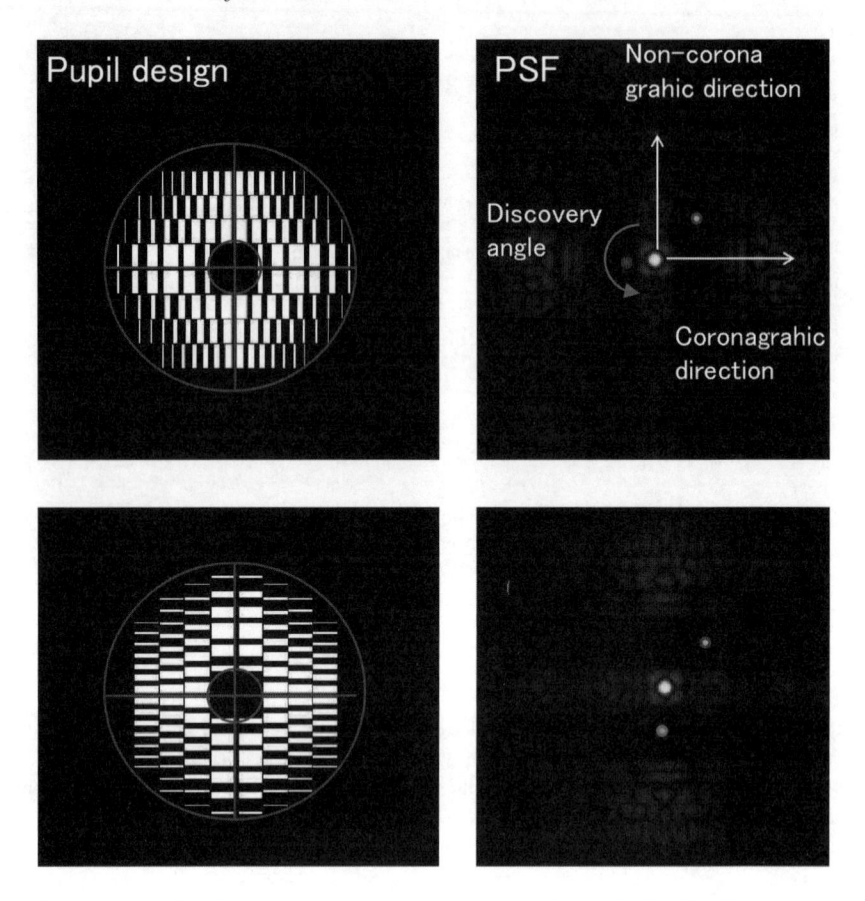

Figure 3. *Top-left*: an example of multi-barcode mask for SPICA pupil. Transmissivity of white and black area is 1 and 0, respectively. The grey circles and cross show constraints by the primary mirror, the secondary mirror, and the support structure of the telescope. *Top-right*: expected PSF. Bottom: figures to show the principle of mask rotation. Bright dots show possible planet positions. Mask rotation makes observation of all them possible.

we selected a checkerboard-type design of pupil mask as presented by Vanderbei, Kasdin, & Spergel (2004). A PSF contrast of 6.7×10^{-8} was achieved without adaptive optics. This value satisfies the requirement of SCI. It should be noted that a software LOQO presented by Vanderbei (1999) was used for optimization of shaped pupil masks.

Mask Design for the Pupil of SPICA

The pupil of SPICA is obscured by the secondary mirror and the support structure of the telescope (Figure 3). So a special new design of shaped pupil mask was needed to adapt to the pupil of SPICA. An example of our solution is shown at the top of Figure 3. This design consists of multi-barcode masks and has coronagraphic power only in one dimension (the horizontal direction in Figure 3).

As a result, the specification on IWA is satisfied and a large opening angle is realized. Furthermore, mask rotation makes it possible to rotate the diffraction tail in the PSF produced by the mask. Owing to two images before and after the mask rotation, the total discovery angle at PSF becomes larger. It should be noted that the principle of a barcode mask was presented by Kasdin et al. (2005).

Cryogenic Active Optics

The specification of the wave-front quality is 350 nm rms for the SPICA telescope, while the requirement of the SCI is higher. Therefore, SCI needs active wave-front correction which works at the temperature of work-surface of SCI (i.e., ~5K). We developed a prototype of the cryogenic deformable mirror (Kasdin et al. 2005). The prototype device with 32 channels consists of Micro Electro Mechanical Systems (MEMS) of Boston Micromachines Corporation (BMC) and special substrates designed to minimize thermal stress caused by cooling. The demonstration of the prototype at cryogenic environment (95K) has been carried out. Development of a larger format deformable mirror (~1000 channels) is ongoing.

Wave-front Correction Algorithm

An important issue in operating a deformable mirror is how to correct the wavefront error. In the context of the study of such an algorithm, we started experiments with a visible laser and a commercially available deformable mirror of BMC. Remarkable improvement of contrast, over 10^{-6}, was confirmed at the area just out of IWA, which is an important area for the exo-planet search (Kotani et al. 2009).

Cryogenic MIR Testbed

Many of our laboratory experiments were performed at room temperature, in atmosphere, at visible wavelengths, using a monochromatic laser as a light source. Contrarily, the SPICA coronagraph will finally have to be evaluated at cryogenic temperature, in vacuum, at infrared wavelengths, using a source with some bandwidth. We are developing a cryogenic vacuum chamber as a testbed of whole MIR coronagraph including cryogenic active optics and free standing mask (i.e., without substrate) of which manufacturing of prototype is ongoing.

4. SPICA Coronagraph as a Precursor toward Terrestrial Planets

The SPICA mission gives us a unique opportunity for high-contrast observations because of the large telescope aperture, the simple pupil shape, and the capability for infrared observations from space (Nakagawa 2009). It should be noted that SCI has potential to provide a unique spectral atlas of various exoplanets in mid-infrared, which could contribute to the characterization of planetary systems and their birth and evolution. A coronagraph mission with SPICA includes advanced technology, (light-weight large telescope made of SiC family material coronagraph optics with wavefront control, and so on), which will be potentially useful for the terrestrial planet search. At both scientific and technical point of view, we believe the SPICA coronagraph can be a quite fruitful precursor toward the study of terrestrial planets in future.

Acknowledgments. We deeply thank to all pioneers relating to this work, especially R. Vanderbei and J. Kasdin. This work is supported by the Japan Aerospace Exploration Agency.

References

Abe, L., Vannier, R., Petrov, R., Enya, K., & Kataza, H. 2009, In SPICA Joint European/Japanese Workshop, eds A. M. Heras, B. M. Swinyard, K. G. Isaak & J. R. Goicoechea, (EDP Sciences), 02005
Burrows, A., Sudarsky, D., & Hubeny, I. 2004, ApJ, 609, 407
Enya, K. & SPICA working group 2010, AdSR, in press, (arXiv:0905.3829)
Fukagawa, M., Itho, Y., & Enya, K. 2009, In SPICA Joint European/Japanese Workshop, eds A. M. Heras, B. M. Swinyard, K. G. Isaak & J. R. Goicoechea, (EDP Sciences), 02006
Kalas, P., Graham, J. R., Chiang, E., Fitzgerald, M. P., Clampin, M., Kite, E. S., Stapelfeldt, K., Marois, C., & Krist, J. 2008, Sci, 322, 1345
Kasdin, N. J., Vanderbei, R. J., Littman, M. G. & Spergel,D. N. 2005, Appl.Optics, 44, 1117
Kotani, T., Enya, K., Nakagawa, T., Miyata, T., Sako, S., Nakamura, T., Haze, K., Higuchi, S., & Tange, T. 2009, this volume
Marois, C., Macintosh, B., Barman, T., Zuckerman, B., Song, I., Patience, J., Lafrenière, D., & Doyon, R. 2008, Sci, 322, 1348
Mayor, M. & Queloz, D. 1995, Nat, 378, 355
Nakagawa, T. 2009, this volume
Swain, M. R., Vasisht, G., & Tinetti, G. 2008, Nat, 452, 329
Tinetti, G., Vidal-Madjar, A., Liang, M.-C., Beaulieu, J.-P., Yung, Y., Carey, S., Barber, R. J., Tennyson, J., Ribas, I., Allard, N., Ballester, G. E., Sing, D. K., & Selsis, F. 2007, Nat, 448, 169
Vanderbei, R. J., Kasdin, N. J. & Spergel, D. N. 2004, ApJ, 615, 555
Vanderbei, R. J. 1999, Optimization Methods and Software, 11, 451
Yamashita, T., Narita, N., & Enya, K. 2009, In SPICA joint European/Japanese Workshop, eds A. M. Heras, B. M. Swinyard, K. G. Isaak & J. R. Goicoechea, (EDP Sciences)

Session VI

Synthesis of Panel Discusssions

Pathways Towards Habitable Planets
ASP Conference Series, Vol. 430, 2010
Vincent Coudé du Foresto, Dawn M. Gelino, and Ignasi Ribas, eds.

Do we Need to Solve the Exozodi Question? If Yes, How to Best Solve It?

O. Absil,[1] C. Eiroa,[2] J.-C. Augereau,[3] C. A. Beichman,[4] W. C. Danchi,[5] D. Defrère,[1] M. Fridlund,[6] and A. Roberge[5]

Abstract. When observing an extrasolar planetary system, the most luminous component after the star itself is generally the light scattered and/or thermally emitted by a population of micron-sized dust grains. These grains are expected to be continuously replenished by the collisions and evaporation of larger bodies just as in our solar zodiacal cloud. Exozodiacal clouds ("exozodis") must therefore be seriously taken into account when attempting to directly image faint Earth-like planets (exoEarths, for short). This paper summarizes the oral contributions and discussions that took place during the Satellite Meeting on exozodiacal dust disks, in an attempt to address the following two questions: Do we need to solve the exozodi question? If yes, how to best solve it?

1. Impact of Exozodis on ExoEarth Imaging with Optical Telescopes

The most obvious effect of exozodiacal dust on optical wavelength direct observations of Earth-like planets is to increase exposure times. Light scattered off the local zodiacal dust and the exozodiacal dust around nearby stars will be mixed with the planet signal in both images and spectra. The larger the final PSF, the more background will be mixed in. This background will likely be the largest source of noise, if the stellar light is canceled by 10^{-10} at the planet pixel.

In the background limited case, the exposure time to image a planet is linearly proportional to the exozodi surface brightness. Obviously, if the exozodi is too bright, the integration time to get the required planet SNR becomes prohibitively long. How do various optical mission concepts perform under varying levels of exozodi? Using publicly available information, it is very hard to come up with estimates that can be reasonably compared. In addition, most groups assume a fixed exozodi level for calculating their mission performance, typically one zodi, which is not tremendously informative. There is a strong need for uniform mission performance analysis that pushes to high exozodi levels.

---

[1]Dept. AGO, Université de Liège, 17 Allée du Six Août, Sart Tilman, Belgium

[2]Departamento de Física Teórica, C-XI, Universidad Autónoma de Madrid, Madrid, Spain

[3]LAOG–UMR 5571, CNRS and Université Joseph Fourier, Grenoble, France

[4]NExScI, California Institute of Technology, 770 S. Wilson Ave, Pasadena, CA, USA

[5]Exoplanets and Stellar Astrophysics Laboratory, NASA-GSFC, Greenbelt, MD, USA

[6]ESA-ESTEC, PO Box 299, Keplerlaan 1 NL, Noordwijk, The Netherlands

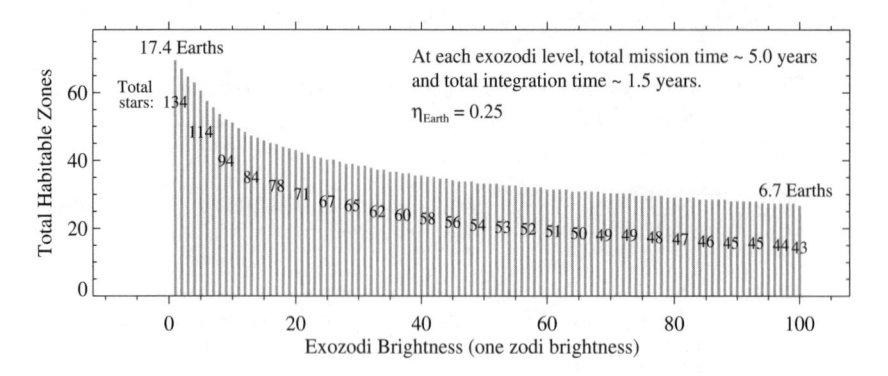

Figure 1. Total habitable zones searched vs. exozodi surface brightness. Each grey bar represents a possible observing program, assuming $\eta_\oplus = 0.25$ and the parameters of the *New Worlds Observer* mission concept (for details, see Turnbull et al. 2010). The numbers superimposed on some of the grey bars are the total numbers of stars observed in the programs. The y-axis is the total habitable zones searched, which is the sum of the completeness values for all the individual stars observed. A star's completeness value is the probability that a habitable zone planet would be detected in a single visit, given the possible range of system inclinations and planet eccentricities. The expected number of exoEarths characterized is the total habitable zones searched times $\eta_\oplus$. To generate a program for each exozodi level, stars were chosen in order of decreasing weighting factor (completeness / exposure time) until the total on-target integration time reached 1.5 years or the total mission time reached 5 years, whichever came sooner. The total integration times were calculated assuming that for $\eta_\oplus \times 100 = 25\%$ of the targets, a spectrum with $S/N \geq 10$ and $R = 100$ was obtained in addition to the imaging observation. Total mission times include 11 days per target for moving the occulter.

That being said, Figure 1 illustrates the decline in mission performance with increasing exozodi brightness for an optical mission. These preliminary calculations were done assuming the parameters of the *New Worlds Observer* mission concept (4-m telescope, broad-band imaging channel covering 500 to 700 nm), but the general behavior should be similar for other optical wavelength missions using either external occulters or internal coronagraphs. However, it is important to note that these calculations include only statistical errors due to local zodiacal background, exozodiacal background, and unsuppressed starlight. Possible systematic errors associated with modeling light scattered off a non-uniform exozodiacal dust distribution and removing it from a planet image have not been thoroughly characterized for optical missions. Furthermore, confusion between planets and exozodiacal dust structures like resonant clumps remains poorly studied, although initial attempts to evaluate its impact on mission performance appear in Savransky et al. (2009) and Turnbull et al. (2010).

2. Impact of Exozodis on ExoEarth Imaging with IR Interferometry

The mid-infrared wavelength range presents several advantages for Earth-like planet characterization. In addition to including spectral bands of water, car-

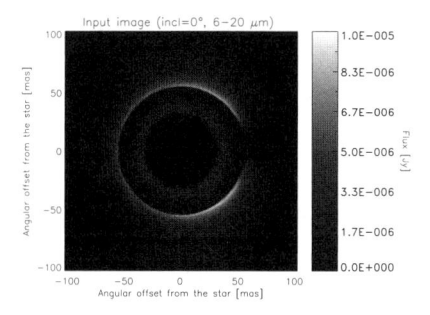

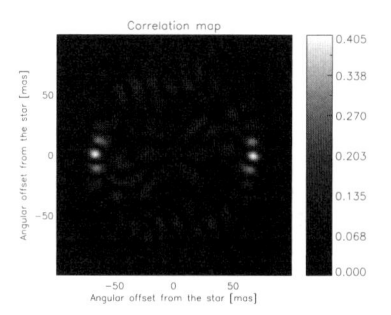

Figure 2. *Left:* Simulated image of the thermal flux produced by a 10-zodi face-on exozodiacal dust cloud in the 6-20 μm range. The simulation includes the presence of an Earth-mass planet located at 1 AU on the x-axis, and assumes a Dohnanyi size distribution for the dust grains (Stark & Kuchner 2008). *Right:* Corresponding dirty map computed for an Emma X-array nulling interferometer, obtained by cross-correlating the measured signal with templates for the expected signal from a point source at each location on the sky plane (Defrère et al. 2010).

bon dioxide and ozone, the contrast between a star and an exoEarth is only $\sim 10^7$ whereas it is $\sim 10^{10}$ in the visible. However, resolving the habitable zone around nearby stars in the mid-infrared would require a gigantic telescope with a diameter up to 100 m. Space-based interferometry is therefore considered as the most promising technique to achieve this goal. A large effort has been carried out during the past decade to define a design that provides excellent scientific performance while minimizing cost and technical risk. This has resulted in a convergence and consensus on a single mission architecture consisting of a non-coplanar X-array, called Emma (see e.g., Cockell et al. 2009), using four collector spacecraft and a single beam combiner spacecraft. Such a design enables the implementation of phase chopping, a technique which suppresses from the final output all sources having point-symmetric brightness distributions.

The impact of exozodis on the mission performance is twofold: on one hand, their point-symmetric component contributes to the overall shot noise and can therefore drive the required integration time to detect exoEarths, while on the other hand, asymmetric structures in exozodis (such as resonant clumps) are not suppressed by phase chopping and thereby contribute as possible biases (or false positives), which could prevent the detection of small planets.

Considering the Emma X-Array design with four 2-m apertures, point-symmetric exozodis of about 100 zodis can be tolerated while preserving 75% of the mission outcome, i.e., surveying the habitable zone of at least 150 targets with > 90% completeness during the 2-yr search program. However, when including the resonant structures created by an exoEarth orbiting at 1 AU around a Sun-like star (see left part of Figure 2), the asymmetry created by the hole in the dust distribution near the planet significantly contributes to the final detected signal (see dirty map in the right part of Figure 2) and can thereby prevent from detecting the planet itself. The tolerable dust density then goes down to about 15 times the solar zodiacal dust density. This upper limit on the tolerable exozodiacal dust density gives an estimation of the typical sensitivity that precursor instruments will need to reach on exozodiacal disks in order to prepare the scientific program of future exoEarth characterization missions.

3. Current Exozodiacal Disk Detection Efforts

Two main techniques are been used to evaluate the amount of exozodiacal dust around main sequence stars. Infrared spectro-photometry can reveal the presence of an excess emission on top of the expected stellar photospheric flux. This requires high-accuracy photometry in the mid-infrared regime, which is generally done from space, e.g., using the various instruments on board the Spitzer Space Telescope. The intrinsic accuracy of Spitzer photometry limits the sensitivity to exozodiacal disks about 1000 times more luminous than the solar zodiacal cloud (assuming the same brightness distribution, see e.g., Bryden et al. 2006). Note that, even with an infinite accuracy on the photometry, this technique would still be limited to an accuracy of about 1% by the capability to predict the mid-infrared stellar photospheric flux.

The second way to detect exozodiacal disks overcomes this limitation by angularly separating the signal of the star from its surrounding dust disk. The requested angular resolution can generally be achieved only with infrared interferometry. Two types of interferometers are contributing to exozodi surveys: high-accuracy near-infrared interferometers such as CHARA/FLUOR and VLTI/VINCI (see e.g., Absil et al. 2006, 2009), and mid-infrared nulling interferometers such as MMT/BLINC and the Keck Interferometer Nuller (see e.g., Liu et al. 2009; Colavita et al. 2009). Both types of instruments reach a typical accuracy of 10^{-3} on the disk/star contrast, which corresponds to roughly 1000 zodis in the K band and 300 zodis in the N band. These observations are however still restricted to a small amount of targets, due either to limited observing time or to the limiting magnitude of the instrument. A new generation of nulling interferometers, such as the LBTI or the ALADDIN project on the ground, or FKSI in space, could significantly improve the current sensitivity to exozodiacal disks, pushing the detection limit down to ~ 30 zodi from the ground or ~ 1 zodi from space (see other paper by Absil et al. in this volume).

4. Current Exozodiacal Disk Modeling Efforts

The current exozodi modeling efforts follow two main paths. On the one hand, radiative transfer codes for optically thin disks are being employed to reproduce the scarce exozodi measurements, in particular near- and mid-infrared interferometric data (see e.g., Figure 3), and show that the solar system zodiacal spectrum does not match the spectral energy distributions (SEDs) of detected exozodiacal disks. The fit to the 2.2 μm excesses with the solar system zodiacal model indeed indicates densities a few thousand times larger than the zodiacal density (e.g., about 3000 zodis for Vega and 5000 zodis for Fomalhaut), but this model predicts much too large flux (by about an order of magnitude) in the mid-infrared compared to the observations. Dust much closer to the star (down to the sublimation radius at a fraction of an AU), and with a sufficient amount of refractory material, are required to shift the spectrum to shorter wavelengths and match the current data sets. Spectral decomposition techniques are also being developed to reproduce extreme Spitzer/IRS spectra showing unusually large amount of warm dust and clear spectral solid-state features (e.g., Lisse et al. 2009).

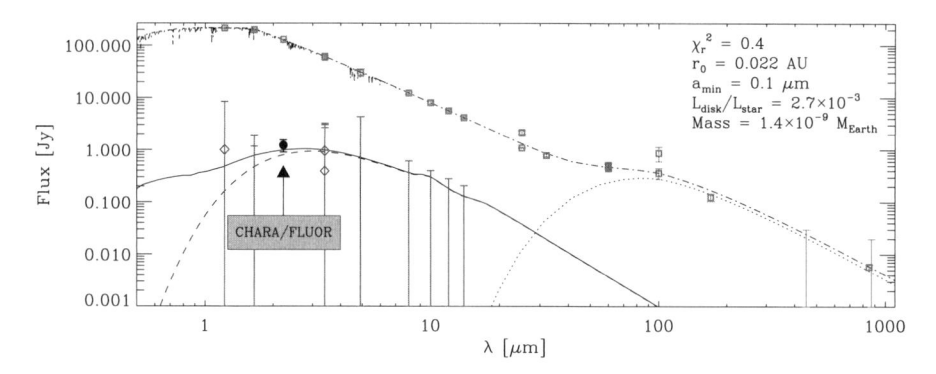

Figure 3. A possible fit to the photometric excesses (represented with diamonds) and interferometric observations (filled circle at $2.2\,\mu$m) of the τ Ceti exozodi (from Di Folco et al. 2007). Dashed line: thermal emission from the best-fit disk model. Solid line: includes the scattered light contribution. Dotted line: approximative blackbody fit to the long-wavelength excesses produced by a Kuiper-like belt. Dashed-dotted line: total emission from the star-disk system.

On the other hand, dynamical models employing classical N-body codes are being used to assess the influence of planets on the shape of exozodiacal disks and discuss the dust production mechanisms. Some models for instance simulate the sculpting of an asteroid/dust belt by a planet due to their capture in mean motion resonances (e.g., Stark & Kuchner 2008). Larger scale models, involving the outward migration of a planet toward a Kuiper belt are currently being developed, basically relying on the assumption that exozodis are fed by the outer, much more massive disk regions, thereby linking the inner and outer regions of planetary systems. This can either be due to a sudden event, e.g., a Nice LHB-like model (Booth et al. 2009), or be more progressive (Augereau et al., in prep.). The amount of dust produced can be assessed for individual systems following this methodology, but it is very sensitive to the assumed planetary system architecture. Finally, collisional models, using statistical approaches, can evaluate the lifetime of asteroid belts due to collisions (e.g., Krivov et al. 2006; Thébault & Augereau 2007). An ISSI working group has been assembled to further discuss the modeling of exozodis (see http://www.issibern.ch/teams/exodust/).

5. Conclusions and Recommendations

This short review illustrates how the final SNR for direct exoEarth detection depends on the quantity of exozodiacal dust around main sequence stars. On one hand, it drives the required integration time to detect the planetary signal as soon as its density reaches a few tens of zodis, and on the other hand, its potential asymmetries induce biases and false positives, which in turn demand the planetary systems to be observed longer in order to extract the actual planetary signal. If space missions had an unlimited lifetime, this wouldn't be a major issue, as one would just skip the inappropriate targets, or integrate longer to eventually reveal their planets. However, space missions are limited in time, and the exozodi issue could thus become a major hurdle in case bright exozodis are common. The sensitivity of current exozodi finders ($\sim$ 300 zodis at best) is not appropriate to assess whether exozodis in the 10–100 zodi range are common or not. A dedicated effort to solve this question is therefore mandatory.

Three possible avenues have been identified to make sure that exoEarth imaging missions will be capable of reaching their goals:

- Perform an exozodi survey with a sensitivity of $\sim$ 30 zodis on a statistically meaningful sample of main sequence stars to constrain the distribution of exozodi brightness down to an appropriate level. Space missions can then be designed (in terms of aperture size and mission lifetime) to cope with the inferred mean exozodi level.

- Measure the exozodi level with an accuracy of $\sim$ 10 zodis on all the candidate mission targets, once the targets have been identified (e.g., Sun-like stars hosting exoEarth(s), detected through high-precision astrometric or radial velocity surveys).

- Use the fact that the statistical distribution of cold debris disks will soon be known at a level similar to the density of the solar Kuiper belt thanks to Herschel. Extrapolating the statistics of cold debris disks towards that of warm exozodis is however not straightforward based on theoretical models, and needs to be backed up by observational data anyway.

The Panel therefore recommends that significant efforts and support be put in next generation exozodi finders, starting with LBTI and continuing with mid-term ground-based, balloon-borne or space projects such as ALADDIN or FKSI. The Panel also underscores the importance of continued modeling efforts to better understand the origin and dynamics of second generation dust grains around main sequence stars.

Acknowledgments. O.A. acknowledges the financial support from an F.R.S.-FNRS postdoctoral fellowship, and from the Communauté Française de Belgique (ARC – Académie universitaire Wallonie-Europe).

References

Absil, O., Di Folco, E., Mérand, A., et al. 2006, A&A, 452, 237
Absil, O., Mennesson, B., Le Bouquin, J.-B., et al. 2009, ApJ, 704, 150
Booth, M., Wyatt, M. C., Morbidelli, A., Moro-Martin, A., & Levison, H. 2009, MNRAS, 399, 385
Bryden, G., Beichman, C. A., Trilling, D. E., et al. 2006, ApJ, 636, 1098
Cockell, C., Herbst, T., Léger, A., et al. 2009, Exp. Astron., 23, 435
Colavita, M. M., Serabyn, G., Millan-Gabet, R., et al. 2009, PASP, 121, 1120
Defrère, D., Absil, O., den Hartog, R., Hanot, C., & Stark, P. 2010, A&A, 509, A9
Di Folco, E., Absil, O., Augereau, J.-C., et al. 2007, A&A, 475, 243
Krivov, A., Löhne, T., & Sremčević, M. 2006, A&A, 455, 509
Lisse, C. M., Chen, C. H., Wyatt, M. C., et al. 2009, ApJ, 701, 2019
Liu, W., Hinz, P., Hoffman, W., et al. 2009, ApJ, 693, 1500

Savransky, D., Kasdin, N. J., & Vanderbei, R. J. 2009, SPIE, 7440, 34
Stark, C. C., & Kuchner, M. J. 2008, ApJ, 686, 637
Thébault, P. & Augereau, J.-C. 2007, A&A, 472, 169
Turnbull, M. C., Roberge, A., Cash, W. C., et al. 2010, in prep

Pathways Towards Habitable Planets
ASP Conference Series, Vol. 430, 2010
Vincent Coudé du Foresto, Dawn M. Gelino, and Ignasi Ribas, eds.

How to Consolidate Efforts within the Community and the Related Agencies?

H. Zinnecker

Astrophysical Institute Potsdam, An der Sternwarte 16, Potsdam, Germany

Abstract. The author was charged by the organisers to run a panel discussion in order to sample opinions from the community and the representatives of the agencies on how to best consolidate efforts towards habitable exoplanets. The conclusions from this panel discussion are given here.

1. Questions

The panel chair, after reminding everyone that ESA did not select the proposed DARWIN mission during the Cosmic Vision I process, raised 3 basic questions in an analysis to appreciate WHY Darwin was NOT chosen. One major issue was the immature technological readiness level (free flyer concept, interferometry in space) but another, equally important, issue was community support. In the long run, we clearly want a DARWIN/TPF type mission, but which probe or M-class mission would lead to it? Thus, to start off, we discussed publicly

1. How to unite the community (behind a habitable planet search and study project)?

2. How to set up collaborations, and ultimately,

3. Do we need to re-organize our efforts, i.e., do we need to found a transnational exoplanet institute?

2. Discussion

In the discussion of the first point, astronomers Hatzes, Schneider, Liseau, and Fridlund stood up and emphasized that we must speak with one voice, but also that there is a general fear of interferometry. In Europe we must teach people about interferometry (VLTI) and prepare the astronomical community for ESA's Cosmic Vision Cycle No. 2, while the decadal review sets the stage in the U.S.

In the discussion of the second point, the chair asked several astronomers to express their opinion, among them Lin (US, China), Nakagawa (Japan), Devirian (NASA), as well as Fridlund (ESA) and Kissler-Patig (ESO). There was general agreement that a NASA/ESA 50:50 mission (instead of 90:10) would be a recipe for disaster, i.e., there must be a lead responsibility by someone.

In addition, it was pointed out that ALMA does not qualify as a role model for a space mission. In the end, Jean Schneider made the point that not only do we need collaboration but even more we need coordination of our efforts!

This then led to the third question whether we need a new organisational vision, i.e., a European or world-wide exoplanet institute? Given the importance of and huge public interest in the habitable planets issue, such a thought has occurred. Of course, one of the immediate subquestions turned out to be whether we would need a real institute (with a new building) or just a virtual institute. The chair asked the US attendees about their experience with NExScI in California and with the NASA Astrobiology Institute (NAI).

How well do the virtual astrobiology nodes work together? The answer was not clear cut. In Europe, one would have to ask whether ESO and ESA would accept such an independent development. Here the answer was that ESO's and ESA's terms of reference are different, so there would not be an obvious conflict of interest. The crucial comment came from Doug Lin who suggested that it is more efficient to spread support over many nodes rather than having a centralized system. He stressed the broad support that is needed to make headway and mentioned the possibility of creating exoplanet fellowships (like the Carl Sagan fellowship). In conclusion, the feeling was that a single new exoplanet institute was not such a good idea in the end.

3. Personal Roadmap

After this, the chairman offered his personal HZ roadmap for a Darwin-like mission, where NASA and ESA in the near future go separate ways: with NASA heading towards SIM, with 15% ESA participation and ESA going for a combined light spectrometer like THESIS (following Mark Swain's idea), with NASA being the minor (15%) partner. [I think I courageously proposed the name SEXOPLANET mission for spectroscopic exoplanet mission or something like that.] The main idea would be to study the atmosphere of habitable planets around M-stars over a wide wavelength range from the far optical to the mid-infared (1–15 microns), a range that not even JWST can cover. This would then be a comparative exoplanetology mission, behind which Europe could unite, exploring cooperation with new non-European partners such as Japan, China, India, and perhaps Russia. Then, by 2025 or beyond, NASA and ESA could converge on one flagship mission of the Darwin/TPF type. At the present time there would simply not be enough money in both agencies to pursue such an idea, let alone the issue of technical readiness.

4. Acknowledgment

Our host, the city of Barcelona, did such a wonderful job, not only offering the world-class Cosmocaixa museum as our meeting place but also sponsoring a superb conference dinner. They truly made us feel "at home" and the city indeed suggested that we, the exoplanet community, should come back in 2–3 years to continue the "Barcelona process" for studying a space mission towards habitable planets.

Pathways Towards Habitable Planets
ASP Conference Series, Vol. 430, 2010
Vincent Coudé du Foresto, Dawn M. Gelino, and Ignasi Ribas, eds.

Designating Habitable Planets for Follow-up Study - What are the Relative Parameter Spaces of Radial Velocities and Astrometry?

I. Ribas[1] and F. Malbet[2,3]

[1] Institut de Ciències de l'Espai (CSIC-IEEC), Campus UAB, Bellaterra, Spain

[2] Jet Propulsion Laboratory/California Institute of Technology, 4800 Oak Grove Drive, Pasadena, CA, USA

[3] Université de Grenoble J. Fourier/CNRS, Laboratoire d'Astrophysique de Grenoble, BP 53, Grenoble, France

Abstract. In this contribution we summarize the satellite meeting (P2 panel) held at the "Pathways towards Habitable Planets" conference about the complementarity of two techniques, namely *astrometry* and *radial velocities*, to detect telluric planets in the habitable zone of solar type stars. This question has a direct impact on the future possibility of launching a characterization mission with the aim of detecting bio-signatures of Earth-like planets. Beyond the level of instrumental noise, the main issue is how to cope with the intrinsic stellar variability due to magnetic activity (i.e., starspots).

The P2 panel session included short presentations by F. Pepe (Obs. Genève), E. Martín (CAB), M. Shao (JPL), F. Malbet (JPL/U. Grenoble), and W. Traub (JPL), and additional discussion time. The text below summarizes the main items addressed and the general conclusions that met with consensus.

1. Performances Needed

The astrometric signal of an Earth-mass planet located at 10 pc around a solar-type star is $0.3\,\mu$as and decreases linearly with the distance of the star. The radial velocity of the same planet is 9 cm s^{-1} (~ 8 cm s^{-1} for the expected value of randomly selected orbital inclinations) and is independent of the distance.

2. Instrumental Limitations

Research and development for the SIM astrometic mission has demonstrated in laboratory experiments that a floor noise level of $0.035\,\mu$as can be reached, thus permitting the detection of astrometric planet signatures down to $0.22\,\mu$as, and therefore either an Earth-mass planet at 15 pc or a $0.7\,M_\oplus$ planet at 10 pc. This corresponds to a signal-to-noise ratio of 6 (Catanzarite et al. 2006). In SIM Lite astrometry, the target star is surrounded by reference stars located within a radius of about 1°. The least-square determination of three parameters requires a minimum of three reference stars. The astrometric precision obtained in this one-dimensional narrow-angle observing sequence is currently specified at 1.0 μas for a star with a group of reference stars (Goullioud et al. 2008). This performance has been demonstrated in the microarcsecond metrology testbed at the Jet Propulsion Laboratory. At 1.0 μas single-measurement accuracy, a differential measurement with 780 s total integration time divided among a 7 mag target star and set of 10 mag reference stars has an astrometric accuracy of 1.4 μas. This includes photon noise, instrument noise, and a multiplier that accounts for the geometric distribution of the reference stars. A two-dimensional narrow-angle observation is a pair of visits with the interferometer baseline oriented along quasi-orthogonal directions on the plane of the sky. An even time sampling for the series of observations is assumed along each of the two baseline orientations, and that observation pairs are quasi-simultaneous. However in reality, scheduling constraints in the mission (including a solar exclusion zone) preclude even sampling. Therefore sampling of SIM planet-search targets will be serendipitous. Yearly gaps in the sampling, ranging from several weeks to several months (for targets near the ecliptic), will occur during times when the target is in the solar exclusion zone.

By comparison, ground-based efforts, such as the STEPS instrument installed on the 5-m Palomar telescope (Pravdo et al. 2004), reach an astrometric precision of 1 mas per a single series of 20–30 exposures. The first astrometric discovery of a giant exoplanet (VB 10b) around a main-sequence star was announced by Pravdo & Shaklan (2009). However, there is debate as to whether the astrometric signal is due to a giant planet or a spurious detection (Zapatero Osorio et al. 2009; Bean et al. 2009b). Given the performances, it is clear that a space-borne high-accuracy astrometry mission is required to detect Earth-type planets.

The ESO/HARPS spectrograph has demonstrated a performance reaching radial velocity residuals down to 0.8 m s^{-1} over 4–5 years (Mayor & Udry 2008). The recent study of Mayor et al. (2009) announced discovery of a 1.9 M$_\oplus$ planet corresponding to a signal of 1.85 m s^{-1} with a residual of 1.53 m s^{-1}. These residuals include instrumental noise (telescope guiding,...), noise from stellar origin (pulsations, activity,...) and also, possibly, smaller planets not yet detected. Several approaches are currently being developed to lower the instrumental noise level. The key elements are the stability and repeatability, the importance of good centering and guiding, and calibration and wavelength solution. The short-term goal is to routinely reach 0.5 m s^{-1} with a single thorium-argon reference source stable to 1 m s^{-1} over 1 month. Note that rms residuals as low as ∼ 0.35 m s^{-1} have already been claimed by Lovis et al. (2006) using a clever averaging strategy (binning over the rotational period) on a very quiet target.

The next generation radial velocity instrument is the so-called EXPRESSO project for the VLT, having a radial precision goal of 0.1 m s^{-1}. First light is expected for 2014. Estimates indicate that an integration time of 900 s will be needed for the VLT 8-m telescope to reach 0.1 m s^{-1} on stars with $V = 8$, thus permitting the detection of Earth-like planets around bright solar-type stars. Further in the future, the CODEX instrument proposed for the European Extremely Large Telescope project (E-ELT) has a radial velocity precision goal of 1 cm s^{-1} over a baseline of a decade. However, questions have been raised as to whether sufficient observing time will be available to carry out an ambitious planet search program. Efforts are also being put into developing high-precision spectroscopy in the near-IR (NIR), thus opening a new area of planet search around very low mass stars (with the increased radial velocity signal of habitable planets and relative immunity to activity effects – see below). Currently, the radial velocity precision achieved is around 5 m s^{-1} on special cases (Bean et al. 2009a; Figueira et al. 2009) but several projects, such as CARMENES, NAHUAL, SpIRou, UPF, are aiming at 1–3 m s^{-1} precision. The main limitation to NIR spectroscopy is the contamination from telluric lines.

3. Stellar Intrinsic Noise

Besides the instrumental effect, one of the main limitations for both techniques at the level of Earth twins is the noise from stellar sources, i.e., biases coming from stellar pulsation, atmosphere granulation, and stellar activity due to stellar spots. The latter has been shown to be the bigger concern given the highly stochastic nature and the potential size of the effect.

The SIM team (Shao et al.) has devised a toy model to evaluate the effect of starspots on both the astrometric and radial velocity signals in the frame of the detection of Earth-like planets around solar-type stars. A first model was defined by considering a Sun-like star located at 10 pc with an area of 0.1% covered by one dark spot. The results indicate that such spots can introduce an astrometric bias of about 0.25 μas and about 1 m s^{-1} in radial velocities. On a second stage, a spot model was build by considering dynamic active dark spots that produce photometric variations matching the solar temporal power spectrum. The results show that the astrometry jitter due to solar-type activity is about 0.08 μas per measurement and the radial velocity jitter is about 0.45 m s^{-1}. In the case of the Sun, this bias becomes random only for epochs separated by more than one week.

A similar study was carried out by the Geneva team (Dumusque et al.) also by considering a toy model with dark spots reproducing the power temporal power spectrum of solar photometric variations. The study is focused on radial velocities only and is based on synthetic velocities from HARPS targets. The simulations indicate that photometric jitter is on the level of 0.4–0.8 m s^{-1} per measurement for relatively quiet stars. However, the study also concludes that an improved observational strategy (i.e., 3 measures per night of 10 minutes, separated by 2 hours, and repeated every 3 nights) could yield precisions of about 0.1–0.2 m s^{-1} when binning over time intervals of about 10 days.

The simulations from both the SIM and radial velocity teams yield comparable results and suggest that both techinques can potentially overcome stellar noise to identify Earth-like planets, athough astrometry seems less affected by this activity jitter (see also Makarov et al. 2009). Note that in the case of radial velocities in the NIR, the activity jitter can be diminished by a factor of $\sim$2 because of the smaller contrast of starspots (Reiners et al. 2009).

4. How Quiet is the Sun? How Quiet are Most Stars?

The SIM team has carried out a preliminary analysis of about 100 stars observed by the CoRoT mission. The results shows that $\sim$40% of the stars are about 10 times or more variable than the Sun and about 10–15% of the stars are quieter than the Sun in terms of photometric variations. The Geneva team reported on the distribution of log R'_{HK} in a limited volume of stars around the Sun observed with HARPS. The statistical analysis shows that 25% of the stars surveyed (already pre-selected) have a Mount Wilson chromospheric activity index below log $R'_{HK} = -5.0$, and 47% of stars have an index below log $R'_{HK} = -4.9$. For comparison, the Sun has log $R'_{HK} = -5.0$ in quiescence and log $R'_{HK} = -4.8$ at the peak of its activity cycle. This is in agreement with past studies in the literature (e.g., Lockwood et al. 2007). The conclusion from the Geneva team is that between 25% (log $R'_{HK} < -5.0$, ideal case) and 50% (log $R'_{HK} < -4.9$) of the dwarf stars (FGK) can be considered as favorable cases for the radial velocity technique in the context of the 0.1–0.2 m s^{-1} precision by 10-day bin obtained for stars with log $R'_{HK} = -4.9$.

The results from the two teams are in apparent disagreement but there are some caveats. First, the noise in the CoRoT data may not be all from stellar origin. Instrumental effects are known to be present in the data and careful screening has to be carried out on a target-by-target basis. Noise estimates from an automatic analysis of the rms CoRoT light curves may lead to an overestimation. Secondly, photometric variations and the chromospheric indices may not be fully correlated. Lockwood et al. (2007) show that the Sun has the smallest photometric variations in spite of having a chromospheric activity index that is not the lowest of the sample. Data from the Kepler mission should provide further indications on the level of photometric variability of solar-type stars.

5. Earth-like Planets in the Habitable Zone of Nearby Stars

An ultimate future goal within exoplanet research is to carry out the detection *and further characterization* of an Earth-mass planet within the stellar habitable zone. For such characterization to be successful, the system should be located at a distance (d) close enough from the Sun so that sufficient photons reach the Earth to perform spectroscopic observations over a reasonable time. Calculations considering a spectroscopic mission (M. Shao, priv. comm.) show that the integration time scales as d^4 for background limited detection/spectroscopy. Table 1 presents the integration time needed to detect (broad band) an Earth clone with SNR = 10 and an instrument with 5% efficiency and to obtain a spectrum with 50 spectral channels. The calculations illustrate that realistic characterization efforts will have to be restricted to targets within the nearest 10 (or perhaps 15) pc. In the case of the 10-pc sample, this implies some 20 G-type stars, some 50 K-type stars and some 240 M-type stars[1]. For a 15 pc volume these numbers could be multiplied by ∼3.

From the numbers above it becomes apparent that M-type stars are prime targets for the detection of nearby planets. Most of the discussion in this paper has been focussed on Sun-Earth analogs and should be generalized to Earth-type planets around M-type stars. The location of the habitable zone scales as the square root of the luminosity and, roughly, as the square of stellar mass (assuming a mass-luminosity relationship). Thus, the astrometric wobble of a

---

[1] http://www.chara.gsu.edu/RECONS/census.posted.htm

Table 1. Integration time needed to detect (broad band) an Earth clone with SNR = 10 and an instrument with 5% efficiency and to obtain a spectrum with 50 spectral channels.

	2.5-m telescope			4-m telescope		
Distance	10 pc	20 pc	25 pc	10 pc	20 pc	25 pc
Detection	10 hours	3 days	16 days	1.5 hours	1 day	2.5 days
Spectrum	21 days	11 months	2.2 years	3 days	51 days	4 months

planet in the habitable zone scales linearly with stellar mass and inversely with the distance. The distance gain for M-stars is therefore partially cancelled out by the lower mass. Radial velocities (in the IR) are in a better situation since the radial velocity wobble of habitable planets is roughly proportional to the inverse of the stellar mass to the power of 1.5, and independent of the distance. In both cases, however, stellar noise may prove to be the limiting factor and this is an aspect that deserves careful study beyond the Sun-Earth paradigm. Besides the detection issue, a more fundamental question is still awaiting an answer, namely the possible habitability of such planets, having suffered strong high-energy irradiation and rotating in orbital synchronism.

6. Conclusions

The conclusions from the P2 panel discussion can be summarized as follows:

- Radial velocity has the capacity with an 8-m telescope to detect several Earth-like planets in the habitable zone if the instrumental noise is decreased to a level of 0.1–0.2 m s^{-1} within $d \leq 50$ pc. The main limitations will be the effect of stellar noise (related with rotational velocity) and the consequently diminished stellar sample available.

- However, only the exoplanets found within 10–15 pc should be considered for further characterization within a spectroscopic study. A rough scaling with the sampled volume implies a reduction by a factor ∼40 in the number of planets detected. Thus, it might well be that at the end, only 1 or 2 Earth-like planets are found by radial velocities that can be further studied spectroscopically. The number could be significantly increased by NIR radial velocities because of the larger nearby sample available.

- Space-based astrometry is a significantly more expensive approach but has the capability of detecting most (if not all) Earth-like planets in the habitable zone around the 60 stars closest to the Sun.

- There is consensus that the radial velocity approach should be followed even if there is a limited chance of finding appropriate habitable Earths at an accessible distance, as these would be extremely valuable systems. However, for the identification of Earth-like systems adequate for the search of bio-signatures via spectroscopic follow-up, astrometry is probably required to attain a full census of suitable targets, although at a significantly higher cost.

References

Bean, J. L., et al. 2009a, arXiv:0911.3148
Bean, J. L., et al. 2010, ApJ, 711, L19
Catanzarite, J., Shao, M., Tanner, A., Unwin, S., & Yu, J. 2006, PASP, 118, 1319
Figueira, P., et al. 2009, arXiv:0912.2643
Goullioud, R., et al. 2008, Optical & Infrared Interferometry, SPIE, 7013, 70134
Lockwood, G. W., et al. 2007, ApJS, 171, 260
Lovis, C., et al. 2006, Nat, 441, 305
Makarov, V. V., et al. 2009, ApJ, 707, L73
Mayor, M., & Udry, S. 2008, Physica Scripta Volume T, 130, 014010
Mayor, M., et al. 2009, A&A, 507, 487
Pravdo, S. H. et al. 2004, ApJ, 617, 1323
Pravdo, S. H., & Shaklan, S. B. 2009, ApJ, 700, 623
Reiners, A., et al. 2009, arXiv:0909.0002
Zapatero Osorio, et al. A&A, 505, L5

Peter Schuller, Richard Barry, and Bill Danchi exchange viewpoints at a coffee break.

Session VII

A Broader Picture

Exoplanets and Earth Sciences

E. Pallé[1,2]

[1] *Instituto de Astrofísica de Canarias. C/ Vía Láctea, La Laguna, Tenerife, Spain*

[2] *Departamento de Astrofísica, Universidad de La Laguna. Av. Astrofísico Francisco Sánchez, La Laguna, Spain*

Abstract. It is foreseen that in the near future we will be able to measure the light from extrasolar planets similar to the Earth. When these data become available, a truly interdisciplinary approach to their analysis will be necessary in order to understand the physical properties of these worlds based on globally-averaged measurements. In this task, observation of the Earth (as the only inhabited planet that we know of) and the rest of the planets of the solar system will be our guide to interpret the observations. Here I make a brief review of the existing observations of the Earth seen as a planet.

1. Introduction

Over the past few years, we have developed the capacity to discover planets orbiting around stars other than the Sun, and the number of detections is increasing exponentially. Even though we are not yet capable of detecting and exploring planets like Earth, ambitious missions are already being planned for the coming decades. In the near future, it is likely that Earth-size planets will be discovered, and efforts will then be directed toward obtaining images and spectra from them.

The first thing an extraterrestrial observer would find out about Earth would be that it is the largest of the rocky planets in the solar system, in both diameter and mass. More detailed observations would identify its minimum (or absolute, depending on the observing techniques) mass, and determine its orbital parameters. The determination of the Sun-Earth distance, the spectral type of the Sun, and the mass of the planet would immediately point out to the possibility of finding life, as Earth lies inside the habitable zone of the Sun. Thus, once curiosity is triggered, more detailed observations of the planet would be in order.

However, what these observations would reveal about our planet will be highly dependent on the techniques used, the temporal sampling, the spectral range and the timing. In reflected light, the observing geometry will determine whether one samples either a full planet hemisphere or just a thin crescent.

When observing an extrasolar planet from an astronomical distance, all the reflected starlight and radiated emission from its surface and atmosphere will be integrated into a single point. Thus, in order to interpret the resulting spatially-integrated observations, we will need to compare them with observations with similar resolution of the only planet that we know to be inhabited, namely

Table 1. A compilation of globally-integrated observations of the Earth from remote satellite platforms (adapted from Vazquez et al. 2010). More details can be found in: [a]Sagan (1994), [b]Sagan et al. (1993), [c]Christensen and Pearl (1997), [d]http://www.esa.int, [e]http://www.esa.int, [f]http://messenger.jhuapl.edu, [g]http://www.nasa.gov, [h]http://www.esa.int, [i]http://epoxi.umd.edu/4gallery/Earth-Moon.shtml, [j]Piccioni et al. (2008).

Mission	Date	Observations
[a] Voyager 1	1989	Pale blue dot image (6.4 billion km)
[b] Galileo	1990	Very low-resolution spectroscopy
[c] Mars Global Surveyor	1996	Infrared spectroscopy
[d] Mars Express	2003	Low-res. VIS and near-IR spectra
[e] Rosetta	2005	High-res VIS and IR imaging
[f] MESSENGER	2005	Earth images during flyby
[g] CASSINI	2006	Image of Earth from Saturn's orbit
[h] Rosetta	2007	Composite imaging, true colors
[i] EPOXI	2008	Composite imaging, true colors
[j] Venus Express	2007-08	Repeated VIS and IR spectroscopy

Earth. However, despite the fact that Earth is our own planet, and that we have a very detailed knowledge of its physical and chemical properties, Earth observations from a remote perspective are limited. They can only be obtained by integrating high-resolution Earth orbiting satellite observations, from a very remote observing platform, or from the ground by observing the earthshine reflected on the dark side of the Moon.

1.1. Observations from Long-range Spacecraft

Globally-integrated observations of the Earth as a point, or as a planet, are sometimes taken by solar system probes, usually as an instrumentation test or as an effective public-relations event. Planetary flybys, while often required to eventually settle in a flying path, are also critical opportunities for the mission teams to test the spacecrafts and their scientific instruments. Earth flybys are often required to achieve a final trajectory to a distant planet and can provide a critical opportunity for a mission team to test the operation of a spacecraft and its scientific instruments. Over the years, a series of Earth images and spectra, mostly as one-time observations, have been compiled. Some of the most significant are listed in Table 1.

These observations offer a unique perspective of our planet, and will be discussed in more detail in the following sections. The first of such observations was taken by the Galileo spacecraft (Sagan 1994), and had a deep impact among both scientists and the the general public. Recently the EPOXI mission has released a set of images revealing changes in the visible and near-infrared reflectance as the Earth rotates as well as the Moon's transit in front of the Earth.

Although almost all of the observations are taken only as one-time efforts or over short periods of time, it is worth noting the effort recently undertaken by the Venus Express Team. Venus Express started a program in May 2007, to continuously monitor the Earth spectra, with approximately monthly data (Piccioni et al. 2008). These data will be extremely useful in the near future to characterize the more important features of the Earth's spectra, and especially their variability with time and orbital configurations.

For extrasolar planets, IR photometry has not been considered so far as a useful tool characterize their atmospheric/suface signature, because the variability is smaller than in the visible range. A planet with an atmosphere and oceans, such as the Earth, exhibits a small range of emission temperatures between day and night, due the the very larger thermal inertia of the ocean and air (Gaidos et al. 2006). Several pictures of the Earth at infrared wavelengths, have been taken by the Mars Odyssey, the MESSENGER and the Rosetta missions. Gomez et al. (2009) have simulated the infrared photometric variability of the Earth observed from an astronomical distance, based on measured hourly data of the broadband outgoing longwave radiation ($\lambda > 4\mu m$) at the top of the atmosphere, and a geometrical averaging model. The radiation data were obtained from the Global Energy and Water-cycle Experiment (GEWEX).

On average, the daily maximum integrated IR flux, in the direction of a distant observer, is found when the Sahara region is within the observer's field of view. The minimum emitted flux is found when the Pacific and Asia/Oceania regions are in the field of view. Thus, IR emission from hot desert surface is the dominant factor, although the infrared flux is heavily altered by the presence of clouds. Gomez et al. (2009) found that the determination of the rotational period of the Earth from IR photometric observations is feasible in general, but it becomes difficult at short times scales if cloud and weather patterns dominate the emission.

In the past, there have been numerous attempts to observe the spectrum of the pale blue dot from the distance in different spectral regions. Among other efforts, observations of the Earth as a planet from afar have been made in the visible range from the Galileo spacecraft (Sagan et al. 1993), and from the Mars Global Surveyor spacecraft, at a distance of 4.7 million km, in the thermal infrared (Christensen and Pearl 1997).

Earth's spectral observations in the thermal infrared are also available. The thermal emission spectrometer (TES) on the Mars Global Surveyor spacecraft acquired observations of the Earth from a distance of 4.7 million km for instrument performance characterization on November 24, 1996 (Christensen and Pearl 1997). An infrared spectrum of the Earth was also taken by VIRTIS on 5 March 2005 after Rosetta's closest approach to Earth. The Earth's IR brightness peaks around 10 μm and then decays slowly. Spectral features of the gases carbon dioxide (CO_2), water vapor (H_2O), and ozone (O_3) dominate the Earth's spectra in this range, as well as methane (CH_4) and several other minor constituents.

1.2. An Indirect View of Earth: the Earthshine

The earthshine, the ghostly glow visible on the dark side of the Moon, is sunlight that has reflected from the dayside Earth onto the dark side of the Moon and back again to Earth. Already in the XVIth century, Leonardo da Vinci correctly deduced that earthshine was due to sunlight reflected from the Earth toward the dark side of the Moon. From ground-based telescopes, earthshine observations can be used to determine the Earth's reflectance, by measuring the brightness ratio between the bright and dark sides of the Moon (Pallé et al. 2004; Qiu et al. 2003).

In the past few years, with the blooming of exoplanet detections and the plans for future missions, there has been a renewed interest in the earthshine measurements. This has lead several research groups to undertake observation campaigns of the earthshine in the visible and infrared ranges. Observations of the earthshine allow us to explore and characterize its photometric, spectral and polarimetric features, and to extract precise information on what are the distinctive characteristics of Earth, and life in particular.

From these measurements the photometric variability of the Earth is found to be of the order of 10-15% from night to night and during one Earth rotation (Qiu et al. 2003), while Goode et al. (2001) have shown that over a year and a half, the Earth's seasonal variation is larger than 10%.

Ford et al. (2001) were the first to point out that the light scattered by a terrestrial planet will vary in intensity and color as the planet rotates, and the resulting light curve will contain information about the planet's surface and atmospheric properties. However, when clouds are added the reflected light curve it is not so straight forward to interpret. Nevertheless, Pallé et al. (2008) found that scattered light observations of the Earth can accurately identify the rotation period of the Earth's surface. This is because large-scale time-averaged cloud patterns are tied to the surface features of Earth, such as continents and ocean currents. Moreover, for an extrasolar Earth-like planet, photometric observations could be used to infer the presence of a 'variable' surface (i.e. clouds), even in the absence of spectroscopic data. This would strongly suggest the presence of liquid water on the planet's surface and/or in the planet's atmosphere, especially if the effective temperature of the planet was also determined by other means.

With the aim of improving the spectral resolution and the sampling of seasonality and phase changes, spectroscopic measurements of the earthshine have also been taken from ground-based observatories and analyzed at visible wavelengths (Woolf et al. 2002; Montañés-Rodríguez et al. 2005, 2006), and in the near infrared wavelengths (Turnbull et al. 2006).

The rotation of a planet with surface features such as continents and oceans should modulate the polarized reflectances in a simple and predictable manner. Lyot & Dollfus (1949) studied the polarization of the earthlight by means of a coronograph and a polarizer at the Pic du Midi Observatory. They concluded that there is a strong polarization which reaches a maximum when the Moon's phase angle is near $80°$ and then it exceeds 10% over the seas. Dollfus (1957) continued with observations of the earthshine polarization, and obtained direct measurements of polarization from the ground from balloon observations. From his measurements, Dollfus concluded that the atmospheric polarization is by far larger than that introduced from the ground. And more recently, based on polarization observations of Earth made from the Polder satellite, Wolstencroft & Breon (2005) determined that the degree of linear polarization, for Earth at 443 nm and $90°$ scattering angle, is 23% for averaged cloud cover conditions (55%).

McCullough (2006) predicted the shape of the phase function of the earthshine's linear polarization observed by Dollfus (1957); the maximum polarization agrees with Dollfus' observations but is approximately twice as large as that predicted by Wolstencroft & Breon (2005). Thus, linear polarization could be a potentially useful signature of oceans and atmospheres of Earth-like extrasolar planets. As with other techniques, global polarimetric observations of our planet will be the guideline.

On the Earth, water surfaces provide the only significantly polarized natural sources of thermal infrared radiation, while emission from the atmosphere and ground is almost always unpolarized to any practical degree (Shaw 2002). For practical purposes, the degree of polarization of the infrared flux emitted from Earth, observed as a planet, is negligible.

Finally it is worth noting that lunar observations also allow us to obtain the transmission spectrum of the Earth. When a planet passes in front of its parent star, part of the starlight passes through the planet's atmosphere and contains information about the atmospheric species it encounters. Transmission spectra (in transmission or emitted light) are very interesting because they provide the only current methodology to characterize exoplanetary atmospheres. The characterization of spectral features in our planet's transmission spectra can be achieved through observations of the light reflected from the Moon during a lunar eclipse, which resembles the observing geometry during a planetary transit (Pallé et al. 2009).

References

Christensen, P. R. and Pearl, J. C., 1997, JGR, 102, 10875
Dollfus, A.,1957, Supplements aux Annales d'Astrophysique, 4, 3
Ford, E. B., Seager, S. and Turner, E. L., 2001, Nat, 412, 885
Gaidos, E., et al., 2006, in IAU Colloq. 200, Direct Imaging of Exoplanets Science Techniques, ed. C. Aime & F. Vakili, (Cambridge: Cambridge University Press), 153
Gómez et al., 2009, in preparation
Goode, P. R., et al, 2001, GRL, 28, 1671
Lyot, B. & Dollfus, A., 1949, Memoires et communications de l'academie des sciences, 222, 1773
McCullough, P. R., 2006, arXiv:astro-ph/0610518
Montañés-Rodríguez, P. et al., 2005, ApJ, 629, 1175
Montañés-Rodríguez, P., et al., 2006, ApJ, 651, 544
Pallé, E., Goode, P. R., Montañés-Rodríguez, P. & Koonin, S. E. 2004, Sci, 304, 1299
Pallé E., et al. 2008, ApJ, 676, 1319
Pallé, E, et al. 2009, Nat, 442, 51
Piccioni, G., Drossart, P., Cardesin, A. & VIRTIS-VenusX team, Venus Express Workshop, La Thuile, Italy
Qiu, J., Goode, P.R., Pallé, E., Yurchyshyn, V., Hickey, J., Montañés Rodríguez, P., Chu, M.-C., Kolbe, E., Brown, C.T., & Koonin S.E. 2003, JGR 108, D22, 4709
Sagan, C., Thompson, W. R., Carlson, R., Gurnett, D. & Hord, C. 1993, Nat, 365, 715
Sagan, C., 1994, Pale Blue Dot. A Vision of the Human Future in Space, (New York: Random House)
Shaw, J. A., 2002, Polarization Measurement, Analysis, and Applications V., SPIE, 4819, 129
Turnbull, M. C., et al., 2006, ApJ, 644, 551

Vazquez, M., Pallé, E., & Montañés-Rodríguez, 2010, The Earth as a Distant Planet: A Rosetta Stone for the Search of Earth-Like Worlds, Springer-Verlag Astronomy and Astrophysics Library, ISBN-10: 1441916830, in press

Wolstencroft, R. D. & Breon, F.-M., 2005, in ASP Conf. Ser. 343, Astronomical Polarimetry: Current Status and Future Directions, eds. A. Adamson, C Aspin, C. J. Davis & R. Fujiyoshi, (San Francisco: ASP), 211

Woolf, N. J., Smith, P. S., Traub, W. A., Jucks, K. W., 2002, ApJ, 574, 430

Pathways Towards Habitable Planets
ASP Conference Series, Vol. 430, 2010
Vincent Coudé du Foresto, Dawn M. Gelino, and Ignasi Ribas, eds.

Life in the Solar System

A. Giménez

Centro de Astrobiología, CSIC-INTA, Instituto Nacional de Técnica Aeroespacial, Ctra. de Torrejón a Ajalvir, Torrejón de Ardoz, Madrid, Spain

Abstract. We have studied for a long time the physical and chemical nature of our solar system, its formation, the evolution of its different bodies and their chemical composition. If the physics and the chemistry we know to be valid in our planet works everywhere in the solar system, the question is, why not biology? In this paper, the aims and methodology of astrobiology to understand the possibilities for life in our solar system are summarized. The objective is to share with astronomers this experience in the exploration of our neighborhood so that it could be extended, obviously using different techniques, to the search for life in extrasolar planetary systems.

1. Habitability

Although we know the structure and composition of the solar system reasonably well, no life has yet been identified outside our own planet. But in order to search for it, we need to know what we are looking for and it is not easy to define life. This difficulty is generally sorted out by describing the properties of what we call life. In other words, we admit we do not know what is life but we can recognize it when we see it. For this purpose, a widely accepted description is that of a self-supported replicating cellular structure subject to Darwinian evolution.

We also think we know what life needs to emerge and survive: organic compounds, liquid water, and a source of energy. The search for life is thus equivalent to exploring the places of the solar system where all these elements could be present. If they exist and life is not found at least in an extant form, we obviously have a problem with such concepts as universality and/or inevitability of life. NASA's "follow the water" has to be therefore extended to the search of the chemical elements and sources of energy. This is done through the concept of habitable environment, i.e., by using "follow habitability".

1.1. Organic Compounds

The most important organic compounds for life are C, H, O, and N, which are also the most abundant, but P and S (thus, CHONPS) are also essential for the building blocks of life. These elements are part of life as we know it, Terran life, and though exotic or weird forms of life can not be discarded, the natural way to search for life is to focus on what could be the result of using the most abundant elements in the solar system.

Organic compounds have now been found in places like the interstellar medium, protoplanetary disks, comets, and asteroids. In fact, the cometary and meteoritic infall of organic material to Earth has been estimated to be quite large, particularly during the late heavy bombardment, around 3.9 Gyr ago, induced by the migration movements of giant planets. This is the time in the evolution of our planet when life emerged. Comets are known to carry possible components for life and a recent analysis of samples from Comet Wild-2 demonstrated the presence of glycine, a simple amino acid (the makers of proteins). On the other hand, carbonaceous meteorites have also shown the existence of extraterresrial amino acids as indicated by the Murchison meteorite. In a recent study, the same meteorite has shown the presence of nucleobases, the key molecular ingredients of DNA and RNA. Carbon isotope ratios for uracil and xanthine showing excess ^{13}C relative to ^{12}C indicated a non-terrestrial origin for these compounds. The demonstration that uracil and xanthine are indigenous to Murchison adds nucleobases to the other key elements of biochemistry (amino acids, sugars, carboxylic acids, etc.) known to be present in meteorites. These results support the idea that external material may have played a key role in the origin of life.

Nevertheless, this does not imply that the actual building blocks of life on Earth came from outside, or at least not all of them. Laboratory experiments have demonstrated that aminoacids can also be synthesized in special environment conditions that prevailed in early Earth, as shown by Miller and Urey in the fifties.

It seems, anyhow, that the material of which life is made of is more common than expected in our solar neighborhood. Life should thus be abundant in the solar system! This is of course not true since it assumes that a necessary condition is also a sufficient one and ignores two additional necessities for life.

1.2. Liquid Water

Liquid water is the best and most abundant solvent known to allow for the above mentioned compounds to build complex structures, as needed by living organisms. Water allows the formation of membranes and its polarity creates forces and dissociation ions that facilitate chemical reactions. As a result, many chemicals dissolve in water allowing them to mix and react.

In addition, liquid water is present in a wider range of temperatures than many other possible solvents. Versus 100°C for water, ammonia is liquid in a range of 45°, and methane in only 18°C. Ethane is close to water, in the sense of a range of 94°C, but at a temperature below 100°C. Indeed water as a liquid element peaks around the temperature at the Earth's surface while sulfuric acid does for Mercury or Venus, and methane plays the same role in the environements of Jupiter and Saturn. Beyond that, temperature drops in such a way that liquid nitrogen is possible and even hydrogen. We could then expect lakes of ethane in places like Titan, but carbohydrates unfortunately do not provide a good environment for complex carbon-based structures to be formed.

Standard theories about the formation of the solar system indicate that, while water ices should have been abundantly accreted by protoplanets in the regions beyond the ice-line, the inner solar system did not have this option. We will come back to this issue.

1.3. A Source of Energy

Chemical reactions need a source of energy to drive them. Photosynthesis absorbs light from the Sun and it is the current major primary source of energy for living organisms. Nevertheless life is not limited to the availability of sun light. There are many other chemical and physical processes that can provide the necessary energy as shown by known Earth species that have no access to sun light. Organisms have been discovered deep beneath the oceans, in black smokers, where they absorb energy from chemicals in the water, and life has also been found below the ices of the Arctic seas.

The study of extreme Earth environments, where life could evolve, has been an area of research in recent years providing new views and inspiration about the limits of life. Organisms have been found in high-temperature environments like in thermal zones at Yellowstone, US, or Gunhuver, Iceland; at high pressures like deep in the oceans; or with very low pH like in Rio Tinto, Spain. Rio Tinto is in fact an acid river with pH 2.3 and a high concentration of heavy metals that provide energy and stability to the observed system.

An interesting example of life without the sun is an ecosystem that has been discovered 2.8 km underground in a gold mine near Johannesburg, South Africa. This system has been isolated from the surface for ∼10 million years and uses energy of naturally occurring radioactivity which splits water into hydrogen and hydrogen peroxide. Hydrogen peroxide reacts with naturally occurring sulfide in the rocks to make sulfate. Microbes reduce sulfate back to sulfide using electrons from hydrogen left over from the splitting of water. This ecosystem raises the possibility that organisms could survive even on planets whose surfaces have long since become lifeless.

2. The Sun's Influence

Our central star certainly affects climate and the environmental conditions for life on Earth. Nuclear evolution of the Sun changes its temperature, radius, and luminosity rather dramatically. Current stellar evolutionary theories predict that the Sun had about 70% of its current luminosity by the time life emerged on our planet. This implies a problem to explain the presence of liquid water on Earth at the time, the so-called Faint Sun Paradox. The temperature of a planet irradiated by the Sun depends on the incident radiation and the reflectivity of the planet. In the case of Earth, this results in an average temperature of $-20°C$. The difference with the more comfortable value of $+15°C$ is due to the greenhouse effect. But with the current atmospheric chemical composition, the temperature of the Earth with a fainter Sun should have been below $0°C$ until around 2000 million years ago, when life was already quite extended. Solutions to the Faint Sun Paradox normally include an increase of the greenhouse-effect gases like carbon dioxide, ammonia, or methane.

Moreover, the long-term nuclear evolution of the Sun indicate that ∼1 Gyr from now the oceans will start to evaporate and ∼4 Gyr from now it will be 40% more luminous and a runaway greenhouse effect will take place due to evaporated water. Fortunately, there is a still a lot of time until this happens!

Another possible effect of the Sun on environmental conditions is the influence of high-energy emissions. They may contribute to the energy balance as well as alter the chemistry of the upper atmosphere. But the old-Sun was much more active than currently, emitting higher fluxes of X-rays (around 1000×) or UV (around 100×) at the time of the emergence of life on Earth.

All these comments are relevant to the issue of habitability within our solar system. Of the three conditions mentioned: components, liquid water, and energy, the second one appears to be the biggest constraint. Liquid water requires the existence of water within a given range of temperature and this may be easily linked to the Sun's distance, i.e., the total irradiance. But reflectivity and the greenhouse effect have to be also taken into account as well as the atmospheric protection to high energy particles, including the presence of a magnetic field. Furthermore, other sources of thermal energy, like internal activity or tidal effects, may provide the necessary temperature for liquid water to be present far from the Sun, below the surface of a given body. Habitability is thus not just a function of the solar irradiance and distance. Other parameters, like the mass of the planet, its atmosphere, tectonic activity, magnetic field, etc. have to be considered. In conclusion, a planet within the habitable zone, as defined by the Star-planet distance, may not be habitable and a body outside it may be habitable.

3. The Outer Solar System

As mentioned before, standard theories of solar system formation predict that water should be everywhere in its outer regions, beyond the ice-line. Ices should have been present during the accretion process of the initial solar nebula and integrated in the structure of giant planets and their satellites. Therefore water is expected to be prominent in the satellites of Jupiter and Saturn as well as in the icy planets beyond, Uranus and Neptune. But is it possible to find water in liquid form? Not on the surface at the expected temperatures of these icy worlds, but it could be very common in the subsurfaces. Europa fulfills the criteria for life to develop since the interaction between Jupiter and its satellite produces a source of internal energy opening the possibility of liquid water below the few kilometers thick crust of ice. It has also been found the presence of salts, magnesium and sodium sulfates, and possibly organic molecules so that and ocean of liquid water below the ice could actually harbor life. Energy is provided by either thermal vents due to volcanism or other mechanisms. The dynamical activity of Europa leads us conclude that an ocean/rock interface is likely and with it, the continuous resupply of potential nutrients. This structure of core, rocky mantle, and rock/ocean interface provides the necessary environment for life. If the compounds of life are really present, the interaction between the mantle and the liquid salty water provide the required nutrients. Moreover no craters or very few have been found on the surface, indicating the recent evolution of Europa and its moving surface.

Another interesting satellite of Jupiter is Ganymede which possesses a magnetosphere produced through convection within its liquid iron core. A joint study of ESA and NASA of a mission to the Jupiter system, with the preliminary name of Laplace or EJSM, focuses its attention on these two satellites. ESA should provide a Ganymede orbiter and NASA the Europa equivalent.

Around Saturn, the main objective has been Titan. This large satellite has a dense atmosphere, rich in organic components and shows some similarities with the prebiotic Earth. Liquid water could be present below the surface, sandwiched between two ice layers. Water ice rocks have been observed by the European probe Huygens at the surface, rounded by flows of liquid methane. Indeed, on Titan methane plays the role of water on Earth, since at the surface temperature, around -180 C, methane is at its triple point and can be found in solid, liquid, and gaseous states. But life is not thought to be possible on Titan. The Cassini mission radar has obtained an altimetry profile of Lake Ontario on Titan. It sits in a 300-m depression and its surface is very flat. Hydrocarbon lakes could exist but complex organic molecules cannot be formed in such an environment. There could be a different chemistry leading to exotic forms of life, based on a liquid solvent other than water. This is a new challenge for astronomers observing exoplanets in connection with the potential impact of life in their atmospheres. The study of observable biomarkers in the case of unusual forms of life will lead to unexpected biomarkers and only the out-of-equilibrium state of the atmosphere may be used.

Enceladus is another satellite around Saturn of interest for astrobiology. The discovery of water ejections from cracks in the ice surface indicated the existence of chimneys through which liquid water could access the surface and revealed the presence of water reservoirs below the crust of ice.

4. The Inner Solar System

Earth is the only planet in the solar system on which liquid water can exist at the surface. The heat capacity of the oceans is very important to keep Earth's temperature relatively stable and liquid water is also responsible for most of the erosion and weathering of the Earth's continents. In Mars, water could have played a similar role in the distant past before it was lost.

The problem with the presence of water, liquid or not, in the inner solar system is linked to the estimated position of the ice-line. No ices could be accreted during the formation of these planets and water had to be somehow carried into the inner regions. Water is abundant on Earth and is present in the atmospheres of Venus and Mars and expected to have been at their surfaces in the past. Water is suspected to be present in their subsurfaces and has been detected, at least in ice form, in Mars and in the Moon. The most accepted way to explain water in the inner solar system is the previously mentioned period, around 3.9 billion years ago, of Late Heavy Bombardment, caused by the rapid migration of the giant planets. The Nice model in fact predicts a scatter of planetesimals throughout the solar system and the dislocation of most asteroid belt bodies producing a violent temporary instability. Of course, many asteroids are hydrated and likely delivered water and other volatile species to the planets of the inner solar system. Carbonaceous asteroids are also the most important source of exogenous organic matter and may have contributed to the origin of life. The amount of cometary material expected to have fallen on Earth explains a significant amount of the current ocean mass and it is consistent with the D/H ratio measurements. The Nice model also explains current planetary orbits and many aspects of the composition distribution and dynamical behavior of solar system bodies.

In Venus, thick clouds of sulfuric acid trap heat and provoke a runaway greenhouse effect responsible for a surface temperature around 500°C at which there cannot be liquid water or life. Venus and Earth do otherwise have similar mass, radius, and even composition. But the evolution of the Venusian atmosphere was very different. Venus possibly had large quantities of water (equivalent to several Earth oceans) in the past, but today is completely lost except for some small remains in the atmosphere. Some scientists even think that life could have actually survived in the clouds.

The case of Mars is much more promising. The possible existence of habitable environments in the past and the probability of some still present today has driven active searches for fossil remains, water-formed and life-modified minerals, and even possible traces of extant life. Mars has a surface with many characteristics typical of erosional processes. Craters as well as canyons and torrents suggests that Mars was a wet planet at some point.

NASA's rovers Spirit and Opportunity have found minerals indicating the existence of liquid water in the past. Orbital missions have also found indications of past surface standing water, and that much of Mars' ancient crust was pervasively altered by water, and shallow ancient groundwater. Evidence for recent water in Mars is provided by many narrow channels running from the rim down the floor of some craters. This is the case of the Newton crater outburst of subsurface water which may occasionally and briefly de-ice and sublimate.

If Mars had liquid water, there was a high potential for a habitable environment. Global climate change affected the Martian hydrological cycle and surface habitability over the planet's history. There are indications of large floods and probably a global ocean. Mars has a major sedimentary record and a variety of local environments for the possible evolution of life. Multiple sedimentary facies have been discovered, sulfates and phyllosilicates, as well as subsurface water. Minerals altered by water or water-deposited minerals have been discovered. Sedimentary rocks could have been formed in acidic conditions of ancient lakes or oceans. The analysis of the Endurance crater showed hematites through Mossbauer spectroscopy and the so-called blueberries. Jarosites have been observed at El Capitan as well as Fe+ sulfates exhumated at crater Gusev or silica of possible hydrothermal origin. Mars also had potentially favorable conditions for the preservation of past life in fossil form, as given by sulfates and clays. Moreover, possible episodic modern liquid water has been identified as well as modern near-surface ice. This is much more than could be expected, but no modern thermal sites have yet been found.

Another interesting aspect of Mars is the detection by ground-based observations, concurrent with space missions, of methane in its atmosphere. When present, methane occurs in extended plumes and the maxima of latitudinal profiles imply that the methane was released from discrete regions. The presence of methane in the atmosphere, together with formaldehyde, an oxidation product of methane that has also been found, would indicate that Mars either bears volcanic activity or, less probably, biological processes today. Clear correlation between water vapor in the boundary layer and methane concentrations observed from orbit, further illustrates the volcanism vs. life debate. The correlation between water vapor, methane, and possible underground acquifers, points to a common underground source for water and methane.

Paleo-oceans have been indicated by gammaray techniques with water near the surface, while radar measurements show water ice in the South pole. Modern habitability potential was recently found by the Phoenix mission at Borealis Planum when near-surface water ice was found together with perchlorate, a possible energy source. Perchlorates are powerful but stable oxidants and are very hydroscopic.

The study of climate change in Mars will also be essential to understand the evolution of the planet. Early warm, wet Mars (Noachian) has evolved to cold, dry Mars (from Hesperian) but also quasi-periodic climate changes could have been present. In terms of evolution, igneous rocks and associated tectonics will have to be studied, as well as multi-year records of CO_2, water and dust, atmospheric dynamics, and SO_2.

Missions in the future will be looking for fossil bacteria, or at least mineralogical biomarkers, in the subsurface. Current conditions at the surface of Mars, like oxidation, intense UV and ionizing radiation, do not seem to be ideal for the development of life. There are many more possibilities for it to survive in extinct or extant form in the subsurface. It is therefore very important to explore these regions of Mars by drilling at least few meters, beyond the penetration of organic destructive agents.

Pathways Towards Habitable Planets
ASP Conference Series, Vol. 430, 2010
Vincent Coudé du Foresto, Dawn M. Gelino, and Ignasi Ribas, eds.

The Extraterrestrial Life Debate in Different Cultures

J. Schneider

LUTh - Paris Observatory, Paris, France

Abstract. Surprisingly, the question "Is there Life in the Universe outside Earth?" has been raised, in rational terms, almost only in the western literature throughout the ages. In a first part I justify this statement. Then I try to develop an explanation of this fact by analyzing the different aspects of the notion of "decentration".

1. Introduction

The question "Are we alone in the Universe?" is one of the main motivations of this Conference "Pathways Towards Habitable Planets." It is often claimed to be "as old as Humanity itself." It indeed looks very natural since Life is spread out over the whole Earth and therefore even a child rising his eyes toward the sky can ask "Is there also Life out there?" But, very surprisingly, there is almost no written occurence of this question in "non western" ancient cultures. In a first part of this paper I justify this statement. Then I will try to understand why it is so. I will thus be led to first clarify what can characterize and delimit "Western" culture. Then I will propose a hypothesis to explain why the question of Life in the Universe has almost never been raised by non-western cultures. Finally I will address the question "why did this movement start in Greece?".

There are generally two ways to consider the question of extraterrestrial life. First, the point of view of living organisms, leading to the question "Is there Life elsewhere in the Universe?", which is the subject of exobiology and extraterrestrial intelligence, leading to the question "Are we alone?" or "Is there anybody out there?", which is subject of SETI (Search for Extra-Terrestrial Intelligence). And a different, but connected, question is the nature of Life: how different can it be from terrestrial life? This question is symbolized by the word "Aliens" often found in the literature. Here I will treat these three questions as if they were only one.

2. Survey of the World-wide Literature and Traditions

The question of extraterrestrial life in the literature since the Greeks has been compiled in the remarkable books "The Extraterrestrial Life Debate 1750-1900 - The Idea of a Plurality of Worlds from Kant to Lowell" (Crowe 1986) and "The Extraterrestrial life debate, antiquity to 1915" (Crowe 2008). They are a must on this topic and represent an almost exhaustive compilation of all authors having expressed an opinion on this debate.

It is remarkable that almost all authors entering the debate have expressed that the existence of extraterrestrial life seemed natural to them. Among the most famous authors, the only few remarkable exceptions are Aristotles, Augustine, Hegel, Schopenauer and to some extent Plato. That means a very few skeptics among hundreds of optimists. Another curiosity is that, while the debate has increased in intensity among the scientific community at the end the the XIXth century, almost no philosopher after Schopenauer was interested in this subject. Only H. Bergson in his *Evolution Créatrice* and more vaguely C.S. Peirce and W. James did mention the question of extraterrestrial life. This is strange because several philosophers at the beginning of the XXth century, like Husserl, Cassirer, Wittgenstein, were well aware of scientific developments of their times. To me it remains a mystery. It cannot be explained by ignorance: many novelists like Charles Cros, H.G. Wells, G. Flaubert[1] A. Strindberg[2], Marconi, Stapledon[3] and Tristan Bernard[4] did contribute to an outreach of the extraterrestrial life debate in the general culture. Only in the second half of the 20th century Paul Watzlawick, from the Palo Alto school in sociology, addressed seriously the question of communication with extraterrestrials (Watzlawick 1976).

The most important, although obvious, observation from Crowe's books is that all authors cited are Europeans and (after 1800) North-Americans. No reference to extraterrestrial life exists in "Astronomy Across Cultures - The History on Non-Western Cultures" (Selin 2000) nor in "L'Astronomie des Anciens" (Nazé 2009). There seem to be a few exceptions. I give here the full quotations because they are not very widely known. In the Jewish literature, the "Guide for the Perplexed" by Moise Maimonides (circa 1135 - circa 1204) says "The whole mankind at present in existence [...] and every other species of animals, form an infinitesimal portion of the permanent universe [...] it is of great advantage that man should know his station, and not erroneously imagine that the whole universe exists only for him." (Chapter XII p. 268). But Maimonid was a European Jew living in Cordoba, Spain. He knew well the ancient Greeks' work and participated in the cultural atmosphere also represented by Michael Scot (1175 - 1235) and Albertus Magnus (1193 - 1280) who were among the Middle Age philosophers supporting the idea of extraterrestrial life. In the Hindu tradition, Bhrigu says in Chapter 9 of the Mahabharata:

"The sky thou seest above is Infinite. It is the abode of persons crowned with ascetic success and of divine beings. It is delightful, and consists of various regions. Its limits cannot be ascertained. The Sun and the Moon cannot see, above or below, beyond the range of their own rays. There where the rays of the Sun and the Moon cannot reach are luminaries which are self-effulgent and which possess splendor like that of the Sun or the fire. Know this, O giver

---

[1] in *Bouvard and Pecuchet* (Penguin Classics)

[2] In his drama "Father", one of the key characters worked on panspermia.

[3] Olaf Stapledon (1886-1950), a British psychologist, envisaged communication with extraterrestrial in his "Last and First Men" (Stapledon 1930)

[4] French humorist, 1866-1947

of honours, that possessed of far-famed splendor, even these last do not behold the limits of the firmament in consequence of the inaccessibility and infinity of those limits. This Space which the very gods cannot measure is fall of many blazing and self-luminous worlds each above the other. Beyond the limits of land are oceans of water. Beyond water is darkness. Beyond darkness is water again, and beyond the last is fire. Downwards, beyond the nether regions, is water. Beyond water is the region belonging to the great snakes. Beyond that is sky once more, and beyond the sky is water again. Even thus there is water and sky alternately without end."

This text, with no clear date of writing (between 400 BC and +400 AD) is interesting, but it is not a rational deduction of extraterrestrial life in the framework of a treatise in Natural Sciences like in Epicure's Letter to Herodotus. In ancient China, there seem to be only two allusions to the plurality of worlds. In the chapter Qianxiang ("The Heavens") of the great encyclopedia *Gujin ushu Jicheng* refers to Yi Shizhen who wrote around 1300 AD:

"Humans and things are without limit, and the same holds for the Earth and the Heavens. As a comparison, when a parasite is in a man's stomach, it does not know that outside this man there are other men; Man being himself in the stomach of the Earth and the Heavens, he does not know that beyond the Earth and the Heavens there are other Earths and other Heavens".

In the book *Po Ya Ch'in*, Deng Mu writes around 1000 AD:

"Heaven and Earth are large, yet in the whole of empty space they are but as a small grain of rice .... It is as if the whole of empty space were a trunk and heaven and earth were one of its fruits. Empty space is like a kingdom and heaven and earth no more than a single individual person in that kingdom. Upon one tree there are many fruits, and in one kingdom many people. How unreasonable it would be to suppose that besides the heaven and earth which we can see there are no other heavens and no other earths!"

Let us note here that the latter text by Deng Mu only refers to other "Earths" and not to extraterrestrial life. In addition, these seem to be of Buddhist inspiration and they are very late compared the the IV - III Century BC Greek literature.

To be complete, one must say that there are references to non-human beings in some African and Hopi cultures, but they are rather of the "supernatural" type, like angels for instance.

In conclusion, the extraterrestrial life debate has by far be initiated mostly in the arc going from India to Greece and later on in Western Europe.

What is curious is that the question "Why is this debate essentially restricted to the western literature?" has never been discussed, at least to my knowledge.

3. Why Does the Question "Are We Alone?" Exist Only in "Western" Culture?

Let us now go beyond a simple historical compilation and try to understand this indubitable dis-equilibrium between western and non-western cultures and the roots of preconditions of the extraterrestrial life debate. Here I will illustrate my argument using historical examples. My purpose is nevertheless not a historical perspective. It is rather a-historical and structural. I will develop a hypothesis which rests on a main guiding principle: "elsewhere" and "aliens" require some distanciation or differentiation. This principle is schematic compared to the complexity of historical situations, much like how the Galilean inertia principle is apparently contradicted by everyday life dominated by dissipative frictions.

There are two types of distanciation: the distanciation of concepts from their empirical objects and spatial distanciation. These two aspects are closely connected and in particular spatial distanciation requires distianciation by concept as a prerequisite. Let us nevertheless shortly discuss them separately.

Distanciation by concepts
The essence of concepts is to introduce a distance between them and what they are about. The idea that life can exist elsewhere requires that the word "Life" is not identical with the living beings with which we have personal relationships. In other words, it requires a concept of "Life". Only concepts can be generalized. This points toward the "universalizing" structure of concepts. What is called "abstraction" is then the result of this universalization.

We can at this point try to characterize "Western culture" as the culture of concepts with their mathematization and the logical constraints that they impose.

Concepts are created by the words naming them. This view is illustrated for instance by the ideas of nominalism (Abelard) and the Berkeleysian so-called idealism[5]. Moreover, what is not subject of language cannot be imagined different: to imagine that things are different one must give them names AND detach the word from the designated object. Hence the above-mentioned conceptual distanciation. An example is given by the idea of "circle": it is an abstraction insofar as there is no perfect circle in nature[6] and a source of universalization since it allows to put all empirical curves resembling a circle into the same single category. Another, less abstract, example of universalization is given by the introduction of the metrical system which abandoned local customs for a "universal" length unit (the corresponding "universe" being the Earth, shared by every country). The latter example is a good transition toward spatial distanciation.

---

[5]The truth is that the so-called materialism is in fact a true idealism as we never experience anything like "matter itself", but only perceptions and what language makes of it.

[6]See *The Origin of Geometry* by E. Husserl.

Spatial distanciation

Euclid's Elements introduced a rigorous structure of spatiality, the realm of potential freedom of motion. An important consequence was Thales' theorem. The latter permitted one to make rigorous statements on objects (their length) inaccessible to direct manipulation. As such, it opened the possibility of *extra*-polation, the possibility of transferring to distant objects characteristics of objects within our reach, like harboring life for "other worlds". It is also worthwhile to note that the idea of proportion underlying Thales' theorem is in Latin the same word as "reasoning" ("ratio"), another aspect of the above-mentioned conceptual distanciation. Moreover Euclid's geometry introduced homogeneity of space, opening the possibility that "here" is not a center, not "the" center. It is not necessary to recall the fortune of this idea with the end of geocentrism introduced by Aristarchus of Samos and Copernicus. About the latter, it is interesting to note that there no reference to extraterrestrial life in his writings. In other words, one is thus led from distanciation to decentration.

This homogeneity underlines the great difference with Aristoteles' conception of space for whom the Universe was divided into the Earth (the sublunar world) and Heavens (the superlunar world), by the way like in ancient Chinese astronomy. Both were very heterogeneous and it would have been illogical to transfer to the Heavens something like terrestrial living organisms.

The rationalized structure of space is significantly opposed to the idea of Yin and Yang where every "yang-like" notion contains some "yin-like" quality and vice versa. This Yin-Yang structure is impossible to express in geometrical terms[7]. This difference in the treatment of space in ancient China and Europe is well illustrated by the difference between Chinese painting and Italian perspective.

To summarize, our hypothesis is that the appearance of the theme of extraterrestrial life in the Indo-European area, and its culmination in Greece, is related to the apparition of Euclidean geometry and of the so-called Greek *logos*.

4. Conceptual Distanciation, a Societal Consideration

In addition to distanciation, another important aspect of concepts is that they are likely to be shared by every individual. Indeed, it belongs to the essence of concepts that they are not the property of a political power. It results that the political power (King, Emperor) cannot be the source of concepts. They are are their own, impersonal, source. To express it in a radical way, they are their own power. In astronomy, things were very different in ancient China where, for instance, the few astronomical knowledge like the prediction of eclipses or even the calendar were the private property of the Emperor, because they did provide some power. In addition, concepts are open to debate. That is why concepts and democracy go together, if here by democracy one means "public debate of ideas" rather than such or such election systems, in other words intellectual

---

[7]It can nevertheless be mathematized in modern terms thanks for instance to "non-well founded" set theory (J. Barwise and J. Etchemendy. "The Liar" Oxford University Press 1987) or to Combinatory Logic (Schneider J. "La non-tratification" in *La psychanalyse et la réforme de l'entendement* available at http://www.obspm.fr/~schneider).

rather than political democracy. And it is a fact that in this sense democracy has appeared in the part of the world in which the *logos* also appeared.

It is also interesting to note that in the European Age of Enlightenment where the extraterrestrial life debate gained in intensity with authors like Bernard le Bovier de Fontenelle[8], the idea of decentration gained also a societal tone. This is for instance witnessed by Montesqieu's and Voltaire's work[9]. There is here a significant contrast with one of the old China's name: "The Empire of the Middle".

One may wonder if such considerations do not lead to a Eurocentrism. Such a potential Eurocentrism seems to culminate with Kant when he writes in his *Idea for a Universal History from a Cosmopolitan Point of View*: ".. our continent [Europe] (which will probably give law, eventually, to all the others)..." (9th Thesis)[10]. But to this real concern one can reply:

- That it is not an ideological position but a matter of fact that the entire world has adopted the scientifico-technical concepts.
- That these concepts are not the only respectable values. Ethical values are as important than scientific rational concepts. And notions like Yin and Yang are more useful in some human affairs than rigid rationality. The German philosopher Martin Heidegger has developed at length in his article *Dialogue with a Japanese* (in *On the Way to Language*) that western philosophy has a great deal to learn from the Japanese notion of *koto ba* (which means something like "gracefulness").
- In another vein, Greeks' literalism missed the kabbalistic approach of the reading of great texts which is undoubtedly one of the sources of psychoanalysis.
- One can finally note that if Chinese did have a somewhat elaborated technology, Greeks did not have a systematic development of technology. For instance, they used steam machines to open the heavy doors of their temples, but did not think of applying it in a systematic way to everyday practical life and therefore missed premises of industrialization.

5. Why Did All This Start Essentially in Greece?

This movement did start in the Indo-European arc (which comprises Arabic countries). But it exploded in Greece a few centuries B. C. One could search for some geographical, economical or climatic reason for that[11]. But my thesis is that this Greek geographical location is causeless. Its origin is pure genuine

---

[8] *Entretiens sur la pluralité des mondes.*

[9] Montesiquieu: "If I knew a thing useful to me but harmful to my family, I would reject it. If I knew a thing useful to my family, but useless to my homeland, I would forget about it. If I knew a thing useful to my homeland or to Europe, but prejudicial to the human gender, I would consider it as a crime." in his *Carnets*. See also Voltaire's "point of view from Sirius" in his *Micromegas*.

[10] He meant ethical laws, pointing toward human rights.

[11] For instance, the French book *Le Secret de l'Occident* (David Cosandey, Editions Arléa, 1997) claims that science started in Greece because international exchanges were facilitated by Mediterranean navigation.

fortuitousness, spontaneous generation. This claim results from the epistemo-analytical view[12] according to which ideas emerge from nowhere. This view is illustrated by the *a priori* essence of concepts pointed out by Kant: concepts do not emerge FROM experience, they are a prerequisite to make it intelligible. In another domain, modern language theories rest on de Saussure's principle of *arbitrariness* of signs: linguistics symbols are also given *a priori* and are not the result of a causal process.

6. Conclusion

The personal views presented here are open to debate. Disagreement with them is of course always possible, but any disagreeing opinion should at least offer an alternative explanation of the fact pointed out here that the extraterrestrial life debate seems to be essentially restricted to "Western" literature. This first attempt is not the last word and deserves further investigations, in particular the search for the possible occurrence of the extraterrestrial life debate in other parts of the world.

Acknowledgments. I am grateful to Anne Cheng, Michel Didier, Subhash Kak, Marc Kalinowski, Jean-Claude Marztloff, and Tsevi Mazeh for discussions.

References

Crowe, M. 1986, *The Extraterrestrial Life Debate 1750-1900* (Cambridge: Cambridge University Press)
Crowe, M. 2008, *The Extraterrestrial Life Debate, antiquity to 1915* (Cambridge: Cambridge University Press)
Nazé, Y. 2009, *L'astronomie des anciens* (Paris: Belin)
Selin, H. 2000, *Astronomy Across Cultures* (Kluwer)
Stapledon, O. 1930, *Last and First Men* (Orion Books)
Watzlawick, P. 1976, *How Real is Real* (New York: Random House)

---

[12] Epistemo-analysis is the psycho-analysis of *episteme*, i.e. of knowledge.

Looking for Life Elsewhere: A Biologist's Perspective

M.-C. Maurel
Acides Nucléiques et Biophotonique, FRE 3207 CNRS, Fonctions et Interactions des Acides Nucléiques, UPMC Université Paris 06, France

Life has developed its processes gradually, never rejecting what it has built, but building over what has already taken place. As a result the cell resembles the site of an archeological excavation with the successive strata on top of one another, the oldest one the deepest.

Szent-Gyorgyi (1972) – The Living State

1. Introduction

Looking for life elsewhere must take into account biological studies of terrestrial life, which might provide clues on how to search for life in the universe. This does not mean that life as we know it today exists elsewhere, but it is the only way to handle the problem. First we will consider how to identify life on Earth. What are the standards and the crucial factors that define the right biomarkers? Then, taking into account geological parameters of the solar system as well as the analysis of several spatial objects, we will ask if it is easy to implement the established criteria to apply them to these objects. Finally is it possible to extend this kind of reasoning in searching for life elsewhere in the universe?

2. What Does Life Need?

The cell is the smallest unit of life on Earth. It is composed of a compartment including macromolecules that perform genetic tasks and metabolic reactions. The main macromolecules that store and transfer genetic information over generations are the nucleic acids, DNA, and RNA. Hence, the capacity to reproduce. Proteins are the other class of macromolecules involved in numerous processes, such as catalysis (enzymes), hormones, immunity etc. They also play crucial roles in transport (cytoskeleton), protection and recognition (membranes), packaging (chromosomes) and are involved in plastic properties of the cell. Other important macromolecules are sugars and lipid components as well as large pigments for instance, which are crucial for the transformation of physical energy into usable biochemical energy. The metabolism which brings into play all the macromolecules is characterized by pathways of enzyme-catalyzed biochemical reactions, producing and consuming energy and nutrients. Within the cell, enzymes are highly specific. This is a structural specificity which means that the enzyme recognizes the shape of the substrate with high precision. The consequence is that in practice there exist nearly as many enzymes as chemical reactions. This criterion is highly restrictive if one wishes to understand the chemical and gradual processes leading to the origins of life. If we try to reproduce biochemical reactions in the laboratory without any enzymes, thereby simulating the initial environment, we must proceed in restrictive physico-chemical conditions (of temperature, pressure, and concentration) which are not encountered in living cells and which are sometimes incompatible with present-day life. Hence the importance of ever more discoveries of life in "extreme conditions" (high or low temperatures, pressures, salts, radiations, etc.) that could help to know more about the behavior of macromolecules in drastic environments as well as to better understand primitive processes.

A living system is an open system capable of exchanging with its environment in both directions through a semi-permeable membrane. Nutrients are taken from the outside, stored and transformed, or degraded and then dumped as waste material. Some complex compounds can also be rejected into the environment. Finally, without entering into details, we can say that life on Earth needs a chemical disequilibrium and physico-chemical conditions such as a range of temperatures from -10°C to 120°C, a range of pressures from 10^{-3} Mpa to 10^3 Mpa, food, that is to say an electron donor (reduced carbon, H_2S, H_2, Fe(II), etc), an electron acceptor, such as O_2, organic compounds, Fe(III), CO_2, SO_4^{2-}, carbon sources, minerals, N, P, H_2O which allow diffusion, electron transfer and energy production. These specific conditions could help define the habitable zone in the solar system.

3. Clues that Demonstrate Life on Earth

We know today that life occupies more and more areas, specific and unexpected habitats: soils, atmosphere, rivers, lakes, oceans (surface, depth and sediments) etc. The microbial biomass corresponds to more than half the total terrestrial biomass and more than 99% of bacteria are unknown! This is a surprising assessment since bacteria are omnipresent, numerous and at times live in particularly inhospitable conditions (Madigan et al. 1997).

Looking for extant life on Earth amounts to achieving the characterization and quantification of the biomass, the identification of new forms of life in several habitats, a priori inhospitable, such as new kinds of extremophiles as well as the recognition and the identification of the standard (or common) building blocks and macromolecules of life.

These biomarkers (bio-signatures of life), amino acids, canonic nitrogenous bases – A,U,T,G,C – and modified bases that are likely to have occurred in the primitive RNA world (Cermakian & Cedregren 1998), sugars, lipids, etc., and the macromolecules themselves (DNA, RNA, carbon chains) (Figure 1), can be identified with the help of several methods and tools: Raman spectroscopy (El Amri et al. 2003) that is highly sensitive and requires the handling of few samples thus at the same time saving time and minimizing the risk of contaminations, chemical analytical instruments (ex: GC-MS), aptamer competitors and/or beacons (that is highly specific recognizing molecules), antibodies, metagenomics (a new high performance sequencing method) etc.

Regarding extinct life, the problem is more difficult because of the lack of molecular fossils due to diagenesis and evolution through natural selection. Current technologies are not sufficiently sophisticated to detect femto traces,

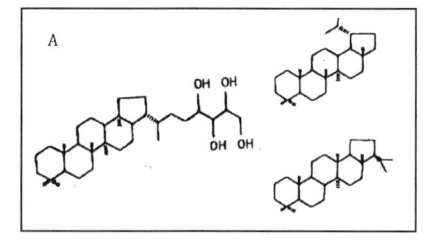

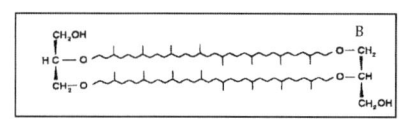

Figure 1. Potential bacterial markers. *A*: Pentacyclic hydrocarbon with hopane skeleton (characteristic constituent of Eubacteria). *B*: Isoprenoid alkyl chains present in ethers from Archaea.

something that might be possible in the future. Nevertheless, following the idea quoted at the beginning of this article (Szent-Gyorgyi 1972), a top-down approach could help to identify the previous and more ancient components at the origin of current metabolic pathways.

4. Life Beyond the Earth

Looking for life in the solar system requires that the same line of investigation be followed. It is necessary to specify what we are looking for, extant or extinct life. The specific geological and environmental conditions of several planets leave little hope of discovering extant life as we know it today. Nevertheless, during the formation of the solar system, some planets and spatial objects followed the same process of formation as the Earth and consequently the presence of organic components of biological interest can be found as traces of extinct life. For instance, the study of numerous spatial objects (comets, asteroids, dusts) or meteorites reveal the presence of soluble organic compounds of biological interest (Stocks & Schwartz 1981) (Figure 2),

Furthermore, based on the presence of past and/or present salts and water ((possibly liquid in depth on Europa and Mars Kargel 1998; Maurel & Zaccaï 2001; Bibring et al. 2006)), it is now common to search in situ or to bring back planet samples to Earth. Several programs are at work in this perspective. The possibility of embarking instruments such as miniaturized Raman spectrometers is particularly interesting since we have shown that adapted Raman techniques allow us to detect subpicomoles of nucleic acids (Percot et al. 2009) and other organic components. We have shown a low detection threshold (5×10^{-3} picomoles) by Raman microscopy, a solid starting point for the development of in situ detection methods of old or modified nucleic acids extracted from mineral samples and that cannot be amplified by PCR (El Amri et al. 2004, 2005). Finally, the aptamer methods could be applied to meteorite samples just as we obtained aptamers for adenine (Meli et al. 2002), a biological molecule of particular interest, easily synthesized in prebiotic conditions from pentamerisation of HCN (Oro & Kimball 1961) and which might be an indication of a prebiotic/biotic primeval pathway. It is now of paramount importance to search for carbon, organic compounds, carbon chains, building blocks etc., given the avail-

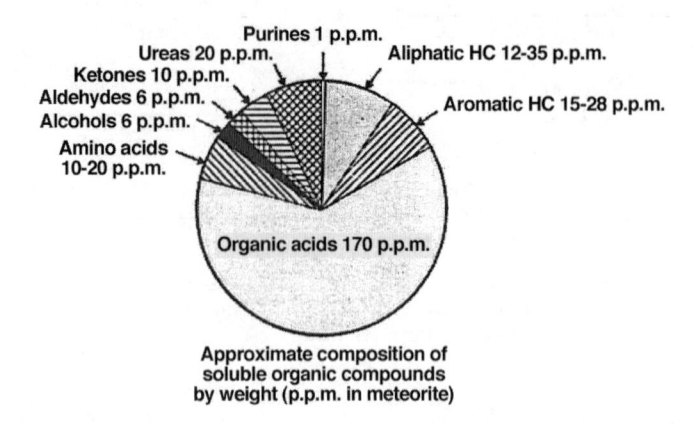

Figure 2. Approximate composition of soluble organic compounds by weight (ppm) in the Murchison meteorite (kindly provided by D. Deamer).

ability of an ever increasing number of highly sensitive techniques that can help to detect and identify these compounds.

5. Life Beyond the Solar System

The impressive discovery of numerous exoplanets has led to the dazzling question of the possibility of life elsewhere in the universe. Are the approaches designed to search for life in the solar system easy to implement for exoplanets? This is a tremendous conceptual and technological challenge.

As stated by Owen (1980) several years ago "simple spectrophotometry can reveal whether or not their atmosphere contain gases such as oxygen, methane and water vapor and/or combinations that would indicate the presence of life." Thus, we need to know more about the characteristics of the atmosphere to correlate the spectroscopic data with established chemical and biochemical parameters. This is a huge program of investigation; it demands working in a systematic way with the help of automated facilities and to make strategic scientific choices.

Finally some authors have hypothesized that since the presence of oxygen and stellar radiation is the prerequisite for life, wherever these two parameters exist, there must be life! This presupposes that the presence of oxygen in a planetary environment is linked to photosynthetic activity. This is a major problem for biologists given the sophistication of photosynthetic phenomena in present-day cells that requires pigments, complex photosystems, specific enzymes, etc. It is clear that life elsewhere – if it exists at all – is not a "clone" of terrestial life, nor is it a genetic or genealogical derivative of life on Earth.

It is also obvious that more than a single trace of life is required. In fact several traces must be considered to propose that some thing is alive. We need more than the detection of something like one biosignature on Earth that would be very specific of Earth-life. We must consider that if life occurred elsewhere, evolution might have led to other paths, that is to say other species and organisms, and also that other kinds of processes must be expected.

If life on Earth were to again take the path it followed 4 Gy ago, no one could be sure that it would take the same path. If life exists — or existed elsewhere — having taken advantage of the same initial planetary conditions, it most likely does not have the same history. Thus, how can one possibly recognize and/or identify something that cannot be conceived? This requires a tremendous amount of imagination, that is to say, great liberty of thought accompanied by rigorous and in-depth studies.

In conclusion, I will introduce an approach often used in biology. It concerns the frontier between what is alive and what is not (for instance, viruses, satellite RNAs and virods). Just as what is pathological in biology provides an insight into what is normal, the space that lies at the border between the living and the non-living will maybe allow us to consider other forms of life (that cannot be imagined today).

Acknowledgments. I gratefully acknowledge Anne-Lise Haenni for her help in the final English version of this paper.

References

Bibring, J.-P., Langevin, Y. , Mustard, J.-F., Poulet, F. , Arvidson, R., Gendrin, A. , Gondet, B. , Mangold, N. , Pinet, P. , Forget, F. and the OMEGA Team, 2006, Sci, 312, 400
Cermakian, N., & Cedregren, R. 1998, in Modification and Editing of RNA, eds H. Grosjean & R. Benne (Washington DC: Am. Soc. for Microbiology Press) Chap. 29.
El Amri, C., Baron, M. H., & Maurel, M. C. 2003, Spectrochim Acta A Mol Biomol Spectrosc., 59, 2645
El Amri C., Maurel, M.-C, Sagon, G., & Baron, M-H. 2005, Spectrochimica Acta, 61, 2049
El Amri C., Baron, M-H., & Maurel, M-C. 2004, J. Raman Spectroscopy, 35, 170
Kargel, J.S. 1998, Sci, 280, 1211
Madigan, M. T., Martinko, J.M., & Parker, J. 1997, Brock Biology of Microorganisms, Eighth Edition, (Prentice-Hall, USA)
Maurel, M-C., & Zaccaï, G. 2001, BioEssays, 23, 977
Meli, M., Vergne, J., & Maurel, M.-C. 2002, in Proc. of the First European Workshop on Exo-Astrobiology, ed. J. Lacoste. ESA SP-518, (Noordwijk, Netherlands: ESA Pub. Div.), 479
Oro, J., & Kimball, A.P., 1961, Arch. Biochem. Biophys., 94, 217
Owen, T. 1980, in Strategies for the Search for Life in the Universe, ed. M.D. Papagiannis, 177
Percot, A., Lecomte, S., Vergne, J., & Maurel, M. C. 2009, Biopolymers, 91, 5, 384
Stocks, P. G., & Schwartz, A. W. 1981, Geochim. et Cosmochim Acta, 45, 563
Szent-Gyorgyi 1972, The Living State (Academic Press: New York)

Pathways Towards Habitable Planets
ASP Conference Series, Vol. 430, 2010
Vincent Coudé du Foresto, Dawn M. Gelino, and Ignasi Ribas, eds.

The ITER Pathway: Useful Lessons for Exoplanet Projects?

J. Jacquinot

Cabinet of the French High Commissioner for Atomic Energy, CEA, Gif-sur-Yvette, France

Abstract. ITER, "the way," in Latin, is an international organisation aimed at the demonstration of fusion energy for peaceful purposes. It has the overall responsibility for constructing in 10 years, and then operating for 20 years, a very large scientific device based on the Tokamak principle. The cost of the construction is estimated to be 5 billion Euros. The treaty-based organisation includes 7 partners (EU, China, India, Japan, Russia, South Korea, and the USA) representing 32 countries. The chosen construction site is Cadarache in the south of France. Key elements necessary for making such an unprecedented collaboration possible are identified: a pre-existing active scientific network acting as a promoter, a in-kind supply by the partners of key high tech components thus ensuring a fair return to them, organisation supported by domestic agencies, intellectual property rights, a win-win agreement for choosing the site of the organisation etc. This article aims at identifying the general aspects of the ITER experience gained so far which could apply to an international exoplanet project.

1. Introduction

In the recent past, the space and astronomy communities have enjoyed considerable success in setting up and operating collaborative ventures. It was therefore a bit surprising to be asked to report on the ITER project (originally the International Thermonuclear Experimental Reactor) and to comment on its relevance to a possible international collaboration in the search for habitable exoplanets. After thinking a bit about it, it became clear that the glorious days of space conquest are over and the emergence of a highly competitive world raises new constraints which must be taken into account in order to increase the chances of success of large international venture projects.

In the first part, we will outline the particular ITER objectives and substance. Then we describe the key dates in the long march starting in 1986 at the Reykjavik Summit which led, in November 2007, to the initial conceptual stage of the project. The process which followed involved an intricate mix of science, engineering and, not surprisingly political aspects, with the latter dominating the final phase. A compromise between numerous constraints was eventually reached, leading to some unusual organisation with a heteroclite combination of highly favourable aspects mixed with undesirable ones, notably a large number of complex interfaces.

2. ITER: Objectives, Substance and Key Dates

On November 21, 2006 in the Paris Elyse Palace (Figure 1), the seven representatives of countries or groups of countries which constitute the ITER partners (Ikeda 2010) signed an agreement to construct together a very large fusion machine in Cadarache, France. From this date, this meant that more than half of the world population from 33 countries representing more than 75% of the world gross product is directly involved in this project.

Figure 1. The representatives of the seven ITER partners together with the French president at the signature of the ITER documents in November 2006.

The long term goal of this project is of huge interest: with less than two kilograms per day of fuel composed of deuterium and tritium, it would be possible to generate 1000 MW of electricity, while 6000 tons of petrol is needed to produce the same energy. The realisation of such an objective requires a very substantial research effort over the long term. However, the reviews held by each partner concluded that, despite the technical complexity, the risk inherent to such a scientific development was well worth taking in view of the potential advantages of fusion. The most appealing of these advantages are the impossibility of a "Tchernobyl effect", the very low impact on the environment and finally the quasi unlimited reserve of fuel. Moreover, this method of energy production does not present a proliferation risk as it does not involve any fissile material required for making bombs, and the tritium used in the process cannot be extracted from the machine without violating its integrity.

It is not necessary here to emphasise the necessity to develop a new large scale energy source which will ineluctably have to replace the sources from fossil origin, not only because of the impact of fossil fuel combustion on the environment, but also because hydro carbonates need to be preserved for use in the chemical industry.

Therefore the perspective of a source of plentiful energy that respects the environment constitutes a strategic stake of the highest order.

The long term perspective for fusion was recognized in 1986 at the Reyjavik Summit when the Heads of States suggested setting up an international team to study, on the conceptual level, the possibility of magnetic fusion for energy production. Substantial progress had just been made in magnetic fusion machines all over the world, notably with the recent successful operation of large devices JET (EU) and TFTR (USA), and it was felt that a new larger device was required to demonstrate, at the appropriate scale, fusion energy production (Jacquinot 2010).

Europe, Japan, the USSR, and the USA agreed to institute a conceptual design activity (CDA) team under the auspices of the IAEA. The CDA phase lasted until 1992. It was hosted by the Max-Planck-Institut für Plasma Physik in Garching, Germany. About 50 professionals worked together for about 6 months on the ITER conceptual design. At the end of this phase, the concept of a D/T burner was proposed and accepted as the basis for a detailed engineering design phase (EDA phase) which took place between 1992 and 2001. The design team was located at three sites: Garching (Germany), Mito (Japan) and San Diego (USA). It worked effectively but the size of the device under study was subject to fluctuating economical/political constraints. Initially aimed at a thermal output of 1.5 GW, it had to be reduced to 0.5 GW after the withdrawal of the USA in 1998, when the three remaining partners required reducing the overall cost by a factor of two. During the EDA phase a substantial ($800M) R&D programme was carried out in order to validate key ITER technologies. Major prototype components of super conducting magnets, remote handling systems, vacuum vessel sectors and high heat tolerant components were constructed and successfully tested (Figure 2).

In 2001, the three remaining partners accepted the engineering design of ITER as submitted by the EDA team. This decision triggered the start of government negotiations on drafting an agreement for site selection and machine construction. By autumn 2002, four candidate sites had been formally proposed with each proposing country interested in the economical and structural appeal of the high technology implied by such a large scale project. The province of Ontario in Canada proposed a site adjacent to its Candu reactors in Clarington near Toronto. It benefitted from a fully operating tritium plant. Japan presented a new site located in Aomori, north of Japan. Europe had to decide between two sites: Vandellos, a fission power plant site near Barcelona, proposed by Spain, and Cadarache, a large CEA R&D centre in the southeast of France, which was proposed by France. In each case, local authorities strongly supported their candidate site. The French local authorities offered 10% of the ITER cost for financing the additional resources that France would have to supply if the Cadarache site was selected. They acted with great success for obtaining the central government support. Site assessment committees were established by the partners. They pondered at length the pros and cons of each site, but at the end of their mission they concluded that all four sites were acceptable.

There was no easy way out of the site selection deadlock and a win-win agreement involving other aspects, preferably in the fusion programme, had to

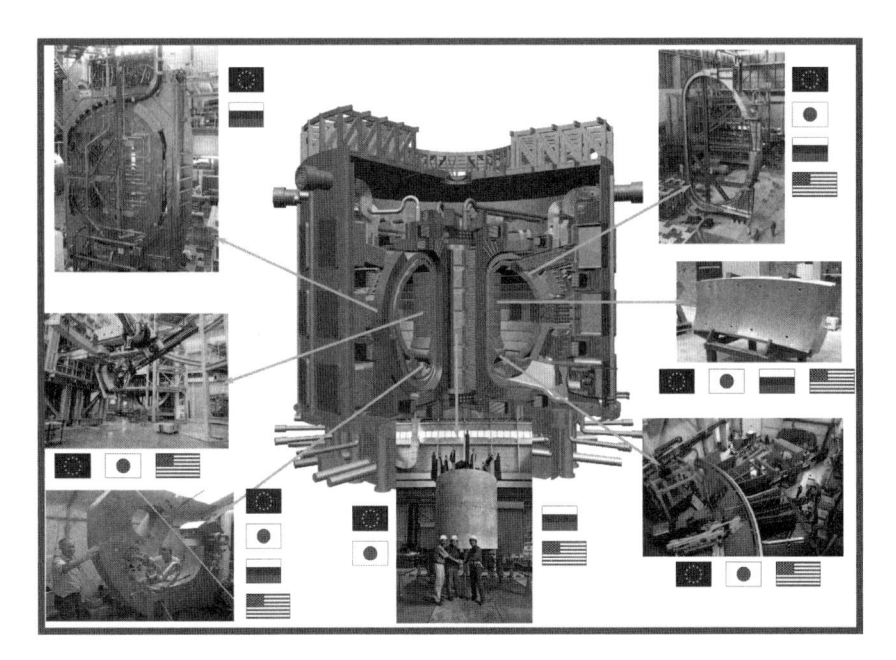

Figure 2. Centre: cut away view of ITER; machine mass: 23350 t (cryostat + Vacuum Vessel + magnets). Photos on the sides: prototypes of main machine components realized during the EDA phase by industrial consortia in order to demonstrate key technologies: superconducting coils (top left and bottom centre), vacuum vessel (top right), remote handling (middle left and bottom right) and divertor module (bottom left).

be found at the political level. It went all the way to the Heads of States. Some of them included ITER high on their agenda, already very full with consequences of the Iraq war. During the site selection process ITER gained high public visibility reminiscent of the selection of sites for the Olympic Games or for the America Cup. This visibility attracted protests against the nuclear aspects of the project but on average, public debates drew a large endorsement by the general public.

During the early phase of the site selection process, the US rejoined the three partners. They were followed by new applications to become full partners from China, South Korea, and finally India all of which were accepted. We now had seven full partners representing more than half the world population.

By the end of 2003, the number of candidate sites had been reduced to two as Ontario dropped out and Europe selected Cadarache, agreeing that the European Domestic Agency (now called "Fusion for Energy") foreseen in the draft ITER statutes would be located in Spain. A year later in June 2005, the seven partners unanimously agreed upon Cadarache. Simultaneously, Europe and Japan concluded a programme of a "broader approach on fusion" which included equal participation and financing in programmes (computing centre, EDA on a material irradiation facility and an upgrade of the large Japanese Tokamak facility) to be carried out in Japan, the total value of which reaches about 14% of the ITER cost. The ITER agreement was signed in Paris on 21 November 2006 by the governments of China, EU, India, Japan, South Korea, the Russian Federation and the US. However the ITER agreement entered into force and the ITER Organization was formally established only on 24 October 2007 after formal ratification by the member states. At the time of writing, the ITER staff exceeds 450 people and construction has started on site.

3. Key Points for Approval of an International Project

3.1. Vision, Scientific Consensus and Opportunism

Not surprisingly, the prerequisites to be met when contemplating a large international scientific project start with a visionary objective which can be shared by the general public and has a solid scientific basis. For ITER, the vision was clean energy and the scientific basis comes mainly from the results of existing large facilities. Then, the project has to be supported by a comprehensive engineering analysis demonstrating its feasibility. Finally, it has to be endorsed by the relevant scientific community. This last point may well be the most difficult to achieve, as we all know that scientists often enjoy being critical. In the case of ITER, the consensus slowly emerged from a long-lasting practice of collaborative programmes. In the case of magnetic fusion, since 1958 Euratom has run a tight network of European laboratories which involves notably the joint construction and exploitation of JET, a precursor of ITER in many respects. On the international scene, the International Energy Agency (IEA) coordinates a number of implementing agreements between facilities operating in its member countries. Lastly, the International Atomic Energy Agency (IAEA) organises a general conference, several technical meetings per year and voluntary physics groups (ITPA) which address theoretical or experimental topics on the critical path for the development of magnetic fusion.

Once the prerequisites are in hand, one needs to be able to grab the right opportunity for putting forward the proposal. The windows of such opportunities are generally short lived and should not be missed. For ITER, the steep cost increase of the barrel of oil provided a good one. Once the project proposal is launched, progress can be slow but all is well as long as steady progress is maintained. Better to make small steps forward than risking a 'NO' from a political instance which may well be final!

3.2. Cost Sharing, Supply in-kind and Intellectual Property Rights

The statutes of the project organisation must take into account the unbalance of the economical returns to the host partner when compared to the other members. Clearly the host will receive more economical returns than the other partners and this fact, well known in international projects, has to be compensated by a series of measures. These can include: proper tailoring of the costs (construction and operation), the methodology for the procurement of the machine, and the intellectual property rights (IPR). In the case of ITER, the ITER partner (EU) will pay 46% of the construction and 34% for operation and dismantling while the other members will each pay only 9% for the construction and 13% (Japan and US) or 10% (China, India, Korea, FR) for operation. Secondly, the IPR agreement states that all members will receive equal rights for using inventions, excluding underlying technology, resulting from ITER for the purpose of fusion energy development. When used for other purposes, rights pro rata to financial contributions will apply. Finally, some 90% of the machine components will be procured "in-kind". We note here that the percentages mentioned above do not correspond to cash money but to "values" estimated by the EFDA team. The distribution of components to be supplied by each partner and the corresponding values are part of the ITER documents ratified by the partners. Figure 3 illustrates the distribution of the in-kind procurement packages. Generally, it reflects the desire of the partners to be involved in the high technology aspects in order to acquire the related know-how. This leads inevitably to complex interfaces between key components of the machine. The ITER organisation (IO) retains the full responsibility for the technical specifications, assembly, acceptance of the components and commissioning and operation.

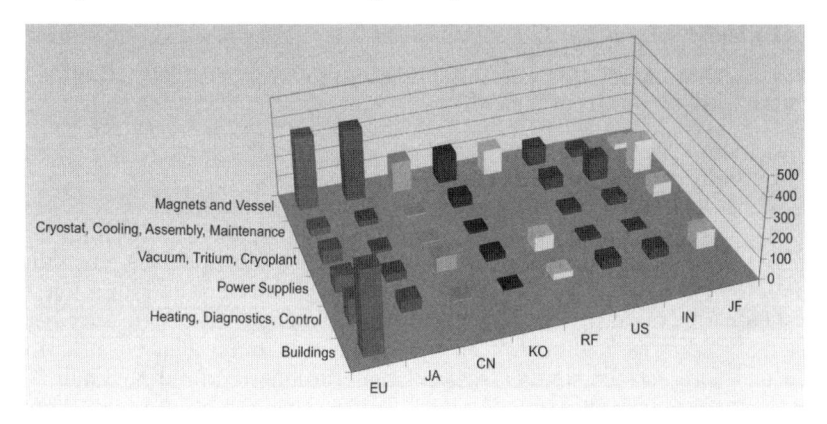

Figure 3. Distribution of the procurement packages to be supplied in-kind by each partner of the ITER project. In total 90% of the machine will be supplied in this way.

4. Pros and Cons of Large Collaborative Projects

Among the pros, the pooling of both the human and financial resources is the dominant advantage. In the case of ITER, this aspect was essential as no single country would have had been able to even consider building such a facility alone. We also note the stability that international collaboration brings to such long term projects. It is a lengthy procedure to establish it, but once launched it is equally difficult to stop. Stability against political and financial crises is mandatory for projects lasting over several decades. A high level international agreement is highly desirable for providing such a durable platform.

Among the cons, it is easy to spot that the necessity to find a political compromise for concluding the negotiation phase may well generate a less-than-ideal organisation. Management at all levels should be primarily appointed on the basis of competence and adapted as the need arises during the lifetime of the project. The procurement method should always keep in view the technical and practical requirements. International collaboration will inevitably generate a lot of technical interfaces between the various partners and the different sectors of the organisation. They can only be resolved if the central organisation is given the full power to manage them. As experience progresses, the good will of all partners will be the essential ingredient to form teams integrating all key players and devoted to realizing highly challenging technological devices. In any case, a totally new organisation is likely to take at least a couple of years before becoming effective. One tends to underestimate the time it takes to recruit and train adequate staff while setting up operating procedures and answering a constant flow of legitimate requests from the partners.

The simple remarks outlined above are drawn from experience of large international collaborations such as JET and CERN. In CERN, the Atlas project (ten Kate 2008) was realized using in-kind procurements similar to what has been foreseen for ITER. The project could be completed in a quite reasonable time and budget. But in order to achieve this, it was emphatically stressed that the central team needed to be strong and empowered to manage all interfaces and conduct acceptance of deliverables.

5. Conclusions

The experience gained from the ITER pathway shows that establishing an international collaboration of such a scale can be a very slow process. Ingredients essential for success have been based on a visionary objective and on a scientific consensus on the value of the project and the choice of its design parameters. Consensus building is a difficult step and requires a comprehensive practice of international projects of smaller scales. Negotiations required to finalize the project statutes may lead to a complex organisation with a very large number of interfaces. A strong central organisation and well integrated partner teams are required to manage these interfaces and avoid the related delays and cost increases.

So why pursue international projects and pay the associated price of a slow start and the increased complexity? Is it worth it? The answer based on past experience is a clear yes. The pulling together of intellectual and material resources is highly stimulating and efficient. Such projects will, after the inevitable slow start, achieve a breadth of results which could not have been conceivable in the frame of pure national projects. Even more importantly, this will contribute to building a worldwide scientific community capable of jointly addressing the global issues facing our planet where resources are becoming scarce. This alone is already an invaluable investment for the future of mankind.

References

Ikeda, K. 2010, Nucl. Fusion 20, 014001 (See also: http://www.iter.org/)
Jacquinot, J. 2010, Nucl. Fusion 50, 014001
ten Kate, H. 2008, ATLAS superconducting toroids completion, lessons learnt and how they can be used for fusion toroids, Invited Lecture SOFT Rostock, http://www.ipp.mpg.de/eng/for/veranstaltungen/soft2008/

Session VIII
Conclusion

A Personal Insight on the Conference

P. J. Léna

Université Paris Diderot & Observatoire de Paris, 5 Pl Jules Janssen, Meudon, France

Abstract. This short contribution is a modest attempt to express some of the obvious convergences of projects presented and opinions expressed in the conference. It emphasizes the clear emergence of an international and enthusiastic community working in the field of exoplanets, and the wish of this community to establish strong links in order to fully optimize, worldwide, the resources which are necessary for the formidable discovery path laying ahead. Hence the building up of a *Barcelona process* appears as a new step, which emerged in the conference and got support from all participants. This process should be made broadly known, since the issue of other Earths and life in the universe concerns so many people on our planet, beyond the scientists themselves.

The Nature of Exoplanet Science

After 14 years of discoveries following the first exoplanet, 51 Pegasi, in 1995, it is possible to outline some of the main characteristics of this new field of research. Over the long term, which may be a matter of years, decades or even more, the ultimate goal is undoubtly to answer the old question: *Is there life elsewhere in the Universe?* Indeed, a negative answer is completely out of reach, while speculations about the likelihood of a positive one cover the full range of possibilities, from almost certain to fully improbable. Yet today, on the basis of an extended Copernican principle, most astrophysicists would adopt the former perspective and be prepared to organize step by step an ambitious process, investing time, energy and money in such a long term search.

Exoplanet science, as it can be described today, is grounded on the discoveries made in the last decade, namely over 300 stars with planets, including systems with several planets. These observations are complemented by a increasingly large amount of theoretical work, going from fundamental biology and biochemistry to the geophysical and geochemical analysis of conditions on primitive Earth, from celestial to fluid mechanics, in order to establish boundary conditions for the apparition and sustainability of life. Many of these avenues have been presented at the conference and exobiology emerges as a new discipline, at the crossing point of many others. Needless to say, there also exists more adventurous speculations on forms of life which would drastically differ from the terrestrial one, but these may not be very useful for designing the discovery steps toward answering the above formulated question.

Observational, modeling, and theoretical strategies are intimately connected, as the latter must constantly guide the former, while observations did, and will continue to reveal a rich and often unexpected range of planet configurations and physical conditions. As an example, let us simply mention the migrations of planets, plate tectonics and its chemical consequences, the ocean planets, the climatic consequences of phase-locking or high ellipticity orbits, etc. Consequently, exoplanet science is by nature an interdisciplinary and expanding domain, where the parameters used in modeling to describe and understand the observations get enriched and modified on short periods of time – namely a few years.

Observational techniques are extremely diverse, and reveal an intense creativity: e.g., the old technique of coronography is now blossoming in all kinds of new concepts and realizations. Whether they involve ground- or space-based instruments, the levels of difficulty, hence of instrument sophistication, cost and implementation, are great. Hence, establishing observational strategies has to be done for a long period ahead, such as 10 to 15 years, while the field itself changes much faster. The mismatch of this time constant with the previous one offers a serious planning challenge.

Long Term Goal and Necessary Steps

Many discussions were held at the conference in order to more precisely define an agreed long term goal. Assuming the commonly accepted definition of *habitability* as referring to the existence of liquid water on the planet, the goal may be formulated as: *Characterizing a set of habitable planets comparable to Earth*. This characterization implies direct imaging and spectral analysis of the planetary light emission, absorption, or scattering. Looking exclusively for Earth-like (same mass and G-type star) objects seems both too difficult and too restrictive, and a consensus seemed to be reached on defining this set as made of objects with masses 1 to 10 $M_\oplus$, with the parent stars of spectral types later than F.

The conference has indeed discussed many aspects of exoplanet science which deal with hot Jupiters or similar mass planets. On the other hand, the long term question clearly deals with Earths and super-Earths, and we shall focus on these. Yet, the continuation of observations and models for the larger masses is essential, as it will make more and more precise the continuity and differences along the mass spectrum, as well as stimulating new observational techniques.

For super-Earths and Earth-like planets, three successive steps are identified:

- Step 1: It is necessary to obtain a good measurement of the statistical occurence of such objects;

- Step 2: A reasonable number of potential targets for spectroscopic analysis must be identified;

- Step 3: Only then comes the detailed characterization requiring the determination of the physical and chemical conditions on the surface (solid, or solid/liquid, or liquid) and in the atmosphere. It also requires obtaining extensive information on the whole system: star, age, other planets, residual disk, etc.

At this point, the ambitious European *Darwin* project ought to be mentioned, since its goal was essentially to cover the two last items, as well as its U.S. counterpart *Terrestrial Planet Finder*. *Darwin*, proposed with great insight in 1993, two years before the 51 Peg discovery, played a fundamental role in establishing a community, in exploring all the difficulties and hopes of life detection, in triggering numerous and fruitful instrumental studies. *Terrestrial Planet Finder* did the same smewhat later.

Space Missions 2010-2020

To possibly reach the ultimate goal stated above, at least one complex and expensive space mission (flagship) will be needed. Planning for it requires more maturation of the field, as today many unknowns remain to be explored and clarified. Space is absolutely necessary to achieve this clarification task, which ought to be pursued in the decade 2010-2020. The COROT, KEPLER, HST missions are already providing data. The GAIA (2012) and JWST (2014) missions will bring more relevant data and are on their way.

A number of new missions, in the cost range 0.5 to 1 G euros, have been presented at the Conference. In Europe, the ESA *Cosmic Vision* process has already selected for study PLATO, a successor of COROT (transits), and SEE-COAST (coronographic imaging) prepares for the second round of selection in 2010, for launches in 2018. In Japan, SPICA (coronographic imaging, 2017) could be decided soon, with possible European participation. In the USA, the process of Decadal Review is going on, with conclusions expected by mid-2010. The SIM-LITE (astrometric interferometer) mission has now been studied for a long time, but the Astrophysics Strategic Mission Concept Studies, set by NASA in 2008, have produced many interesting concepts, such as THEIA and New World Observer (free-flying occulter), THESIS (spectroscopy of atmospheres), ACCESS (active coronograph). Other missions might have an interesting impact on statistical aspects of exoplanet science, such as the study at galactic scale of microlensing events with EUCLID.

Discussions at the Conference have not elaborated on the existing complementarity of these missions with ground-based observations, such as radial velocities, astrometric interferometry, coronography, and imaging. This task remains to be done. It seems agreed that several of these missions will happen, following independent choices made by national or regional agencies. Nevertheless, it is wished that crossed minority participations be implemented, demonstrating the emergence and value of the exoplanet community. Steps 1 (statistics), 2 (targets designation) and partially 3 (characterization) will then be well underway with some of the following missions:

- COROT, KEPLER and radial velocities done on the ground in the visible and soon in the infrared (for M stars). These transit missions may be pursued and extended to a larger part of the Galaxy with PLATO, TESS and others.
- The exploration of the full mass spectrum of planets, providing a global picture of their formation and occurrences;
- An even broader statistical picture provided by microlensing events (EUCLID).

In parallel with these missions, it appears that the astrometry mission SIM-LITE can identify a complete set of reasonably close targets (distance ca. 10 pc), i.e. fullfilling step 2, in cooperation with observations done from the ground with radial velocities (see below) and astrometry (VLTI-PRIMA). The conclusion seems to be a recommendation for the NASA mission SIM-LITE, with some minority participation from ESA.

To achieve step 3, where spectroscopy is mandatory, transit (primary and/or secondary) and off-transit observations from space on already defined targets are likely to be the best option. Such a mission should account for the following points:

- Broad spectral coverage (0.5 to 20 μm or even more) will be necessary to remove ambiguities in the interpretation of spectra, as already shown by simulations;

- Even the JWST may not be sufficient to achieve this, because of its limitations in observing time dedicated to exoplanet programs, operational pointing constraints and possibly an insufficient S/N ratio over this broad spectral range;

- Step 2 is already achieved on massive planets, and one can be confident that it will soon extend to objects of smaller masses in the habitable zone for some targets; the feasibility of such a mission is already established; the comparison with models becomes straightforward and highly conclusive;

On this basis, at least one medium-class (M) mission, in the 500 Meuros range, should be planned to happen before 2020 (ESA, NASA), with supporting ground programs (see below).

While spectroscopy remains the main tool to achieve characterization, one should not underestimate the rapid progress made by coronographic and interferometric imaging techniques, making them capable to contribute to step 3. For this reason the SPICA mission, as well as a possible NASA Probe mission, would demonstrate the concepts. Nevertheless, this path requires clarification of the issue of exozodi emissions, to establish whether and in which cases these emissions would present insuperable obstacles to full characterization. Double blind studies may provide precious support for mission planning and data analysis. In any case, JWST will also contribute significantly to the progress in this field.

Once these three steps are mastered, a flagship mission, possibly with global resources, in the decade 2020-2030 will fully make sense: one will be equipped to plan a sound observing program and to optimally exploit the data, especially the spectroscopic ones.

Programs from Ground-based Instruments

Ground-based programs are obviously confined to larger planetary masses, although efforts are clearly recommended to lower the mass limit (super-Earths) which the VLT or ELTs could reach.

- Radial velocities ought to be extended to the infrared (M stars), with HARPS, VLT/EXPRESSO and others;

- New instrumentation is coming for 10-m class telescopes and ELTs, such as extreme coronography and extreme adaptive optics (EXAO);

- Optical interferometry is emerging as an additional tool for astrometry (VLTI/PRIMA), discs (KI, LBT) and exozodi (ALADDIN in Antarctica) studies;

- Microlensing is also possible from the ground (MOA).

Building-up an International Community

Numerous presentations have clearly demonstrated the great degree of prospective analysis and organisation achieved in the USA, as demonstrated by the Decadal Review, the NASA Exoplanet Exploration Program, the Exoplanet Science Institute, the Exoplanet Forum, and the Sagan Fellowships.

In Asia, Japan appears also as well organized, with its long term space strategy for exoplanet science and the synergy with the SUBARU telescope. India is progressively getting interested in the domain, while China has not yet choosen to invest significant efforts into it.

European strategies appear less structured than the American ones. ESO and ESA do not have a common policy analysis. The ASTRONET prospective exercize, an European Union-funded review published in 2008, does not focus much on exoplanet science, which occupies at most 8 pages out of 126 in the report. Its recommandations are sound, but very general, such as: high contrast and high resolution infrared interferometry in space; high precision photometry($\leq$ 0.01 mag) from space and ground; high spectral resolution on E-ELT with adaptive optics in the infrared; a next generation of radial velocities instruments, reaching 0.1 m/s.

Russia at the moment seems quite absent from the domain.

While the conference clearly manifested a pressing need and a strong will to exchange and collaborate on a global scale, the structures to do so are not readily available.

Establishing a *Barcelona Process*

Having agreed that exoplanet science may represent one of the key developments of science and astronomy during the $21^{s}t$ century, and considering that the efficient use of resources worldwide is necessary to progress towards answering the fundamental question, the proposal is made to initiate a process, the name of which could be *The Barcelona Process*. This is in reference to the generous host city of this conference as well as to the other convergence processes which mark successive steps in the construction of the European Union (Lisboa Process, Bologna Process).

To be more specific, this Process may include at the global level:

- A regular *Pathways* Conference, e.g., every 3 years;

- An identified place for regular interactions at global scale, possibly on the model of the highly successful International Center for Theoretical Physics (ICTP) at Trieste (Italy);

- A dedicated, long-term action for outreach, possibly using the remarkable model of the HST Baltimore Center. Such an outreach strategy will foster the public interest for the fundamental question tied to exoplanet science;

- A vision, placing this science in a broader context with ethical and epistemological dimensions.

Focusing more specifically on European issues, it could be proposed to:

- Establish a permanent committee representing the community, in order to present formally the goal and its steps to the European Union; to get adequate support (in the Framework Program FP-8, to begin in 2014) ; to be recognized by ESO and ESA;

- Establish a dedicated fellowship program;

- Implement the Barcelona process in Europe.

Finally, regarding China, India, and Russia, the conference concluded on the need for its participants to contribute to the emergence of a community in these countries, e.g., by adequate workshops and schools, then to establish as soon as possible progressive collaboration programs, such as the one currently initiated between the Thirty Meter Telescope and China.

Conclusion

The question *Is there life elsewhere in the Universe?* may or may not find an answer during the next decades, and nobody is able to predict the outcome of the quest. But one can agree on the fact that many inhabitants of the Earth and many cultures are supporting this quest as a fundamental one for the representation of the place of Man and mankind in the universe. Astronomers and scientists from other disciplines feel this deep wish to know, this *natural desiderio di sapere*, as it was expressed by Federico Cesi while founding in 1603 the Academia de Lincei in Rome. And the Barcelona Process may be an important step to pursue the goal.

Acknowledgments. It is a pleasure to thank here Ignasi Ribas and Vincent Coudé du Foresto for having proposed this personal insight to be presented, as well as the several participants who kindly have communicated their presentation to the author.

Session IX

Special Session: Advanced Strategic Mission Concept Studies

The New Worlds Observer: The Astrophysics Strategic Mission Concept Study

W. Cash

University of Colorado Boulder, CO, USA

and the New Worlds Study Team (see Table 1)

Abstract. We present some results of the Astrophysics Strategic Mission Concept Study for the New Worlds Observer (NWO). We show that the use of starshades is the most effective and affordable path to mapping and understanding our neighboring planetary systems, to opening the search for life outside our solar system, while serving the needs of the greater astronomy community. A starshade-based mission can be implemented immediately with a near term program of technology demonstration.

1. The Mission

The New Worlds Observer (NWO) is a mission concept for direct observation and study of exoplanets all the way down to Earth-like objects. Its goal is very simple: map the planetary systems of the nearby stars, search for habitable planets, and look for signs of life. An Astrophysics Strategic Mission Concept Study (ASMCS) has just been completed, and this paper summarizes some of the results of that study (NWO ASMCS Study Report) which the reader is encouraged to consult for material backing up the assertions contained herein. The many people of the New Worlds study team who contributed to this study and thus to this paper are listed in Table 1, along with an indicator of their roles.

Is Earth a unique outpost of life in a vast and empty Universe? How did planets come into being and why are they in their current state? What are the circumstances under which life arises, and how common is it? NASA can definitively address these questions in the coming decade with the New Worlds Observer (NWO).

Hundreds of giant exoplanets have now been detected and improvements in technology are moving the detection limits to smaller and smaller masses. NWO can discover Earth-like planets, but detecting their existence is just the beginning: only spectroscopy of planets in the habitable zones of dozens of stars can answer the question of how common life is in the Universe. A facility capable of finding and characterizing terrestrial planets requires that the starlight be suppressed by a factor of at least 10^{10} to enable the planet's light to be seen against the light of its host sun. This suppression needs to be confined within tens of milliarcseconds (mas) so that the planet's light is not blocked. Direct imaging with NWO will reveal most of the planets in an extrasolar system in

354 Cash et al.

just a single exposure. Through spectroscopy, we can determine the nature of each planet discovered.

Table 1: The New Worlds Study Team

Co-author	Institution	Study Role	Co-author	Institution	Study Role
Steve Kendrick	Ball	Optics and Telescope Design	June Tveekrem	GSFC	Optics, Stray Light
Charley Noecker	Ball	Mission Design/Telescope Lead	Bruce Woodgate	GSFC	Instruments
John Bally	CU	Science Team	Maggie Turnbull	GSI	Science Team Lead
Julia DeMarines	CU	Outreach/Education	Dean Dailey	NGC	Mechanism and Structures Lead
James Green	CU	Science, Instrument Design	Kent Decker	NGC	Power Subsystem Lead
Phil Oakley	CU	Science Team	Reza Dehmohseni	NGC	ACS/GN&C Lead
Ann Shipley	CU	Starshade Laboratory Testing	Brian Gaugh	NGC	Thermal Control Subsystem Lead
Scott Benson	GRC	NEXT Propulsion System Lead	Tiffany Glassman	NGC	Science team/ Starshade Analysis
Steve Oleson	GRC	NEXT Propulsion System	Mickey Haney	NGC	Comm System Lead
David Content	GSFC	Laboratory Testing	Reem Hejal	NGC	Dynamics Lead
Dave Folta	GSFC	Flight Dynamics	Chuck Lillie	NGC	Senior Advisor
Sharon Garrison	GSFC	Study Management	Amy Lo	NGC	Deputy PI, Mission Design
Keith Gendreau	GSFC	Science System Engineer	David O'Conner	NGC	Starshade ACS Lead
Kate Hartman	GSFC	Study Manager	Gina Oleas	NGC	System Engineer
Joe Howard	GSFC	Optical Design	Ron Polidan	NGC	Senior Advisor
Tupper Hyde	GSFC	Technology Lead	Rocco Samuele	NGC	Starshade Laboratory Testing
Darryl Lakins	GSFC	Lead System Engineer	Stephen Shields	NGC	Mechanical Design Integration
Jesse Leitner	GSFC	Formation Flying Systems	James Shirvanian	NGC	Formation Flying Conops
Doug Leviton	GSFC	Optics, Instrument, Lab Testing	David SooHoo	NGC	Propulsion
Rich Luquette	GSFC	Formation Flying Systems	Giovanna Tinetti	UCL	Science Team
Bill Oegerley	GSFC	Science Team	Bryan Dorland	USNO	JMAP Instrument PI
Karen Richon	GSFC	Orbits, Navigation lead	Rachel Dudik	USNO	JMAPS Instrument
Aki Roberge	GSFC	Science Team	Ralph Gaume	USNO	Astrometry
Steve Tompkins	GSFC	Ground Systems/Mission Ops	Brian Mason	USNO	Science Team, astrometry

The NWO mission concept (Figure 1) can do all of this and more. Full suppression of the starlight before it enters the aperture relieves the telescope of demanding requirements such as ultra-high quality wave front correction and stray light control. The NWO telescope requires only diffraction-limited wavefront quality. This design results in a clean separation of light suppression and light collection. The starshade is a passive mechanical structure that only has centimeter-level requirements on the edge, but not over the surface. Integrated development of NWO could start today.

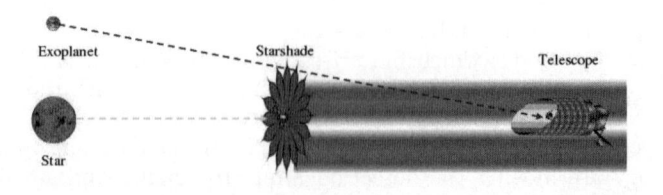

Figure 1. NWO's cost-effective starshade shadows the telescope from the star, while light from a terrestrial exoplanet passes the edge of the starshade unimpeded.

The NWO mission is illustrated in Figure 2. Two launch vehicles take the 50 m starshade and the 4 m telescope to L2, where they enter a halo orbit. The two spacecraft are separated by ∼80,000 km. The starshade moves relative to

The New Worlds Observer Concept Study 355

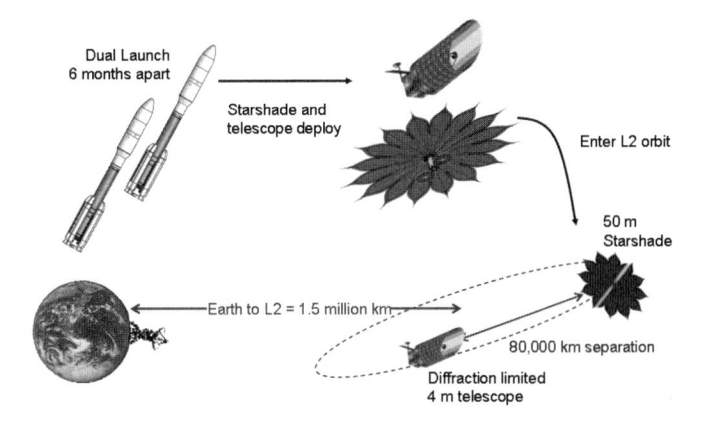

Figure 2. Employing existing technology, NWO uses a 4 m telescope and a 50 m starshade orbiting around the Sun-Earth L2 point to image and characterize terrestrial planets.

the telescope to occult target stars. The average exoplanet observing cycle is 2 weeks per star, with the capability of more than 100 cycles over a 5 year mission.

The study has shown that starshades are an extendable technology. Larger starshades can accommodate successively larger telescopes, freeing the telescope to adopt architectures such as segmented mirrors and central obstructions that are not available to other direct-imaging techniques. We took a detailed look at the technology necessary to make the New Worlds Observer a reality: most of this technology already exists. What is new for NWO is that these existing technologies have not been combined in this particular way before. Technology development for NWO can be rapidly implemented to lead to a flight program within a few years, with well-understood and controlled risks. The launch of a starshade can be envisioned to happen in as little as six years. The cost of a starshade-only mission is in the medium cost category. The cost of a 4 m UVOIR observatory is, unsurprisingly, in the flagship range.

The idea of a starshade is not new (Spitzer 1962), but eliminating light diffracting around an external occulter for imaging Earth-like planets has been impractical Marchal (1985); Copi & Starkman (2000). Recently, Cash (2006) found an apodization function that made such a system practical with todays technology. Shown in Figure 1, the starshade is an opaque screen that sits in the line of sight from the telescope to the star. If the starshade is sufficiently distant, it will subtend a small angle, blotting out the star's light while allowing the exoplanet light to pass unobscured past the edge. A detailed requirements document for the Starshade was generated. Our science goals are taken from the TPF STDT report, and the NWO mission is designed to fulfill and exceed those requirements.

Cash's offset hyper-Gaussian apodization function reduces diffraction by many orders of magnitude; Figure 3 shows the parameters of this apodization function. A starshade with $2(a + b) = 50$ m (the effective diameter), operating ≈80,000 km from a 4 m telescope is capable of 10^{10} starlight suppression within 50 mas for wavelengths from 0.1 to 1 μm. Our studies of the starshade in the past four years have shown that the optimal petal number for NWO is $P = 16$, a balance between starshade mass and shadow diameter. The hypergaussian parameter is optimized at $n = 6$ over the wavelength range. Four independent software codes have been developed to simulate starshade performance. Figure 4 shows the suppression efficiency of the baseline starshade design as a function of both shadow radius and angular offset for two representative wavelengths.

We have chosen to baseline a 4 m telescope for its ability to resolve the exoplanet from the background, and because it is the breakpoint for a monolithic mirror fitting inside existing launch vehicles and using existing facilities. While it is larger than HST, this telescope has roughly the same tolerances. The optical design of the telescope is straightforward. This is primarily because the starlight from the target stars will be fully suppressed, so there are no special requirements on the optical train as there are for internal coronagraphic techniques. For example, segmented mirrors and any mirror coating may be used. The primary mirror configuration is still being studied (monolith vs. segmented) but our current baseline is a monolith. A 4 m UV/Optical/near-IR telescope is within the state of the art for space telescopes.

We chose a modified Three Mirror Anastigmat (TMA) as the baseline optical design. This design allows wide-field imaging for General Astrophysics applications and a high-quality narrow field at the Cassegrain focus. The various instrument apertures are spread around the focal plane and the light is sent into a given instrument by steering the telescope in the manner of HST.

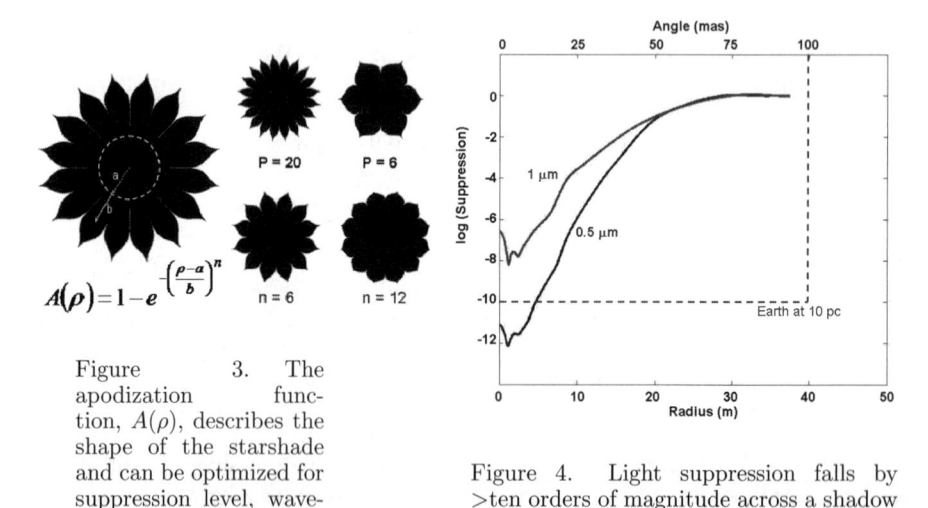

Figure 3. The apodization function, $A(\rho)$, describes the shape of the starshade and can be optimized for suppression level, wavelength range, shadow size, and IWA.

Figure 4. Light suppression falls by >ten orders of magnitude across a shadow radius of just 20 m, allowing observations of planets as close as 50 mas.

2. Key Science Goals

The science enabled by the New Worlds Observer is extensive and groundbreaking. With current and near-term technology, we can make great strides in finding and characterizing planets in the habitable zones (HZs) of nearby stars. The key science goals of NWO are: 1) discover dozens of Earth-like planets in the HZs of nearby stars with a total search completeness of 30; 2) characterize the planets we find using time-resolved photometry, spectroscopy, and polarimetry, giving us information such as atmospheric conditions, internal structure, mass estimates, and signs of life; 3) study other aspects of the extrasolar system including giant planets, planetesimal belts, and exozodiacal dust; and 4) conduct a large range of astronomical research ≈70% of the time, while the NWO starshade is moving from target to target.

Once exoplanets have been discovered, detailed observations such as time-resolved photometry, spectroscopy, and polarimetry will reveal the true nature of these planets and the systems in which they were born. The physical properties of exoplanets can be characterized using visible-band, reflected starlight which depends on the size of the planet, the distance between the planet and the star, the composition and structure of the planet's atmosphere and surface, the wavelength of the observation, etc.

Spectroscopy of terrestrial exoplanets will quickly reveal a wealth of information about the planet's atmospheric and surface conditions including habitability or even the presence of life. Water, carbon dioxide, oxygen, methane, ozone, and ammonia give the key signatures. Water is the necessary ingredient for the types of life found on Earth and it has played an intimate, if not fully understood, role in the origin and development of life on Earth. The presence of carbon-dioxide would indicate (1) that carbon is available for the biosphere, (2) a greenhouse effect, and (3) the possibility of climate regulation via carbon cycling between the atmosphere and hydro/geosphere. A large amount of oxygen in a terrestrial atmosphere would be extremely interesting; oxygen is so chemically reactive that it must be continuously produced at enormous rates to persist. O_2 in the Earth's atmosphere is the result of continuous input from the biosphere (Lovelock 1979).

A simulated spectrum of the Earth at 10 pc, viewed for 50 hours by NWO, is shown in Figure 5. All known sources of noise are included. Clearly visible in the spectrum is the rise to short wavelength, indicating Rayleigh scattering. Toward the red end are strong absorption features of water, indicative of oceans and clouds. Most exciting is the presence of biomarkers such as absorption lines from molecular oxygen and an absorption edge from ozone in the near ultraviolet. These features in the spectrum of the Earth arise solely as a byproduct of plant life.

An analysis of a planet's color, brightness variability, and spectrum provides an estimate of the planet's reflectivity, or albedo. From this, the planetary radius can possibly be derived as well as an estimate of its density (rocky planets tend to have much lower albedo than gas giants). This classification system provides a method to estimate planetary mass. While measuring the mass of the planet is an important parameter for detailed modeling, the most important information regarding habitability is gained through direct observation. Measurement of mass should follow planet detection and classification, as opposed to being a necessary first step.

The full suite of astrophysical techniques will be available for exoplanet observations. We can make rough measurements of atmospheric density from Raleigh scattering. Photometric monitoring could reveal surface variations for planets with relatively transparent atmospheres (Oakley, Cash & Turnbull 2008). A high-resolution spectrometer might be used to capture a detailed spectrum of a particularly interesting planet. Similarly, other general astrophysics (GA) instruments might be used to characterize planets in special circumstances.

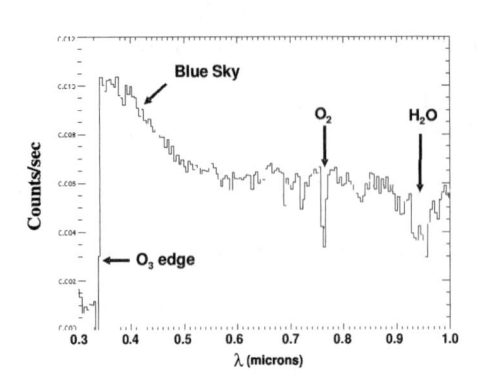

Figure 5. The spectrum of the Earth at 10 pc as seen by NWO. Note the prominent water and oxygen absorption lines and the ozone edge in the near UV.

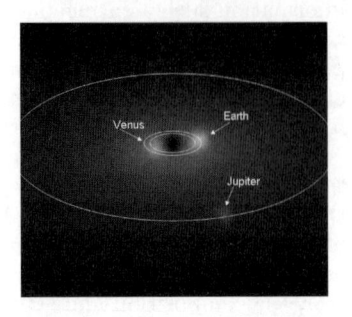

Figure 6. The exozodiacal light can help estimate the inclination of the system and therefore constrain planet orbits. This image shows a simulation of a hypothetical system with three planets Venus, Earth and Jupiter. The exozodiacal light has total brightness equal to our own, but is more spatially extended.

Since NWO will have a large field of view of ≈0.2 square arcminutes, we will discover outer planets and diffuse emission while searching the HZ of the star. The detection, characterization, and orbit determinations of gas and ice giants in the outer parts of planetary systems will provide important clues about the system's long-term dynamical evolution. NWO will provide reliable statistics on the presence of ice and gas giants in long-period orbits in mature planetary systems, estimates of disk lifetimes, etc. Given the parameters observable with NWO, it will be possible to differentiate between and constrain models of planet formation and evolution.

We must also carefully consider diffuse emission from interplanetary dust in the extrasolar systems. This exozodiacal dust (or "exozodi") is crucial, both for its science return and as a source of background noise.

The amount of exozodi is typically quantified by the fractional infrared luminosity (L_{IR}/L^*) which is proportional to the dust mass, though other factors like grain properties affect it. Currently known exozodi disks (better known as debris disks) have fractional infrared luminosity (Lovelock 1979) (L_{IR}/L^*) $\sim$ $10^3 - 10^5$. The zodiacal dust interior to our asteroid belt has (L_{IR}/L^*) $\simeq 10^7$, which we call 1 "zodi". We are not currently able to detect this amount of dust around other stars; this can only be done with high-contrast direct imaging. Since NWO has no outer working angle and produces zero distortions in the field, exozodiacal light and debris disks will be optimally imaged by this system.

The distribution of exozodiacal light is a sensitive tracer of the system's orbital dynamics. Planetary orbital resonances will be displayed as gaps and enhancements in the dust. Tiny planets, too small to be seen directly, should leave distinct marks. The observed dust distribution gives us critical information like the inclination of the system's ecliptic plane (Figure 6). By eye, one can place an ellipse over the system, estimating the orientation of the plane. Then, concentric ellipses may be drawn about the central star and those that pass through a planet show the orbit of that planet under the assumption of circularity. Exozodiacal light has the potential to give us a first estimate of the orbit of each planet from a single image. Revisits will determine other planet orbit parameters.

Zodiacal and exozodiacal dust also create a background flux that is mixed with the planet signal in both images and spectra. Even if nearby systems have exozodi levels no greater than the solar system level, the zodiacal and exozodiacal background will be the largest source of noise for most terrestrial planet targets, assuming the starlight is suppressed to $\sim 10^{-10}$. The surface brightness of the exozodi is the main factor controlling how long it takes to detect an exoplanet that is buried in it. We know very little about exozodi levels in nearby stellar systems. However, NWO is quite robust against the presence of many zodis of dust in the extrasolar system. A useful benchmark goal is S/N = 10 on an Earth-like planet in a solar system twin at 10 pc viewed at a inclination, which NWO can achieve in 3.3 hours. Even if there is 10 zodi in this system ($\sim$ 19 mag/arcsec^2 at the planet location), NWO can image the Earth twin in less than a day.

3. Summary of Findings

- NWO is capable of studying thirty complete habitable zones given a contrast limit of 10^{-11} and inner working angle of 50mas.
- Spectra of all major planets in a solar system twin at 10 pc can be obtained in less than 24 hours at $R \approx 100$ over a wavelength range of 0.1 to 1.1 μm, due to the high throughput of the NWO system.
- The wavelength range of NWO is capable of obtaining the ozone edge, water lines, and some methane lines, making the detection of life possible.
- The design of the telescope is independent of the design of the starshade. A diffraction-limited UV/O/IR telescope like HST can be used, with no additional constraints from the starshade. 70% of the telescope time is available for general astrophysics during the time that the starshade is traveling to the next target star.
- NWO does not require invention of new technology: all the elements of the technologies used exist, including the deployment and telescope-to-starshade alignment systems, but they need to be integrated into a working system.
- Starshades provide a versatile architecture. They may be expanded, extended, and upgraded for future missions.
- A starshade can effectively be used with the upcoming James Webb Space Telescope. This represents the fastest, lowest cost route to acquiring spectra of Earth-like planets.

References

Cash, W. 2006, Nat, 442, 51
Copi, C.J. & Starkman, G. 2000, ApJ, 532, 581
Lovelock, J. E., 1979, "A New Look at Life on Earth," (Oxford: Oxford University Press
Marchal, C. 1985, Acta Astronautica, 12, 195
Oakley, P., Cash, W. & Turnbull, M. 2008, SPIE, 7010, 70103Y
Spitzer, L. 1962, American Scientist, 50, 473
The NWO ASMCS Study Report & Appendices, http://newworlds.colorado.edu

Advanced Technology Large-Aperture Space Telescope (ATLAST): Characterizing Habitable Worlds

M. Postman,[1] W. A. Traub,[2] J. Krist,[2] K. Stapelfeldt,[2] R. Brown,[1] W. Oegerle,[3] A. Lo,[4] M. Clampin,[3] R. Soummer,[1] J. Wiseman,[3] and M. Mountain[1]

[1] *Space Telescope Science Institute, Baltimore, MD, USA*

[2] *Jet Propulsion Laboratory, California Inst. of Technology, Pasadena, CA, USA*

[3] *NASA/Goddard Space Flight Center, Greenbelt, MD, USA*

[4] *Northrop Grumman Aerospace Systems, Redondo Beach, CA, USA*

Abstract. The Advanced Technology Large Aperture Space Telescope (ATLAST) is a set of mission concepts for the next generation UV-Optical-Near Infrared space telescope with an aperture size of 8 to 16 meters. ATLAST, using an internal coronagraph or an external occulter, can characterize the atmosphere and surface of an Earth-sized exoplanet in the habitable zone of long-lived stars at distances up to ∼45 pc, including its rotation rate, climate, and habitability. ATLAST will also allow us to glean information on the nature of the dominant surface features, changes in cloud cover and climate, and, potentially, seasonal variations in surface vegetation. ATLAST will be able to visit up to 200 stars in 5 years, at least three times each, depending on the technique used for starlight suppression and the telescope aperture. More frequent visits can be made for interesting systems.

1. Overview of the ATLAST Concept

We are at the brink of answering two paradigm-changing questions: Do Earth-sized planets exist in the habitable zones (HZs) of their host stars? Do any of them harbor life? The tools for answering the first question already exist (e.g., Kepler, CoRoT); those that can address the second can be developed within the next 10-20 years (Kasting, Traub et al. 2009). ATLAST is our best option for an extrasolar life-finding facility and is consistent with the long-range strategy for space-based initiatives recommended by the AAAC Exoplanet Task Force (Lunine et al. 2008). ATLAST is a NASA Astrophysics Strategic Mission Concept for the next generation flagship UVOIR space observatory (wavelength coverage: 110 nm – 2400 nm), designed to answer some of the most compelling astronomical questions, including "Is there life elsewhere in the Galaxy?" The ATLAST team investigated two different observatory architectures: a telescope with an 8-m monolithic primary mirror and two variations of a telescope with a large segmented primary mirror (9.2 m and 16.8 m). The two architectures span the range in viable technologies: e.g., monolithic vs. segmented apertures, Ares V launch vehicle vs. EELV, and passive vs. fully active wavefront control. This approach provides several pathways to realize the mission that would be narrowed to one as the needed technology development progresses and the availability of launch vehicles is clarified.

While ATLAST requires some technology development, all the observatory concepts take full advantage of heritage from previous NASA missions, as well as technologies available for missions in development. The 8 m monolith architecture is similar to the Hubble Space Telescope (HST), although the optical design is different. The 8 m mirror has an areal density exceeding that of the HST mirror, providing superb stiffness and thermal inertia. The use of such a massive mirror provides excellent wavefront quality and is made possible given a launch vehicle with the capabilities of the proposed Ares V. The 9.2 m and 16.8 m segmented mirror concepts rely heavily on design heritage from the James Webb Space Telescope (JWST) and work at JPL and ITT (e.g. Hickey et al. 2002; Strafford et al. 2006) in development of lightweight optics, the OTA deployment mechanics, and wavefront sensing and control. The 9.2 m segmented concept can be launched on a Delta IV Heavy launch vehicle with a modified 6.5 m fairing; the 16.8 m concept requires an Ares V. The non-cryogenic nature of ATLAST makes the construction and testing of the observatory much simpler than for JWST. We have also identified departures from existing NASA mission designs to capitalize on newer technologies, minimize complexity, and enable the required improvements in performance.

Four significant drivers dictate the need for a large space-based telescope if one wishes to conduct a successful search for biosignatures on exoplanets. First, and foremost, Earth-mass planets are faint – an Earth twin at 10 pc, seen at maximum elongation around a G-dwarf solar star, will have V $\sim$ 29.8 AB mag. Detecting a biosignature, such as the presence of molecular oxygen in the exoplanet's atmosphere, will require the ability to obtain direct low-resolution spectroscopy of such extremely faint sources. Second, the average projected angular radius of the HZ around nearby F,G,K stars is less than 100 milli-arcseconds (mas). One thus needs an imaging system capable of angular resolutions of $\sim$10 to 25 mas to adequately sample the HZ and isolate the exoplanet point source in the presence of an exo-zodiacal background. Third, direct detection of an Earth-sized planet in the HZ requires high contrast imaging, typically requiring starlight suppression factors of 10^{-9} to 10^{-10}. Several techniques (c.f. Levine et al. 2009) are, in principle, capable of delivering such high contrast levels but all require levels of wavefront stability not possible with ground-based telescopes because the timescale of wavefront variations induced by the Earth's atmosphere is shorter than the time to measure the wavefront error to the required precision. A space-based platform is required to achieve the wavefront stability that is needed for such high contrast imaging.[1] Lastly, biosignature-bearing planets may well be rare, requiring one to search tens or even several hundred stars to find even a handful with compelling signs of life. The number of stars for which one can obtain an exoplanet's spectrum at a given SNR and in less than a given exposure time, scales approximately as D^3, where D is the telescope aperture diameter. This is demonstrated in Figure 1 where we have averaged

---

[1] Such stability is achievable regardless of the space telescope's aperture (within reasonable limits) – so this driver is primarily for the space locale of the telescope rather than its aperture.

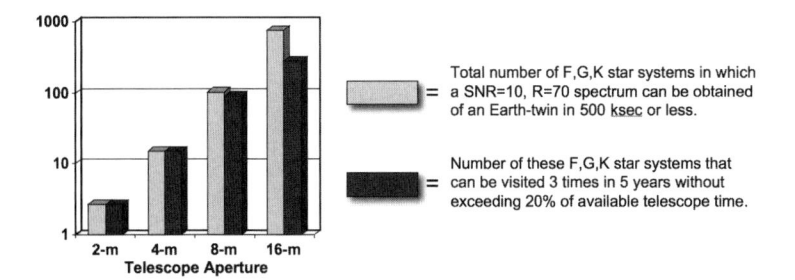

Figure 1. The number of F,G,K stars as a function of telescope aperture where an R=70 SNR=10 spectrum could be obtained of an Earth-twin in the HZ in less than 500 ksec. These counts assume every star has an Earth twin.

different simulations done using various starlight suppression options (internal coronagraphs of various kinds as well as an external occulter). To estimate the number of potentially habitable worlds detected, one must multiply the numbers in Figure 1 by the fraction of the stars that have an exoplanet with detectable biosignatures in their HZ ($\eta_\oplus$). The value of $\eta_\oplus$ is currently not constrained but it is not likely to be close to unity. One must conclude that to maximize the chance for a successful search for life in the solar neighborhood requires a space telescope with an aperture size of at least 8 meters.

All ATLAST concepts require many of the same key technologies. We believe these designs compose a robust set of options for achieving the next generation of UVOIR space observatory in the 2020 era. There are several fundamental features common to all our designs. All ATLAST concepts are designed to operate in a halo orbit at the Sun-Earth L2 point. The optical designs are diffraction limited at 500 nm (36 nm RMS WFE) and the optical telescope assembly (OTA) operates near room temperature (280 K – 290 K). All OTAs employ two, simultaneously usable foci: a three-mirror anastigmat channel for multiple, wide field of view instruments, and a Cassegrain channel for high-throughput UV instruments and instruments for imaging and spectroscopy of exoplanets (all designs have an RMS WFE of < 5 nm at < 2 arcsec radial offset from Cass optical axis). The ATLAST concept study is available on line and on the astro-ph archives (Postman 2009a,b).

2. Simulated Data and Exoplanet Characterization Performance

Estimates of the SNR of habitability and biosignature features in an Earth-twin spectrum, achievable with ATLAST, are shown in Table 1. For these calculations we use a fully validated model of the Earth's spectrum (Marais et al. 2002; Woolf et al. 2002), in combination with the observed visible reflection spectrum of the present Earth. We assume that the exoplanet is at maximum elongation and that the planet is observed for a length of time sufficient to achieve an SNR of 10, at a spectral resolution $R = 70$, in the red continuum. The vegetation signal is the enhanced albedo of the Earth from land plants, for wavelengths longer than 720 nm (Seager et al. 2005), with a modest SNR. Column 3 gives the width of the spectral feature. All of these SNR values can easily be improved with re-visits.

Table 1. Habitability and bio-signature characteristics

Feature	λ (nm)	$\Delta\lambda$ (nm)	SNR	Significance
Reference Continuum	750	11	10	
Rayleigh Scattering	<500	100	4	Protective atmosphere
Ozone (O_3)	580	100	5	Source is O_2; UV Shield
Oxygen (O_2)	760	11	5	Plants emit; Animals inhale
Cloud/Surface reflection	750	100	30	Rotation indicator
Land Plant reflection	770	100	2	Vegetated land area
Water vapor (H_2O)	940	60	16	Needed for life

In addition, ATLAST will allow us to glean substantial information about an exo-Earth from temporal variations in its features. Such variations inform us about the nature of the dominant surface features, changes in climate, changes in cloud cover, and potentially, seasonal variations in surface vegetation (Ford et al. 2003). Constraints on variability require multiple visits to each target. The 8-m ATLAST (with internal coronagraph) will be able to observe ~100 different star systems 3 times each in a 5-year interval and not exceed 20% of the total observing time available to the community. The 16-m version (with internal coronagraph) could visit up to ~250 stars three times each in a five-year period. The 8-m or 16-m ATLAST (with a single external occulter) can observe ~85 stars 3 times each in a 5-year period, limited by the transit times of the occulter. Use of two occulters would remove this limitation.

Figure 2 shows two simulated ATLAST spectra for an Earth-twin at 10 pc, one at R=100 and one at R=500, taken with sufficient exposure to reach SNR=10 at 750 nm in the continuum. A 3-zodi background was used (local plus exosolar). The R=100 exposure times are 46 ksec and 8 ksec, respectively, for an 8-m and 16-m space telescope. The corresponding exposure times for the R=500 spectrum are 500 ksec and 56 ksec, respectively, for the 8-m and 16-m telescopes. The reflected flux from an Earth-like rocky planet increases as $M^{2/3}$, where M is the exoplanet mass. Hence, the exposure times for a $5M_\oplus$ super-Earth would be ~3 times shorter. At both resolutions, the O_2 features at 680 nm and 760 nm are detected, as are the H_2O features at 720, 820, 940, 1130, 1410, and 1880 nm. Rayleigh scattering is detected as an increase in reflectivity bluewards of 550 nm. The higher spectral resolutions enabled by large-aperture space telescopes enable the detection of molecular oxygen in exoplanets with lower abundances than those on Earth and provide constraints on the kinematics and thermal structure of the atmosphere that are not accessible at lower resolution.

For a 16-m class space telescope, time-resolved spectroscopy over intervals of a few hours may reveal surface composition variations, if the planet is not cloud dominated, as the exoplanet rotates. However, even broadband photometry can be used to detect short-term variations in albedo that can determine the rotation period and constrain the amount of cloud cover. Ford et al. (2003) generated model light curves for the Earth over 6 consecutive days using data from real satellite observations. Photometric variations of 20 – 30% on timescales of 6 hours were typical in the B,V,R,I passbands. Their models are shown in Figure

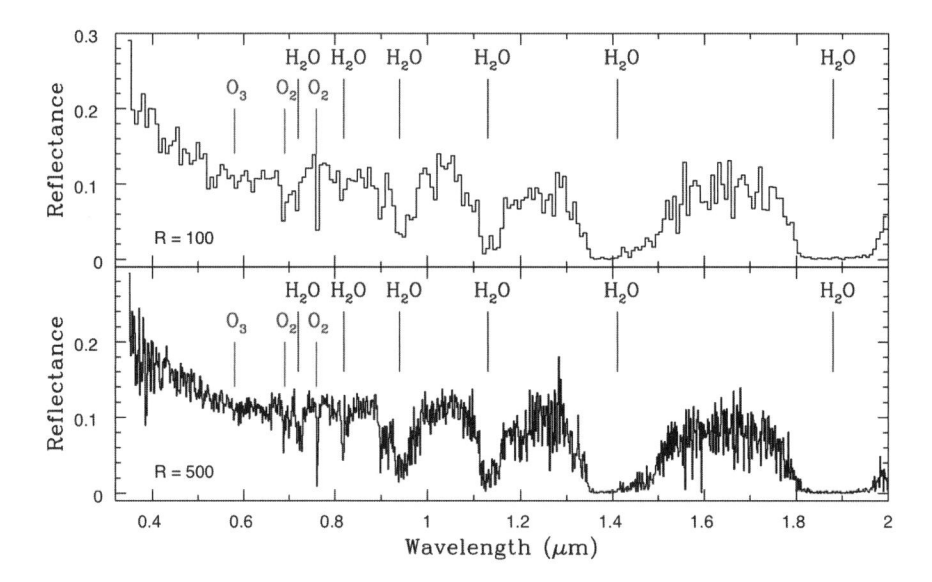

Figure 2. Simulated ATLAST spectrum of an Earth-twin at 10 pc, shown at R=100 in the upper plot and R=500 in the lower plot. The SNR=10 at 750 nm in both cases. Key O_2, O_3, and H_2O features are shown. The increased reflectance at the blue end of the spectrum is due to Rayleigh scattering. An 8-m ATLAST obtains the R=100 and R=500 spectra in 46 ksec and 500 ksec, respectively.

3. On top of these light curves we show grey bands whose height represents the ±5% uncertainty for SNR=20 broadband (R=4) photometry and whose length represents the time it would take to perform such an observation with a 4-m, 8-m, and 16-m space telescope. We show the results of such calculations for an Earth twin at 10 pc and 20 pc. The 4-m has marginal capability to study such photometric variations at 10 pc but telescopes with apertures of 8-m or larger would be able to perform the measurements well as the integration time is less than the typical period between significant albedo changes. At 20 pc, even an 8-m telescope reaches its limits but a 16-m telescope is still able to acquire the needed accuracy in photometry in less than 4 hours.

Transit spectroscopy with ATLAST will permit characterization of super-Earth mass exoplanets. Figure 4 shows two simulated ATLAST transit spectra for planets around a 6th magnitude M2 star where the orbital period in the HZ is ∼20 – 30 days. Such observations are time consuming but do not require the use of a coronagraph or occulter.

In summary, with ATLAST, we will be able to determine if HZ exoplanets are indeed habitable, and if they show signs of life as evidenced by the presence of oxygen, water, and ozone. ATLAST will also provide useful information on the column abundance of the atmosphere, the presence of continents and oceans, the rotation period, and the degree of daily large-scale weather variations.

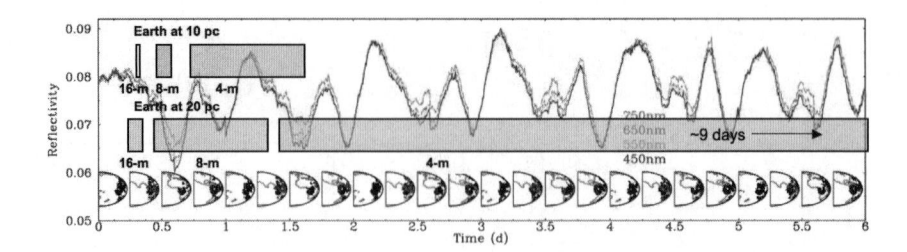

Figure 3. Model light curves of Earth for a six day interval (Ford et al. 2003). Superposed are grey bands whose length indicates the time required (in days) to achieve SNR=20 broadband photometry of an Earth twin at 10 pc (top bands) and 20 pc (bottom bands) for space telescopes with apertures of 4, 8, and 16 meters. The height of the grey bands corresponds to the ±5% photometric error.

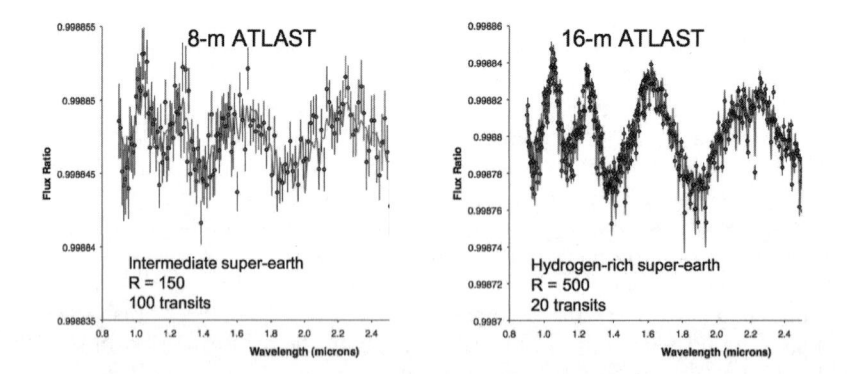

Figure 4. Simulated ATLAST transit spectra for two super-Earth exoplanets around a K=6 mag M2 star. Instrumental effects are modeled assuming JWST NIRSpec performance. The transit period is ∼22 days. The broad features are water absorption bands.

References

Des Marais, D. J., et al. 2002, Astrobiology, 2, 153
Ford, E. B., Seager, S., & Turner, E. L. 2003, in ASP Conf. Series 294, Scientific Frontiers in Research on Extrasolar Planets, eds. D. Deming & S. Seager, (ASP: San Francisco), 639
Hickey, G. S. et al., 2002, SPIE, 4849, 63.
Kasting, J., Traub, W. et al. 2009, arXiv:0911.2936
Levine, M., et al. 2009, arXiv:0911.3200
Lunine, J. I., et al., 2008, arXiv:0808.2754
Postman, M. 2009a, ATLAST website: http://www.stsci.edu/institute/atlast
Postman, M. 2009b, arXiv:0904.0941
Seager, S., Turner, E. L., Schafer, J., & Ford, E. B. 2005, Astrobiology, 5, 372
Strafford, D. N. et al., 2006, SPIE, 6273, 0R
Woolf, N. J., Smith, P. S., Traub, W. A., & Jucks, K. W. 2002, ApJ, 574, 430

Pathways Towards Habitable Planets
ASP Conference Series, Vol. 430, 2010
Vincent Coudé du Foresto, Dawn M. Gelino, and Ignasi Ribas, eds.

Dilute Aperture Visible Nulling Coronagraphic Imager (DAViNCI)

M. Shao and B. M. Levine

California Institute of Technology, Jet Propulsion Laboratory, Pasadena, CA, USA

1. Introduction

The DAViNCI concept was one of the concepts studies funded in 2008 by NASA in its call for future science missions. It is unique in its ability to search over 100 candidate stars for Earth clones and measure the planet spectra at a cost that is dramatically less expensive than competing approaches. DAViNCI is a multiple telescope interferometer composed of four 1.1 m telescopes on a two position baseline that uses a nulling interferometer/coronagraph to achieve an inner working angle of ~40 milliarcsec (mas) at its longest baseline (Figure 1). With collecting area roughly equal to the Hubble telescope, it will detect an Earth-Sun at 10 parsecs in approximately 10 hours of integration. DAViNCI has an inner working angle (IWA) equal to a filled aperture coronagraph with an 8m coronagraph (IWA ~ 2 λ/D) and an exoplanet science capability of a ~ 7m coronagraph. With a wavelength range from 550–1700 nm, it will measure spectra at R ~ 80 and SNR=10 on a number of critical biomarkers (Shao et al. 2009a).

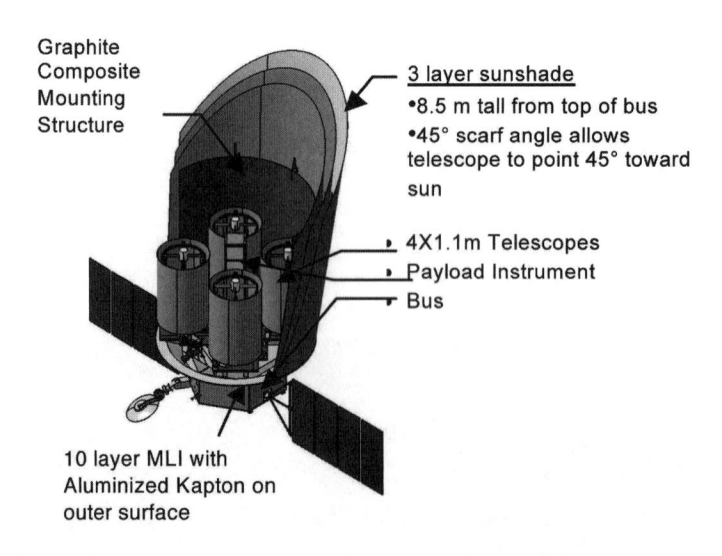

Figure 1. Principal components of the DAViNCI observatory.

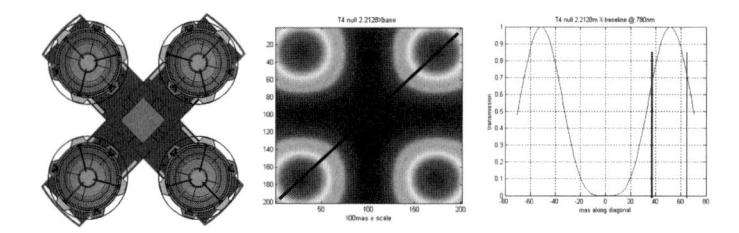

Figure 2. (Left) Downward view of DAViNCI at maximum baseline 2.2m. (Center) The transmission of this four telescope array configuration. (Right) A diagonal slice through the transmission pattern showing an inner working angle of 38 mas.

DAViNCI is a nominal 5-year mission to be launched on an Atlas V 521 in an earth-trailing orbit, with its payload fitting within a 5m fairing. The mission cost estimated by JPL TEAM-X is $1.2B. Costs have been moderated through the maximum use of existing commercial technology, namely, ITT Space Systems LLC NextView telescope, Northrop-Grumman Aerospace Systems thermal control systems and spacecraft, and SIM laser metrology. DAViNCI's other major partnership includes Lockheed-Martin for instrument design and construction. Technology will be developed to demonstrate the high contrast imaging needed for imaging Earth-like planets, which is 10^{-10} contrast over a 2as field of view. Future technology development will produce components such as 1000 element segmented deformable mirrors and 1000 subaperture single mode fiber arrays (Shao et al. 2009a). These will be integrated into the APEP test bed for technology readiness demonstrations.

2. Instrument and Mission Overview

DAViNCI is a phased dilute aperture telescope array with a nulling interferometer, imager, and spectrometer designed to detect and characterize extrasolar planets performing imagery and R ~ 80 spectroscopy over the wavelength range from $\lambda = 550$ to 1700nm. All major optical components are reflective in the nulling interferometer and beam train (except for a few components in the nulling interferometer and in the spectrometer). This affords wavelength coverage and simplicity of design and alignment (Shao et al. 2009a).

Four afocal collection telescopes form the dilute aperture along with a compressor telescope that feeds the instrument. The 110 mm diameter parallel beam exiting each collector telescope is directed to the science instrument (SI) by four flat mirrors to the smaller shared compressor telescope. In each of the four paths, the first two of the four flats are delay line mirrors for wavefront phasing, the third mirror is a fine steering mirror for line of sight control and stabilization, and the fourth flat mirror directs the beams into the compressor telescope aperture. The geometry of these sixteen flat mirrors compensates polarization, permits phase and pointing control, and permits modification of the aperture baseline in-orbit with simple linear motions of the collector telescopes (Figure 2, left) (Shao et al. 2009a).

3. Team X Study (Shao et al. 2009a)

JPLs TeamX was used to do the cost estimation of DAViNCI, which is categorized as a class A mission. Team X guidelines for this study were to provide independent design and costing analysis for each mission concept. Project-provided designs were used, but not project-provided cost estimates. The cost estimates summarized in this document were generated as part of a Pre-Phase-A preliminary concept study, are model-based, and were prepared without consideration of potential industry participation. The accuracy of the cost estimate is commensurate with the level of understanding of the mission concept, typically Pre-Phase A, and should be viewed as indicative rather than predictive. Team X typically adds 30% reserves for development (Phases A-D) and 15% for operations (Phase E). Project Cost is roughly $1.2B FY09 with appropriate contingencies, including five years of science operations. The cost breakdowns (rounded to two significant digits) are can be seen in Table 1.

Table 1. Team X DAViNCI mission cost summary

Item	Cost ($M FY09)*	Notes
Management, Systems Engr., Mission Assurance	84	
Payload System – Instrument	390	1
Flight System	190	
Mission Ops Preparation/Ground Data System	58	
Launch Vehicle	170	Atlas V 521
Assembly, Test, Launch Operations	23	
Science	40	
Education and Public Outreach	8	
Mission Design	7	
Reserves	230	
Total Project Cost	1200	

Notes:
* Individual WBS elements have been rounded to 2 significant digits.
1. Payload system includes instrument.
2. Four telescopes, metrology package, thermal shield, mounting structure.

4. DAViNCI Science

Direct imaging of an exo-Earth can be used to measure a number of important planetary parameters, 1) planetary orbit, is it in the habitable zone? 2) planet flux versus orbital phase, is the planet a Lambertian scatterer? 3) spectra of its atmosphere in the visible and near-IR, is there oxygen or water in the atmosphere? and 4) possible seasonal changes in albedo, and spectra. If the IWA is 2 × smaller, on average we will be able to detect Earth-like planets around stars twice as far away, or roughly an 8 fold increase in the number of targets (Figure 3).

The major advantage of DAVINCI is that it has an inner working angle of ~ 38 mas at 780 nm (oxygen line) equal to an 8 m coronagraph working at 2 λ/D. The dilute aperture approach breaks the cost paradigm of large monolithic

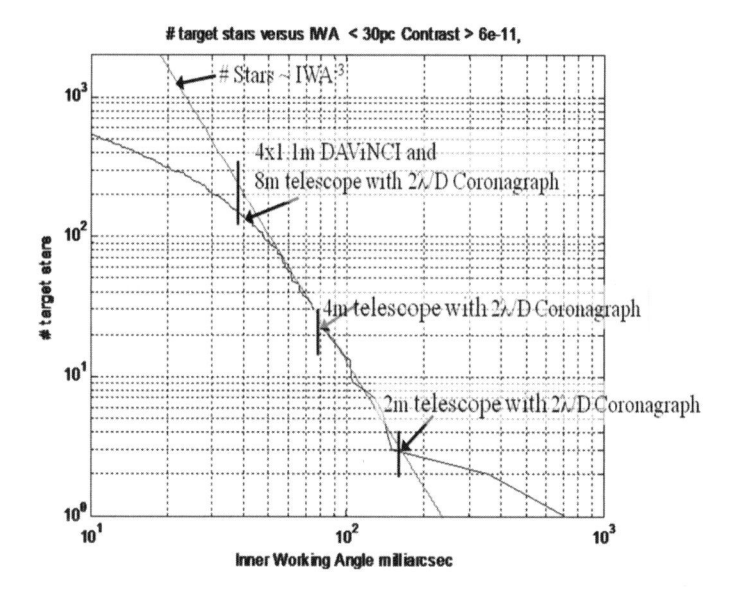

Figure 3. The relationship between coronograph IWA and the number of potential target stars. IWA and planet orbit is for a maximum star-planet separation > 1.2 .

mirrors, requiring four telescopes of modest aperture (1.1m ITT, Space Systems LLC NextView-class). The instrument includes a R ~ 80 spectrometer with spectral range of 0.5μm to 1.6μm. The size of this range requires detection of discrete wave bands each of approximately 25% bandwidth.

When a potential exo-Earth is detected, we first want to know, "is this an Earth?" and "is it in the habitable zone?" Measuring the orbit of a planet in our solar system is pretty straight forward because we can observe the planet over approximately 90% of its orbit. In Figure 4, the largest circle is the IWA and the small grey arc segments are the parts of the orbit when the planet is observable. The planet is not always observable even outside the IWA, when the bright side of the planet is facing away from us. With a coronagraph whose IWA is only slightly smaller than the maximum star-planet separation, some orbital parameters like orbit inclination cant be measured. The planets apparent brightness can vary by a factor of 3 from the "full moon" phase to the half moon phase. In multiple planet systems two images with one planet in each image leaves open the possibility that there are two separate planets only one of which is outside the IWA at a time. As the IWA becomes substantially smaller (e.g. 50%) of the maximum star-planet separation, the planet becomes observable over most of its orbit (Figure 4). In this case it will be possible to look for seasonal variations in albedo. Such variations may be because the surface is a non-lambertian scatterer. But there could also be real seasonal variations, such as the top layer of a water world freezing over in winter time (Shao et al. 2009b).

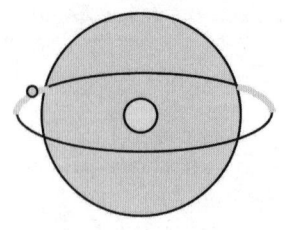

Figure 4. Observable IWA and planet orbit.

5. Technology Needs and Status

The technology development tasks are divided into 1) wide band starlight suppression and demonstration of high contrast imaging using the VNC over an extended field of view, and 2) the development of segmented deformable mirrors and fabrication of coherent single mode fiber arrays. Previous development with a nulling interferometer and a single optical fiber resulted in demonstration of 10-9 equivalent white light nulls and 10-10 laser light nulls. These levels are sufficient for imaging exo-Jupiters and exo-Earths, respectively (Samuele et al. 2007). Future demonstrations will be made on the APEP test bed. It will be capable of imaging over an extended field of view and also be capable of demonstrating all technical milestones necessary to retire technical risk for DAViNCI

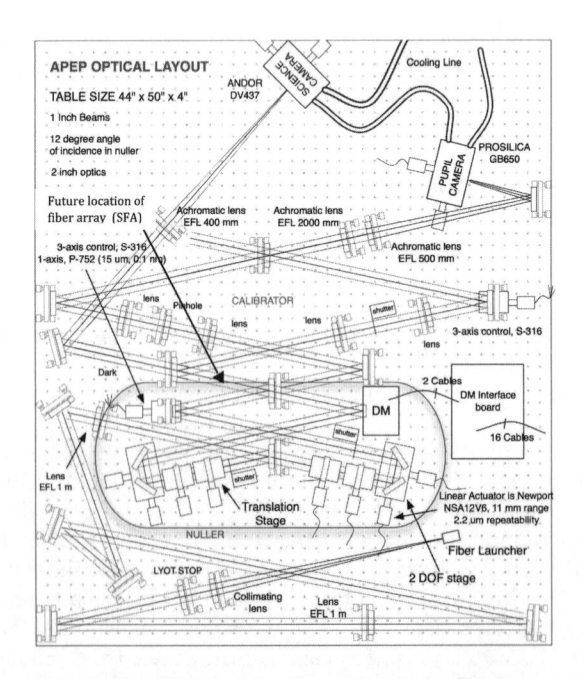

Figure 5. JPL Nulling Test bed for demonstration of high contrast imaging using a nulling interferometer, DM, fiber array, and with a post-coronagraphic wavefront sensor.

Figure 6. 331 Segmented DM technology demonstrator.

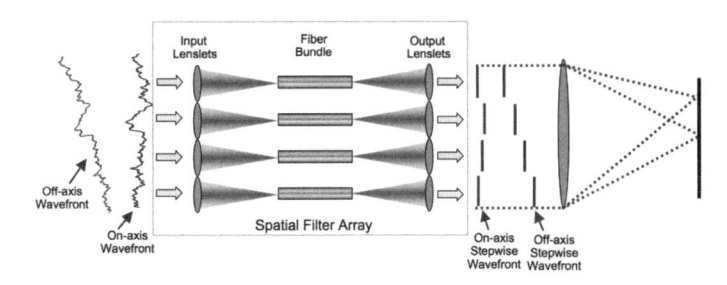

Figure 7. Principal of a spatial filter array. Input wavefront and amplitude aberrations are converted into intensity variations and stepwise phase on output. (Figure from R. Lyon, NASA/GSFC).

(Figure 5). Not mentioned is the use of the wavefront Calibrator to sense and to eliminate residual speckles in the image plane. See Shao (2007) for further description and explanation.

Under SBIR funding, Boston Micromachines Corporation (BMC) has delivered a 331 segment deformable device (Figure 6 and Stewart et al. 2007). They have demonstrated stroke of over 1.5mm, tilts of over 5mrad, all with segment flatness to 4m radius of curvature and 5nm rms roughness. To achieve a deep null, both the amplitude and phase of the interfering wavefronts must be matched to high tolerance in the pupil plane. A deformable mirror (DM) is placed in one arm of each of the nulling interferometers. At the output of the nuller, we use an array of single mode fibers (Figure 7). Within each fiber the electric field is fully specified by two numbers, the phase and amplitude. Each nuller uses a segmented DM, with piston as well as tip/tilt control of each segment. Piston motion controls the phase of the light, while tip/tilt changes the coupling of the light into the single mode fiber. This combination allows us to control both amplitude and phase of the interfering beams in the fiber. At the output of the single mode fiber array, the light from each fiber is collimated. The array of collimated beams is then focused onto a detector (or spectrometer slit). The fiber array is called coherent because the optical paths through all the fibers are matched to 1/4 of a wave. Fiberguide Industries is currently working to realize an array with 217 fibers. Future development will work toward realizing 1000 segment deformable mirrors with comparable 1000 channel fiber arrays.

6. Summary

DAViNCI is a phased 4 × 1.1m telescope dilute aperture with a nulling coronagraph front end to imaging camera and an R ~ 80 spectrometer. DAViNCI fits into a 4.5m fairing and at maximum baseline, it is capable of achieving an IWA of 38mas at 780nm wavelength. This compact instrument has an IWA that yields a potential target list of around 150 stars. The spectral range of the spectrometer goes from the 0.5μm-1.6μm which provides on the order of 20 targets.

Although light suppression has been demonstrated at a level sufficient (10^{-9} to 10^{-10} equivalent) for imaging exo-earths, a full imaging demonstration has begun to achieve the required contrast levels over an extended field of view. In order to achieve this, development is required to produce coherent fiber arrays and segment deformable mirror components with 1000 channels. Vendors have been found and are working to realize these components. In addition a new approach to speckle subtraction is in development that reduces the requirement on telescope stability without the need for extreme telescope stability over long integration times.

DAViNCI is a mission that has the potential for realization at a lower cost than a comparable coronagraph behind a 7m – 8m telescope working at a $2\lambda/D$ IWA. The nulling coronagraph architecture is also compatible with large diameter on-axis and segmented telescopes. So in the absence of a dedicated planet-finding mission, a nulling coronagraph instrument could be flown on a large general purpose telescope to meet high contrast imaging requirements and at a cost that is inexpensive when compared to addition, say, of an external occulter.

Acknowledgments. The authors would like to acknowledge the contributions of the entire DAViNCI ASMCS team. The individuals are listed in Shao et al. 2009a,b. This work was performed at the Jet Propulsion Laboratory, California Institute of Technology, under contract to the National Aeronautics and Space Administration.

References

Samuele, R., Wallace, J. K., Schmidtlin, E., Shao, M., Levine, B. M. & Fregoso , S. 2007, IEEE Aerospace Conference, 1, 10.1109/AERO.2007.353078
Shao, M. 2007, Comptes Rendus Physique, 8, 340
Shao, M. et. al. 2009a, DAVINCI: Dilute Aperture Visible Nulling Coronagraphic Imager, Astro2010 Request for Information White Paper 36, http://sites.nationalcademies.org/BPA/BPA_049855
Shao, M. et. al. 2009b, Direct Detection and Spectroscopy of Exo-Earths; The Need for High Angular Resolution and Other Observational Requirements, Science White Paper 68 submitted to the Astro2010 review Committee, http://sites.nationalcademies.org/BPA/BPA_050603
Shao, M. et. al. 2009c, Technology Development Toward Imaging and Spectroscopy of Exo-planets via Visible Nulling Coronagraphic Techniques, Astro2010 Technology White Paper 52, http://sites.nationalcademies.org/BPA/BPA_049522
Stewart, J. B., Bifano, T. G., Cornelissen, S., Bierden, P., Levine, B. M. & Cook, T. 2007, Sensors & Actuators, 138, 230

ACCESS: A Concept Study for the Direct Imaging and Spectroscopy of Exoplanetary Systems

J. Trauger,[1] K. Stapelfeldt,[1] W. A. Traub,[1] J. Krist,[1] D. Moody,[1]
E. Serabyn,[1] D. Mawet,[1] L. Pueyo,[1] S. Shaklan,[1] C. Henry,[1] P. Park,[1]
R. Gappinger,[1] P. Brugarolas,[1] J. Alexander,[1] V. Mireles,[1] O. Dawson,[1]
O. Guyon,[2,3] J. Kasdin,[4] B. Vanderbei,[4] D. Spergel,[4] R. Belikov,[5]
G. Marcy,[6] R. Brown,[7] J. Schneider,[8] B. Woodgate,[9] G. Matthews,[10]
R. Egerman,[10] P. Voyer,[10] P. Vallone,[10] J. Elias,[10] Y. Conturie,[10]
R. Polidan,[11] C. Lillie,[11] C. Spittler,[11] D. Lee,[11] R. Hejal,[11]
A. Bronowick,[11] N. Saldivar,[11] M. Ealey,[12] and T. Price[12]

Abstract. ACCESS is one of four medium-class mission concepts selected for study in 2008/9 by NASA's Astrophysics Strategic Mission Concepts Study program. In a nutshell, ACCESS evaluates a space telescope designed for extreme high-contrast imaging and spectroscopy of exoplanetary systems. An actively-corrected coronagraph is used to suppress the glare of diffracted and scattered starlight to the levels required for exoplanet imaging. The ACCESS study asks: What is the most capable medium-class coronagraphic mission that is possible with telescope, instrument, and spacecraft technologies available today?

1. Overview

Our science objective is the direct observation of exoplanetary systems, possibly dynamically full, that harbor exoplanets, planetesimals, and dust/debris structures. Direct coronagraphic imaging at visible (450–900 nm) wavelengths and low-resolution (R=20) spectroscopy of exoplanet systems in reflected starlight

---

[1] Jet Propulsion Laboratory, Caltech, Pasadena, CA, USA

[2] Subaru Telescope, Hilo, HI, US

[3] University of Arizona, Tucson, AZ, USA

[4] Princeton University, Princeton, NJ, USA

[5] NASA Ames Research Center, Moffett Field, CA, USA

[6] University of California, Berkeley, CA, USA

[7] Space Telescope Science Institute, Baltimore, MD, USA

[8] Paris Observatory, Meudon, France

[9] NASA Goddard Space Flight Center, Greenbelt, MD, USA

[10] ITT Space Systems Division, Rochester, NY, USA

[11] Northrop Grumman Corporation, Redondo Beach, CA, USA

[12] Northrop Grumman Xinetics, Devens, MA, USA

Figure 1. The ACCESS observatory, an actively corrected coronagraphic space telescope for the study of exoplanetary systems.

enables a broad science program that includes a census of nearby known RV planets in orbits beyond ∼1AU; a search for mature exoplanet systems beyond the RV survey limits including giant planets, super-Earths, and possibly a dozen Earth-mass planets; observations of debris structures as indicators of unseen planets and planetesimals; and imaging of dust structures in circumstellar environments as a probe of the life cycle of planetary systems from young stellar objects to proto-planetary nebulae.

The ACCESS study compares the performance and readiness of four major coronagraph architectures. ACCESS defines a conceptual space observatory platform as the "level playing field" for comparisons among coronagraph types. And it uses laboratory validation of four representative coronagraph types as a second "level playing field" for assessing coronagraph hardware readiness. The "external occulter" coronagraph is not considered here, on the presumption that a concept requiring two spacecraft is beyond the bounds of a medium-class mission. ACCESS identifies a genre of scientifically compelling mission concepts built upon mature subsystem technologies, and evaluates science reach of a medium-class coronagraph mission.

2. Performance Assessment

The observatory architecture represents the "best available" for exoplanet coronagraphy within the scope (cost, risk, schedule) of a NASA medium-class mission. Visible wavelengths (450–900 nm) are selected for a minimum inner working angle (IWA). All coronagraphs require an observatory system with exceptional pointing control and optical stability, with deformable mirrors (DMs) for active wavefront control. ACCESS requires systems with high technology readiness (near or above TRL6) for reliable estimates of science capabilities and reliable determinations of cost and schedule. The baseline observatory architecture defines a capable platform for meaningful comparisons among coronagraph types.

The ACCESS observatory (Figure 1) is comprised of a Gregorian telescope with an unobscured 1.5 meter diameter aperture, end-to-end system design for alignment stability, thermal isolation of the telescope secondary mirror and all downstream optics, an precision pointing control system, and an actively-corrected coronagraph for the suppression of diffracted and scattered light. The observatory orbits at L2 halo for a baseline mission of five years.

Figure 2. The development of the monolithic PMN deformable mirrors. From left to right: a 32 × 32 mm array (1024 actuators), of the type used for all HCIT demonstrations to date; a 64 × 64 mm array (4096 actuators) first installed on HCIT in 2009; a 48 × 48 mm array (2304 actuators) to be used to demonstrate TRL6 flight-readiness; and the 48 × 48 array on the JPL shake table.

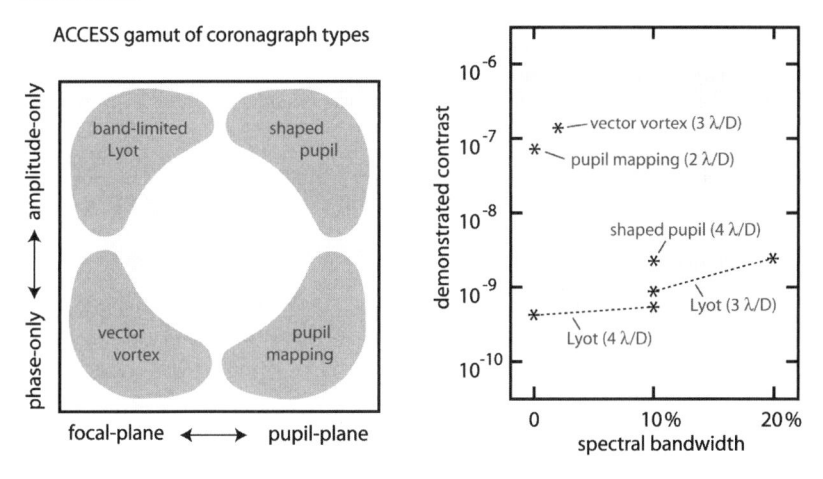

Figure 3. *Left:* The coronograph types in the ACCESS study. *Right:* The best contrast demonstrated in the laboratory to date (September 2009) (see text for details).

High-order wavefront control is provided by a pair of deformable mirrors. The evolution of precision deformable mirrors based on monolithic PMN electroceramic actuator arrays is illustrated in Figure 2. Mirror facesheets are fused silica, with surfaces polished nominally to $\lambda/100$ rms. Surface figure is settable and stable (open loop) to 0.01 nm rms over periods of 6 hours or more in a vacuum testbed environment. All DMs have been manufactured and delivered to JPL by Xinetics Inc.

The gamut of coronagraph types in the ACCESS study is indicated in Figure 3 (at left). The four major coronagraph types perform starlight rejection with combinations of phase and amplitude elements placed in focal and pupil planes.

The best demonstrated laboratory contrast to date (September 2009) for each type is plotted in Figure 3 (at right), as follows. Lyot data at $4\lambda/D$ are TPF performance milestones demonstrated on the High Contrast Imaging Testbed

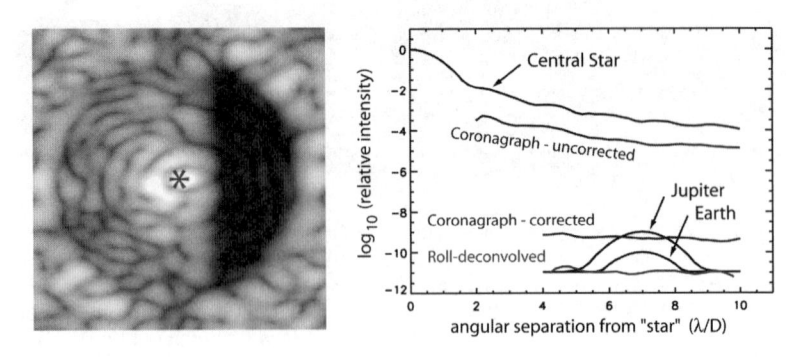

Figure 4. *Left:* The high-contrast dark field (D-shaped) created by a single DM in the laboratory experiments. *Right:* A comparison of the azimuthally averaged PSFs of (a) the star with focal plane mask offset and Lyot stop in place; (b) the coronograph field with all DM actuators set to equal voltages; (c) the coronograph with DM set for a dark half-field and (d) the result of simulated roll deconvolution with the set of 480 consecutive coronograph images. PSFs of a nominal Earth and Jupiter and also indicated (Trauger & Traub 2007).

(HCIT) (Trauger et al. 2006, 2007; Kern et al. 2008) with band limited masks (Kuchner & Traub 2002). Lyot data at $3\lambda/D$ were achieved on the HCIT in the course of the ACCESS study with hybrid Lyot masks (Moody et al. 2008). Shaped pupil (Spergel 2000) data were obtained on the HCIT with masks designed at Princeton (Belikov et al. 2004). Vortex result was demonstrated on the HCIT during the ACCESS study with a vector vortex mask (Mawet et al. 2010). The result for pupil mapping (Guyon et al. 2006) came from the new Ames testbed (Belikov et al. 2009). We note that post-observation data processing methods can be expected to improve the threshold for exoplanet detection by an order of magnitude compared to the raw contrast values plotted in Figure 3, for all coronograph types, and as illustrated in Figure 4 for the case of a Lyot coronograph. We further note that significant improvements are expected in the coming months and years as an outcome of active laboratory developments with well-understood technologies.

Coronograph contrast and stability have been demonstrated in the laboratory at the levels required to detect exoplanets. Figure 4 shows the high-contrast dark field (D-shaped) created by a single DM in laboratory experiments (a pair of DMs clears a full, two-sided dark field). At right in the figure is a comparison of azimuthally averaged PSFs of (a) the star, with focal plane mask offset and Lyot stop in place; (b) the coronograph field with the DM set to a flat surface figure; (c) the coronograph with DM set for a dark half-field; and (d) the result of simulated roll deconvolution with the set of 480 consecutive coronograph images. PSFs of a nominal Earth and Jupiter are also indicated (Trauger & Traub 2007).

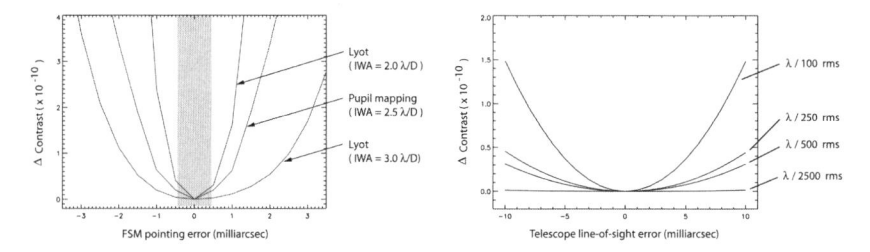

Figure 5. *Left:* Contrast deltas at the IWA for representative coronographs designed for 2.0, 2.5 and 3.0 λ/D. *Right:* Contrast deltas (vs. rms surface figure of the optical elements following the primary mirror) due to beamwalk on the optics upstream of the fine steering mirror.

Structural and thermal models guide the observatory design and inform the optical performance models with estimates of structure dynamics, vibration isolation, pointing control, thermal gradients across the primary mirror and forward metering structures, alignment drift in response to telescope slews and roll.

Telescope body pointing (i.e., line of sight) is stabilized to 1 milliarcsec (3 σ) with an active jitter control system. Figure 5 shows the contrast deltas (vs. rms surface figure of the optical elements following the primary mirror) due to beamwalk on the optics upstream of the fine steering mirror. The telescope attitude control system, augmented by a fine steering mirror within the coronagraph, stabilizes the star image on the coronagraph occulting mask (all four coronagraphs have an occulting mask) to 0.45 milliarcsec (3 σ), as required for high contrast at inner working angles as small as 2 λ/D (Figure 5).

3. Science Program

A baseline minimum science mission has been developed in terms of end-to-end optical models (e.g., Krist 2007) that incorporate the baseline observatory architecture and laboratory-validated estimates of coronagraph performance. A number of results are collected here.

Figure 6 depicts the ACCESS discovery space, which lies above the labeled curve at lower right in the diagram. A 1.5 meter coronagraph in space offers significant contrast advantages over even the largest current and future observatories on the ground.

Figure 7 gives two representations of the completeness in an ACCESS survey for exoplanets. At left are the detections of Jupiter-twins within 45 ° of elongation from their parent stars, to S/N = 10, using the ACCESS Lyot coronagraph with an IWA = 2 λ/D for a number of integration times. Note that the probability that an exoplanet will have a star-planet separation greater than that at 45 ° elongation is 50% or more. At right, the number of planets, in four mass categories, detectable to S/N = 10 in integration times of one day or less, using the ACCESS Lyot coronagraph with an IWA = 2.5 λ/D.

Figure 8 tabulates of the number of nearby stars that could be searched with various ACCESS coronagraphs to the depth of 10-σ detections of Jupiter

380   Trauger et al.

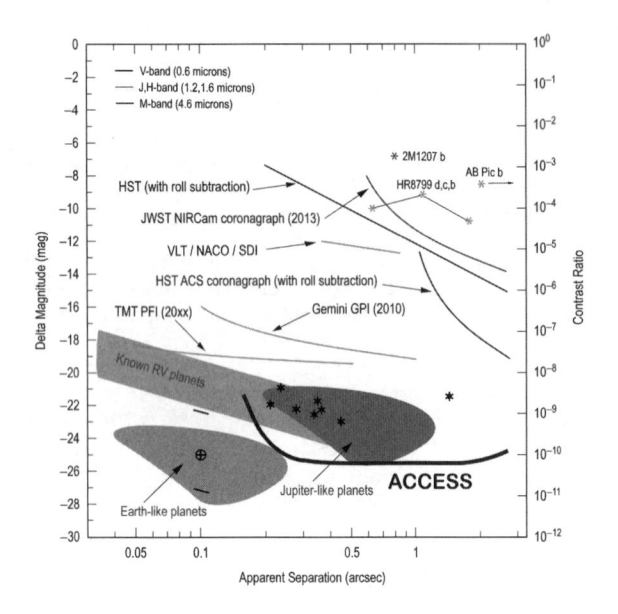

Figure 6. The ACCESS discovery space. Sensitivity for exoplanet detections is compared with current and future observatories in terms of brightness relative to the central star vs. apparent separation. Known exoplanets are shown as asterisks. Shaded areas indicate the regions of high probability of detecting planets orbiting the nearest 100 AFGK stars (for Jupiter-twins in 5AU orbits and Earth-twins in 1 AU orbits, respectively). The detection range for ACCESS is the area above the bold curve at bottom right.

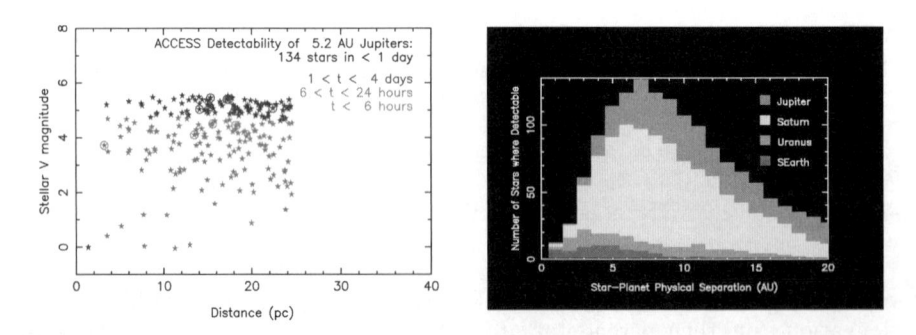

Figure 7. Two representations of the completeness in an ACCESS survey for exoplanets. *Left:* Detections of Jupiter-twins within 45deg of elongation from their parent stars to S/N = 10, using the ACCESS Lyot coronagraph with an IWA = 2 λ/D for a number of integration times. *Right:* The number of planets, in four mass categories, detectable to S/N = 10 in integration times of one day or less using the ACCESS Lyot coronagraph with an IWAof 2.5 λ/D.

ACCESS 381

Table 1: Number of nearby stars that can be surveyed for 5.2 AU Jupiters, IWA 3.0 λ/D

Coronagraph Type	Planet 45° from max elong	Planet 15° from max elong
→ Lyot	117	175
PIAA	166	278
Vortex	135	204

Table 2: Number of nearby stars that can be surveyed for 5.2 AU Jupiters, IWA 2.5 λ/D

Coronagraph Type	Planet 45° from max elong	Planet 15° from max elong
Lyot	153	218
PIAA	178	267
Vortex	154	228

Table 3: Number of nearby stars that can be surveyed for 5.2 AU Jupiters, IWA 2.0 λ/D

Coronagraph Type	Planet 45° from max elong	Planet 15° from max elong
Lyot	170	230
Vortex	164	241

Note: Accurate PIAA wavefront control solution not available for IWA = 2.0 λ/D

Figure 8. The number of nearby stars that could be searched with various ACCESS coronagraphs to the depth of 10-σ detections of Jupiter-twins in each of six visits to the star over a period of 2.5 years. The arrow corresponds to the ACCESS minimum science program based on current demonstrated technologies. Ongoing developments are expected to bring the demonstrated readiness of other coronagraph configurations to the search sensitivities shown in the table.

twins in each of six visits to the star over a period of 2.5 years. The row indicated by the arrow is an estimate based on coronagraph performance demonstrated in the laboratory at 3.0 λ/D with the Lyot coronagraph. The other rows represent coronagraph performance that may be achieved with further development of known technologies in the near future. The column for 45° from maximum elongation corresponds to an observational completeness of 50% or more in each visit, approaching 100% after six epochs spread over several years.

4. Summary

The ACCESS study has considered the relative merits and readiness of four major coronagraph types, and hybrid combinations, in the context of a conceptual medium class space observatory.

Using demonstrated high-TRL technologies, the ACCESS minimum science program surveys the nearest 120+ AFGK stars for exoplanet systems, and surveys the majority of those for exozodiacal dust to the level of 1 zodi at 3 AU. Discoveries are followed up with R=20 spectrophotometry.

Ongoing technology development and demonstrations in the coming year are expected to further enhance the science reach of an ACCESS mission, in advance of a NASA AO for a medium class mission. The study also identifies areas of technology development that would advance the readiness of all major coronagraph types in the coming 5 years.

Acknowledgments. The research described in this paper was carried out at the Jet Propulsion Laboratory, California Institute of Technology, under a contract with the National Aeronautics and Space Administration.

References

Belikov, R., et al. 2009, in Techniques and Instrumentation for Detection of Exoplanets IV, ed. Shaklan, S. B., SPIE, 7440, 74400J
Belikov, R., Kasdin, N.J., & Vanderbei, R.J. 2006, ApJ, 652, 833
Giveón, A., et al. 2007, in Astronomical Adaptive Optics Systems and Applications III, ed. Tyson, R. K. & Lloyd-Hart, M., SPIE, 6691, 66910A
Guyon, O., et al. 2005, ApJ, 622, 744
Guyon, O., et al. 2008, in Space Telescopes and Instrumentation 2008: Optical, Infrared, and Millimeter, ed. Oschmann, et al., SPIE, 7010, 70102Z
Kern, B., et al. 2008, JPL Document D-60951
Krist, J.E. 2007, in Optical Modeling and Performance Predictions III, ed. Kahan, M. A., SPIE, 6675, 66750P
Kuchner, M.J., & Traub, W.A. 2002, ApJ, 570, 900
Mawet, D., et al. 2010, ApJ, 709, 53
Moody, D., Gordon, B., & Trauger, J. 2008, in Space Telescopes and Instrumentation 2008: Optical, Infrared, and Millimeter, ed. Oschmann, et al., SPIE, 7010, 701081
Spergel, D.N. 2000, arXiv:astro-ph/0101142v1
Trauger, J., et al. 2006, JPL Document D-35484
Trauger, J., et al. 2007, SPIE, 6693, 66930X
Trauger, J.T., & Traub, W.A. 2007, Nat, 446, 771
Trauger, J., et al. 2009, ACCESS final report to NASA (copies available on request)

Pathways Towards Habitable Planets
ASP Conference Series, Vol. 430, 2010
Vincent Coudé du Foresto, Dawn M. Gelino, and Ignasi Ribas, eds.

The ExtraSolar Planetary Imaging Coronagraph

M. Clampin and R. G. Lyon

Code 667, Exoplanets and Stellar Astrophysics Laboratory, Goddard Space Flight Center, Greenbelt, MD, USA

and the EPIC Science Team

Abstract. The Extrasolar Planetary Imaging Coronagraph (EPIC) is a 1.65-m telescope employing a visible nulling coronagraph (VNC) to deliver high-contrast images of extrasolar system architectures. EPIC will survey the architectures of exosolar systems, and investigate the physical nature of planets in these solar systems. EPIC will employ a Visible Nulling Coronagraph (VNC), featuring an inner working angle of $\leq 2\lambda/D$, and offers the ideal balance between performance and feasibility of implementation, while not sacrificing science return. The VNC does not demand unrealistic thermal stability from its telescope optics, achieving its primary mirror surface figure requires no new technology, and pointing stability is within state of the art. The EPIC mission will be launched into a drift-away orbit with a five-year mission lifetime.

1. Introduction

The Extrasolar Planetary Imaging Coronagraph (EPIC) is a Probe-class mission concept designed to survey the architectures of exosolar systems, and investigate the physical nature of planets in these solar systems (Clampin et al. 2009) EPIC was recently funded for an Astrophysics Strategic Mission Concept (ASMC) study. Addressing EPIC's science goals is an important precursor to future large-scale missions that will image and characterize terrestrial planets. EPICs strategy is to focus initially on the detection and characterization of planetary systems identified from radial velocity surveys. EPIC employs a Visible Nulling Coronagraph (VNC) to detect and characterize extrasolar planets. The VNC is primarily a hybrid coronagraph, since it employs interferometric nulling, but is implemented on a single, filled-aperture telescope. The primary benefit afforded by a VNC is its inner working angle of $\leq 2\lambda/D$ that takes full advantage of EPICs 1.65-m telescope aperture to maximize the discovery space for extrasolar planets. Recent performance comparisons of coronagraphic techniques (Guyon et al. 2006) have focused on performance metrics such as contrast and inner working angle. Many coronagraphic techniques claim impressive performance, but demand challenging performance from the observatory. Key observatory requirements that have to be considered for a coronagraphic mission include pointing, stability (structural, thermal and optical), the optical figure of the telescope, performance verification, and the ability to sense and control wavefront and amplitude errors at levels significantly below 1 nm rms. The EPIC team has adopted the VNC as it achieves a balance between coronagraphic performance and the complexity of implementation.

2. Scientific Objectives

EPIC will survey the architectures of exosolar systems, conducting a census of planets in these systems and setting limits on the zodiacal dust emission. EPIC will investigate the physical nature of planets detected in these solar systems. EPICs science goals can be divided into two areas of focus, the detection and characterization of extrasolar planets, and the characterization of zodiacal dust structures within extrasolar planetary systems.

The number of exoplanets detected by radial velocity surveys that currently lie within the discovery space of EPIC given its $\leq 2\lambda/D$ inner working angle (IWA) continues to grow and is currently ~ 55. The discovery space enabled by EPIC's survey of nearby stars is shown in Figure 1, and illustrates that EPIC can readily detect gas giant and intermediate mass planets in extrasolar systems. The achievable contrast ratio ultimately depends on the systems stability level. Sensing and control of wavefront and amplitude errors is repeated every 7,000 secs to prevent contrast falling below 10^9. However, EPIC can reach $\sim 5 \times 10^{10}$ for specific targets by null phasing on a shorter timescales (~ 2000 secs) allowing the detection of some superearths or terrestrial planets. EPIC will conduct its science investigations using two operational modes: (1) discovery, and (2) characterization. In discovery mode, a high contrast image of the entire field of view

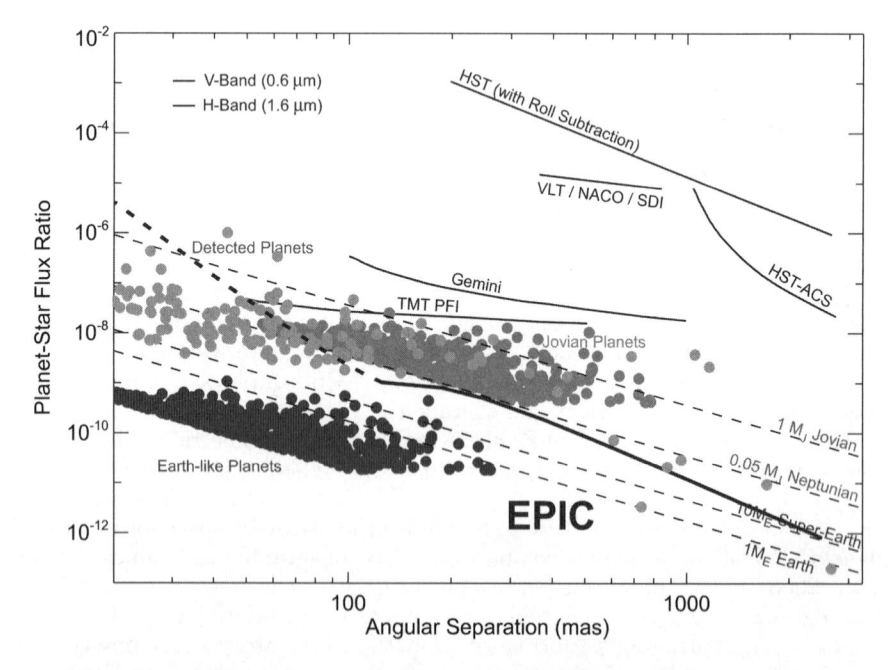

Figure 1. With an inner working angle of $\leq 2\lambda/D$ the VNC provides the maximum discovery space for a given telescope aperture, as illustrated by the 1.65-m Extrasolar Planetary Imaging Camera (EPIC) mission concept. Shown are $1M_J$ Jupiters at 3 AU (luminosity corrected) around potential EPIC survey targets, $1M_E$ Earths at 1 AU (luminosity corrected) and all currently detected planets (RV, Imaging, transit).

surrounding the nulled starlight is built from a sequence of observations (see Figure 2). This survey mode is used to locate and track any possible planets in the stellar system. Once a planet is unambiguously located, spectroscopic characterization of the planet's atmosphere can be conducted efficiently without the preliminary mapping sequence. Spectra and images of extrasolar giant planets help to constrain their masses, atmospheric composition, and particularly their cloud decks. Because the atmospheric temperature of a giant planet depends on a combination of its age, mass, composition, photochemistry, and the incident flux, the EPIC mission will measure an illuminating assortment of planetary spectra and albedos. With a typical stellar age of 2 Gyr the survey will encompass effective temperatures ranging from 100 to over 400 K, which includes a wide variety of atmospheres, from cold, ammonia-cloud dominated planets to planets with water clouds, to cloudless worlds.

EPIC will also study planetary systems by observing circumstellar disks. Debris dust around stars, generated by the collisions and evaporation of planetesimals, is both a source of noise against which planets must be detected and a signal of the architecture of planetary systems. The location and structure of dust bands around other stars will reflect their planetary system architectures. EPIC will be sensitive to 2× zodi of dust in its habitable zone with 8 hours of integration. If structures such as that seen around Eps Eri are present in the inner (1-5 AU) regions around other stars, EPIC will image them, even if they are created by much lower mass planets (Stark & Kuchner 2008). The simulation shown in Figure 3 shows what a resolved disk due to a 10 M_{earth} planet might look like. The planet carves a hole and generates clumps, both of which will orbit with it.

3. The Visible Nulling Interferometer

The VNC is a hybrid coronagraph that employs a combination of active and passive components to achieve the desired contrast. The VNC works by projecting contrast maxima and minima onto the sky, such that placing a star on a null (contrast minima) eliminates its light, while the light for a planet falling on a maxima is transmitted. The overall concept of the VNC is detailed in Figure 2. The VNC's inner working angle of $\leq 2\lambda/D$ is especially important since it maximizes the discovery space for imaging exoplanets, for a given telescope aperture size. The VNC also delivers a lower but useful contrast in to $\sim 0.5\lambda/D$, since it does not employ a coronagraphic mask. While the VNC appears complex it typically contains a similar number of optical elements to Lyot coronagraphs once optics for pointing control, and wavefront sensing systems are considered. The VNC is self contained because it provides via its interferometric stages all of the outputs required to track pointing jitter, and sense and control wavefront error (WFE). The VNC is implemented as a compact, modular package, designed to provide thermal and mechanical stability; see Lyon et al. (2009).

One of the most important advantages of the VNC is that *it can null to a contrast of* $\geq 10^9$ *using the starlight of the centrally nulled star*, while other coronagraphs have to null using the residual light within the dark hole. The VNC is an interferometer that separates on-axis stellar starlight into one output channel and off-axis planet light into another channel, i.e. into different and spatially

386 Clampin and Lyon

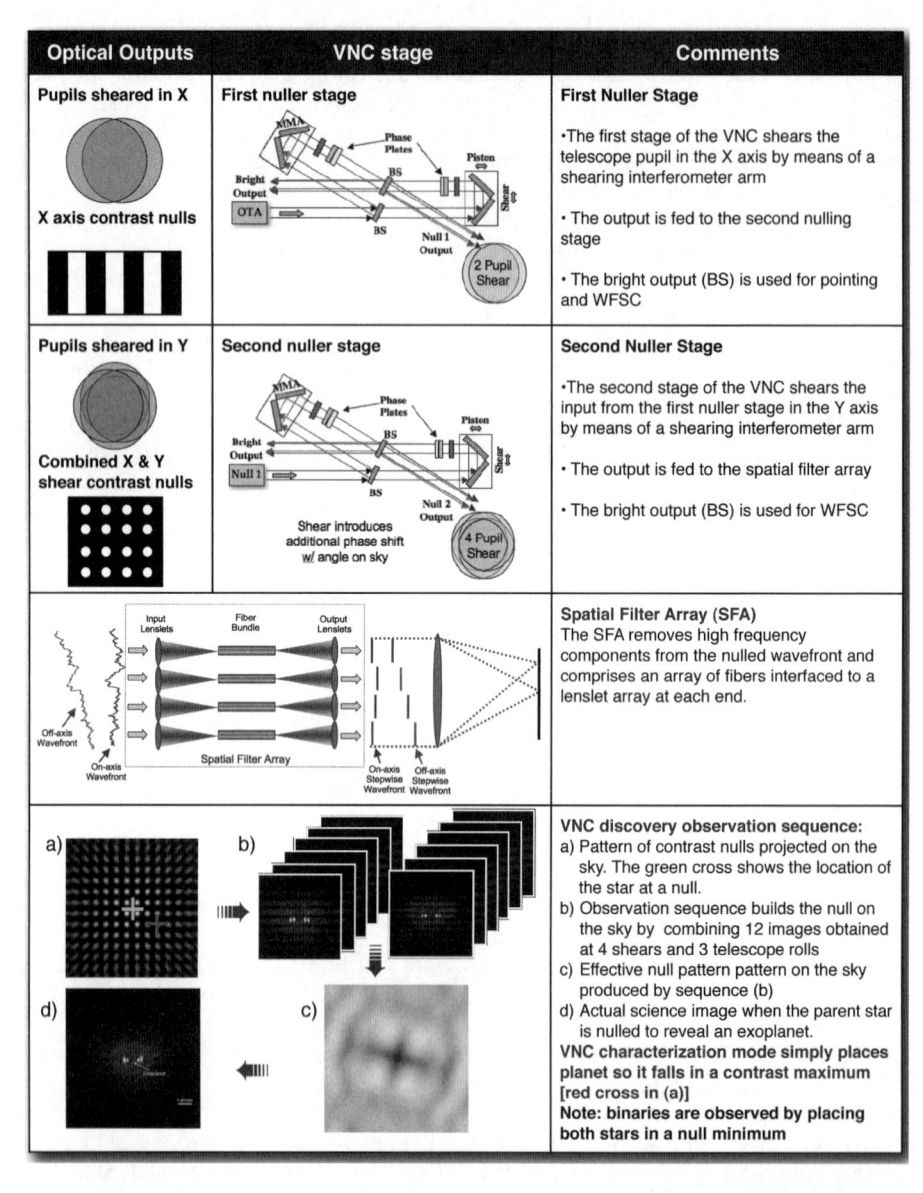

Optical Outputs	VNC stage	Comments
Pupils sheared in X X axis contrast nulls	**First nuller stage**	**First Nuller Stage** • The first stage of the VNC shears the telescope pupil in the X axis by means of a shearing interferometer arm • The output is fed to the second nulling stage • The bright output (BS) is used for pointing and WFSC
Pupils sheared in Y Combined X & Y shear contrast nulls	**Second nuller stage**	**Second Nuller Stage** • The second stage of the VNC shears the input from the first nuller stage in the Y axis by means of a shearing interferometer arm • The output is fed to the spatial filter array • The bright output (BS) is used for WFSC
	Spatial Filter Array diagram	**Spatial Filter Array (SFA)** The SFA removes high frequency components from the nulled wavefront and comprises an array of fibers interfaced to a lenslet array at each end.
a) b) d) c)		**VNC discovery observation sequence:** a) Pattern of contrast nulls projected on the sky. The green cross shows the location of the star at a null. b) Observation sequence builds the null on the sky by combining 12 images obtained at 4 shears and 3 telescope rolls c) Effective null pattern pattern on the sky produced by sequence (b) d) Actual science image when the parent star is nulled to reveal an exoplanet. VNC characterization mode simply places planet so it falls in a contrast maximum [red cross in (a)] Note: binaries are observed by placing both stars in a null minimum

Figure 2.   Operation and concept of the Visible Nulling Coronagraph

separated output channels of the interferometer. The rejected starlight is used for null control (wavefront and amplitude), first during initial coarse nulling, and then during closed-loop WFC throughout the science observation. Thus sensing the wavefront from its three output ports is independent of how dark the science channel becomes, since as the dark science channel becomes successively darker with each control step the bright channels become successively brighter. WFC

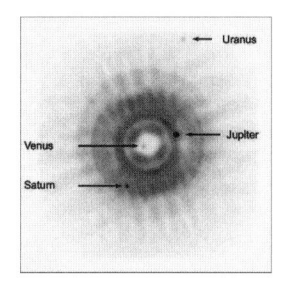

Figure 3. EPIC will image extrasolar planets, and zodiacal dust structures. A simulated 48,000 second, V-band EPIC observation of an inclined debris disk at 10 pc, dynamically sculpted by an embedded Jupiter mass planet at 5.2 AU.

control is independent of how the photons are distributed between the output channels and is only driven by the aggregate sum of the two. This is not true for other coronagraph options, where only the dark science channel is available for WFC. Thus as other coronagraphs converge to the desired contrast level it takes successively longer to achieve sufficient signal to noise to make a WFC step. If the system is not stable on timescales longer than the wavefront sensing time, an unstable control system will result and the system cannot converge to the required contrast. However, the VNC converges extremely rapidly since it uses the rejected stellar photons, a wavefront sensing signal $\sim 10^9$ -10^{10} brighter than a traditional coronagraph. This capability will prove to be an important consideration in terms of the cost and feasibility of an Exoplanet Probe mission, but will be more important as a driver for future large flagship missions to image and characterize terrestrial planets.

The VNC is scalable, both to larger aperture telescopes e.g. ATLAST (Postman et al. 2008), and to large baseline, dilute-aperture telescopes e.g. DaVinci (Shao et al. 2008). Technical demonstration of the VNC for a Probe-Class Mission would provide the technical foundation for a larger aperture missions in the future. The VNC operates seamlessly with segmented mirror telescopes, where it is able to null the individual mirror segments to a common solution, an important point, since it is very likely that future large aperture, space-telescopes will have segmented mirror architectures.

4. The EPIC Mission Concept

The EPIC science team has worked with NASA/GSFC, LMCO, BATC, and SAO/CfA to develop a mission concept with carefully assigned risks. For a Probe class mission we chose to implement our science requirements by means of a modestly sized (1.65-m) telescope feeding a Visible Nulling Coronagraph, and set our contrast requirement at $\geq 10^9$. EPIC employs a heritage spacecraft bus, since it can accommodate up to a 1.65-m telescope without redesign, providing a real cost saving. EPIC will be launched into a drift-away orbit to make high contrast observations of extrasolar planetary systems. A summary of the mission is shown in Table 1. The spacecraft comprises a 1.65-m tele-

Table 1. Overview of the EPIC's Mission Parameters

Science Payload	Performance
Telescope Optics	1.65 meter aperture off-axis primary
Coronagraph Design	Visible Nulling Coronagraph
	-Instrument throughput: 18%
	-Instantaneous Bandpass: 20% - 50%
Science Modes	
• Discovery mode	$\geq 10^9$ contrast: $2\lambda/D$: $4.5'' \times 4.5''$ FOV
• Discovery mode	$\geq 10^7$ contrast: $2\lambda/D$: $10'' \times 10''$ FOV
	(w/o Spatial Filter array)
• Characterization	R = 20-50 spectroscopy

scope, feeding a visible nulling coronagraph, mounted on a Kepler heritage bus. The Kepler bus already meets the mission-level requirements for EPIC and does not require a redesign to accommodate the 1.65-m primary mirror. The Kepler bus, in combination with a fine steering mirror in the optical chain, can also meet the pointing requirements imposed by the EPIC optical error budget. The EPIC payload will be launched into a heliocentric Earth trailing orbit, similar to Spitzer, which provides excellent sky coverage, combined with thermal stability. Since EPIC has a very modest data rate requirement (0.5 Gbits/day), the drift-away orbit does not drive communications costs. EPIC is designed to be a Class-B mission with redundancy on all major systems. We have designated a nominal mission lifetime of 5 years.

EPIC's science payload consists of an off-axis, optical-telescope element (OTE) coupled to a visible nulling coronagraph (VNC). The science payload mounts to a spacecraft bus that provides guidance, navigation, fine pointing, power, communications, and telemetry. The OTE is a two-mirror cassegrain telescope with metering truss, support structure, pupil relay and fine steering mirror (FSM); the pupil relay and FSM physically reside within the VNC enclosure to insure stability but are considered as part of the OTE. The VNC contains fore optics which relays the OTE entrance pupil onto the fine steering mirror (FSM) and a pupil relay to relay the pupil at the into the VNC. The VNC has three output ports that function as part of the control system, these are: (i) the X-bright object sensor (XBOS), (ii) Y-bright object sensor (YBOS) and (iii) nulled beam which feeds into the science instruments. The XBOS is used during coarse phasing of the system and continually in closed-loop to feedback to the pointing control system (PCS). The YBOS operates as part of the null control system to monitor and control the MMAs during science operations. In addition, during the setup for each new target a pupil plane null sensor is used to fine-tune the MMAs to achieve the contrast of $\geq 10^9$. The two science instruments are the imaging camera and a dispersing spectrometer and are physically mounted to the VNC enclosure. The imaging camera operates in two modes: narrow field-of-view (FOV) with the Spatial Filter Array (SFA), and wide FOV-field of view without the SFA.

The VNC has capability to sense and correct OTE wavefront errors at the outset of a science observation. During an observation these errors drift, and are continually sensed and corrected. The contrast becomes limited by the drift between successive null control steps and the correctable range and resolution of the DMs. The VNC is a differential approach, and less driven by absolute tolerances on the optics, relaxing OTE requirements. It employs a direct approach which senses and controls only that which is important to the science, i.e., null depth, and directly controls it both passively (via the SFA), and actively (via the MMAs), rather than directly sensing and controlling the wavefront and amplitude errors.

5. Summary

The Extrasolar Planetary Imaging Coronagraph (EPIC) is a Probe-class mission concept designed to survey the architectures of exosolar systems, and investigate the physical nature of planets in these solar systems. EPIC's ASMC study has demonstrated the technical feasibility of EPIC and validated the design concept of avoiding challenging observatory level requirements that drive the cost and complexity of the mission. EPIC has achievable observatory requirements such as pointing jitter and OTE stability. The VNC is the optimum choice for a small aperture (1.65 m) coronagraphic mission. Technology development for the VNC is advancing rapidly with recent investment at both the Goddard Space Flight Center and the Jet Propulsion Laboratory (Lyon et al. 2009).

References

Clampin, M. and the EPIC Science Team, 2009, "The Extrasolar Planetary Imaging Camera", Final Report: Astrophysics Strategic Mission Study.
Guyon, O., et al., 2006, ApJS, 167, 81
Lyon, R. G., et al., 2009, SPIE 7440, 11
Postman, M., et al., 2008, SPIE 7010, 21
Shao, M., et al., 2008, SPIE 7013, 32
Stark, C., & Kuchner, M. 2008, ApJ, 686, 637

The Pupil Mapping Exoplanet Coronagraphic Observer (PECO)

O. Guyon,[1,2] T. Greene,[3] K. Cahoy,[3] S. Shaklan,[4] D. Tenerelli,[5] and the PECO Team

[1] *Steward Observatory, University of Arizona, Tucson, AZ, USA*

[2] *Subaru Telescope, Nation. Astron. Observ. of Japan, Hilo, HI, USA*

[3] *NASA Ames Research Center, Moffett Field, CA, USA*

[4] *NASA Jet Propulsion Laboratory, 4800 Oak Grove Drive, Pasadena, CA, USA*

[5] *Lockheed Martin Space Corporation*

Abstract. The Pupil-mapping Exoplanet Coronagraphic Observer (PECO) is a mission concept using a coronagraphic 1.4-m space-based telescope to both image and characterize extra-solar planetary systems at optical wavelengths. PECO delivers 10^{10} contrast at 2 λ/D separation (0.15″) using a high- performance Phase-Induced Amplitude Apodization (PIAA) coronagraph which remaps the telescope pupil and uses nearly all of the light coming into the aperture. PECO's heliocentric drift-away orbit provides a stable thermal environment for wavefront control. PECO acquires narrow field images simultaneously in 16 spectral bands over wavelengths from 0.4 to 0.9 μm, utilizing all available photons for maximum wavefront sensing and sensitivity for imaging and spectroscopy. The optical design is optimized for simultaneous low-resolution spectral characterization of both planets and dust disks using a moderate-sized telescope. PECO will image the habitable zones of about 20 known F, G, K stars at a spectral resolution of R ~15 with sensitivity sufficient to detect and characterize Earth-like planets and to map dust disks as small as a fraction of our own zodiacal dust cloud brightness. An active technology development program, including coronagraph and wavefront control laboratory demonstrations at NASA Ames and NASA JPL, is currently addressing key technology needs for PECO.

1. Overview of PECO Mission Concept

PECO is a 1.4-m off-axis telescope operating at room temperature. The mission will last three years, with an option for extension to five years. Key features of this design are the PIAA coronagraph system for diffraction suppression, active wavefront control, photon counting focal plane science detectors, extremely low vibration and high pointing stability (< 10 mas rms for OTA, < 1 mas rms within the instrument). In order to meet the desired wavefront stability requirements, the design uses a drift-away heliocentric orbit, in conjunction with an internal thermal control system.

The PECO science instrument is designed to take full advantage of the high throughput and small inner working angle (IWA) of the PIAA coronagraph. PECO has four parallel coronagraph channels to allow simultaneous

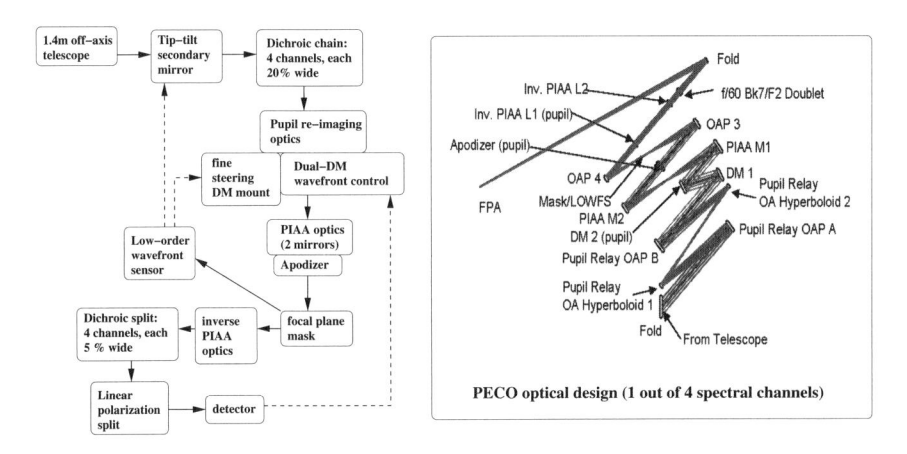

Figure 1. Block diagram of the PECO optical layout. Solid arrows show the light path; dashed arrows show wavefront control signals. The 1.4-m off-axis telescope's beam is apodized by the PIAA mirrors. A set of dichroics splits light in 4 spectral channels. The optical layout within one of these spectral channels is shown in the right part of the figure.

acquisition of all photons from 0.4 to 0.9μm. A functional block diagram of the PECO instrument is shown in the left part of Figure 1. Light is split by dichroics into four nearly identical coronagraph channels. The right side of the figure depicts the optical design of the instrument for one of these channels. The wavefront control subsystem consists, in each spectral channel, of two deformable mirrors (DMs; required for correction in a 360 deg field around the star) and a coronagraphic low order wavefront sensor (Guyon et al. 2009a) for accurate measurement of pointing errors and low order aberrations. The two DMs per channel provide the degrees of freedom to correct both phase and amplitude errors in the pupil (Shaklan et al. 2009) so that the primary mirror quality requirements do not exceed those already proven for the Hubble Space Telescope. A set of two aspheric PIAA mirrors (PIAA M1 and M2 on Figure 1) and a conventional pupil apodizer are used in each channel to fully apodize the telescope beam. The DMs are placed upstream of the PIAA optics allowing them to correct a dark hole extending to roughly 20 λ/D, or about 1.6 arcsec in the visible. An off-axis parabola (OAP3) focuses the apodized beam onto a coronagraph mask (labeled "Mask/LOWFS" in Figure 1) that blocks the central beam and reflects an annular beam extending to 2 λ/D to the LOWFS camera (not shown in the figure). A simpler, lower quality inverse PIAA system (Inv. PIAA M1 and M2 in Figure 1) reverses the coma-like aberration in the beam to form a sharp image of the planet across the field (Guyon et al. 2003; Vanderbei et al. 2005; Guyon et al. 2005; Lozi et al. 2009). Before reaching the detector, the beam is split into two linear polarizations by a Wollaston prism and four spectral channels with dichroics (these optics are not shown in Figure 1). In all there are 32 separate images formed on the detectors in the four channels (16 spectral bands x 2 polarizations).

PECO is thus designed to make optimal use of incoming photons, with a large spectral coverage (0.4 μm to 0.9 μm). Maximizing the total number of

Guyon et al.

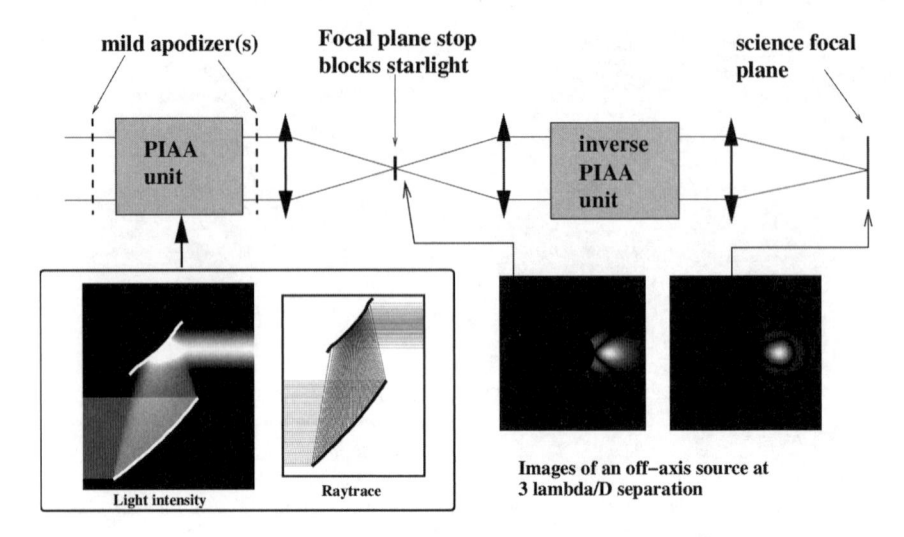

Figure 2. Schematic representation of the PIAAC. The telescope light beam enters from the left and is first apodized by the PIAA unit. Mild apodizer(s) are used to perform a small part of the apodization, and are essential to both mitigate chromatic diffraction propagation effects and allow for the design of friendly aspheric PIAA mirrors. A high contrast image is then formed, allowing starlight to be removed by a small focal plane occulting mask. An inverse PIAA unit is required to remove the off-axis aberrations introduced by the first set of PIAA optics.

photons transmitted to the detectors is essential for both science (exoplanets are faint, and the total observing times required for detection with a 1.4-m telescope are long – typically a day per target – even in broadband) and wavefront control (with more photons detected per unit of time, the wavefront control system is better able to track and correct aberrations). The number of PECO spectral channels (four in the baseline concept) is driven by the spectral bandwidth over which high contrast can be achieved, and can be traded against wavefront control agility (for example, number and location of DMs in each channel).

2. PIAA Coronagraph

PECO's Phase-Induced Amplitude Apodization (PIAA) coronagraph efficiently suppresses stellar light while preserving most of the planet's light. This coronagraph offers simultaneously a high throughput (close to 100%) and a small inner working angle (2 λ/D). A detailed description of this coronagraph technique can be found in previous publications (Guyon et al. 2003; Traub et al. 2003; Guyon et al. 2005; Vanderbei et al. 2005; Martinache et al. 2006; Vanderbei 2006; Pluzhnik et al. 2006a,b), and we briefly summarize in this section the principle and performance of PIAA-type coronagraphs.

The coronagraph architecture is shown in Figure 2. The telescope beam is apodized by two aspheric mirrors which reshape the telescope beam, and a conventional mild apodizer which can be located before or after the PIAA mirrors. In addition to making PIAA M1 manufacturable, sharing the apodization with a conventional apodizer also greatly improves the chromaticity of the PIAA coronagraph. Thanks to the apodization, a high contrast image of the star is produced in the coronagraphic focal plane: a small occulting mask can therefore block starlight while having little effect on off-axis sources. The beam shaping performed by the PIAA optics introduces strong off-axis aberrations which limit the useful field of view to about 8 λ/D. The PIAAC shown in Figure 2 therefore includes an inverse PIAA to recover a wider unaberrated field of view in the science focal plane.

Thanks to the lossless apodization performed by the PIAA optics, the PECO coronagraph simulatneously offers high contrast, nearly 100% throughput, small IWA and full 360 degree discovery space. At 10^{-10} contrast, the PIAA IWA, defined as separation at which the throughput is 50% of its maximum, is slightly under 2 λ/D.

3. Wavefront Control

Pointing: The instrument pointing tolerance during observations is no more than 1% of the diffraction width (i.e., <1 mas). The PECO strategy to meet this stability relies on creating a stable environment (heliocentric orbit and vibration isolation of reaction wheels) and deriving an accurate control signal from the bright target star image reflected by the coronagraphic stop (this is done with sub-mas accuracy at >100 Hz). Measurement of pointing error with light reflected from the central focal plane mask has been recently demonstrated in the laboratory to a precision equivalent to 0.1mas on PECO (Guyon et al. 2009a).

Control of mid-spatial wavefront aberrations: The goal of the wavefront correction algorithm is to command the DM to cancel out mid-spatial frequency wavefront errors, manifested as scattered light (speckles) and measured in the science focal plane. The DMs are commanded by a phase diversity signal to modulate the speckle intensity, allowing recovery of both amplitude and phase of the speckles without being affected by the incoherent planet light and the zodiacal light. Wavefront sensing is achieved by monitoring scattered light in the PECO science frames, which are acquired with photon-counting CCD read every few seconds. The DMs are continuously updated to correct for varying wavefront aberrations and provide the phase diversity signal necessary to measure the complex amplitude of starlight scattered by aberrations. Signals from all spectral channels can be added to improve wavefront sensing sensitivity in order to optimally track time-variable aberrations which are expected to be mostly achromatic. The Electric Field Conjugation (EFC) algorithm (Giveón et al. 2007) baselined for PECO, has been used on the High Contrast Imaging Testbed (HCIT) at JPL, using one DM and a band-limited coronagraph, achieving contrasts of 6 10^{-10} in 10% broad light at $4\lambda/D$. Similar approaches have already been successfully used as close as 2 λ/D in the Subaru PIAA testbed.

4. Coronagraph System Technology Development

PIAA coronagraph optics design and manufacturing: At the core of the PECO coronagraph are the PIAA mirrors which are a pair of aspheric high quality optics which perform the apodization. Tinsley, under contract to NASA ARC has recently delivered a pair of high fidelity PIAA-generation 2 mirrors to the NASA JPL High Contrast Imaging Testbed for testing of a PIAA coronagraph to broadband milestone levels.

Ongoing PIAA laboratory testing: Starlight suppression with PIAA coronagraph was first tested in the Subaru Telescope laboratory with a monochromatic testbed operated in air. The testbed achieved a 2.27×10^{-7} raw contrast between 1.65 λ/D (inner working angle of the coronagraph configuration tested) and 4.4 λ/D (outer working angle) (Guyon et al. 2010). The Subaru Telescope PIAA testbed activity has now ended and efforts at Subaru Telescope are focused on deploying already validated technology on the ground-based telescope. PIAA technology is now actively developed in two laboratories which are more capable than the Subaru lab was:

- The NASA Ames Research Center PIAA coronagraph laboratory is a highly flexible testbed operating in air (Belikov et al. 2009, Greene et al. 2009). It is dedicated to PIAA technologies and is ideally suited to rapidly develop and validate new technologies and algorithms. It uses MEMS-type deformable mirrors for wavefront control.

- The NASA JPL High Contrast Imaging Testbed is a high stability vacuum testbed facility for coronagraphs. PIAA is one of the coronagraph techniques tested in this lab, which provides the stable vacuum environment ultimately required to validate PIAA for flight (Kern et al. 2009).

5. Expected Science Return

The efficient coronagraphic approach adopted for PECO offers optimal science return for a given aperture size. PECO can therefore address some of the key science goals of previous mission designs that had planned to use larger telescopes but with less efficient coronagraphs (e.g., Terrestrial Planet Finder's Flight Baseline 1). Most importantly, our study shows that detection and characterization (low-resolution spectroscopy) of planets as small as Earth in the habitable zones of nearby stars is possible even with PECO's 1.4-m aperture.

PECO's science goals are to (1) image and characterize rocky planets in the habitable zones of ≈20 nearby stars, (2) image and characterize a large sample of giant planets and (3) map exozodiacal disks around nearby stars. PECO devotes a large fraction of its 3-year mission to the observation of 20 high priority targets, which are the stars for which it has the sensitivity to image a super-Earth in the habitable zone. These stars are revisited at least 10 times to maximize detection probability and minimize orbit uncertainties and possible confusion with other planets or exozodi structures.

PECO's PIAA coronagraph offers a 2 λ/D IWA, allowing PECO to image the habitable zones of main sequence stars at up to 5 to 10pc, depending on stellar luminosity and wavelength. PECO's blue channels offer the best IWA and are therefore essential for identification and orbit determination of previously unknown planets.

Detailed performance simulations have been performed to identify stars around which PECO could detect planets of different types (Cahoy et al. 2009).

We assume here that a planet is detectable if a SNR of 5 can be reached within 12 hr exposure time at a spectral resolution of 5 at λ=550nm along 20% of its orbit. With this detection limit, simulations show that:

- Super-Earths (assumed to be twice the diameter of Earth) can be imaged in the habitable zones of $\approx$ 20 nearby stars;

- Earths can be detected around 9 stars;

- At least 13 known radial velocity planets can be imaged;

Much of the design and analysis that went into PECO could be readily applied to a larger, more expensive telescope, resulting in enhanced science return.

References

Belikov, R., Pluzhnik, E., Connelley, M. S., Lynch, D.H., Witteborn, F. C., Cahoy, K., Guyon, O., Greene, T.P. & McKelvey, M.E. 2009, SPIE, 7440, 74400J
Cahoy, K., Guyon, O., Schneider, G. H., Marley, M.S., Meyer, M. R., Ridgway, S. T., Kasting, J., Traub, W.A. & Woolf, N.J. 2009, SPIE, 7440, 74400G
Giveón, A., Kern, B., Shaklan, S., Moody, D. C. & Pueyo, L. 2007, SPIE, 6691, 66910A
Greene, T., Belikov, R., Connelley, M., Guyon, O., Lynch, D., McKelvey, M., Pluzhnik, E. & Witteborn, F. 2009, this volume
Guyon, O. 2003, A&A, 404, 379
Guyon, O., Pluzhnik, E. A., Galicher, R., Martinache, F., Ridgway, S. T. & Woodruff, R. A. 2005, ApJ, 622, 744
Guyon, O., Matsuo, T. & Angel, R. 2009a, ApJ, 693, 75
Guyon, O., Pluzhnik, E.A., Martinache, F., Totems, J., Tanaka, S., Matsuo, T., Blain, C.& Belikov, R. 2010, PASP, 122, 71
Kern, B., Belikov, R., Give'on, A., Guyon, O., Kuhnert, A. C, Levine, M. B., Moody, D. C., Niessner, A. F., Pueyo, L.A., Shaklan, S. B., Traub, W. A. & Trauger, J. T. 2009, SPIE, 7440, 74400H
Lozi, J., Martinache, F., & Guyon, O., submitted to PASP
Martinache, F., Guyon, O., Pluzhnik, E.A., Galicher, R., & Ridgway, S. T. 2006, ApJ, 639, 1129
Pluzhnik, E. A., Guyon, O., Ridgway, S.T., Martinache, F., Woodruff, R. A., Blain, C. & Galicher, R. 2006a, ApJ, 644, 1246
Pluzhnik, E. A., Guyon, O., Warren, M., Woodruff, R.A., & Ridgway, S.T. 2006b, SPIE 6265, 62653S
Shaklan, S. B. & Green, J. J. 2006, Appl.Optics, 45, 5143
Traub, W.A. & Vanderbei, R. J. 2003, ApJ, 599, 695
Vanderbei, R.J. & Traub, W. A. 2005, ApJ, 626, 1079
Vanderbei, R.J. 2006, ApJ, 636, 528

Franck Selsis during a panel session.

Session X
Poster Presentations

Pathways Towards Habitable Planets
ASP Conference Series, Vol. 430, 2010
Vincent Coudé du Foresto, Dawn M. Gelino, and Ignasi Ribas, eds.

Interactions between a Massive Planet and a Disc

P. Amaro-Seoane,[1,2] I. Ribas,[2] U. Löckmann,[3] and H. Baumgardt[3]

[1] *Max-Planck Institut für Gravitationsphysik (Albert Einstein-Institut), Am Mühlenberg 1, Potsdam, Germany*

[2] *Institut de Ciències de l'Espai, Campus UAB, Torre C-5, Bellaterra, Barcelona, Spain*

[3] *Argelander-Institut für Astronomie, Auf dem Hügel 71, Bonn, Germany*

Abstract. We analyse the potential migration of massive planets forming far away from an inner planetary system. For this, we follow the dynamical evolution of the orbital elements of a massive planet undergoing a dissipative process with a gas disc centred around the central sun. We use a new method for post-Newtonian, high-precision integration of planetary systems containing a central sun by splitting the forces on a particle between a dominant central force and additional perturbations. In this treatment, which allows us to integrate with a very high-accuracy close encounters, all gravitational forces are directly integrated, without resorting to any simplifying approach. After traversing the disc a number of times, the planet is finally trapped into the disc with a non-negligible eccentricity.

1. Introduction

It has been suggested that the most massive planets in a planetary system can be formed by a process of gas collapse, independently of metallicity, whilst the lighter components would have formed via core accretion. This can lead to a situation in which massive planets that have originated relatively far away from the inner system, migrate inwards into it (see Font-Ribera, Miralda-Escudé & Ribas 2009). In this process, these more massive planets with larger semi-major axis, cross the gas disc centred around the central sun. When going through this dissipative process, the planets loses kinetic energy because of the friction with the gas. As a consequence, the inner disc is heated up and the semi-major axis of the planet shrinks. After some passages, the planet is trapped in the disc with a residual eccentricity. We propose this scenario as a plausible way of explaining the existence of massive planets distributed around a sun with non-zero eccentricities. We give results based on high-accurate dynamical simulations about the distribution of the orbital elements of the trapped objects in the disc. We find that the massive planets are typically captured after some $\sim 10^5$ yrs and the final eccentricity is non-negligible ($e \sim 0.1$).

2. Method, Results, and Conclusions

We have recently developed an integrator bhint specialised for dynamical processes in the vicinity of a very massive particle, which relies in the assumption that the very massive particle dominates the motion of the smaller ones (Löckmann & Baumgardt 2008). For this new mechanism, we retain the Hermite scheme as a basis. Collisions are taken into account.

We initially set a disc made out of 10^3 small particles which is "hosting" a sun in the centre and follows a simple $1/r$ profile. The integrated mass is of $0.1\,M_\odot$ and the radius of some 30 AU. The thickness of the disc is about the diameter of the central sun and has a gap around the central sun which extends 0.1 AU. The mass of the central sun is $1\,M_\odot$, the mass of the planet of $10^{-2}\,M_\odot$ and the mass of the particles in the disc of $10^{-4}\,M_\odot$, so that they are planetesimals. The mass forming the disc are all single-mass. The massive planet, a massive particle of 5 Jupiters is set in an orbit such that the initial eccentricity with the sun is of $e = 0.95$. Initially, the particle is 100 AU away from the sun and the inclination angle is $i = 90$ degrees. The system (disc plus interloper) is integrated until the interloper is trapped by the disc. In the figure we show the evolution of the orbital parameters. Whilst we cannot discuss them in detail because of the publication limits, we note that after some 4×10^5 years the inclination has almost not changed as compared to its initial value. Then, after a short time of 75×10^3 years elapses, it abruptly decays from almost 80 degrees to a very small number, to be finally trapped in the disc after 5×10^5 years. In the figure the third panel from the top shows the evolution of the energy normalised to the initial energy. The eccentricity, whilst it decays from the initial high value of 0.95, is of $e \sim 0.1$ when the massive planet is totally trapped in the disc, within a final semi-major axis which is well within the range of expectation. In the particular simulation presented here the particles did not accrete mass (neither the planet nor the planetesimals). An extended and detailed scrutiny of the parameter space of this capture process we propose will be soon published elsewhere (Amaro-Seoane et al. 2010). The addressing of this scenario has direct bearing on our understanding of planetary dynamics and migration mechanisms.

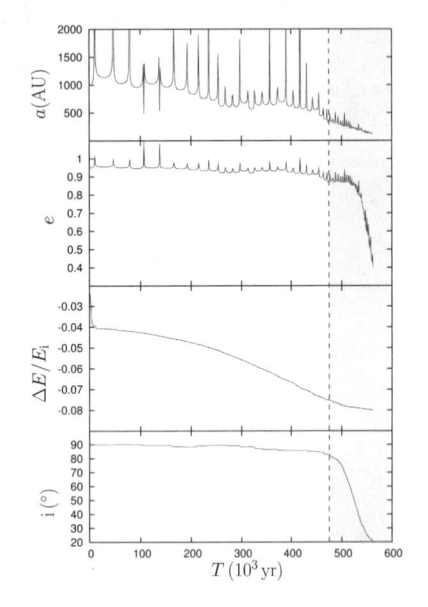

References

Amaro-Seoane P., Ribas I., Löckmann U., & Baumgardt H. 2010, in preparation
Font-Ribera A., Miralda Escudé J., & Ribas I. 2009, ApJ, 694, 183
Löckmann U. & Baumgardt H., 2008, MNRAS, 384, 323

Ground-based Astrometric Planet Searches using Medium-sized Telescopes

G. Anglada-Escudé, A. P. Boss, and A. J. Weinberger

Carnegie Institution of Washington/Department of Terrestrial Magnetism, 5421 Broad Branch Rd., Washington, DC, USA

Abstract. Ground based astrometry has not been very successful in detecting extrasolar planets. Some reasons are the relatively long time baselines required and instrumental stability requirements. Also, the number of free parameters is large compared to other methods (such as Doppler spectroscopy) and additional information is often required to constrain the true nature of the candidate signals. An example is the recently announced astrometric detection of a planet around the low mass star VB 10, where a careful reanalysis of the astrometric data casts some doubts on the true nature of the announced low mass companion. The Carnegie Astrometric Planet Search Program (CAPS), is focused on the detection of gas giant exoplanets around nearby low mass stars. We show that accuracies at the level of 0.4 mas can be reached on time-scales of years with a 2.5 class meter telescope given a sufficiently stable and optimized camera (CAPScam-S). This accuracy enables the detection of Jupiter-sized planets around nearby cool stars providing at the same time, accurate measurements of their distances and spatial motion.

1. The Carnegie Astrometric Planet Search

The Carnegie Astrometric Planet Search(CAPS) uses the du Pont 2.5m telescope at Las Campanas Observatory (Chile) with an especially designed camera for astrometric observations. With a FoV of 6 × 6 arcmin and a natural narrow wavelength band (800-930 nm), CAPScam is optimized to observe nearby M-dwarfs which are the optimal targets for planet detection by astrometry(see Boss et al. 2009, for more detailed information). We have developed a fully automatic data reduction pipeline to extract the star positions, calibrate for geometric distortions and produce an astrometric catalog for all the objects in the field (available under request at anglada@dtm.ciw.edu).

2. Least Squares Periodograms

Astrometry only. In order to maximize the sensitivity of our planet search, we have developed a least-squares periodogram approach which is the natural generalization of the so-called Lomb-Scargle Periodogram. It solves simultaneously for the periodic signal and the classic astrometric parameters. The χ_P^2 of a given solution is compared to the null-hypothesis to compute the power $z(P)$ that is then used to estimate the false alarm probabilities. We have applied it to the astrometry of the planet candidate around VB 10 recently announced by Pravdo & Shaklan (2009) identifying 2 significant periods: one at 50 d and the

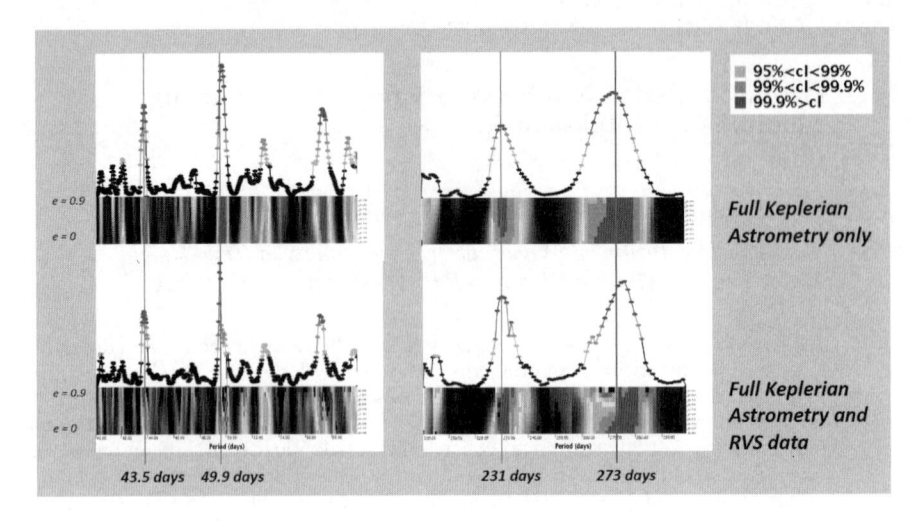

Figure 1. Confidence level maps for VB10b. They represent the single frequency confidence level for different eccentricity-period pairs. The vertical plots show the higher power obtained for a given period by fitting all the Keplerian parameters. The 50 d signal is clearly still dominant after including the available RV data.

published one at 270 d. Since the eccentricity is not constrained by the current astrometry, additional data is required to further constrain the orbit.

Including RV data. The least-squares approach, can be easily generalized to include non-linear parameters and radial velocity data. We have developed our own software to perform such analysis and the results applied to the VB10b candidate are shown in Figure 1. The colored maps in Figure 1 show the confidence level of the best fit solution given a period-eccentricity combination. The vertical periodogram-like plots show the maximum power obtained for each period. We show how the confidence level areas around the 49.9 (left) and the 273 days (right) period are updated when RV data (Zapatero-Osorio et al. 2009) is added. The eccentricity is still unconstrained and more data is required to confirm the existence of VB10b and/or constrain the allowed orbital solutions.

Conclusions. The analysis of astrometric data for planet searches requires a revision of the classical methods to properly identify the relevant signals. As an example, the dominant 50.0 days signal in the astrometry of VB 10 was missed in the original analysis of Pravdo & Shaklan (2009), and still provides the most likely solution given the available data (by September 2009). Further observations using the MIKE spectrograph at Clay/Magellan Telescope are ongoing.

References

Boss, A. P. et al. 2009, PASP, 121, 1218
Pravdo, S. H. & Shaklan, S. B. 2009, ApJ, 700, 623
Zapatero-Osorio, M. R., et al. 2009, A&A, 505, L5

Pathways Towards Habitable Planets
ASP Conference Series, Vol. 430, 2010
Vincent Coudé du Foresto, Dawn M. Gelino, and Ignasi Ribas, eds.

The Fourier-Kelvin Stellar Interferometer: Exploring Exoplanetary Systems with an Infrared Probe-class Mission

R. K. Barry,[1] W. C. Danchi,[1] B. Lopez,[2] S. A. Rinehart,[1] O. Absil,[3] J.-C. Augereau,[4] H. Beust,[4] X. Bonfils,[4] P. Bordé,[5] D. Defrère,[3] P. Kern,[4] A. Léger,[5] J.-L. Monin,[4] D. Mourard,[2] M. Ollivier,[5] R. Petrov,[2] F. Vakili,[2] and the FKSI Consortium

[1] *NASA Goddard Space Flight Center, Greenbelt, MD, USA*

[2] *Observatoire de la Côte d'Azur (OCA), Département Gemini UMR 6023, Nice, France*

[3] *Université de Liége and the FKSI Consortium*

[4] *Laboratoire d'Astrophysique de Grenoble (LAOG), Grenoble, France*

[5] *Institut d'Astrophysique Spatiale (IAS), Orsay, France*

Abstract. We report results of a recent engineering study of an enhanced version of the Fourier-Kelvin Stellar Interferometer (FKSI) that includes 1-m diameter primary mirrors, a 20-m baseline, a sun shield with a ±45° Field-of-Regard (FoR), and 40K operating temperature. The enhanced FKSI is a two-element nulling interferometer operating in the mid-infrared (e.g. ∼ 5-15 μm) designed to measure exozodiacal debris disks around nearby stars with a sensitivity better than one solar system zodi (SSZ) and to characterize the atmospheres of a large sample of known exoplanets. The modifications to the original FKSI design also allows observations of the atmospheres of many super-Earths and a few Earth twins using a combination of spatial modulation and spectral analysis.

1. Design Evolution of FKSI

During the last few years, considerable effort has been directed towards flagship-scale missions to detect and characterize Earth-radius down to Mars-radius planets around nearby stars in the mid-infrared (∼ 7 − 20 μm), namely the Darwin and Terrestrial Planet Finder Interferometer (TPF-I) missions. However, cost and technological issues such as formation flying and control of systematic noise sources are likely to prevent these missions from entering Phase A until the 2020+ time frame. Presently more than 400 exoplanets have been discovered, and little is known about the majority of them other than their approximate mass, semi-major axes, and eccentricities.

FKSI was originally conceived of as a Discovery-class mission for an imaging and nulling interferometer for the near- to mid-infrared spectral region (3 − 8 μm) (Danchi et al. 2003; Barry et al. 2006; Danchi & Lopez 2007), with a 12.5 m baseline, cooled to 65K, and with an FoR of ±20 degrees. The shorter wavelength operation allowed for a reduction in costs primarily through simpli-

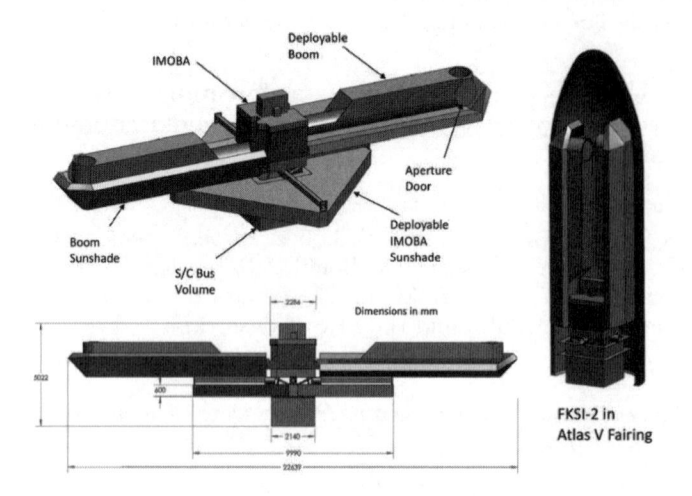

Figure 1. Engineering realization of the enhanced Fourier-Kelvin Stellar Interferometer (FKSI) spacecraft and instrument.

fied cooling and sunshade requirements. This version of FKSI had as it primary science requirement the measurement of exozodiacal emissin levels to 1 SSZ for nearby FGK stars, spectroscopic characterization of known exoplanets, and other scientific activities requiring high spatial resolution. However, in order to detect and characterize many super-Earths, and to have enough sensitivity to detect and characterize a modest number of Earth-twins, some modifications to the original design needed to be investigated. In particular recent performance simulations (Danchi et al. 2009) allowed us to show that, with a science band from 5 to 15 μm, 1-m primary mirrors, a 20-m baseline, a sunshield with a $\pm 45°$ FoR, and 40K operating temperature, this enhanced version is able to detect and characterize ~ 5 Earth-twins and about $30 \sim 2R_{Earth}$ super-Earths.

Recently, we conducted a substantial engineering effort to explore the discovery space afforded by the introduction of a new mid-size Probe-class mission category. The result of the engineering study (Figure 1) was that such an enhanced version of FKSI could be built well within the expected Probe cost cap and that technology from JWST could be directly incorporated into FKSI including detectors, cryocoolers, sunshade and other deployable technologies and precision cryogenic structures. The budget for the instrument package roughly doubled in cost.

References

Barry, R. K., et al. 2006, SPIE, 6265, 1L
Danchi, W. C., Deming, D., Kuchner, M. J., & Seager, S. 2003, ApJ, 597, L57
Danchi, W. C. & Lopez, B. 2007, Compte Rendu Physique, 8, 396.
Danchi, W. C., et al. 2009, Sf2A Proceedings, 317

Towards Astrometric Detection of Neptune- to Earth-Mass Planets around M-Stars

C. Bergfors,[1] W. Brandner,[1] M. Janson,[2] N. Kudryavtseva,[1] S. Daemgen,[3] S. Hippler,[1] F. Hormuth,[1] and T. Henning[1]

[1] *Max-Planck-Institute für Astronomie, Königstuhl 17, Heidelberg, Germany*

[2] *University of Toronto, Dept. of Astronomy, 50 St George Street, Toronto, ON, Canada*

[3] *European Southern Observatory, Karl-Schwarzschild Strasse 2, Garching, Germany*

Abstract. Astrometric planet searches with the GRAVITY VLTI instrument will be able to detect planets in the range from a few Earth-masses to Neptune-masses around nearby low-mass stars. High-resolution Lucky Imaging with *AstraLux* has provided a sample of nearby M dwarf binaries which defines a potential target list for astrometric planet detection with GRAVITY. We present the GRAVITY planet detection limits for one of the M dwarf binaries detected in the *AstraLux* survey.

1. Introduction

The intrinsic faintness of M dwarfs makes them difficult targets for today's exoplanet searches using transits and direct imaging. To date we know only of a few planets belonging to these stars. However, the low mass of M dwarfs makes them good targets for astrometric planet detection. While it appears from observations so far that giant planets are more scarce around M dwarfs than around solar-type stars, core-accretion planet formation models predict that Neptune-mass and terrestrial-type planets should be common around low-mass stars (Laughlin, Bodenheimer & Adams 2004). A few Neptune and sub-Neptune mass planets belonging to M dwarfs have indeed already been discovered (GJ 436b, Gl 581b,c,d,e, Gl 876d, etc.). To understand how the large variety of planets form and evolve, it is necessary to include a census of the occurrence and properties of the planets of M dwarfs, the most numerous stars in our neighbourhood.

2. Detecting Planets around Nearby Binary M Dwarfs With GRAVITY

GRAVITY will be a second generation VLTI instrument which will make use of all four VLT 8 m Unit Telescopes. Using differential astrometry, a positional accuracy of 10 μas can be achieved for targets brighter than m_K ~15 mag (see Eisenhauer et al. (2008) for a description of GRAVITY). The GRAVITY detec-

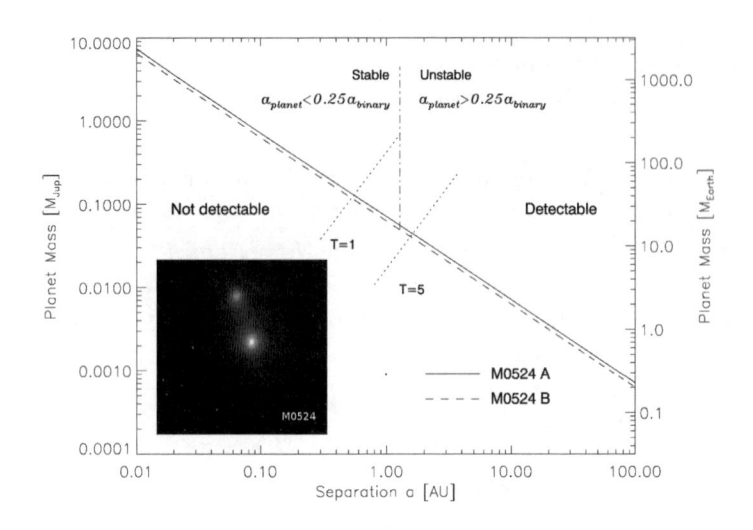

Figure 1. GRAVITY planet detection limits for one of the newly discovered binaries in the *AstraLux* M Dwarf Survey. The inset shows the system as observed with *AstraLux*. T is the orbital period of the planet in years.

tion limit corresponds to planets of 3 $M_{\rm Earth}$ around an M5V star at a distance of 5 pc, or planets less than 2 $M_{\rm Neptune}$ around an M3V star at 25 pc.

The *AstraLux* M dwarf binary survey is a high-resolution Lucky Imaging survey with *AstraLux* at NTT, La Silla (Hippler et al. 2009) and the 2.2 m telescope at Calar Alto (Hormuth et al. 2009). The survey is monitoring a sample of ~ 800 nearby M dwarfs from the Riaz, Gizis & Harvin (2006) catalogue of young, nearby late-type stars in the search for close low-mass companions. The binaries will define a sample for astrometric exoplanet searches with GRAVITY. The GRAVITY planet detection limits are depicted in Figure 1 for one of the M dwarf binary systems recently discovered with *AstraLux* Lucky Imaging. The system consists of stars with photometrically estimated spectral types M4.5 and M5 at a distance of 8 pc from the Sun. The angular separation is 0.639″, corresponding to a projected separation of 5.1 AU. In this system, planets with masses $M_p > 1 M_{\rm Neptune}$ in stable orbits within ~1.3 AU will be detectable with GRAVITY over a time span of 1 to 5 years.

References

Eisenhauer, F. et al., 2008, SPIE, 7013, 70132A
Hippler, S. et al., 2009, Msngr, 137, 14
Hormuth, F., Brandner, W., Janson, M., Hippler, S., & Henning, Th. 2009, in AIP Conf. Ser. 1094, Cool Stars, Stellar Systems and the Sun, ed. E. Stemplels (New York: AIP), 935
Laughlin, G., Bodenheimer, P. & Adams, F. C. 2004, ApJ, 612, L73
Riaz, B., Gizis, J. E & Harvin, J. 2006, AJ, 132, 866

How Common are Extrasolar, Late Heavy Bombardments?

M. Booth,[1] M. C. Wyatt,[1] A. Morbidelli,[2] A. Moro-Martín,[3,4] and H. F. Levison[5]

[1] *Institute of Astronomy, Madingley Rd, Cambridge, UK*

[2] *Observatoire de la Côte d'Azur, Nice, France*

[3] *Centro de Astrobiologia - CSIC/INTA, 28850 Torrejón de Ardoz, Madrid, Spain*

[4] *Department of Astrophysical Sciences, Peyton Hall, Ivy Lane, Princeton University, Princeton, NJ, USA*

[5] *Department of Space Studies, Southwest Research Institute, Boulder, CO, USA*

Abstract. The habitability of planets is strongly affected by impacts from comets and asteroids. Indications from the ages of Moon rocks suggest that the inner solar system experienced an increased rate of impacts roughly 3.8 Gya known as the Late Heavy Bombardment (LHB). Here we develop a model of how the solar system would have appeared to a distant observer during its history based on the Nice model of Gomes et al. (2005). We compare our results with observed debris discs. We show that the solar system would have been amongst the brightest of these systems before the LHB. Comparison with the statistics of debris disc evolution shows that such heavy bombardment events must be rare occurring around less than 12% of Sun-like stars.

In addition to evidence for the LHB on the Moon (Tera et al. 1974), there is also evidence that the rest of the solar system also went through a period of increased bombardment including the Earth (Kring & Cohen 2002; Graae Jorgensen et al. 2009). Impacts on the Earth from comets and asteroids may have caused mass extinctions like that which may have wiped out the dinosaurs. However, the earliest evidence for life on Earth comes from just after the time of the LHB (Mojzsis et al. 1996). Since large impacts have the potential to cause the formation of deep sea vents, which may be the sites where life first developed, it is plausible that the LHB was, in fact, necessary for life (Kring 2003).

Whether the LHB helped or hindered life, it almost certainly had an effect on life and so it is interesting to see whether other planetary systems would also have gone through a late phase of heavy bombardment like the solar system. To answer this question, we need to know what observable properties of the solar system would change during and after an LHB event.

Planetesimals in any planetary systems are constantly undergoing collisional evolution. This produces dust and can be observed as a debris disc. Gomes et al. (2005) model the dynamics of planetesimals during the LHB. By making some assumptions about the planetesimals and their size distribution, we use

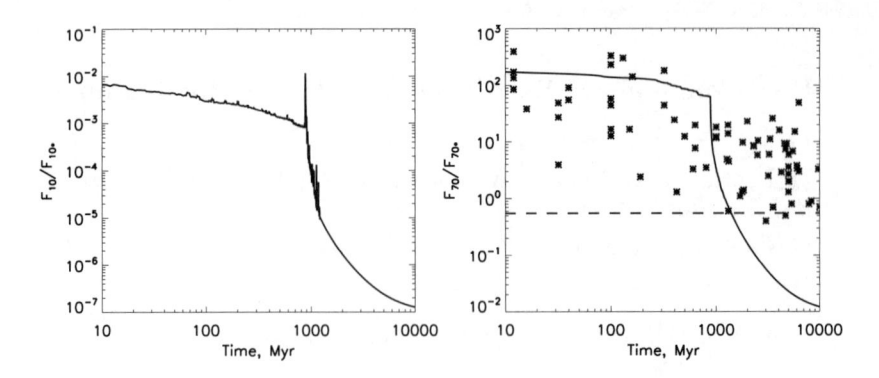

Figure 1. Excess ratio versus time for 10 μm (left) and 70 μm (right). The solid line represents the emission of our Solar system from our model. The asterisks are observed discs and the dashed line shows the approximate observational limit.

the evolution of the mass distribution from this model to find out how the flux from the Kuiper belt changes due to the LHB (Booth et al. 2009).

Figure 1 shows the emission from the solar system at 10 and 70 μm. The figure also shows observed debris discs from the literature. This shows that the solar system would have been amongst the brightest discs before the LHB and that its emission rapidly dropped after the LHB. There is also a rise in mid-IR emission during the LHB, which could explain some observed debris discs that have mid-IR excesses above what would be expected from steady state evolution of debris discs (Wyatt et al. 2007). Trilling et al. (2008) survey debris discs and find that 16% of stars are observed with 70 μm and that this number remains approximately constant with age. By assuming that any star that is observable at 70 μm when it is young and not observable when it is old has undergone an LHB-like event, we can constrain the number of stars that would have undergone an LHB to 12%.

Acknowledgments. MB acknowledges the financial support of a PPARC / STFC studentship.

References

Booth, M., Wyatt, M. C., Morbidelli, A., Moro-Martín, A., & Levison, H. F. 2009, MNRAS, 399, 385
Gomes, R., Levison, H. F., Tsiganis, K., & Morbidelli, A. 2005, Nat, 435, 466
Graae Jorgensen, U., Appel, P. W. U., Hatsukawa, Y., Frei, R., Oshima, M., Toh, Y., & Kimura, A. 2009, arXiv:0907.4104
Kring, D. A., & Cohen, B. A. 2002, JGR (Planets), 107, 5009
Kring, D. A. 2003, Astrobiology, 3, 133
Mojzsis S. J., Arrenhius G., McKeegan K. D., Harrison T. M., Nutman A. P., & Friend C. R. L. 1996, Nat, 384, 55
Tera, F., Papanastassiou, D. A., & Wasserburg, G. J. 1974, EPSL, 22, 1
Trilling, D. E., et al. 2008, ApJ, 674, 1086
Wyatt, M. C., Smith, R., Greaves, J. S., Beichman, C. A., Bryden, G., & Lisse, C. M. 2007, ApJ, 658, 569

Pathways Towards Habitable Planets
ASP Conference Series, Vol. 430, 2010
Vincent Coudé du Foresto, Dawn M. Gelino, and Ignasi Ribas, eds.

Occurrence, Physical Conditions, and Observations of Super-Ios and Hyper-Ios

D. Briot and J. Schneider

Observatoire de Paris-Meudon, 61 avenue de l'Observatoire, Paris, France

Abstract. If a volcanic activity similar to that of Io occurs in Exo-Earths or super-Earths, these objects would correspond to super-Ios and even more hyper-Ios. We study extreme physical conditions of these objects and their very special observational features.

1. Introduction

In the solar system, volcanic activity is observed in the larger "solid" bodies, as well as in some satellites, demonstrating that this process can happen in bodies of various sizes and masses. However, Io remains a very special object because of its very intense volcanic activity. We research if conditions which originate in the very intense activity of Io can exist at a larger scale for an extra-solar planet, exo-Earth or super-Earth. We shall call such a planet a super-Io, in case of a terrestrial mass and even more, a hyper-Io in the case of a mass corresponding to a super-Earth. We explore the very special physical conditions and properties of these objects.

2. Possibility of Volcanic Activity in Super-Earths

The volcanism of Io is attributed to the tidal forces exerted by Jupiter. These forces are important because of both the closeness of Jupiter and Io, and the excentricity of Io's orbit. Io's orbit is made eccentric by resonance with the orbits of Europa and Ganymede. Actually, extrasolar planets are frequently very close to their star, their trajectories are often eccentric, and in case of a system of several planets, cases of resonance of orbits are frequently observed. So the probability that there exists volcanic activity similar to Io's in terrestrial extra-solar planets or in super-Earths may be very high. The system of Jupiter and its larger satellites maybe exists on a much larger scale in some stars and their planetary system. It appears that a high proportion of super-Earths can possibly display volcanic activity (see e.g., Jackson et al. 2008).

3. Some Physical Properties of Super-Ios and Hyper-Ios

Some fundamental parameters are quite different in Io and in an extrasolar planet with very intense volcanic activity. The main differences are temperature, mass, and diameter of the planet. Because of the vicinity of the star, instead of Jupiter, the temperature on a super-Io would be much higher than on Io. It can be likely supposed that, as Io is, this object would be co-rotating, which would imply an extreme hotness on the hemisphere facing the star. Super-Ios and hyper-Ios diameters and masses would be those of rocky planets, i.e. much larger than for Io. A very thick atmosphere could be retained around the planet. This atmosphere is fed by material ejected by volcanoes and would contain sulphur dioxide. If this atmosphere is very thick, it could be opaque which would drastically change the observational features.

4. Observations of Super-Ios and Hyper-Ios

Primary Transit - The much larger size of a super-Earth with volcanic activity, as compared with the real Io, implies that a super-Io would have a very much thicker atmosphere than Io. The spectrum of the sulphur dioxide would be observed in the case of transit.

Secondary Transit - The volcanic activity of Io is especially important in the near infrared, in the 3 - 5 μm range. If we suppose, as a first approximation, that the chemistry involved is the same, the radiation of a super-Io makes it an excellent candidate for observations of secondary transits at these wavelengths, especially in the case of a cool star. Moreover, the volcanic activity of Io appears in hot spots and the volcanoes of Io can be detected in the case of an occultation by Europa (Descamps et al. 1992). If the volcanic activity is Io-like on a much larger scale, then in the case of a transparent atmosphere, hot spots could be detected in the first part and the last part of the secondary transit. In this case, some small irregularities could be observed during the decrease and increase of luminosity.

Visibility - If a super-Io emits its own radiation, the ratio between the luminosity of the star and the luminosity of the planet, especially in the case of a cool star, becomes smaller than in the case of a "non-volcanic" planet and a solar-type star. So these hypothetical objects would become very interesting objects to be observed by imaging in the infrared.

5. Conclusion

Tidal effects could be very important in extrasolar planets and occurrence of extrasolar planets undergoing volcanic activity similar to Io is really likely. These super-Ios and hyper-Ios would be both much hotter and more massive than Io. The physical conditions of these objects could be extreme and not correspond to any currently known object. It is important to study these physical conditions to determine observational features. The present paper is a preliminary study, a more detailed one will be published.

Acknowledgments. It is a pleasure to thank Valéry Lainey and Emmanuel Lellouch for very helpful discussions.

References

Descamps, P., Arlot, J.E., Thuillot, W., Colas, F., & Vu, D.T. 1992, Icarus, 100, 235
Jackson, B., Barnes, B., & Greenberg, R. 2008, MNRAS, 391, 237

First Steps on the Design of the Optical Differentiation Coronagraph

M. P. Cagigal,[1] V. F. Canales,[1] P. J. Valle,[1] E. Sanchez-Blanco,[2] M. Maldonado,[2] and M. L. Garcia-Vargas[2]

[1] *Dept. Fisica Aplicada, Univ. Cantabria, Santander, Spain*

[2] *FRACTAL SLNE, Tulipan 2, p 13, 1A., Las Rozas, Madrid, Spain*

Abstract. We present the UC coronagraph, an instrument designed for exoplanet detection. It is based on a standard coronagraph where the occulting disc is substituted by a differentiation mask. The instrument includes the coronagraph itself, a star generator, and a telescope simulator. The coronagraph is mechanically independent of both auxiliary subsystems, to allow its use in an actual astronomical telescope. The turbulence generation will be simulated with a spatial light modulator (SLM) placed at the nominal pupil.

Project Scheme

The development of the UC coronagraph (Oti, Canales & Cagigal 2005) was divided into three stages. The first stage (conceptual, preliminary, and detailed design) has been accomplished and we are currently completing the second one, instrument manufacture and the corresponding checking proofs. The final stage is the design, manufacture, and checking of masks. At this moment we are checking different procedures to mask binarization, their performances, and their capability of being manufactured.

Main Results

The results can be divided into those related with the instrument (star generator, telescope simulator and coronagraph) and those related with the masks.

The star generator artificially produces a double star with adjustable distance and brightness. It starts with a single source (LED + pinhole) that is divided with a pellicle beam splitter (Figure 1). A path with no optics is the reference star, while the second path goes through a wheel with neutral density filters. Both beams are again combined and sent towards the telescope simulator.

The telescope chosen for simulation is the 3.5m telescope at Calar Alto Observatory (CAHA) at Cassegrain focus. This telescope has a Ritchey Chretien optical design with a hyperboloid secondary and a f10 beam at the Cassegrain focus. The pupil is at the primary mirror. The simulator reproduces the optical parameters of the telescope, including the F number, the required FOV, 4.57″, considering the plate scale 169.7 $\mu m''$ and the exit pupil. Image quality is diffraction limited in the FOV central 3″, with a fast degradation to the edge.

Nevertheless, aberrations are compensated at the coronagraph first stage, thus the focal plane for the mask is aberration free.

Since the coronagraph focal and pupil planes are related through Fourier transforms (Oti, Canales & Cagigal 2007), a 4f system will be used with the f collimator distance from the focal plane to the collimator, and f camera to detector plane. The reference wavelength will be 650nm although the goal is a fully achromatic instrument in the widest wavelength range. The instrument will be fed by the 3.5m telescope at CAHA. The instrument FOV will correspond to 60 diffraction rings. Therefore, the focal plane where the occultation mask is placed should accommodate 60 diffraction rings in 15mm approximately. The pupil plane will allow the introduction of a Lyot mask. The pupil size has been chosen to allow simple manufacturing. The detector focal plane will accommodate a 1024x1024 pixels CCD with 24x24 μm pixel size. The coronagraph design is fully reflective, thus achromatic. The functionality comes in two stages. The first stage scales the nominal (telescope) focal plane, 20 times to provide a second focal plane where the occulting mask will be placed. The second stage will follow a collimator-camera configuration while supplying a pupil image of the system.

With respect to the mask we have checked different procedures to binarize the amplitude masks. We have seen that the procedure based on error dispersion, together with a symmetry process, provides the best result for non-circularly symmetric masks. For circularly symmetric masks we are trying an iterative Fourier transform procedure.

Figure 1. Scheme of the UC coronagraph.

References

Oti, J. E., Canales, V. F., & Cagigal M. P. 2005 ApJ, 630, 631
Oti, J. E., Canales, V. F., & Cagigal M. P. 2007 ApJ, 662, 738

Pathways Towards Habitable Planets
ASP Conference Series, Vol. 430, 2010
Vincent Coudé du Foresto, Dawn M. Gelino, and Ignasi Ribas, eds.

The Search for Exoplanets in India

A. Chakraborty,[1] B. G. Anadarao,[1] and S. Mahadevan[2]

[1] *Physical Research Laboratory, Ahmedabad, India*

[2] *The Pennsylvania State University, University Park, PA, USA*

Abstract. The first dedicated program to search for exoplanets has been planned and pursued at Physical Research Laboratory. The Program is called PARAS which stands for PRL Advanced Radial-velocity All-sky Search. The Search will be conducted by an efficient optical fiber-fed Echelle Spectrograph attached to a 1.2 m telescope at the Mt. Abu Observatory in India. Using the simultaneous ThAr calibration technique we plan to achieve RV precision of 3 - 5 m/s on a 10.5 magnitude star. The spectrograph will be commissioned by early 2010 and the search is expected to begin by the end of 2010. The present search will look for Neptune-size planets around G, K, and M type dwarfs. Follow-up precision photometric observations of the prospective candidates will be done using a wide field 50 cm telescope at the same Observatory. By 2015, we have plans to have a new 2.5 m-class telescope at the same place and attach the spectrograph to it, and achieve less than 1 m/s RV precision and look for super Earths.

1. Introduction

Today, astronomy and astrophysics is primarily driven by two topics: a) cosmology and b) exoplanet sciences. While cosmology requires large telescope apertures to do effective science, exoplanet science can be achieved using ground-based small aperture telescopes. Thus, there is a clear scientific need for high-resolution spectrographs with high efficiency and wavelength stability for precise radial velocity measurements (3 to 5 m/s) of stars for various astrophysical studies like planet searches, stellar pulsations etc. Recent success of such spectrometers coupled with small telescopes (1-2 m) proves the case. In India there are many 1 to 2 m class small telescopes which can be very effectively utilized for exoplanet studies.

India is relative newcomer in the field of modern exoplanet research. In 2007 at Physical Research Laboratory, AC started a program called PARAS (PRL Advanced Radial-velocity All-sky Search) for exoplanet searches using a very stable fiber fed optical Echelle spectrograph. In the last few years we have been designing and building the spectrograph (see Chakraborty et al. 2008). By the end of 2010 we will also have a 50 cm telescope equipped with a $4k \times 4k$ front-illuminated CCD with an un-vignetting field of view of 37 arcmin by 37 arcmin for photometric studies of the prospective planet candidate stars.

2. Near Term Goals Up to Five Years

The Physical Research Laboratory (PRL) owns a 1.2 m telescope at Gurushikar, Mt. Abu which is about 230 km from the Institute at Ahmedabad. The observatory site is about 5800 feet above mean sea level and enjoys a median seeing of 1.2 arcsecs with about 150 photometric nights and a total of 250 observing nights in a year. The observatory remains closed during the monsoon from June 15th to the end of August and reopens by the second week of September. During other times, the site is mostly very dry (less than 20% relative humidity; 1 to 2 mm of water vapor column density) and largely cloud free and photometric sky. The observatory is self-sufficient in taking care of telescope mirror optics and includes an aluminum coating facility for the main telescope mirror, a process usually done once every 2 years. For exoplanet searches under the PARAS project, we are guaranteed about 80 to 90 nights in a year.

Considering the fact that the Echelle spectrograph will be attached to a 1.2 m telescope, for the PARAS project we focused and put efforts on maximum possible efficiency. It is a white pupil design of 100 mm beam width and consisting of a $R = 3.75$ Echelle, a single large prism of material PBM8Y producing a cross dispersion of 60 mm between wavelengths 370 nm and 850 nm, highly reflective off-axis parabolas and an F5 camera lens system. The spectrograph is expected to work at a resolution of $R \approx 70000$ and will be kept inside a low pressure vacuum chamber ($\approx 0.01 mbar$) under a constant temperature environment down to the precision of 10 to 20 milli-Kelvin at 27 C. It is expected to be up to 35% efficient from the fiber exit to the CCD detector, and the overall spectrograph efficiency including the telescope and the fiber optics on a good photometric night can be about 10%. Thus, the spectrograph will be ideally suited for exoplanet searches using the ThAr simultaneous referencing technique.

We are in the finishing stages of building the spectrograph. It will be commissioned by early 2010 and science verifications are expected to start immediately after that. By the end of 2010, the actual science program under the PARAS project will start. SM is developing the RV data pipeline for PARAS. The main science goals under the PARAS program are listed below:

1. To go down to 3 m/s precision on a 10th mag star and demonstrate the instrument stability by achieving 1 m/s to 3 m/s on bright stars.

2. To search for planets with a precision of 3 m/s around at least 1000 G, K, and M dwarfs that are brighter than 10.5 mag in V-band.

3. To evolve an observing strategy that will focus on longer period planets (one to a few months periods). Depending upon the spectrograph stability we achieve, we may look for even longer period planets (up to six months).

4. To look for planets around G, K Giants with lower precision of 10 m/s to 30 m/s. A good suite to confirm candidate exoplanets around giant stars detected with ongoing surveys like that at the HET.

We are in the process of installing a 50 cm robotic telescope at the same site on Mt. Abu. This telescope is expected to be operational by the end of 2010, and the PARAS project is guaranteed its use about 50% of the time. We plan to do precision differential photometry ($\approx 0.002mag$) in the R-band on stars with prospective planet candidates. The 50 cm aperture telescope will be equipped with the standard U, V, R, and I filters and the 4K CCD. A data pipe line for photometric reduction is also under development.

Complementary photometric capability along with PARAS enables scientific programs like the search for transits around long period exoplanets by refining their transit windows with RV and then observing with the 50 cm telescope, and with PARAS to detect the transit with photometry and the Rossiter McLaughlin effect. PARAS red wavelengths are useful to test concepts of radial velocity extraction in the presence of telluric features and are an essential test for future NIR spectrographs.

PARAS will provide possible follow-up and confirmation of candidates from ongoing multi-object RV surveys like SDSSIII MARVELS and long term monitoring of bright stars to discover low mass exoplanets. Stellar astrophysics studies will also be possible.

3. Long Term Goals Beyond Five Years

PARAS can be coupled to a future 2.5 m telescope with very little additional slit losses. Our goal is to have a dedicated or at least 50% time on a 2.5 m robotic telescope for PARAS and for RV searches by 2015.

The Mt. Abu site has reasonably dark sky, typically about 21 mag/arcsec^2 in the V band on a moonless photometric night even though it suffers from some light pollution from near-by towns. The sky brightness in the K-band is about 12.5 mag/arcsec^2. Thus the site is suitable for telescope apertures up to 3.5m and especially suited for ground based near-IR observations up to 2.5 microns.

By the advantage of more photon flux we plan to achieve $\sim$1 m/s on stars brighter than 12th mag and less than 1 m/s on bright stars. This will enable us to look for super-Earths around G and K dwarfs and Earth-sized planets around M dwarfs. Our observing strategy will focus on long-period planets and will make efforts to look for planets in the habitable zones of stars.

Acknowledgments. The PARAS project is supported and financed by the Physical Research Laboratory, which is a unit of the Department of Space, Govt. of India. The authors would like to thank the staff and faculty at PRL whose continuing valuable support has made this project possible. AC would like to thank Prof. J.N. Goswami, the director PRL for his continuing support and encouragement for the PARAS project. AC would also like to thank Dr. Francesco Pepe of the Geneva Observatory, Prof. Larry Ramsey of Penn State University, and Mr. Subrahmunium of SAC, ISRO for many valuable discussions.

References

Chakraborty, A. et al. 2008, Proc. of SPIE, Ground-based and Airborne Instrumentation for Astronomy II. eds. I. McLean & M. M. Casali, 7014, 70144G

Characterization of Extrasolar Planets with High Contrast Imaging

R. Claudi,[1] M. Bonavita,[1] R. Gratton,[1] S. Desidera,[1] G. Tinetti,[2] J.-L. Beuzit,[3] and M. Kasper[4]

[1] *INAF Astronomical Observatory of Padova, Vicolo dell'Osservatorio 5, Padova, Italy*

[2] *Dept. of Physics and Astronomy, University College of London, Gower St., London, UK*

[3] *Laboratory d'Astrophysique de l'Observatoire de Grenoble, 414 Rue de la Piscine, Domaine Universitaire, Saint-Martin d'Heres, France*

[4] *European Southern Observatory, Garching, Germany*

Abstract. High contrast imaging appears to be not only the most promising technique to explore the outer regions of extrasolar planetary systems, giving us a view of a complementary separation domain, but will also give us information on their atmospheres. The possibility of coupling an integral field spectrograph to a module for extreme adaptive optics and a 8m class telescope is already on the way (SPHERE for VLT and GPI for South Gemini), focusing on young giant planets. The possibility for similar instrumentation on ELTs (EPICS) is under study, and this will give us a much more complete view of extrasolar planetary systems, at different ages and, in some peculiar cases, reaching the habitable zone. A second order characterization of planets already discovered with other techniques will also be possible. In this framework we present the advantages and limits of the high contrast imaging technique and the potential of EPICS in characterization of atmosphere of both Jupiter-like and Earth-like planets using high resolution spectroscopy (R$\sim$ 3000 and R$\sim$ 20000).

1. Spectroscopic Characterization with EPICS

The EPICS concept is under development with preliminary plans to include an XAO module, able to provide high Strehl ratio (> 90% in H) down to 600 nm (Strehl > 60% in R for bright sources), and ideally down to 350 nm. The reference source should have I < 9, with a goal of I < 13. In order to allow efficient suppression of the central diffraction peak and observation at inner working angle as small as possible, EPICS will have also a diffraction suppression module. Two scientific instruments, a polarization analyzer (EPOL) and a BIGRE based Integral Field Spectrograph (IFS), are the core of EPICS. Not included in the baseline design, EPICS might also include in the future a Self Coherent Camera (SCC) based on speckle correction and differential imaging. The IFS should have at least a 2 arcsec (goal 4 arcsec) diameter field of view. The spectra will cover the wavelength region between 0.95 and 1.7 μm with a spectral resolution larger than 50. It should also have a high spectral resolution mode with

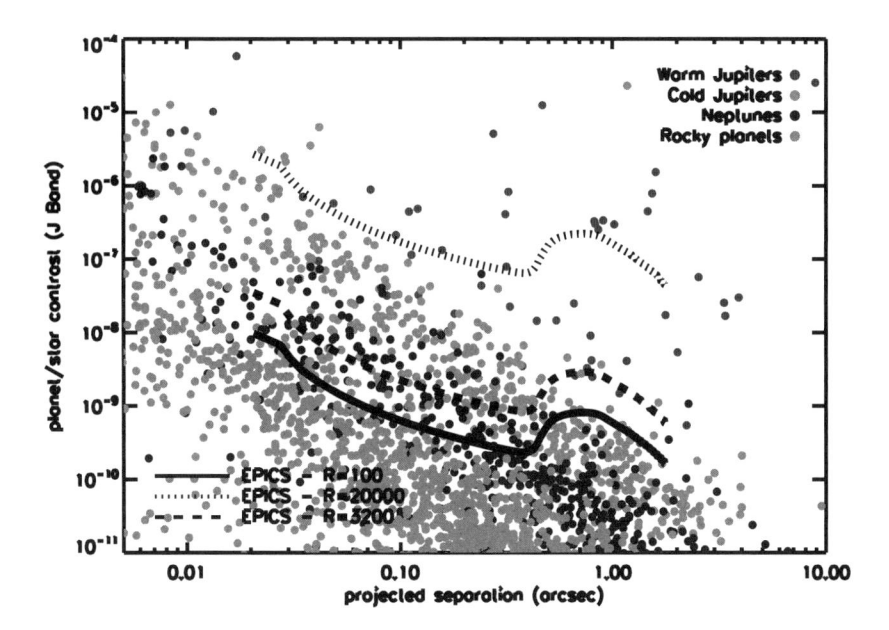

Figure 1. The detectability function of EPICS for the different resolution modes foreseen.

resolution of 3,000 and 20,000. EPICS will be used both in survey mode to search for and detect a large number of planets, and in characterization mode to probe (for planets detected with high enough S/N) the atmospheric composition and structure of a limited number of planets achieving the following main science aims: identification of spectral features, determination of physical parameters (temperature, gravity, chemical composition), cloud processes and their variation with time (e.g., for planets in eccentric orbits). A large number of planets can be observed with EPICS at resolution R=3,000, and even a few at high spectral resolution of R=20,000 (see Figure 1). Observable objects include warm and cold Jupiters, Neptunes, and even a few rocky planets.

Neptunes and Super-Earths are detectable only if they are not too far from the star because in this case they are expected to be warmer, with (equilibrium) temperatures up to ∼ 200-300 K and as consequence brighter. Spectra of giant and Neptune-like planets are expected to be dominated by methane bands; these are resolved at a resolution R >15. Spectra with resolution R > 30 are then not required in order to detect the planets, at least insofar as the band contrast is considered.

As pointed out by the recent discoveries, direct detection seems destined to unveil different kinds of extrasolar planetary systems, with respect to those discovered up to now via indirect methods. The availability of a 40m class telescope will not only allow us to take advantage of the synergy of different methods but also to make a first order characterization of planetary atmospheres, using high resolution spectroscopy, down to masses comparable with Neptune.

ALADDIN: An Antarctic-based Nulling Interferometer for Exozodi Characterization

V. Coudé du Foresto

Observatoire de Paris – LESIA, 5 Place Jules Janssen, Meudon Cedex, Paris, France

Abstract. ALADDIN is an infrared (L-band) nulling interferometer optimized for the study of exozodiacal dust. Although relatively modest in size (two 1-meter class telescopes on a maximum baseline of 32 m), it takes advantage of the privileged atmospheric conditions on the Antarctic plateau to achieve a sensitivity better than what can be obtained with a pair of 8 m class telescopes on a temperate site. Beyond its main mission, the science potential of ALADDIN extends to the study of all kinds of faint circumstellar material (dust and/or molecules) around young, old, or main sequence stars.

The Need for an Exozodi Survey

Exozodiacal dust clouds are interesting in their own right due to their role in defining the boundary conditions for the formation and evolution of planetary systems. But they also play an important role to study the feasibility and scope of future imaging space missions dedicated the spectroscopic analysis of the atmosphere of habitable planets (see summary of Panel Meeting on Exozodiacal Dust Disks discussion by Absil et al. in this volume). Indeed, if one takes the solar system as an example, the intensity of the infrared flux emitted by zodiacal dust (1 zodi by definition) is 300 times brighter than Earth. Fluctuations of this radiation can be a significant noise source in the survey of exoearths, and may even jeopardize the detection if the exozodiacal light is brighter than $\simeq 30$ zodis. This calls for a survey of dust clouds around potential targets, down to that sensitivity level, in order to reduce the risk on the space mission and not waste time on sources where exoearths cannot be detected. Similar studies (e.g., Beichman et al. 2006) have shown that noise due to exozodiacal dust radiation is significant for a visible light coronagraph as well.

The Need for a Dedicated Nuller

While cold debris disks have been detected and sometimes imaged far away from the central star by ground- and space-based instruments, measuring exozodiacal dust near the habitable zone remains elusive. These areas cannot be detected as IR excesses as they contribute to less than a few percent of the photospheric flux of the central star. Their characterization requires a combination of high angular resolution (1 AU at 20 pc is 50 mas) and high dynamic range that only infrared nulling interferometry can achieve. Yet the performance of ground-based nulling interferometers is strongly affected by atmospheric turbulence (phase noise) and thermal background.

Nulling interferometry would be particularly efficient from the high Antarctic plateau, where the French-Italian Concordia station at Dome C and the future Chinese Kunlun station at Dome A are located. Low environmental and sky temperatures (195–235 K) result in a low and stable emission from the thermal background. Extremely dry air improves infrared transmission, reduces dispersion variations, and extends atmospheric windows. Most of the turbulence is concentrated in the ground layer (whose typical height is 30 m), above which the seeing has been found to be much lower than at any other measured site (0.27 arcsec median value). Besides, long coherence times (≥ 7 ms in normalized conditions) have strong benefits for the performance (lower residuals) of real time control loops (tip-tilt, fringe tracking).

A concept for an Antarctic nulling interferometer for exozodis studies is (ALADDIN (Antarctic L-band Astrophysics Discovery Demonstrator for Interferometric Nulling Coudé du Foresto et al. 2006). It has a dedicated architecture optimized at the system level. An initial concept for ALADDIN (Barillot et al. 2006) consists of two 1 m siderostats that can be relocated along a 32 m long rotating beam, with the nulling beam combiner in the center. Using proven modeling tools it was possible to demonstrate (Absil et al. 2007) that at Dome C, an optimized interferometer located above the turbulent layer (or equipped with adequate adaptive optics) would be more sensitive with a pair of 1 m collectors than a nulling instrument with the VLT's UTs (8 m) at Paranal, and be compliant with the 30 zodis requirement.

After its primary mission is complete, ALADDIN could either be dismantled or used for more general science, if adequate support for extended operations can be found. Its observational niche is the characterization of faint environments around central point sources, a common feature of many astrophysical objects. It would be, for example, an exceptional tool to characterize circumstellar disks around YSOs, circumstellar matter ejected by stars in the late stages of their evolution, or even the K- or L-band spectroscopy of a few close-in giant exoplanets around very nearby stars.

Conclusion

The Antarctic plateau is a very favorable site for high dynamic range IR stellar interferometry, and provides the opportunity to carry out an exozodi survey of nearby stars at a fraction of the cost of a space mission that would normally be needed for that purpose. A modest nulling interferometer on the Antarctic plateau would have performance compliant with the exozodi sensitivity required to prepare future space missions dedicated to the spectroscopic characterization of habitable exoplanet atmospheres.

References

Absil O., Coudé du Foresto V., Barillot M., & Swain M. R., 2007, A&A, 475, 1185
Barillot M., et al., 2006, in SPIE 6268, Advances in Stellar Interferometry, ed. Monnier, J., Schöller, M., Danchi, W. C., , 62682Z
Beichman, C. A., et al., 2006. ApJ, 652, 1674
Coudé du Foresto V., et al., 2006, in SPIE 6268, Advances in Stellar Interferometry, ed. Monnier, J., Schöller, M., Danchi, W. C., 626810

Pathways Towards Habitable Planets
ASP Conference Series, Vol. 430, 2010
Vincent Coudé du Foresto, Dawn M. Gelino, and Ignasi Ribas, eds.

A Photometric Transit Search for Planets around Cool Stars from the Italian Alps: Results from a Feasibility Study

M. Damasso,[1] P. Calcidese,[1] A. Bernagozzi,[1] E. Bertolini,[1] P. Giacobbe,[2] M. G. Lattanzi,[3] R. Smart,[3] and A. Sozzetti[3]

[1]*Astronomical Observatory of the Autonomous Region of the Aosta Valley, Loc. Lignan 39, Nus (Aosta), Italy*

[2]*Dept. of Physics, University of Turin, Via Giuria 1, Turin, Italy*

[3]*INAF-Astronomical Observatory of Turin, Strada Osservatorio, Pino Torinese, Turin, Italy*

Abstract. A feasibility study was carried out at the Astronomical Observatory of the Autonomous Region of the Aosta Valley demonstrating that it is a well-poised site to conduct an upcoming observing campaign aimed at detecting small-size (R≤$R_{Neptune}$) transiting planets around nearby cool M dwarf stars. Three known transiting planet systems were monitored from May to August 2009 with a 25 cm f/3.8 Maksutov telescope. We reached seeing-independent, best-case photometric RMS less than 0.003 mag for stars with V≤13, with a median RMS of 0.006 mag for the whole observing period.

1. Introduction

The Astronomical Observatory of the Autonomous Region of the Aosta Valley (OAVdA; 45.7895°N, 7.47833°E) has been identified as a potential site for hosting a photometric transit search for low-mass, small-size planets around nearby cool M dwarf stars. We carried out a study to gauge the near-term and long-term photometric precision achievable, observing known transiting systems in a range of transit depths under variable seeing and sky transparency conditions.

2. Instrumentation and Methodology

We used a 25 cm Maksutov f/3.8 reflector telescope equipped with a CCD Moravian G2-3200ME and an R filter centered at 610 nm (field of view: 52.10 x 35.11 arcmin^2; plate scale: 1.43 arcsec/pix; QE≈87% at 610 nm). The seeing was monitored each night using a Hartmann mask and the DIMM technique. It varied in the range 1-3 arcsec (median ≈1.7 arcsec). Data reduction and ensemble differential aperture photometry (≈100 reference objects) were carried out with the version 1.0 of an automated, IDL-based pipeline we have developed.

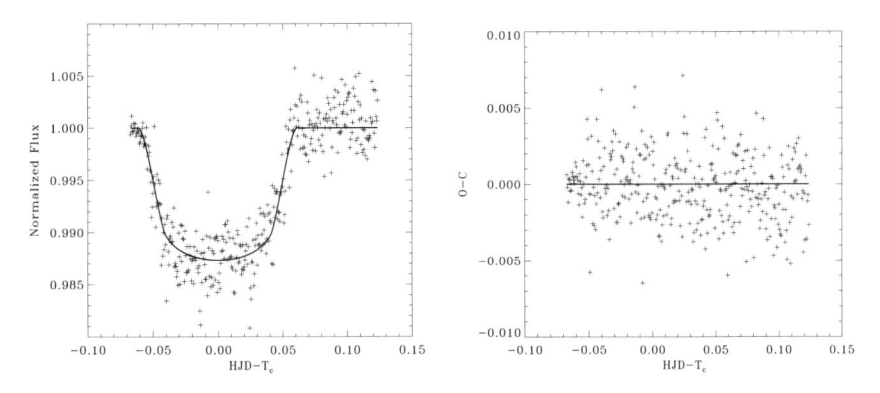

Figure 1. *Left:* WASP-3b transit light curve (28-29 July 2009; expos. 35 s; T$_C$=55041.411) and best-fit curve. *Right:* residuals of the fitted curve.

3. Selected Results

In May-August 2009 we monitored three transiting systems: WASP-3 (Pollacco et al. 2008), HAT-P-7 (Pál et al. 2008) and Gliese 436 (Butler et al. 2004). WASP-3, a 1.24 M$_{Sun}$ star (V=10.64) hosting a 1.76 M$_{Jupiter}$ planet (period=1.846834 days), was monitored for 19 nights, often reaching a photometric RMS below 0.003 mag for stars with V≤13. Figure 1 shows one of the observed WASP-3b transits and the best-fit model is based on the formalism of Mandel & Agol (2002), assuming quadratic limb darkening. We got: b=0.665±0.003 (impact parameter), i=81.24±0.06 degrees (orbital inclination), r=0.1091±0.0006 (planet-to-star radius ratio), T$_C$=55041.411±0.029 HJD (time of mid transit). The median RMS for the whole observing period is 0.006 mag (V≤13).

4. Conclusions and Future Work

The feasibility study demonstrated that the OAVdA is a promising site for a long-term photometric survey aimed to detect transiting low-mass, small-size planets. We are planning to set up a system of five 40 cm identical telescopes for a high-precision observing campaign focused on several hundreds nearby M dwarfs, improving the longitudinal coverage of similar ongoing programs, such as the Arizona-based MEarth project (Nutzman & Charbonneau 2008).

Acknowledgments. M.D., P.C. and A.B. benefit of a grant provided by the European Union, the Autonomous Region of the Aosta Valley and the Italian Department for Work, Health and Pensions.

References

Butler, R. P. et al. 2004, ApJ, 617, L580
Mandel, K. & Agol, E. 2002, ApJ, 580, L171
Nutzman, P. & Charbonneau, D. 2008, PASP, 120, 317
Pál, A. et al. 2008 ApJ, 680, 1450
Pollacco, D. et al. 2008, MNRAS, 385, 1576

Pathways Towards Habitable Planets
ASP Conference Series, Vol. 430, 2010
Vincent Coudé du Foresto, Dawn M. Gelino, and Ignasi Ribas, eds.

Influence of Exozodiacal Dust Clouds on Mid-IR Earth-like Planet Detection

D. Defrère,[1] O. Absil,[1] R. den Hartog,[2] C. Hanot,[1] and C. Stark[3]

[1]Département d'Astrophysique, Géophysique & Océanographie, Université de Liège, 17 Allée du Six Août, Sart Tilman, Belgium

[2]Netherlands Institute Space Research, SRON, Sorbonnelaan 2, Utrecht, The Netherlands

[3]Department of Physics, University of Maryland, Box 197, 082 Regents Drive, College Park, MD, USA

Abstract. The characterization of Earth-like extrasolar planets in the mid-infrared is a significant observational challenge that could be tackled by future space-based interferometers. A large effort has been carried out the past two decades to define a design that provides the necessary scientific performance while minimizing cost and technical risks. These efforts have resulted in a consensus on a single mission architecture consisting of a non-coplanar X-array (the so-called *Emma* configuration), using four collector spacecraft and a single beam combiner spacecraft. The ability to study distant planets with an X-array interferometer will however depend on exozodiacal dust clouds, the counterparts of the solar zodiacal disk. In this paper, we briefly discuss the impact of exozodiacal clouds on the performance of an Emma X-array interferometer dedicated to Earth-like planet characterization.

1. Impact of Exozodiacal Dust Density

An exozodiacal dust cloud similar to the solar zodiacal disk is approximately 400 times more luminous than an Earth-like planet at 10 μm. Thanks to phase chopping enabled by the X-array configuration, the symmetric component of exozodiacal disks are suppressed and only contributes to shot noise (which drives the integration time). The number of targets that can be surveyed during the mission lifetime depends therefore on the amount of exozodiacal dust in the habitable zone of nearby main sequence stars. This is illustrated by Figure 1 showing the number of extrasolar planetary systems that can be surveyed during a 2-year detection phase with respect to the exozodiacal dust density. For 2-m diameter telescopes, about 200 targets could be surveyed if exozodiacal dust clouds are as dense as the solar zodiacal cloud. Exozodiacal dust disks 100 times denser than the solar zodiacal would reduce this number to approximately 150.

2. Impact of Resonant Dust Structures

Although phase chopping suppresses the symmetric component of exozodiacal disks, it has no effect on asymmetric features, such as clumpy resonant dust

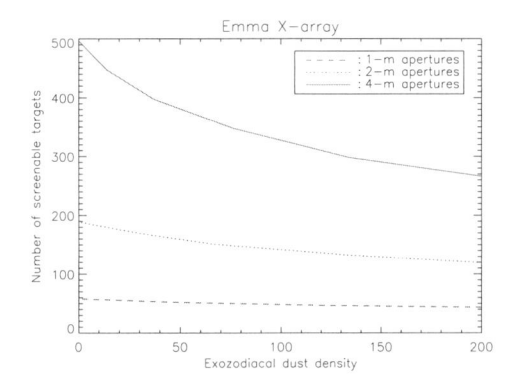

Figure 1. Number of extrasolar planetary systems that can be surveyed for Earth-like planet detection during a 2-year detection phase, assuming a space-based nulling interferometer in the Emma X-array configuration, point-symmetric exozodiacal disks, and various telescope diameters.

rings created by the gravitational influence of embedded planets. For a resonant structure created by an Earth-like planet orbiting at 1 AU around a Sun-like star in a cloud of ∼100-μm dust grains (Stark & Kuchner 2008), models suggest that the tolerable dust density is reduced to approximately 15 times the solar zodiacal density (Defrère et al. 2010). This constraint might be relaxed since this study considered a resonant ring model that does not include dust from highly eccentric or inclined parent bodies, the effects of grain-grain collisions, or perturbations by additional planets, all of which can reduce the density of the resonant ring structures (Stark & Kuchner 2009).

3. Conclusion

Whereas the disk brightness only affects the integration time, the presence of clumpy resonant rings is more problematic and may hamper planet detection. Current models of debris disk structures suggest that in the worst-case scenario, the upper limit on tolerable exozodiacal dust density may be as little as 15 times the density of the solar zodiacal cloud to detect an Earth. This gives the typical sensitivity that we will need to reach on exozodiacal disks in order to prepare the scientific program of future Earth-like planet characterisation missions.

Acknowledgments. DD and CH acknowledge the support of the Belgian National Science Foundation ("FRIA"). OA acknowledges the support from a F.R.S.-FNRS Postdoctoral Fellowship. DD, CH and OA acknowledge support from the Communauté française de Belgique ("ARC").

References

Defrère, D., Absil, O., den Hartog, R., Hanot, C., & Stark, C. 2010, A&A, 509, 9
Stark, C. C. & Kuchner, M. J. 2008, ApJ, 686, 637
Stark, C. C. & Kuchner, M. J. 2009, ApJ, 707, 543

Pathways Towards Habitable Planets
ASP Conference Series, Vol. 430, 2010
Vincent Coudé du Foresto, Dawn M. Gelino, and Ignasi Ribas, eds.

Future Technical Developments Concerning Nulling Interferometry at the Institut d'Astrophysique Spatiale (Orsay)

O. Demangeon

Institut d'Astrophysique Spatiale, Bâtiment 209G, Orsay, France

Abstract. Technical research in the field of nulling interferometry at the Institut d'Astrophysique Spatiale (IAS) is following two principal paths. The first path concerns technology studies for the Darwin project with the test bench NULLTIMATE, and the second one is the study of precursors to the Darwin project which appear compelling and necessary.

1. Darwin Project: NULLTIMATE Test Bench

The Darwin project (like TPF-I) aims at using nulling interferometry to perform a low-resolution spectral study of telluric exoplanets atmospheres. The objective of the NULLTIMATE test bench is to reach the technical specifications for this project regarding nulling ratio (10^{-5}) and its stability (3.10^{-9} over 10 days) with polychromatic light in the K band. Future developments of the bench concern:

Test of Achromatic Phase Shifters (APS)

We possess two types of hopeful APS that we will try to characterize: focus crossing and field reversal.

Stability

The rejection ratio $\rho = \frac{I_{max}}{I_{min}}$ depends on optical path difference (OPD) and flux balance ϵ_I[1] and in order to reach a stability of the rejection ratio of 3×10^{-9} over 10 days, one needs to dither both quantities.

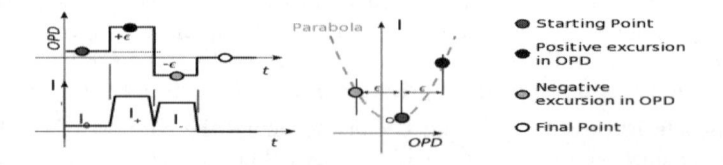

Figure 1. Principle of optical path difference dithering. The final position is computed from measurements of the three other points.

---

[1] ϵ_I is the ratio of fluxes in both arms of the interferometer

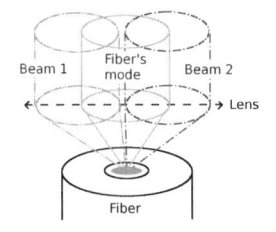

Figure 2. Flux balance dithering by movement of fiber head.

We want to realize the OPD dithering by modulation of the position of the delay line. We work directly on scientific signal which implies a better accuracy than a metrologic system (Figure 1).

As the two beams of the interferometer do not have exactly the same direction and the same position relative to the mono-mode fiber (modal fiber). They are not coupled into the fiber with the same efficiency. Thus, moving the fiber head allows us to balance the flux in both arms (Figure 2) and realize flux balance dithering.

Polarization

Nulling measurements can be performed in linearly polarized light not only oriented parallel to the test bench polarization plane of maximum transmission but also perpendicular to it. Such an experiment could be the first direct demonstration, in the context of nulling interferometry, of the difference in optical path length between the two polarizations.

2. Precursor Experiments to Darwin Project

Some of the technologies for the Darwin mission such as formation flying are not yet mature. Furthermore the prize of the project is actually dissuasive. Therefore precursors to Darwin with less stringent specifications appear necessary and beneficial for the field of nulling interferometry. Scientific purposes of such precursors are numerous:

- Validation of interferometry in space conditions;
- Developments in high angular resolution;
- Studies and characterization of zodiacal environment.

Several concepts are currently under study and IAS want to be involved in such projects. We are currently particularly interested in two possible precursors:

- ALADDIN: Antarctic L-band Astrophysics Discovery Demonstrator for Interferometric Nulling. A ground precursor in quasi-space conditions.
- A balloon experiment like FITE: the Far-infrared Interferometric TElescope. A flying precursor with rigid baseline.

Pathways Towards Habitable Planets
ASP Conference Series, Vol. 430, 2010
Vincent Coudé du Foresto, Dawn M. Gelino, and Ignasi Ribas, eds.

Ground-based Search of Earth-mass Exoplanets using Transit-Timing Variations

J. M. Fernandez

Institute for Astrophysics, University of Goettingen, Goettingen, Germany

Abstract. This work presents recent results from a ground-based transit follow-up program of the extrasolar planet XO-2b in order to find Earth-mass companions. It also introduces the future use of the MONET 1m-class robotic telescopes as part of the effort to overcome the difficulties of this kind of project.

1. Search for Transit-timing Variations

The gravitational interaction between a transiting planet and a secondary object can introduce variations in the expected time of transits for a perfectly periodic orbit. If the system is composed of a short-period Jupiter-mass transiting planet and an Earth-mass object orbiting in mean motion resonance, the variation in the expected time of transit can be larger than a minute (see Holman & Murray 2005; Agol et al. 2005). Photometric observations of six transits of the extrasolar planet XO-2b were obtained with KeplerCam on the 1.2m telescope in Mount Hopkins, Arizona, and with the Centurion 0.46 m (18-inch) telescope, at the Wise Observatory, in Israel. Transit timing errors smaller than 25 sec were derived for two events. If previous published data (Burke et al. 2007) are considered, the transit timings are marginally inconsistent with a constant orbital period. For a detailed description of the observations and analysis, see Fernandez et al. (2009).

2. MONET Robotic Telescopes

Systematic ground-based transit follow-up is a challenging task: it requires permanent good weather conditions and a considerable amount of telescope time. Even when bad weather and telescope time are not a problem, many transits will not be observed (generally) if only one telescope is used because of the rotation of the Earth. The MONET project consists of two robotic telescopes located in Texas and South Africa, built for educational and research purposes. Both telescopes have 1.2m primary mirrors, and can be fully controlled by remote observers, with basically no need of local assistance. CCD cameras with broad and narrowband filters are available to perform high precision photometry. The difference in longitude and latitude between the two telescopes makes them an ideal tool for exoplanet transits follow up. During 2009 and 2010, MONET-north will perform systematic observations of known transiting planets. The flexibility of the observing schedule allows the study of new transiting systems as soon as they are announced.

Acknowledgments. This work was suported by NASA Origins Grant No. NNX09AB33G, by Grant No. 2006234 from the United States Israel Binational Science Foundation (BSF), and is currently suported by the DFG Graduiertenkolleg (GrK) 1351 "Extrasolar Planets and their Host Stars". KeplerCam was developed with partial support from the Kepler Mission under NASA Cooperative Agreement NCC2-1390 and the Keplercam observations described in this paper were partly supported by grants from the Kepler Mission to SAO and PSI.

References

Agol, E., Steffen, J., Sari, R., & Clarkson, W. 2005, MNRAS, 359, 567
Burke, C. J., et al. 2007, ApJ, 671, 2115
Fernandez, J. M., et al. 2009, AJ, 137, 4911
Holman, M. J., & Murray, N. W. 2005, Sci, 307, 1288

Pathways Towards Habitable Planets
ASP Conference Series, Vol. 430, 2010
Vincent Coudé du Foresto, Dawn M. Gelino, and Ignasi Ribas, eds.

The Fabra-ROA Baker-Nunn Camera at Observatori Astronòmic del Montsec: A Wide-field Imaging Facility for Exoplanet Transit Detection

O. Fors,[1,2] J. Núñez,[1,2] J. L. Muiños,[3] F. J. Montojo,[3] R. Baena,[2] M. Merino,[1,2] R. Morcillo,[3] and V. Blanco[3]

Abstract. A number of Baker-Nunn Camera (BNC) were manufactured by Smithsonian Institution during the 60's as optical tracking systems for artificial satellites with optimal optical and mechanical specifications. One of them was installed at the Real Instituto y Observatorio de la Armada (ROA). We have conducted a profound refurbishment project of the telescope to be installed at Observatori Astronòmic del Montsec (OAdM) (Fors 2009). As a result, the BNC offers the largest combination of a huge FOV (4.4°x4.4°) and aperture (leading to a limiting magnitude of V$\sim$20). These specifications, together with their remote and robotic natures, allows this instrument to face an observational program of exoplanets detection by means of transit technique with high signal-to-noise ratio in the appropiate magnitude range.

Refurbishment Project

The BNC was designed as a f/1 0.5m photographic wide field (5°x30°) telescope with a spot size smaller than 20μm throughout the FOV.

Among some others, the BNC has been refurbished following these phases: mechanical modification of the mount into equatorial and motorization of the two axes, optical refiguring of the originally photographic curved FOV to enable the use of a 4kx4k 9-μm custom-designed FLI ProLine CCD camera (see Figure 1) and to comply the Baker's original design spot size diagram, manufacture of a tip-tilt adjustable spider vanes assembly and athermal CCD focus system (see Figure 2), mirror realuminization and outermost 50cm lens repolishing to increase the throughput of the system, construction of a reinforced glass-fiber enclosure with sliding roof which will host the BNC at OAdM (see Figure 3), development of an XML-based messaging protocol and Java GUIs software, named Instrument-Neutral Distributed Interface (INDI), to control every device of the observatory and schedule its operation both in remote and robotic modes (Downey 2009).

On 23 Sep 2008, the BNC successfully saw first technical light at ROA testing site (see Figure 4). Note this image was still taken with the unpolished 50cm lens and non-realuminizated mirror. The definitive commissioning at OAdM is expected by early Spring 2010.

---

[1] Observatori Fabra, Reial Acadèmia de Ciències i Arts de Barcelona, E-08002 Barcelona, Spain

[2] Departament d'Astronomia i Meteorologia and Institut de Ciències del Cosmos (ICC), Universitat de Barcelona (UB/IEEC), E-08028 Barcelona, Spain

[3] Real Instituto y Observatorio de la Armada, San Fernando, Spain

Figure 1. Custom-designed FLI CCD with field flattener.

Figure 2. Spider vanes assembly and focus system for the CCD.

Figure 3. Reinforced glass-fiber enclosure at OAdM.

Figure 4. First technical light of M31 at ROA on 23 Sep 2008.

Transit Exoplanet Detection

The robotic nature of the BNC, its huge FOV and its considerable aperture, enables the telescope to succesfully detect transits of exoplanets. This expectation is supported by the fact that the Automatic Patrol Telescope (APT), originally a BNC twin of ours and, which after a similar refurfishment (although with an smaller FOV and less sensitive CCD), has succesfully compiled the UNSW Extrasolar Planet Search 2004-2007 catalogue of exoplanets candidates (Christiansen et al. 2008). This catalogue shows that BNC-based cameras can accomplish millimagnitude photometry at least up to V$\sim$14 magnitude.

Acknowledgments. This work was supported by the Ministerio de Ciencia e Innovación of Spain and by Departament d'Universitats, Recerca i Societat de la Informació of the Catalan Goverment.

References

Christiansen, J. L., et al. 2008, MNRAS, 385, 1749
Downey, E. C. 2009, Clear Sky Institute, Inc.: INDI Control System Architecture in
    http://www.clearskyinstitute.com/Company/consulting/consulting.html
Fors, O., et al. 2009, Fabra-ROA Baker-Nunn at Observatori Astronòmic del Montsec
    in http://www.am.ub.es/bnc

Pathways Towards Habitable Planets
ASP Conference Series, Vol. 430, 2010
Vincent Coudé du Foresto, Dawn M. Gelino, and Ignasi Ribas, eds.

Dynamical Stability in the Habitable Zones of Nearby Extrasolar Planetary Systems

B. Funk,[1] Á. Süli,[1] E. Pilat-Lohinger,[2] R. Schwarz,[2] and S. Eggl[2]

[1]*Department of Astronomy, Eötvös Loránd University, Budapest, Hungary*

[2]*Institute for Astronomy, University of Vienna, Vienna, Austria*

Abstract. Discoveries of super-Earths raise our hopes to find terrestrial planets moving in the habitable zone (HZ) of Sun-like stars. In the near future astronomers will try to find such an "exo-Earth" orbiting within the HZ of other stars having various ground-based and space missions (e.g., CoRoT, Kepler, Gaia) at their disposal. It is well known, that the evolution of a biosphere is a process covering stellar timescales, therefore one of the basic requirements for habitability is the long-term orbital stability of planets in the HZ. This investigation tackles the dynamical stability of potential additional terrestrial planets in nearby (within 30 pc) extrasolar planetary systems. Global studies do exist, that provide stability maps for systems consisting of a star, a gas giant and mass-less test-planets, e.g., the so-called "Exocatalogue" (see http://astro.elte.hu/exocatalogue/). Additionally we provide the tool "ExoStab" (see http://www.univie.ac.at/adg/exostab/), which allows a more interactive application of the Exocatalogue data. Utilities like these are used to verify the long-term stability of an additional low-mass planet moving in a particular single-star single-planet system. A comparison of the results of numerical simulations including also massive test-planets with the Exocatalogue will be shown. Since some orbital elements, like e.g., the orbital eccentricities, the mass of the gas giant or its semi-major axis are afflicted with errors, we will also perform a parameter study, to ensure the validity of our results.

1. Introduction

Several satellite missions have currently been or will be launched within the next years (for details see e.g., Fridlund (2008), ESA's and NASA's homepages) sharing the aim of detecting extrasolar planets. Since most satellite missions can only observe a restricted sample of stars during their finite time span of operation, it is quite important to choose these target systems very carefully. The goal of this work is to investigate possible target systems of upcoming space missions in which already one or more extrasolar planets have been found. We will be able to exclude systems in which additional terrestrial planets are unlikely within the HZ.

In all cases, the motion of the planets were considered in the framework of purely Newtonian gravitational forces and all the celestial bodies involved were regarded as point masses. For all single star single planet systems, the elliptic restricted three-body problem proves to be a valid model, a fact that was also found by Sándor et al. (2007). For the up to now 13 investigated systems we used the catalogue to determine the stable regions within the HZ. For the terrestrial planets (TP) we did numerical computations using a fine grid of initial conditions. We varied the semi-major axis covering the complete HZ, and we also conducted computations within a certain range of initial inclinations ($i \leq 60°$) and masses of the TPs. A major point of our study was the inclusion of uncertainties of the determined orbits (and masses) of the observed gas giant (GG):

1. The masses of the GG will be varied, because observers can determine just a minimum mass,

2. The eccentricities of the known planets will be changed, because there are uncertainties associated with the observations,

3. The semi-major axis of the GG will also be slightly changed.

2. Results

The list of all up to now investigated systems can be found under: http://www.univie.ac.at/adg/hzcat/. The systems are sorted by their Hipparcos number. In the 2nd and 3rd columns the mass and the spectral-type of the host star are shown. After the name, (4th column as in www.exoplanet.eu) we provide the mass, semi-major axis and eccentricity of the known planet, as well as the width of the HZ (for the calculation of the width of the HZ, see Kaltenegger et al. 2008). In the last column one can find a link, which leads to a more detailed view of each system, displaying the result plots for all different initial conditions. Regarding just the observational given initial conditions, we could find 2 stable, 9 partly stable and 2 completely unstable systems.

Acknowledgments. B. Funk wants to acknowledge the support by the Austrian FWF Erwin Schrödinger grant no. J2892-N16. E. Pilat-Lohinger wants to acknowledge the support by the Austrian FWF project no. P19569. Á. Süli wants to acknowledge the support by the Hungarian Scientific Research Fund PD-75508. R. Schwarz wants to acknowledge the support by Europlanet NA1 and by the ÖFG project 06/11206. S. Eggl wants to acknowledge the support by the Austrian FWF project no. P-20216.

References

Fridlund, M. 2008, Space Sci.Rev., 135, 355
Kaltenegger, L., Eiroa, C., & Fridlund, M. 2008, Target star catalog for Darwin: Nearby stellar sample for a search for terrestrial planets, arXiv:0810.5138
Sándor, Z., Süli, Á., Érdi, B., Pilat-Lohinger, E., & Dvorak, R. 2007, MNRAS, 375, 1495

Pathways Towards Habitable Planets
ASP Conference Series, Vol. 430, 2010
Vincent Coudé du Foresto, Dawn M. Gelino, and Ignasi Ribas, eds.

The NULLTIMATE Testbed: A Progress Report

P. Gabor

Institut d'Astrophysique Spatiale, Université Paris XI, Orsay, France

Abstract. Nulling interferometry has been suggested as the underlying principle for an instrument which could provide direct detection and spectroscopy of Earth-like exoplanets, including searches for potential biomarkers (*Darwin*/TPF-I). Several aspects of this method require further research and development. The NULLTIMATE testbed at the *Institut d'Astrophysique Spatiale* in Orsay, France, is a new instrument, built in late 2008. It is designed to test different achromatic phase shifters (focus crossing, field reversal, dielectric plates) at 300 K using various sources ranging from 2 to 10 μm, with special attention to stabilization (optical path difference and beam intensity balance). Its operational parameters (null depth and stability) were tested with a monochromatic laser sources at 2.32 and 3.39 μm and with a supercontinuum source in the K band. This poster presents a progress report on its performance with a focus crossing achromatic phase shifter.

The *Institut d'Astrophysique Spatiale* in Orsay, France, has been involved in nulling interferometry for more than ten years (Ollivier 1999; Brachet 2005; Chazelas 2007; Gabor 2009). This poster presents the latest of our testbeds. Its objectives are broadband nulling research and development, tests of achromatic phase shifters (APS), and stability studies. The testbed was designed by B. Chazelas and M. Decaudin, set up by P. A. Schuller, P. Duret and P. Gabor, and its preliminary tests were performed by O. Demangeon and P. Gabor.

Optical Layout. NULLTIMATE's source array (Figure 1) currently comprises two monochromatic sources (3.39 μm HeNe laser, and 2.32 μm laser diode) as well as a supercontinuum source used in conjunction with interference filters, mainly used in the K band (2 – 2.5 μm). The flux is injected into the interferometer via a single-mode optical fiber (fluoride glass).

The interferometer is based on a simple Mach-Zehnder architecture. The beam splitter and beam combiner are ZnSe uncoated plates with a wedge angle of 1 degree. Compensator plates, mounted antisymmetrically, eliminate the angular dispersion. One of them also compensates for chromatism: it is translatable in order to adjust ZnSe thickness in the two interferometer arms. Each arm contains a delay line, and an APS module.

The flux, combined for geometrical reasons by the beam combiner, interferes when coupling with the exit single mode fiber. Its primary role is modal wavefront filtering. Two polarizers can be mounted, one immediately in front of the beam splitter, and the other after the beam combiner.

The testbed is installed in a quiet clean room. The AC system maintains the temperature between 20 and 21°C. The table is mounted on active pneumatic legs, and the interferometer is covered by a Plexiglas enclosure. This permits a stable operation with few drifts. All mounts use a line-point-plane system

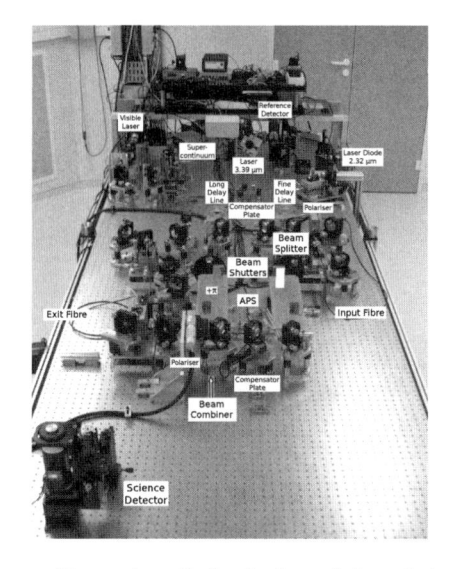

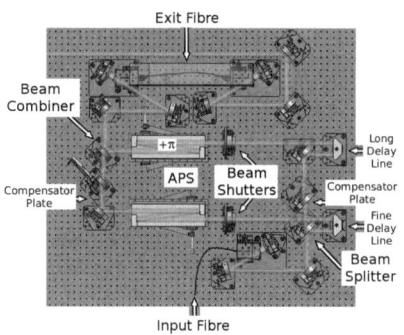

Figure 1. Left: A view of the whole tabletop with the sources on top, the interferometer in the centre and the detector bottom left. Right: The optical layout of the interferometer. The beam enters the setup from the bottom right, and exits top left.

with a reproducibility of < 2 arcsec. The interferometer is built around the two modules of an APS prototype. The APS currently imbedded in the layout is a focus crossing (a.k.a. through focus; produced by Observatoire de Côte d'Azur) setup. NULLTIMATE's mechanics are designed to be easily modified in order to accept another APS prototype, the field reversal (a.k.a. periscope; produced by Max-Planck-Institut für Astronomie Heidelberg).

Preliminary results. The nulling depths reached during commissioning were 4×10^{-5} with the 3.39 μm laser, 3×10^{-4} with the 2.32 μm laser diode, and 1×10^{-3} with the supercontinuum source in the K band (2.0-2.5 μm). It is likely that the performance is limited by the fibers. Changing the operating wavelength to the J band may improve the situation.

References

Brachet, F. 2005, PhD Thesis, Université Paris XI
Chazelas, B. 2007, PhD Thesis, Université Paris XI
Gabor, P. 2009, PhD Thesis, Université Paris XI
Ollivier, M. 1999, PhD Thesis, Université Paris XI

Pathways Towards Habitable Planets
ASP Conference Series, Vol. 430, 2010
Vincent Coudé du Foresto, Dawn M. Gelino, and Ignasi Ribas, eds.

Low-mass Objects in Moving Groups

M. C. Gálvez-Ortiz,[1] J. R. A. Clarke,[1] D. J. Pinfield,[1] S. L. Folkes,[1]
J. S. Jenkins,[2] A. E. García Pérez,[1] B. Burningham,[1]
A. C. Day-Jones,[2,1] and H. R. A. Jones[1]

[1] *Centre for Astrophysics Research, Science and Technology Research Institute, University of Hertfordshire, Hatfield, Hertfordshire, UK*

[2] *Department of Astronomy, Universidad de Chile, Santiago, Chile*

Abstract. We present here the kinematic study of part of a 132 target sample of low-mass objects, previously selected by photometric and astrometric criteria as possible members to five young moving groups (MGs): Local Association (Pleiades moving group, 20 - 150 Myr), Ursa Mayor group (Sirius supercluster, 300 Myr), Hyades supercluster (600 Myr), IC 2391 supercluster (35 Myr) and Castor moving group (200 Myr). We calculated their kinematic galactic components (U,V,W) and apply simple kinematic criteria of membership. The confirmed members will provide a new and substantial population of age and metallicity benchmark ultra cool dwarfs (UCDs). This will give a valuable set to test atmosphere and evolutionary models of ages below 1 Gyr and suitable targets for use in adaptive optic (AO) imaging to search for sub-stellar companions/exoplanets.

1. Sample Selection and Kinematic

Nearby moving group (MG) members will make ideal targets for AO searches seeking faint companions as they provide higher probability of detection compared to typical older field objects. In addition, with well constrained age, composition, and distances, our MG members would provide an ideal population to test atmosphere and evolutionary UCD models.

We compiled our red object sample by cross matching an extended version of the Liverpool-Edinburgh High Proper Motion survey (Pokorny et al. 2004) and Southern Infrared Proper Motion Survey (Deacon et al. 2005) with the Two Micron All Sky Survey and applying colour (for select objects with spectral type >M6) and proper motion cuts (by studying the movement of the objects respecting the convergent point of each MG). We found 132 objects possible members of a MG, see Clarke et al. (2009).

We computed the galactic space velocity components (U, V, W) of the sample taking into account their possible membership to one or several MGs, plotted them (see Figure 1) and applied a simple criterion that consists of identifying a possible member of one of the five MGs by its relative position in the UV and VW diagrams with respect to the boxes (dashed lines) that marked the velocity ranges of different MGs. The boxes are defined by studies with higher mass member samples from the literature (Montes et al. 2001; Barrado Y Navacuès 1998). According to their kinematic behaviour, 49 objects lie inside young disk

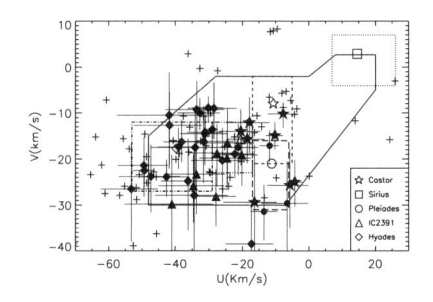

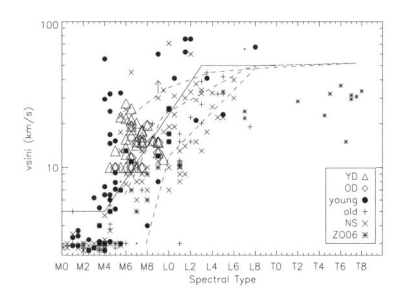

Figure 1. *Left*: UV space motion diagrams for our sample. We have included the boundaries (solid line) that delimit the young disk population as defined by Eggen (1984). The centre of the five main MGs are marked with large open symbols (squares for Sirius, diamonds for Hyades, triangles for IC 2391, circles for Pleiades and stars for Castor), the boxes (dashed line) mark the velocity range of difference MGs and our kinematic members are shown as the corresponding filled symbols. *Right*: Spectral type versus $v \sin i$. See Section 2.

area (YD) and 35 of them are candidates to belong to one of the MGs (see Figure 1 and caption).

2. Youth Constraints: Rotational Velocity

It is known that there is contamination of old objects that share kinematic properties with the genuine young MG members. To provide robust age constraints and assess the membership of our kinematic candidates, we will combine several methods (e.g. lithium I doublet at 6708Å, activity/age relation, etc).

We study here the projected rotational velocity ($v \sin i$) criteria. Following Reiners & Basri (2008), we use $v \sin i$ as an excellent way to discriminate young and old M type UCD populations. We measured the $v \sin i$ of 55 of the total 69 sample and plot them (see Figure 1) in a $v \sin i$-spectral type diagram from Reiners & Basri (2008), where young and old populations occupy distinct regions. Circles are supposed to be young, pluses old and crosses have unknown age. Asterisks come from Zapatero-Osorio et al. (2006). Our data sample is overplotted in different symbols depending on whether they have been classified as YD or old disk (OD). We mark the ages of 2, 5 and 10 Gyr (from upper left to lower right) as dashed lines and plotted the "apparent" separation between the considered "young" and "old" objects as a solid line, and used it as a criterion for determining the possible membership of our targets. From the 42 kinematic YD candidates, 31 present a $v \sin i$ in agreement with being young and 13 can not be dismissed from the $v \sin i$ information.

References

Barrado Y Navacués, D. 1998, A&A, 339, 831
Clarke, J. R. A., Pinfield, D. J., et al. 2009, MNRAS, 402, 575
Deacon, N. R., Hambly, N. C., & Cooke, J. A. 2005, A&A, 435, 363

Eggen, O. J. 1984, AJ, 89, 1358
Montes, D., López-Santiago, J., Gálvez, M. C., et al. 2001, MNRAS, 328, 45
Pokorny, R. S., Jones, H. R. A., Hambly, N. C., & Pinfield, D. J. 2004, A&A, 421, 763
Reiners, A., & Basri G. 2008, AJ, 684, 1390
Zapatero-Osorio, M. R., Martín E. L., Bouy, H., et al. 2006, ApJ, 647, 1405

Characterization of High-Energy Emissions of GKM Stars and the Effects on Planet Habitability

A. Garcés,[1] I. Ribas,[1] and S. Catalán[2]

[1] Institut de Ciències de l'Espai (CSIC-IEEC), Barcelona, Spain

[2] University of Hertfordshire, Hatfield, Hertfordshire, UK

Abstract. An important factor in determining if an Earth-like planet can evolve into a habitable world is the radiation coming from its host star. Magnetic activity in the Sun, and low-mass stars in general, manifests itself in the form of high energy and particle emissions. Such emissions may exert an important influence on planetary atmospheres. The long-term evolution of stellar emissions is of special relevance to understand the overall evolution of a (potentially habitable) planet. A key element for the characterization is a reliable age determination. Here we present our results on the determination of the time-evolution of high energy emissions for a sample of late G, K, and M stars.

Stellar astrophysics becomes a critical element when developing physical and chemical models for planetary atmospheres. The host star is, by far, the main source of energy for an exoplanet's atmosphere, and characterizing such emissions is of central importance. Although the time evolution of solar-type stars is already well constrained from *The Sun in Time* project (Ribas et al. 2005), that of late G-, K- and M-type stars still needs much improvement. Our objective is to carry out an extension for stars with masses below that of the Sun. This is justified by the many differences in the high energy emissions between, e.g., G-type and M-type stars and by the interest of the latter as hosts to exoplanets.

We have started by collecting data on the evolution of X-ray emissions for the younger stars. Their ages have been determined from cluster and moving group membership, and the $\log(L_X/L_{bol})$ values have been obtained from a thorough list provided by Pizzolato et al. (2003) and complemented in a few cases with values estimated directly from ROSAT measurements. The results can be seen in Figure 1. With this sample of stars, ages up to ~ 0.7 Gyr are covered, which is a key point, since it represents the time at which life is supposed to have appeared on Earth's surface. It is thus very important to obtain data for stars with ages > 0.7 Gyr for a full understanding of the overall properties and evolution of planetary atmospheres.

A preliminary calibration for a few stars older than 0.7 Gyr has been obtained but a more complete sample is needed to get more reliable results. To carry out our purpose of extending the sample for this age domain, a new age calibration method based in the properties of white dwarfs (WDs) as stellar chronometers is proposed.

We selected a sample of 40 wide binaries composed of a WD member of DA type (with only Balmer lines in the spectrum) and a G-, K- or M-type companion (2 G, 8 K and 30 M). It is reasonable to assume that the members of a wide binary were born simultaneously and with the same chemical composition, and,

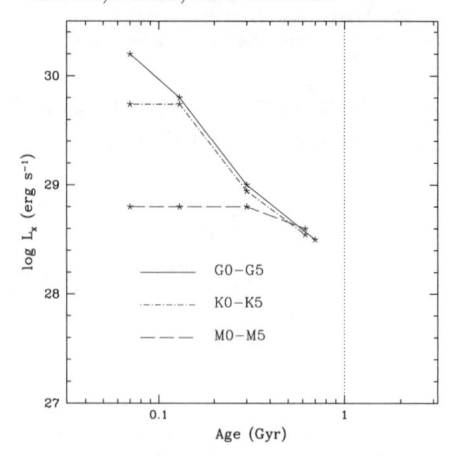

Figure 1. Preliminary results for X-ray luminosity vs. age. Results for stars younger than 0.7 Gyr.

since the components are well separated, they have essentially evolved as isolated stars. The WD member can be used as an age calibrator, and once the age of the system is determined, a full characterization of the low mass star can be carried out.

We have already obtained low-medium resolution and high signal-to-noise spectroscopic observations of the WDs with ISIS and TWIN spectrographs, at WHT at Roque de los Muchachos and 3.5m telescope at CAHA observatory, respectively. We derived the atmospheric parameters (T_{eff} and $\log g$) from this observations by performing fits to the observed Balmer lines with synthetic models (Koester et al. 2001). From these properties, the WD mass and its cooling time can be determined via interpolation in the appropiate cooling sequences (Salaris et al. 2000). On the other hand, from the resulting value of the WD mass, the initial-final mass relationship described in Catalán et al. (2008) and the evolutionary tracks of Dominguez et al. (1999) are used to derive progenitor's lifetime. This yields the total age of the WD, which in turn, is the age of its companion.

Once the ages are determined, we will collect all the possible activity indicators of the GKM companions to calibrate the activity-age relationship. The results from our work should provide critical input information to any effort of understanding the long-term evolution of the atmospheres of planets around low-mass stars, including their potential habitability.

References

Catalán, S., Isern, J., García-Berro, E., & Ribas, I. 2008, MNRAS, 387, 1693
Dominguez, I., Chieffi, A., Limongi, M., & Straniero, O. 1999, ApJ, 524, 226
Koester, D., Napiwotzki, R., Christlieb, N., et al. 2001, A&A, 378, 556
Pizzolato, N., Maggio, A., Micela, G., Sciortino, S., & Ventura, P. 2003, A&A, 397, 147
Ribas, I., Guinan, E. F., Güdel, M., & Audard, M. 2005, ApJ, 622, 680
Salaris, M., García-Berro,E., Hernanz, M., Isern, J., & Saumon, D. 2000, ApJ, 544, 1036

Looking for Transits of Jupiter-Size Planets Orbiting Stars in Habitable Zones

E. García-Melendo[1] and I. Ribas[2]

[1]*Esteve Duran Observatory Foundation, Seva, Spain*

[2]*Institut de Ciències de l'Espai (CSIC-IEEC), Barcelona, Spain*

Abstract. Owing to detection bias, most of the nearly 400 discovered planets thus far are Jupiter-mass bodies. According to the extreme Venus and Mars criteria for the limits of the habitable zone (HZ) around stars with detected exoplanets (Selsis et al. 2007), a few of these Jupiter-size worlds orbit completely inside their parent star's HZ. These planets have orbital semi-major axes between 1 and 4 AU and orbital periods around one year or longer. The discovery of a transiting "warm" Jupiter will provide valuable information on its atmosphere, as well as offer the possibility of detecting Earth and super-Earth type satellites (potentially habitable) by using a variety of techniques such as ultra-high precision photometry or long-term transit timings. An evaluation of the transit probability will depend on a careful study of available and new photometric and spectroscopic data to characterize the host stars and to determine improved ephemeris of the planet-star conjunction time. Transit events, with a duration between seven and ten hours, and photometric depths in excess of 1%, might be easily detected from two or three independent ground-based telescopes as shown recently during the discovery of the optical transit of HD80606b.

1. General Considerations on Transit Data

The discovery of the 111-day period HD80606b transit (Moutou et al. 2009; García-Melendo & McCullough 2009; Fossey et al. 2009), showed that it is possible to detect transits about 12 hours long (photometric depth ∼1%) if independent instruments observe the same event simultaneously. Planets with periods between 300 and 500 days are estimated to display transits on the order of 13-15 hours long. Such estimation can be substantially different for planets with highly eccentric orbits. According to the geometry of transit prediction (see Kane 2007), when $\omega + \nu = 90$ deg, where ω is the star's longitude of the periastron and ν its true anomaly, we may expect a transit if the orbital plane of the system is very close to 90 deg. Radial velocity data should give precise values for the orbital parameters to build an accurate ephemeris. Uncertainties in actual radial velocity data tell that for most of the detected long period planets, it is necessary to improve the quality of the data as well as the span of the time series. Usually, radial velocity data for long period planets only cover not much longer than one cycle. With increasing planetary periods, star's radial velocities decrease and a precise determination of orbital elements (and therefore transit times) becomes more difficult.

2. Three Examples

HD 28185b (Minniti et al. 2009), HD 114783b (Vogt et al. 2002), and HD 221287b (Naef et al. 2007) are three planets with periods of 383, 501, and 456 days respectively, possibly orbiting inside their respective habitable zones, which illustrate the difficulty of computing accurate transit ephemerides due to orbital parameter uncertainties. For example, if the rest of the parameters are assumed to be exact, eccentricity errors may propagate as transit time errors as long as 7 days. Regarding other orbital parameters, Figure 1 (left panel) shows how errors in ω affect predicted times for transits. In this case, observation windows can be considerably wider than those due to eccentricity uncertainties. Another parameter which introduces long observation windows is the planet's period. The usually short data series do not allow an accurate estimate of the period. The right panel of Figure 1 shows that after only two cycles after the planet discovery, the accumulated uncertainty in transit predictions may be as high as 40 days. Additional spectroscopic observations after a few more planet revolutions may notably reduce the error in period determination and narrow down observation windows.

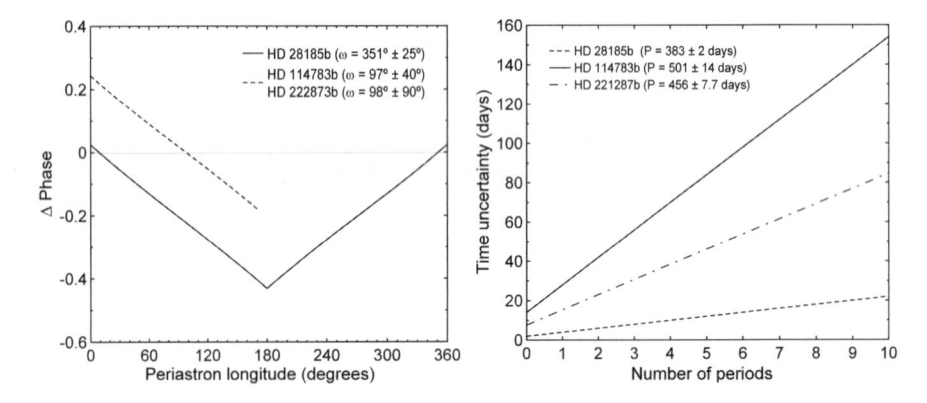

Figure 1. Errors in the time of minimum due to uncertainties in the longitude of the periastron ω (left panel), and cumulative period uncertainties (right panel), assuming that in each case the rest of the orbital parameters are exact.

References

Fossey S. J., Waldmann I. P., & Kipping D. M. 2009, MNRAS, 396, L16
García-Melendo E. & McCullough P. 2009, ApJ, 698, 558
Kane S. R. 2007, MNRAS, 380, 1488
Minniti D. et al., 2009, ApJ, 693, 1424
Moutou C.et al., 2009, A&A, 498, L5
Naef D. et al., 2007, A&A, 470, 721
Selsis F., Kasting F. K., Levrard B., Paillet J., Ribas I., & Delfosse X., 2007, A&A, 476, 1373
Vogt S. S., Butler, R. P., Marcy, G. W., Fischer, D. A., Pourbaix, D., Apps, K., & Laughlin, G., 2002, ApJ, 568, 352

Pathways Towards Habitable Planets
ASP Conference Series, Vol. 430, 2010
Vincent Coudé du Foresto, Dawn M. Gelino, and Ignasi Ribas, eds.

Modeling the Atmospheres of Earth-like Planets

A. García Muñoz,[1,2] E. Pallé,[1,2] and E. L. Martín[3]

[1] *Instituto de Astrofísica de Canarias, La Laguna, Spain*

[2] *Universidad de La Laguna, La Laguna, Spain*

[3] *Centro de Astrobiología, INTA-CSIC, Madrid, Spain*

Abstract. A lunar eclipse offers a unique opportunity to investigate the terrestrial atmosphere. A recent work (Pallé et al. 2009) has published the Earth's transmission spectrum from 0.36 to 2.40 μm obtained during the partial lunar eclipse of August 16th 2008. We have set out to interpret the measured spectrum, and assess to what extent the conclusions can be extrapolated to an exoplanet's in-transit transmission spectrum.

1. Lunar eclipses. Refraction

Before reaching the Moon's surface in umbra, solar photons traverse the terrestrial atmosphere, where they undergo refraction, single and multiple scattering, and partial attenuation by the atmospheric constituents. At the Moon, the photons can be either absorbed at the surface or reflected back into any outward direction. Some of the latter will reach an observer at Earth, which explains the faint but distinct brightness of the Moon when observed from our planet during a lunar eclipse.

Evaluation of the Sun-to-Moon transmission requires the opacities along all photon trajectories reaching the lunar surface in the observer's field of view. Photons that traverse the atmosphere following refracted trajectories without undergoing scattering provide the direct solar contribution to the disk-integrated transmission. Photons that do undergo one or more scattering events provide the diffuse solar contribution.

In polar coordinates (r, θ), the equation for tracing the photon trajectories in a medium with varying refractive index $n(r)$ is (Born & Wolf 1980):

$$\frac{d(\log r)}{d\theta} = \mp\sqrt{\left(\frac{rn}{c}\right)^2 - 1}. \tag{1}$$

Here, c ($\equiv rn(r)\cos\beta$) is the invariant in Snell's law in a spherically symmetric medium, and β is the angle between the light ray direction and the local horizon. c is a measure of the photon incidence angle at the top of the atmosphere.

In a planetary atmosphere, the refractive index typically decreases with altitude. Thus, the product $rn(r)$ goes through a minimum, $r_*n(r_*)$, at a distance r_*. Both r_* and $r_*n(r_*)$ are fully determined by the refractive properties of the atmosphere. The ratio $r_*n(r_*)/c$ dictates the type of photon trajectory, as seen in Equation 1. For ratios > 1, the trajectory monotonically approaches, and

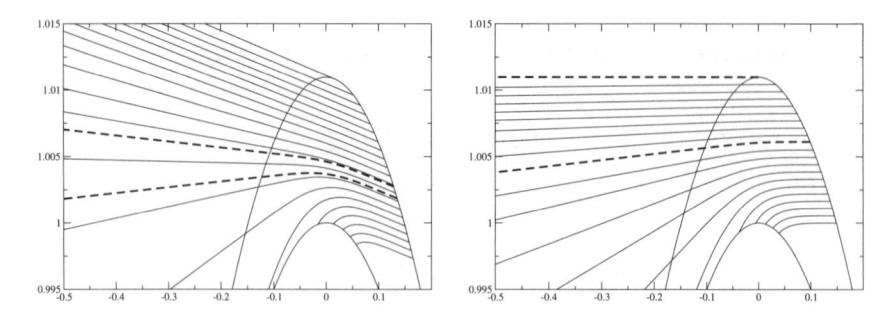

Figure 1. *Left:* Photon trajectories originating from a distance $d/Rp=60$ to the right of the graph. The Sun is left of the graph. Coordinates normalized by R_p. *Right:* Same as *Left* for $d/Rp \to \infty$.

eventually hits, the planet's surface. For ratios < 1, the trajectory becomes parallel to the local horizon at the radial distance that makes singular the right hand side of the equation, thereon following an ascendent trajectory. The condition $r_*n(r_*)/c=1$ separates the two families of trajectories. Only light rays with ratio $R_p n(R_p)/c < 1$ can contribute to the direct radiance in a lunar eclipse.

Figure 1 (*left*) shows a set of photon trajectories originating from the axis Sun-Earth-Moon at a distance $d/R_p \sim 60$ from the Earth's centre, where R_P means one planetary radius. The figure is representative of the middle point of a central eclipse. The curves are found by integrating the system of ODEs for r, θ and β set out by van der Werf (2008), that utilises the path length as the independent variable of integration. The integration is carried out from the Moon towards the Sun, i.e. inversely to the actual direction of propagation of photons, by means of a 4th order Runge-Kutta method. Because of the finite angular width of the Sun, not all photons can be traced back to the star. The rays above the upper dashed curve miss the upper limb of the Sun, whereas the rays below the lower dashed curve miss the lower limb. In our approach, solar limb darkening is easily taken into account. Figure 1 (*right*) is analogous to Figure 1 (*left*), but in the $d/R_p \to \infty$ limit appropriate to the observation of a transiting exoplanet. A major difference between both figures is the range of altitudes probed, broader in the case of a transit. Although not discussed here, other relatively narrow ranges of altitude can be probed at different phases of the lunar eclipse.

References

Born, M. & Wolf, E. 1980, Principles of Optics, (Oxford: Pergamon Press)
Pallé, E., Zapatero Osorio, M.R., Barrena, R., Montañés-Rodríguez, P., & Martín, E.L. 2009, Nat, 459, 814
van der Werf, S. Y. 2008, Appl.Optics, 47, 153

On the Detectability of Biomarkers in Extrasolar Super-Earth Atmospheres

S. Gebauer,[1] M. Godolt,[1] J. L. Grenfell,[1] P. Hedelt,[2,3] P. von Paris,[2] H. Rauer,[1,2] F. Selsis,[3] and A. Belu[3]

[1] *Zentrum für Astronomie und Astrophysik, Technische Universität Berlin (TUB), Hardenbergstr. 36, Berlin, Germany*

[2] *Institut für Planetenforschung, Deutsches Zentrum für Luft- und Raumfahrt (DLR), Rutherfordstr. 2, Berlin, Germany*

[3] *LAB: Laboratoire d'Astrophysique de Bordeaux, CNRS, Université Bordeaux 1, Floirac, France*

Abstract. The presence of biomarker molecules in the atmospheres of terrestrial planets is usually interpreted within the context of biological activity. The instrumental design requirements for the detection of such species are demanding because of the weak signals. In this contribution we present contrast spectra of super-Earth planets.

Among the more than 400 extrasolar planets found in the last decade, most have masses greater than Jupiter. But with better detection methods, more and more low mass planets are being found, like the very interesting object Corot-7b (Léger et al. 2009). Proposed satellite missions such as JWST (Clampin 2008) and THESIS (Swain et al. 2009) will provide us with the first possibility to study spectra of Earth-like planets.

We apply a global column model with a coupled chemistry and radiation scheme from the ground up to the mid-mesosphere. The model calculates globally, diurnally-averaged atmospheric temperature, pressure, water and chemical species profiles for cloud-free conditions for an Earth-like planet in the Habitable Zone. The original code has been described in detail by Kasting, Pollack & Crisp (1984) and was further developed by Segura et al. (2003) and Grenfell et al. (2007). We varied the central star (Sun, AD Leo) and the planetary gravity from 1 to 3 times Earth gravity (9.8 to 29.4 ms^{-2}). All runs employed the same, initial, modern-day Earth-like atmospheric composition (i.e. biogenic fluxes, N_2, O_2 and CO_2 content). The resulting temperature, pressure and concentration profiles are fed into the line-by-line radiative transfer model SQuIRRL (Schwarzschild Quadrature InfraRed Radiation Line-by-line; Schreier & Böttger 2003) to calculate synthetic emission and transmission spectra in the infrared from 1.8 to 25 μm. Additionally, H_2O continuum absorption corrections are performed. Local thermodynamic equilibrium (LTE) is assumed. The database for cross sections and line parameters is HITRAN 2004 and its updates. The emission spectra are calculated using a pencil beam looking downwards with a zenith angle of 38°, to account for the average illumination.

The resulting contrast spectra for emitted infrared flux of an Earth and a super-Earth around the Sun and the M dwarf AD Leo are shown in Figure 1.

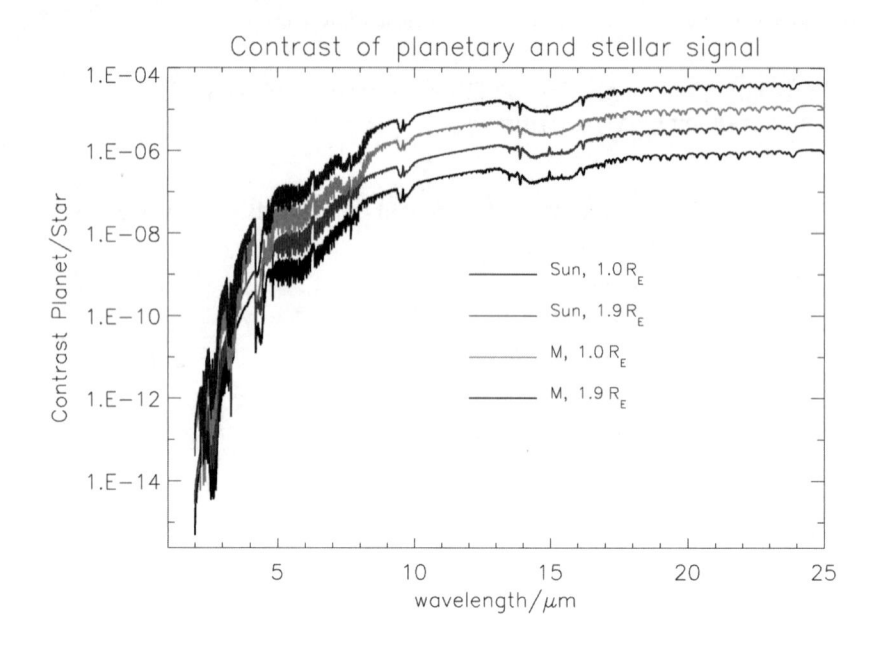

Figure 1. Contrast spectra for an Earth and a super-Earth around the Sun and the M dwarf AD Leo.

One can clearly see the absorption bands of e.g. H_2O (2.7 μm, the broad 6.3 μm band and the rotation bands longwards of 15 μm), CH_4 (3.3 and 7.7 μm), O_3 (9.6 μm) or CO_2 (2.7, 4.3 and 15 μm). The contrast between stellar and planetary flux is very low, reaching almost 10^{-4} in the mid-IR for the Super-Earth around AD Leo. The calculated transmission spectra, resolution and signal-to-noise studies for the different proposed satellite missions can be found in the forthcoming paper.

Acknowledgments. This research has been supported by the Helmholtz Gemeinschaft through the research alliance Planetary Evolution and Life.

References

Clampin, M. 2008, Adv. Space Res., 41, 1983
Grenfell J. L., Stracke, B., von Paris, P., Patzer, B., Titz, R., Segura, A., & Rauer, H. 2007, Planet. Space Sci., 55, 661
Kasting, J. F., Pollack, J. B., & Crisp, D. 1984, J. Atmos. Chem., 1, 403
Léger, A. et al. 2009, A&A, 506, 287
Schreier, F. & Böttger, U. 2003, Atmospheric and Oceanic Optics, 16, 262
Segura, A., Krelove, K., Kasting, J. F., Sommerlatt, D., Meadows, V., Crisp, D., Cohen, M., & Mlawer, E. 2003, Astrobiology, 3, 689
Swain, M. R., Deming, D., Vasisht, G. Henning, T., Bouwman, J. & Akeson, R. 2009, Astro2010: The Astronomy and Astrophysics Decadal Survey, Technology Development Papers, No. 61

Pathways Towards Habitable Planets
ASP Conference Series, Vol. 430, 2010
Vincent Coudé du Foresto, Dawn M. Gelino, and Ignasi Ribas, eds.

Influence of the Spectral Stellar Flux Distribution on Atmospheric Dynamics of Extrasolar Earth-like Planets

M. Godolt,[1] J. L. Grenfell,[1] A. Hamann-Reinus,[2] M. Kunze,[2] U. Langematz,[2] and H. Rauer[1,3]

[1]Zentrum für Astronomie und Astrophysik, Technische Universität Berlin (TU), Berlin, Germany

[2]Institut für Meteorologie, Freie Universität Berlin(FU), Berlin, Germany

[3]Institut für Planetenforschung, Deutsches Zentrum für Luft- und Raumfahrt (DLR), Germany

Abstract. Our ultimate goal is to investigate possible Earth-like exoplanet scenarios and resulting spectra taking into account 3-D atmospheric dynamics and chemistry. As a first step we use the state-of-the-art 3-D general circulation model EMAC to study the influence of spectral flux distributions corresponding to central stars of different spectral type upon Earth-like exoplanets. We focus on those atmospheric responses related to surface habitability such as surface temperature, surface wind, and precipitation. Preliminary results, as well as a comparison with a 1-D radiative-convective model, will be presented.

Three dimensional (3-D) model studies of exoplanetary atmospheres have shown that atmospheric circulation can be crucial for the determination of surface conditions, especially for extreme scenarios (see e.g., Joshi ei al. 1997; Joshi 2003). For less extreme cases, calculating the atmospheric response with a 3-D model also has advantages, since one can investigate the influence of e.g., orbital parameters (Williams & Pollard 2002, 2003) or rotation rate and study climatic feedbacks of clouds or albedo changes due to e.g., snowfall which are not easily incorporated into a 1-D model.

As a first step we used the state-of-the-art 3-D General Circulation Model ECHAM/MESSy Atmospheric Chemistry (EMAC) model, which is a numerical chemistry and climate simulation system that includes sub-models describing tropospheric and middle atmosphere processes and their interaction with oceans, land, and human influences (Jöckel et al. 2006). It uses the first version of the Modular Earth Submodel System (MESSy1) to link multi-institutional computer codes. The core atmospheric model is the 5th generation European Centre Hamburg general circulation model (ECHAM5 Roeckner et al. 2006). For the present study we applied EMAC (ECHAM5 version 5.3.01, MESSy version 1.7) in the T31L39MA-resolution, i.e., with a spherical truncation of T31 (corresponding to a quadratic gaussian grid of approx. 3.75 by 3.75 degrees in latitude and longitude) with 39 vertical hybrid pressure levels up to 0.01 hPa in the middle atmosphere. The applied model setup comprised the submodels CLOUD, CONVECT in the Tiedtke-Nordeng configuration, RAD4ALL and H2O. The radiative transfer is calculated in four spectral bands in the shortwave regime and in 16 spectral bands in the IR regime.

Model calculations were done for three different spectral flux distributions of the incident stellar radiation corresponding to different central stars (K-type, F-type star and Sun for comparison). The spectra were scaled to a total flux of solar value (1366 Wm^{-2}) at the top of the atmosphere. This corresponds to star-planet distances of 0.6 AU for the K-type star and 1.89 AU for the F-type star. Note that no change of orbital period was taken into account. We used a present Earth model setup with climatological sea surface temperatures and sea ice concentrations, climatological ozone and present day volume mixing ratios of the radiative gases (348 ppm CO_2, 1.65 ppm CH_4, 306 ppb N_2O, but no CFCs).

Since we are especially interested in surface habitability we focused on surface conditions. We found that globally and annually averaged changes are small, as expected, since we kept approximately 70% of the surface temperatures fixed. Still some responses could be observed such as an increase (decrease) of one per mill of surface temperature for the K-type spectral distribution (F-type spectral distribution) resulting in an increase (decrease) of the total water vapour column. Local temperature differences were more pronounced, reaching up to 8K for the Northern Hemisphere summer at midlatitudes for the K-type spectral distribution. Statistically significant differences were mainly observed for the Northern Hemisphere in summer, since surface temperatures here were calculated over a larger percentage of landmass (as opposed to the fixed sea surface temperatures). Analysis of the impact on precipitation, windspeed, and cloud cover were also carried out.

A comparison of the global and annual averaged temperature profile with a cloud free 1-D radiative-convective model, originally from Kasting et al. (1984), in an updated version (see Grenfell et al. 2007) confirms the temperature tendencies observed (see e.g., Segura et al. 2003) though they are more pronounced in the 1-D model. For the K-type spectral distribution lower stratospheric temperatures arose since less UV radiation was available resulting in less heating by ozone and higher stratospheric temperatures for the F-type spectral distribution respectively. Tropospheric temperatures were higher for the K-type spectral distribution due to less Rayleigh scattering and more absorption of near-IR radiation by H_2O and CO_2.

Changing the stellar flux distribution resulted in little impact on tropospheric and surface conditions, as expected. Still, small responses for the temperature, the water vapour column, precipitation, cloud coverage, and wind speeds could be observed. Tendencies in the temperature profile are similar to 1-D model results.

Acknowledgments. This work has been supported by the Forschungsallianz *Planetary Evolution and Life* of the Helmholtz Gemeinschaft (HGF).

References

Grenfell, J. L. et al. 2007, Planet. Space Sci., 55, 661
Jöckel, P. et al. 2005, Atmos. Chem. Phys., 5, 433
Jöckel. P. et al. 2006, Atmos. Chem. Phys., 6, 5067
Joshi, M. 2003, Astrobiology, 3, 415
Joshi. M., Haberle, R. M., & Reynolds, R. T. 1997, Icarus, 129, 450
Kasting, J. F. et al. 1984, J. Atmos. Chem., 1, 403
Roeckner, E. et al. 2006, J. Climate, 19, 3771

Segura, A. et al. 2003, Astrobiology, 3, 689
Williams, D. M. & Pollard, D. 2002, Int. J. Astrobiology, 1, 61
Williams, D. M. & Pollard, D. 2003, Int. J. Astrobiology, 2, 1

Pathways Towards Habitable Planets
ASP Conference Series, Vol. 430, 2010
Vincent Coudé du Foresto, Dawn M. Gelino, and Ignasi Ribas, eds.

Exoplanetary Systems with SAFARI: A Far Infrared Imaging Spectrometer for SPICA

J. R. Goicoechea[1] and B. Swinyard[2]
on behalf of the SPICA/SAFARI science teams

[1] Centro de Astrobiología (CSIC-INTA), Madrid, Spain

[2] Rutherford Appleton Laboratory, Chilton, Didcot, UK

Abstract. The far-infrared (far-IR) spectral window plays host to a wide range of spectroscopic diagnostics with which to study planetary disk systems and exoplanets at wavelengths completely blocked by the Earth atmosphere. These include the thermal emission of dusty belts in debris disks, the water ice features in the "snow lines" of protoplanetary disks, as well as many key chemical species (O, OH, H_2O, NH_3, HD, etc.). These tracers play a critical diagnostic role in a number of key areas including the early stages of planet formation and potentially, exoplanets. The proposed Japanese-led IR space telescope SPICA, with its 3m–class cooled mirror ($\sim$5 K) will be the next step in sensitivity after *ESA's Herschel Space Observatory* (successfully launched in May 2009). SPICA is a candidate "M-mission" in *ESA's Cosmic Vision 2015-2025* process. We summarize the science possibilities of SAFARI: a far-IR imaging-spectrometer (covering the $\sim$34 – 210 μm band) that is one of a suite of instruments for SPICA.

1. A New Window in Exoplanet Research

The study of exo-planets (EPs) requires many different approaches across the full wavelength spectrum to both discover and characterize the newly discovered objects in order that we might fully understand the prevalence, formation, and evolution of planetary systems. The mid–IR and far–IR spectral regions are especially important in the study of planetary atmospheres as it spans both the peak of thermal emission from the majority of EPs thus far discovered (up to $\sim$1000 K) and is particularly rich in molecular features that can uniquely identify the chemical composition, from protoplanetary disks to planetary atmospheres and trace the fingerprints of primitive biological activity. In the coming decades many space- and ground-based facilities are planned that are designed to search for EPs on all scales from massive, young "hot Jupiters", through large rocky super-Earths down to the detection of exo-Earths. Few of the planned facilities, however, will have the ability to characterize the planetary atmospheres which they discover through the application of mid-IR and far-IR spectroscopy. SPICA will be realized within $\sim$10 years and has a suite of instruments that can be applied to the detection and characterization of EPs over the $\sim$5–210 μm spectral range (see e.g., our *White Paper*; Goicoechea et al., 2008).

The SAFARI instrument (Swinyard et al. 2009) will provide capabilities to complement SPICA studies in the mid-IR (either coronagraphic or transit studies). Indeed, SAFARI could be the only planned instrument able to study

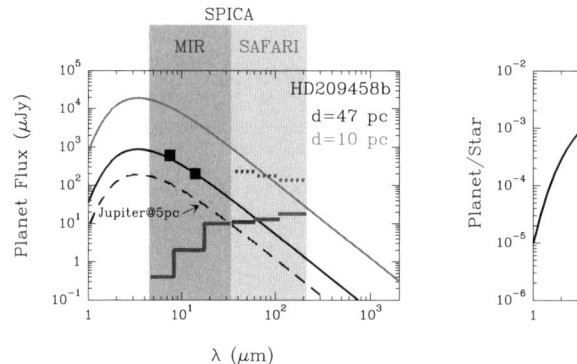

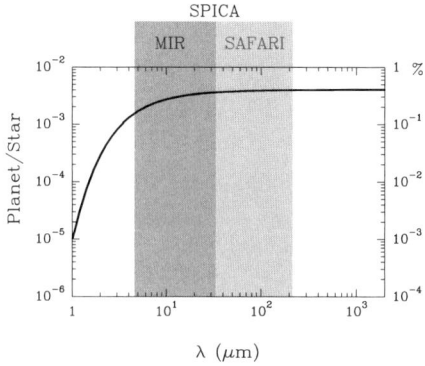

Figure 1. **Left**: Fit to HD 209458b "hot Jupiter" ($T_{eff} \simeq 1000$ K) fluxes inferred from a secondary transit with *Spitzer* (Swain et al. 2008) around a G0 star ($d \sim 47$ pc, in black) and interpolation to $d = 10$ pc (gray). The emission of a cooler Jupiter-like planet at 5 pc is shown in dashed line (reflected emission neglected). Thick horizontal stepped lines are the 5σ-1hr photometric sensitivities of SPICA mid–IR instruments and SAFARI. Short horizontal dotted lines show sensitivities in spectrophotometric mode ($R \simeq 25$). SPICA will observe similar transits of *inner* "hot Jupiters" routinely and will potentially extract their IR spectrum (rich in H_2O, O_3, CH_4, NH_3 and HD features as in Solar System planets). **Right**: Increasing planet-to-star contrast at longer mid– and far–infrared wavelengths (Goicoechea et al. 2008).

EPs in a completely new wavelength domain (for SAFARI's band 1) not covered by *JWST* nor by *Herschel* (Figure 1). This situation is often associated with unexpected discoveries. Since cool EPs show much higher contrast in the far–IR than in the near–/mid–IR (e.g., Jupiter's effective temperature is ~ 110 K), if such EPs are found in the next 10 years, their transit studies with SPICA will help to constrain their main properties, which are much more difficult to infer at shorter wavelengths. Note that following *Infrared Space Observatory* (ISO) observations, Jupiter seen at 5 pc will produce a flux of ~ 35 μJy at 37 μm, but less than ~ 1 μJy at 15 μm. SAFARI band 1 (~ 34 –60 μm) hosts a variety of interesting atmospheric molecular features (e.g., H_2O at 39 μm, HD at 37 μm and NH_3 at 40 and 42 μm). Strong emission/absorption of these features was first detected by ISO in the atmospheres of Jupiter, Saturn, Titan, Uranus and Neptune (Feuchtgruber et al. 1999).

Acknowledgments. We warmly thank the SPICA/SAFARI science teams for fruitful discussions. JRG is supported by a *Ramón y Cajal* research contract.

References

Feuchtgruber, H., Lellouch, E., Bezard, B. et al. 1999, A&A, 341, L17
Goicoechea, J. R., Swinyard, B., Tinetti, G. et al. 2008, *A White Paper for ESA's Exo-Planet Roadmap Advisory Team* (2008 July 29), astro-ph/0809.0242
Swain, M. R., Bouwman, J., Akeson, R. L. et al. 2008, ApJ, 674, 482
Swinyard, B., Nakagawa, T., et al. 2009, Exp. Astron., 23, 193

Pathways Towards Habitable Planets
ASP Conference Series, Vol. 430, 2010
Vincent Coudé du Foresto, Dawn M. Gelino, and Ignasi Ribas, eds.

SIM Lite Instrument Description, Operation, and Performance

R. Goullioud

Jet Propulsion Laboratory, California Institute of Technology, 4800 Oak Grove Dr., Pasadena, CA, USA

Abstract. The SIM Lite Astrometric Observatory, a space-borne mission to be located in a solar Earth-trailing orbit, will conduct high precision differential astrometry (Marr, Shao & Goullioud 2008). During the 5-year mission, SIM Lite will search about 60 nearby Sun-like stars for exoplanets of mass down to one Earth mass, in the habitable zone of those stars. In addition, SIM Lite will survey the architecture of planetary systems of about 1000 stars of all ages and types. SIM Lite will also build a 4 micro-arc second astrometric grid and perform global astrometry on a variety of astrophysical objects (Unwin et al. 2008). The instrument consists of two Michelson stellar interferometers and a precision telescope. The science interferometer is composed of two 50 cm collectors separated by a 6 meter baseline. The technology development was completed in 2005 (Laskin 2006). The mission is now ready for full flight implementation.

The basic elements of a stellar interferometer are shown in Figure 1. Light from a distant source is collected at two points and combined using a beam splitter, where interference of the combined wavefronts produces fringes when the internal pathlength difference (or delay) compensates exactly for the external delay. The astrometric angle, α, between the interferometer baseline and the star can be found using the measured internal optical path difference, x, and the length of the baseline, B. The external metrology system measures the distance between two fiducials (each made of corner cubes) forming the baseline and the internal metrology measures the optical path difference to the beam combiner from the two fiducials.

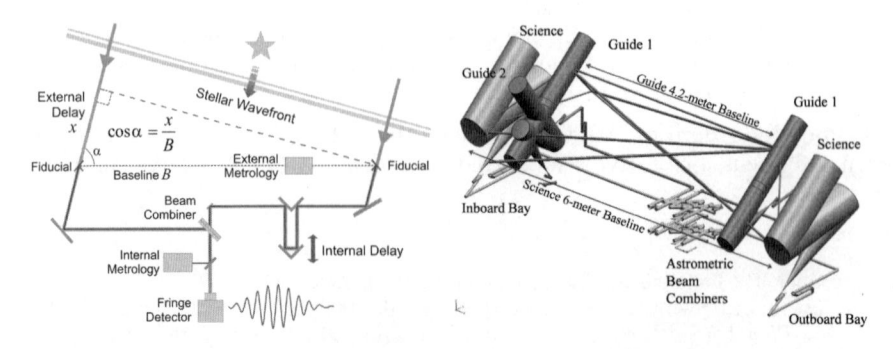

Figure 1. Astrometry with a Michelson stellar interferometer (left picture). Optical configuration of the SIM Lite instrument (right picture).

Narrow angle differential astrometry error budget

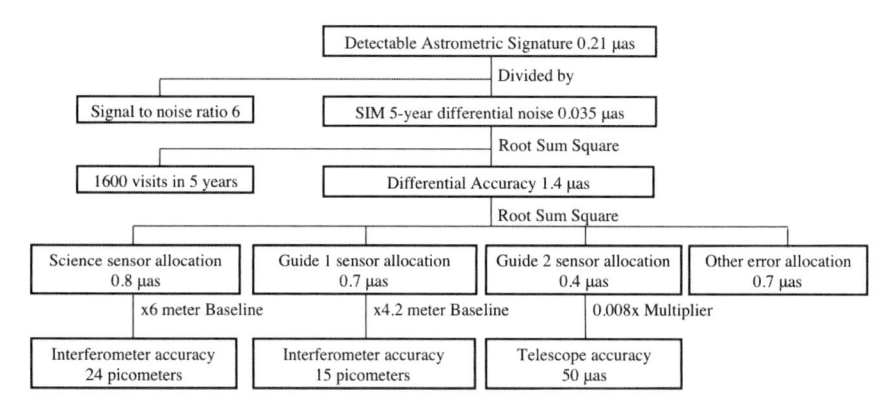

```
                    ┌─────────────────────────────────────────┐
                    │ Detectable Astrometric Signature 0.21 µas│
                    └─────────────────────────────────────────┘
                                    │ Divided by
        ┌───────────────────────┐   │   ┌──────────────────────────────────┐
        │ Signal to noise ratio 6│   │   │ SIM 5-year differential noise 0.035 µas│
        └───────────────────────┘       └──────────────────────────────────┘
                                           │ Root Sum Square
        ┌───────────────────────┐          ┌──────────────────────────────┐
        │ 1600 visits in 5 years│          │ Differential Accuracy 1.4 µas │
        └───────────────────────┘          └──────────────────────────────┘
                                           │ Root Sum Square
┌────────────────┐ ┌────────────────┐ ┌────────────────┐ ┌─────────────────┐
│Science sensor  │ │Guide 1 sensor  │ │Guide 2 sensor  │ │Other error      │
│allocation      │ │allocation      │ │allocation      │ │allocation       │
│0.8 µas         │ │0.7 µas         │ │0.4 µas         │ │0.7 µas          │
└────────────────┘ └────────────────┘ └────────────────┘ └─────────────────┘
│ x6 meter Baseline │ │ x4.2 meter Baseline │ │ 0.008x Multiplier │
┌────────────────┐ ┌────────────────┐ ┌────────────────┐
│Interferometer  │ │Interferometer  │ │Telescope       │
│accuracy        │ │accuracy        │ │accuracy        │
│24 picometers   │ │15 picometers   │ │50 µas          │
└────────────────┘ └────────────────┘ └────────────────┘
```

Figure 2. Narrow angle differential astrometry error budget.

SIM Lite simultaneously employs two stellar interferometers and one telescope to perform astrometry. Precision astrometry requires knowledge of the baseline orientation to the same order of precision as the astrometric measurement. The second interferometer and the precision telescope acquire and lock on bright "guide" stars, keeping track of the uncontrolled rigid-body motions of the instrument, while the main interferometer switches between science targets, measuring projected angles between them. The guide interferometer measures the baseline orientation in the most sensitive direction while the high-precision telescope measures the baseline orientation in the other two directions.

Figure 2 shows the error budget for the planetary survey. 1.4 micro-arcsecond (μas) differential measurement accuracy between the target star and a set of reference stars can be achieved in 15 minutes in the narrow angle observation mode. The observation sequence starts with 15 seconds of observation time on the target star, T, during which interference fringes are collected. The observation is followed by about 15 seconds to slew and reposition the two siderostats and the optical delay line to acquire fringes on the first reference star, R1. After 30 seconds of observation on R1, the interferometer is slewed back to the same target star to be re-observed. Then, the instrument continues slewing and observing between the other reference stars and the target star. Once all the reference stars have been observed, the observation sequence is repeated from the beginning. After 1600 visits to the target star, sampled over 5 years, the astrometric noise is below 0.035 μas, enabling detection of astrometric signatures of 0.2 μas with a signal to noise ratio of 6. As a reference, the signature of the Earth is 0.3 μas for an observer 10 parsecs away. The required performance of 24 and 15 picometers for the two interferometers and 50 μas for the telescope has been experimentally demonstrated (Goullioud et al. 2006; Hahn et al. 2008).

Acknowledgments. This work was conducted at the Jet Propulsion Laboratory of the California Institute of Technology, under contract with the National Aeronautics and Space Administration.

References

Goullioud, R., Shen, T. J., Catanzarite, J. H., 2004, SPIE, 5491, 965
Hahn, I. et al., 2008, SPIE, 7013, 70134W
Laskin, R. A., 2006, SPIE, 6268, 626823
Marr-IV, J. C., Shao, M., & R. Goullioud, 2008, SPIE, 7013, 70132M
Unwin, S. C. et al., 2008, PASP, 120, 38

Pathways Towards Habitable Planets
ASP Conference Series, Vol. 430, 2010
Vincent Coudé du Foresto, Dawn M. Gelino, and Ignasi Ribas, eds.

Testing PIAA Coronagraphs at NASA Ames

T. Greene,[1] R. Belikov,[1] M. Connelley,[1] O. Guyon,[2] D. Lynch,[1]
M. McKelvey,[1] E. Pluzhnik,[1] and F. Witteborn[1]

[1] *NASA Ames Research Center, Moffett Field, CA, USA*

[2] *National Astronomical Observatory of Japan/Subaru Telescope, Hilo, HI, USA*

Abstract. We report on our tests of phase induced amplitude apodization (PIAA) coronagraphs in a new laboratory at NASA Ames. This laboratory is operated in a highly stabilized air environment, and it complements existing efforts at NASA JPL's high contrast imaging testbed. At the time of writing, we are operating a refractive PIAA system and have achieved imaging contrast of 7×10^{-8} in a dark zone from 2.0 to 4.8 λ/D from an artificial star. We are also directing L3 Tinsley Laboratories in the fabrication of a next generation polychromatic PIAA coronagraph mirror set.

1. Introduction and Current Results

Direct imaging of extrasolar planets, and Earth-like planets in particular, is an exciting but difficult problem requiring a telescope imaging system with 10^{-10} contrast at separations of 100 mas and less. The high throughput, small inner working angle, and few practical constraints of the phase induced amplitude apodization (PIAA) coronagraph (e.g., Guyon 2003; Vanderbei & Traub 2005; Pluzhnik et al. 2006) make it promising for use in moderate-sized ($\sim$ 2-m aperture) space telescopes that are capable of imaging Earth-like, super-Earth, and gas giant planets and circumstellar dust disks around nearby stars.

We are developing a new laboratory facility at NASA Ames to test PIAA coronagraphs with a wavefront control system employing micro-electro-mechanical systems (MEMS) deformable mirrors. We will also use this facility to rapidly prototype coronagraph instrumentation architectures that are appropriate for space missions. The laboratory optics operate in air, and we are working towards achieving a contrast goal of 10^{-9}, suitable for observing exozodiacal dust and some planets. Once we meet this goal, we plan to continue testing our coronagraphs and systems in the JPL High Contrast Imaging Testbed (HCIT) vacuum facility at higher contrast levels. Details of the Ames laboratory, its optics, and its performance can be found in Belikov et al. (2009).

As of September 2009, we have achieved monochromatic contrast of 7×10^{-8} in a dark zone from 2.0 to 4.8 λ/D from an artificial star using a refractive PIAA system as shown in Figure 1. This is about a factor of two improvement over Belikov et al. (2009), largely achieved by employing spatial filtering to reduce ringing when reimaging the hard edge of the focal plane occulter (Gibbs effect).

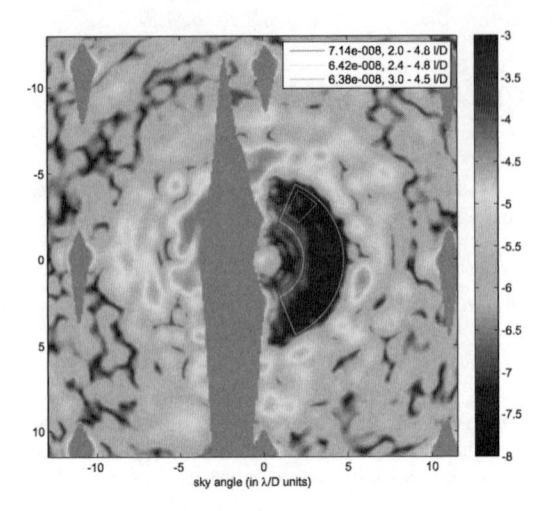

Figure 1. Current monochromatic ($\lambda = 650$ nm) contrast achieved in the Ames PIAA Coronagraph Laboratory as of September 2009. The mean contrast in a nearly semi-annular dark zone is 7×10^{-8} from 2.0 to 4.8 λ/D radius from the central star.

2. Manufacturing the Next Generation PIAA Coronagraph

Tinsley Laboratories has manufactured a second generation PIAA mirror pair that have been designed to provide 10^{-10} contrast over the entire visible wavelength band with 85% system throughput when used with a modest conventional apodizer. The mirrors have been manufactured with conventional computer-controlled optical polishing techniques to a surface quality of approximately 3 nm RMS when filtered to 20 cycles per aperture. Our models indicate that higher frequency errors present on these surfaces will limit polychromatic contrast to several 10^{-9} contrast, so we have contracted Tinsley to improve the surface roughness and then finalize the highly aspheric PIAA surface using a narrow footprint (~ 1 mm) ion beam figuring technique. Once complete, we plan to test these mirrors in the JPL HCIT as well as in the Ames testbed.

Acknowledgments. This work has been greatly aided by collaboration and support from the staff of the JPL HCIT facility, including John Trauger, Brian Kern, Andy Kuhnert, Marie Levine, and others. We also sincerely thank NASA Ames, the NASA Innovative Partnerships Program, and the NASA Exoplanet Exploration Program for providing funding and encouragement for this work.

References

Belikov, R., et al. 2009, SPIE, 7440
Guyon, O. 2003, A&A, 404, 379
Pluzhnik, E. A., Guyon, O., Ridgway, S. T., Martinache, F., Woodruff, R. A., Blain, C., & Galicher, R. 2006, ApJ, 644, 1246
Vanderbei, R. J. & Traub, W. A. 2005, ApJ, 626, 1079

Pathways Towards Habitable Planets
ASP Conference Series, Vol. 430, 2010
Vincent Coudé du Foresto, Dawn M. Gelino, and Ignasi Ribas, eds.

Transfer of Meteorites from Earth to the Interesting Objects within the Solar System and the Extrasolar Planets

T. Hara, T. Takagi, and D. Kajiura

Department of Physics, Kyoto Sangyo University, Kyoto, Japan

Abstract. The probability is investigated that meteorites of Earth origin are transferred to the interesting objects which are supposed to have seas under the icy surface such as Enceladus, Europa, Ceres and dwarf planet Eris and the extrasolar planets. We take the ejection process in collision, such as the Chicxulub crater event, from Earth. If we assume the appropriate size of meteorites as 1cm in diameter, the number of meteorites reaching the interesting objects and the extrasolar planet system could be much greater than one. So we should consider the panspermia theories more seriously as organisms disperse.

It has been established that rocks can be ejected from planetary surfaces by colliding asteroids and comets. The Chicxulub crater event 65 Myr ago provides evidence of the ejection process in collision. The meteorite's size is estimated to be about 10 km in diameter. We estimate the transfer probability of Earth origin rocks to the solar system and nearby star systems.

We assume that N_0 is the number of rocks that are ejected from the Earth, the distance to the object is s(cm), and the cross-section of the object is σ (cm^2). Then the number of captured rocks is $N_{cap} = N_0\,\sigma/(4\pi s^2)$

When the Chicxulub meteorite collided with Earth, it could be estimated that almost the equivalent mass was ejected from Earth. Then it is assumed that the ejected mass from the Solar system is $f_1 \times M_0$, where M_0 is the mass of the Chicxulub meteorite and the factor $f_1(\sim 0.1)$ denotes the fraction of the mass ejected from Earth. Taking that the mean diameter of rocks is r_1 (cm) and the estimated diameter of the Chicxulub meteorite as R_1, the number N_0 of ejected rocks from Earth is

$$N_0 = f_1 \left(\frac{R_1}{r_1}\right)^3 \simeq 10^{17} \left(\frac{f_1}{0.1}\right) \left(\frac{R_1}{10\text{ km}}\right)^3 \left(\frac{r_1}{1\text{ cm}}\right)^{-3}$$

The distance to the interesting icy bodies within the solar system is on the order of an astronomical unit, so we take the representative value for s as $s \sim 1\text{AU} = 1.5 \times 10^8$ km. The problem is the cross-section, σ.

Model A: The cross-section, σ, for the direct collision of the meteorites with the icy body is of the order $\sigma \sim \pi R_0^2$, where R_0 is the radius of the icy body. Then if $R_0 \sim 10^3$ km and $s_0 \sim 1(\text{AU})^2$ the number of ejected rocks hitting the body is estimated to be

$$N_A \sim N_0(\sigma/(4\pi s_0^2)) \sim 10^6 (f_1/0.1)(R_1/10\text{km})^3(r_1/1\text{cm})^{-3}(R_0/10^3\text{km})^2(s_0/1\text{AU})^{-2}$$

This model corresponds to the high velocity case of ejected rocks.
Model B: The gravitational infall to the object could be inferred by the accretion radius, $r_0 \sim Gm_0/v_0^2$, where m_0 is the mass of the object. Then σ

Table 1. The number of Earth-origin meteorites estimated to hit various solar system objects

Object	s_0(AU)	R_0(km)	m_0(kg)	N_A	N_B
Enceladus	10	2.5×10^2	7×10^{19}	6×10^2	5×10^2
Europa	5	1.6×10^3	5×10^{22}	10^5	2×10^8
Ceres	3	5×10^2	9×10^{20}	3×10^4	8×10^4
Eris	100	1.3×10^3	2×10^{22}	2×10^2	4×10^7

is estimated as $\sigma \sim \pi r_0^2$ which is proportional to m_0^2. As the dominant planet is Jupiter, the infall rate is roughly proportional to $(m_0/M_J)^2$ where M_J is Jupiter's mass. Then the number of infalling rocks is estimated for the case of $m_0 \sim 10^{20}$kg as

$$N_B \sim N_0(m_0/M_J) \sim 10^3 (f_1/0.1)(R_1/10\text{km})^3 (r_1/1\text{cm})^{-3} (m_0/10^{20}\text{kg})^2 (M_J/2 \cdot 10^{27}\text{kg})^{-2}$$

This model corresponds to the low velocity case of ejected rocks. The values for Enceladus, Europa, Ceres, and Eris are presented in Table 1.

For every object, the numbers N_A and N_B are much greater than one. Although it is uncertain how rocks enter the presumed sea under the surface, the probability is high that the lives originated from Earth are adapted and growing there.

To extend the above consideration to the extrasolar planets is almost straightforward. We introduce the factor, f_2, which denotes the fraction of rocks ejected from the solar system.

Model C: The cross-section, σ, could be enlarged including the effect of the gravitational interaction such as swingby within the extrasolar system. If rocks can be deccelerated by gravitational interaction, they are captured to the star system. After a few Myr, some fraction of rocks could fall into the planet (Melosh 2003). Then the cross-section goes as $\sigma \sim f_3(r/1\,\text{AU})^2$, where the uncertainty factor f_3 is included. Then the number of falling rocks becomes

$$N_C \sim 10^2 \times \left(\frac{f_1 f_2 f_3}{10^{-3}}\right) \left(\frac{R_1}{10\,\text{km}}\right)^3 \left(\frac{r_1}{1\,\text{cm}}\right)^{-3} \left(\frac{r}{1\,\text{AU}}\right)^2 \left(\frac{s}{20\,\text{lyr}}\right)^{-2}$$

From the simulations of (Melosh 2003), the factors f_1, f_2 and f_3 are ~ 0.1.

Although there are many uncertain factors, the probability of rocks originating from Earth and reaching interesting objects and extrasolar planets is not so small. However, it is not certain that the micro-organisms within small (≥ 1 cm) meteorites are still viable for millions of years, or that Earth-origin meteorites could be transferred to the nearby objects. If we assume they are viable, we should consider the Panspermia theories more seriously as organisms disperse.

References

Melosh, H. 2003, Astrobiology, 3, 207

Pathways Towards Habitable Planets
ASP Conference Series, Vol. 430, 2010
Vincent Coudé du Foresto, Dawn M. Gelino, and Ignasi Ribas, eds.

A Coronagraph Experiment in a High Thermal Stability Environment with a Binary Shaped Pupil Mask

K. Haze,[1,2,3] K. Enya,[3] T. Kotani,[3] L. Abe,[4] T. Nakagawa,[3,5]
S. Higuchi,[3,5] T. Sato,[6] T. Wakayama,[6] and T. Yamamuro[7]

Abstract. We present our results from a laboratory experiment on a binary-shaped pupil mask coronagraph in the context of instrumentation R&D for the direct observation of exo-planets. The aim of this work is testing an axiom of the coronagraph using the method employing the PSF (point spread function) subtraction in thermally-stable condition. Both the raw coronagraphic contrast and the contrast after the PSF subtraction are evaluated. A contrast of 2.6×10^{-7} was achieved for the raw coronagraphic images by analyzing the areal mean of the observed dark regions. A contrast of 1.8×10^{-9} was achieved for the PSF subtraction by the areal variance (1σ) of the observed dark regions. Application of the PSF subtraction to the coronagraph can ease the requirements for the raw contrast.

1. Introduction

The enormous contrast in luminosity between a central star and an associated planet is the primary difficulty in direct observation (Traub & Jucks 2002). We currently promote the development of a coronagraph which improves the contrast between the central star and the planet, with the aim of the direct observation of extra solar planets.

2. Configuration of the Coronagraph Experiment

We used a checkerboard pupil mask (Vanderbei et al. 2004) which could reduce the area of diffracted light from the central star, called "the dark region" in Figure 1. The required contrast in the design was $\times 10^{-7}$ (Enya et al. 2007). A coronagraph optical system was set up on an optical bench in a vacuum chamber,

---

[1] Research Fellow of Japan Society for the Promotion of Science, 8 Ichibancho, Chiyoda, Tokyo, Japan

[2] The Graduate University for Advanced Studies, 3-1-1 Yoshinodai, Sagamihara, Kanagawa, Japan

[3] JAXA/ISAS, 3-1-1 Yoshinodai, Sagamihara, Kanagawa, Japan

[4] Universite de Nice-Sophia Antipolis, Parc Valrose, Nice, France

[5] University of Tokyo, 7-3-1 Hongou, Bunkyo, Tokyo, Japan

[6] AIST, 1-1-1 Azuma, Tsukuba, Ibaraki, Japan

[7] Optcraft, 3-26-8 Aihara, Sagamihara, Kanagawa, Japan

as shown in Figure 2. The vacuum chamber was covered with aluminum-foam polyethylene layers, and active PID temperature control was applied to the optical bench to reduce the instability caused by the thermal deformation. Stability is an important requirement to utilize PSF subtraction with maximum effectiveness. PSF subtraction is a method for subtracting the PSF of one coronagraphic image from another one to improve the contrast.

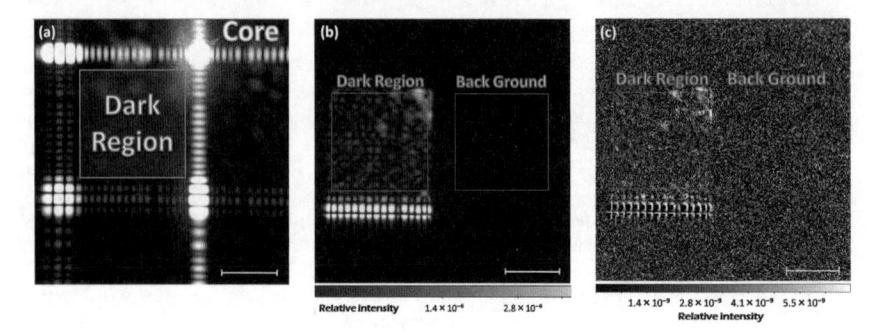

Figure 1. The scale bar is 10/D. (a) Observed coronagraphic images including the core of the PSF. (b) An image of dark region obtained with a black square mask. (c) The PSF subtracted image.

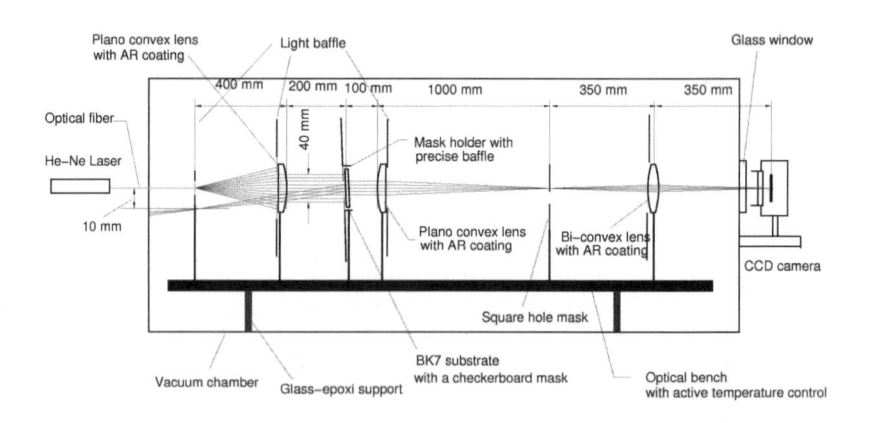

Figure 2. This figure shows the configuration of the experimental optics.

3. Results and Discussion

The core was observed with several exposure times (0.03, 0.3, 3sec) with an optical density of 4. The dark region was observed with a 200s exposure. The contrast between the mean value of the dark region and the peak of the core was 2.6×10^{-7} in a raw coronagraphic image. PSF subtraction used two images of the dark region. The dark regions of the coronagraphic images were observed with a 3600s exposure ×2, and the PSF subtracted image was obtained by subtracting the image of one hour previous from a current image. The standard deviation (1σ) of the fluctuation is given by $\sigma = \sqrt{\sigma_{DarkRegion} - \sigma_{BackGraund}}$, where $\sigma_{DarkRegion}$ is the standard deviation of the dark region and $\sigma_{BackGraund}$ is the standard deviation of the back ground, as shown in Figure 1. A contrast between σ and the peak of the core was 1.8×10^{-9} for the PSF subtraction. The contrast was improved by around two orders of magnitude.

This result shows that the requirements for the raw contrast are significantly relaxed by using PSF subtraction technique. This means that the requirement for the wave-front quality of the Space Infrared telescope for Cosmology and Astrophysics (SPICA) (Nakagawa et al. 2004) and other platforms is relaxed.

References

Enya, K., Tanaka, S., Abe, L., & Nakagawa, T. 2007 A&A, 461, 783
Nakagawa, T., & SPICA working group, 2004, ASR, 34, 645
Traub, W. A., & Jucks, K. W., 2002, Astro-ph/0205369,
Vanderbei, R. J., Kasdin, N. J., & Spergel, D. N. 2004, ApJ, 615, 555

Pathways Towards Habitable Planets
ASP Conference Series, Vol. 430, 2010
Vincent Coudé du Foresto, Dawn M. Gelino, and Ignasi Ribas, eds.

Performance of the Cophasing System of Persee, a Dynamic Nulling Demonstrator

K. Houairi,[1,2,3] F. Cassaing,[1,3] J. M. Le Duigou,[2] J. P. Amans,[4]
M. Barillo,[5] V. Coudé du Foresto,[3,6] F. Hénault,[7] S. Jacquinod,[8]
J. Lozi,[1,2,3] J. Montri,[1,3] M. Ollivier,[8] J. M. Reess,[3,6] and B. Sorrente[1,3]

Abstract. Spectral characterization of exoplanets can be made by nulling interferometers. In this context, several projects have been proposed such as DARWIN/TPF-I, FKSI, PEGASE, and ALADDIN. To stabilize the beams with the required nanometric accuracy, a cophasing system is required, made of a piston/tip/tilt actuator on each arm and a piston/tip-tilt sensor. The demonstration of the feasibility of such a cophasing system is a central issue. In this goal, a laboratory breadboard named PERSEE is under integration at Observatoire de Paris-Meudon. This paper describes the current status of PERSEE. We show that a cophasing at a subnanometric level has been reached, which allowed us to reach a monochromatic null depth of N= $6.2 \times 10^{-5} \pm 6.3 \times 10^{-6}$.

1. Introduction

In the last few years, several nulling interferometry projects have been proposed such as DARWIN (Léger & Herbst 2007), FKSI (Danchi et al. 2006), PEGASE (Le Duigou et al. 2006) and TPF (Lawson et al. 1999), space-based, and ALADDIN (Coudé du Foresto et al. 2006), ground-based. These projects are based on the Bracewell interferometric principle which is made technologically possible thanks to the ability of measuring and correcting the optical path difference (OPD) between the beams. This principle requires a control of the OPD with an accuracy of a few nanometers despite the perturbations during the exposure time mainly arising from the mechanical motions of the spacecrafts (space-based interferometers) or from the atmospheric turbulence (ground-based interferometers). This OPD stability cannot be met without a cophasing system.

A laboratory breadboard called PERSEE (Cassaing et al. 2008; Houairi et al. 2008), the goal of which is to show the feasibility of such missions, is

---

[1]ONERA, Optics Department, BP 72, Châtillon, France

[2]CNES, 18 Avenue Édouard Belin, Toulouse, France

[3]PHASE, the high angular resolution partnership between ONERA, Observatoire de Paris, CNRS and University Denis Diderot Paris 7, France

[4]GEPI, Observatoire de Paris, CNRS UMR 8111, 5 place Jules Janssen, Meudon, France

[5]Thales Alenia Space, 100 Bd du Midi, Cannes-la-Bocca, France

[6]LESIA, Observatoire de Paris, CNRS UMR 8109, Meudon, France

[7]Observatoire de la Côte d'Azur, Boulevard de l'Observatoire, Nice, France

[8]Institut d'Astrophysique Spatiale, Centre universitaire d'Orsay, Orsay, France

a) Residual OPD (nm)   b) Monochromatic null depth

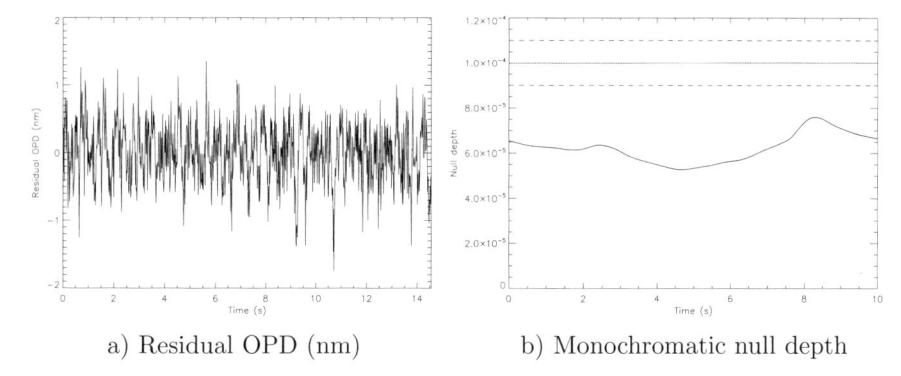

Figure 1. Experimental results obtained with PERSEE cophasing system in closed loop. The solid and dashed straight lines of the right-hand side figure represent the null depth specification.

under integration. PERSEE is built by a consortium including CNES, IAS, LESIA, OCA, ONERA, and TAS. Its main goals are the demonstration of a polychromatic null from 1.6 μm to 3.2 μm with a 10^{-4} rejection rate and a 10^{-5} stability despite the introduction of realistic perturbations.

2. Current Status of PERSEE

By waiting for the final integration of PERSEE, a reduced setup has been carried out in autocollimation: the light will be injected through an output of the Modified Mach-Zehnder interferometer which will thus be used both in order to divide and to recombine the beams.

With this configuration, a cophasing with a residual OPD as low as 0.45 nm rms has been reached, what allowed to reach a monochromatic ($\lambda = 2.32\ \mu m$) null depth of N= $6.2 \times 10^{-5} \pm 6.3 \times 10^{-6}$, as shown in Figure 1.

The large-band nulling is under integration and the complete optical train with the introduction and correction of typical disturbances should end in 2010.

References

Cassaing, F. et al. 2008, SPIE, 7013, 70131Z
Coudé du Foresto, V., et al. 2006, SPIE, 6268, 626810
Danchi, W. C., et al. 2006, SPIE, 6268, 626820
Houairi, K., et al. 2008, SPIE, 7013, 70131W
Lawson, P. R., Dumont, P. J., & Colavita, M. M. 1999, BAAS, 31, 835
Le Duigou, J. M. et al. 2006, SPIE, 6265, 62651M
Léger, A & Herbst T. 2007, arXiv:0707.3385

Search for Exoplanets using TTVs in the Southern Hemisphere

S. Hoyer,[1] P. Rojo,[2] and M. López-Morales[3]

[1] Universidad de Chile, Astronomy Department, Santiago, Chile

[2] Universidad de Chile, Astronomy Department, Santiago, Chile

[3] Department of Terrestrial Magnetism, Carnegie Institution of Washington, Washington D.C., USA

Abstract. The method of transit timing variations (TTVs) is sensitive to detect additional low mass planets in transiting exoplanetary systems that are otherwise undetectable by other methods like RVs. In 2008 we started a homogeneous monitoring of transiting planets in the Southern Hemisphere with observations of cadence of 20 to 50 seconds. By carefully measuring the central time we will be able to detect long- and short-term variations of the orbital period of the primary transit. In this contribution we will present results of the analisys of TTVs interesting case: the transiting hot Jupiter OGLE-TR-111.

1. The Project

Our group is performing a homogeneous monitoring of a set of transiting planets, which will allow us to detect possible variations in the central time of their primary transits. Those time variations that can be produced by the presence of additional orbiting bodies in the systems. We have undertaken careful follow-ups of transiting planets from the Southern Hemisphere making use of SOI (SOAR Telescope), GMOS (Gemini South Observatory) and 4K Optical Imager (1.0 m CTIO Telescope) with cadences of 20 to 50 seconds. Since 2008, we have observed 31 transits events of eight different exoplanets. We are currently performing the reduction and analysis of these data in order to obtain the physical parameters of the transiting exoplanets, and to detect any possible short- or long-term TTVs in the central time of the transits.

2. OGLE-TR-111: An Interesting Case

We have studied the case of the exoplanet OGLE-TR-111b where Díaz (2008) reported differences in the central time of the transits of this planet. In order to confirm this detection we monitored 5 consecutive transits with FORS1 and FORS2 at the ESO Very Large Telescope. However, only four of them were successfully covered and used in this work; one was observed with the V-high band filter while the other three were observed with the Bessell I band filter. Using the DIAPL code (Wozniak 2000), to perform differential photometry we built the light curves, and using the JKTEBOP code (Southworth 2004) to fit

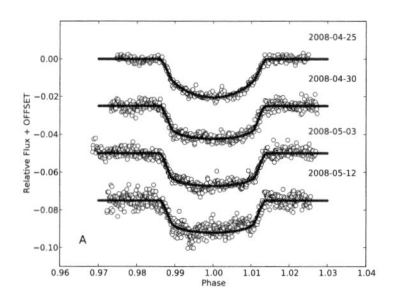

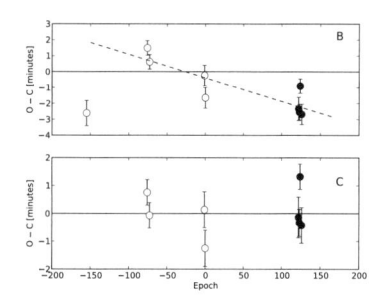

Figure 1. *A:* Light curves of 4 transits of OGLE-TR-111b. The light curve on the top corresponds to the observation in V-high filter while the other three curves correspond to observations with the Bessell I filter. The filled dots correspond to the theoretical fitted curve obtained with JKTEBOP code from our observations (empty dots). *B:* Observed minus Calculated (O-C) central times for the transits of OGLE-TR-111b. The empty points represents the values by Minniti et al. (2007), Winn et al. (2007) and Díaz et al. (2008). The filled dots correspond to the values obtained from the transits presented in this work. The dashed line represent the fitted straight line excluding the value of Minniti et al 2007. *C:* O-C residuals after correcting times with a straight line. These residuals are lower than 1.5 minutes.

a model (Figure 1a), we were able to obtain the inclination, the central time of the transit and the coefficients of limb darkening for each dataset.

Combining previous published values for central times of transits of this exoplanet (Winn 2007; Minniti 2007; Díaz 2008) with the values obtained in this work we built the *Observed minus Calculated* (O-C) diagram (Figure 1b). Excluding the first data point, which is currently being revised as that dataset is suspected to have a timing problem, we can attribute the slope of the other points to residual errors in the ephemerides equation. After correcting for that slope (Figure 1c), we find no TTVs for this system with amplitudes larger than 1.5 min.

Acknowledgments. S. H. acknowledges support by GEMINI # 32070020 and FONDECYT # 11080271 grants.

References

Díaz, F.R., Rojo, P., Melita, M., Hoyer, S., Minniti, D., Mauas, P., & Ruíz, M.T. 2008, ApJ, 682, 49
Minniti, D., Fernández, J.M., Díaz, R.F., Udalski, A., Pietrzynski, G., Gieren, W., Rojo, P., Ruíz, M.T., & Zocalli, M. 2007, ApJ, 660, 858
Southworth, J., Maxted, P.F.L., & Smalley, B. 2004, MNRAS, 351, 1277
Winn, J., Holman, M.J., & Fuentes, C.I. 2007, AJ, 133, 11
Wozniak, P.R. 2000, AcA, 50, 421

Pathways Towards Habitable Planets
ASP Conference Series, Vol. 430, 2010
Vincent Coudé du Foresto, Dawn M. Gelino, and Ignasi Ribas, eds.

First Results from the Transit Ephemeris Refinement and Monitoring Survey (TERMS)

S. R. Kane,[1] S. Mahadevan,[2,3] K. von Braun,[1] G. Laughlin,[4] A. Howard,[5] and D. R. Ciardi[1]

[1] NASA Exoplanet Science Institute, Caltech, Pasadena, CA, USA

[2] Dept of Astronomy & Astrophysics, Penn State University, University Park, PA, USA

[3] Center for Exoplanets and Habitable Worlds, Penn State University, University Park, PA, USA

[4] UCO/Lick Observatory, Univ. of California, Santa Cruz, CA, USA

[5] Department of Astronomy, Univ. of California, Berkeley, CA, USA

Abstract. Transiting planet discoveries have yielded a plethora of information towards understanding the structure and atmospheres of extra-solar planets. These discoveries have been restricted to the short-period or low-periastron distance regimes due to the bias inherent in the geometric transit probability. Through the refinement of planetary orbital parmaters, and hence reducing the size of transit windows, long-period planets become feasible targets for photometric follow-up. Here we describe the TERMS project which is monitoring these host stars at predicted transit times.

1. Introduction

Monitoring known radial velocity planets at predicted transit times presents an avenue through which to explore the mass-radius relationship of exoplanets into regions of period/periastron space beyond that which is currently encompassed (Kane 2007). This is particularly true for those planets in relatively eccentric orbits (Kane & von Braun 2008, 2009), as demonstrated by the variation in transit probability with the argument of periastron (see Figure 1, left). Here we describe techniques for refining ephemerides and performing follow-up observations (Kane et al. 2009). These methods are being employed by the Transit Ephemeris Refinement and Monitoring Survey (TERMS).

2. Transit Ephemerides

The transit window as described here is defined as a specific time period during which a complete transit (including ingress and egress) could occur for a specified planet. The size of a transit window depends upon the uncertainties in the fit parameters. They also deteriorate over time, motivating follow-up of the transit window as soon as possible after discovery. Figure 1 (right) shows the size of the first transit window (after the fit time of periastron passage, t_p) for a sample

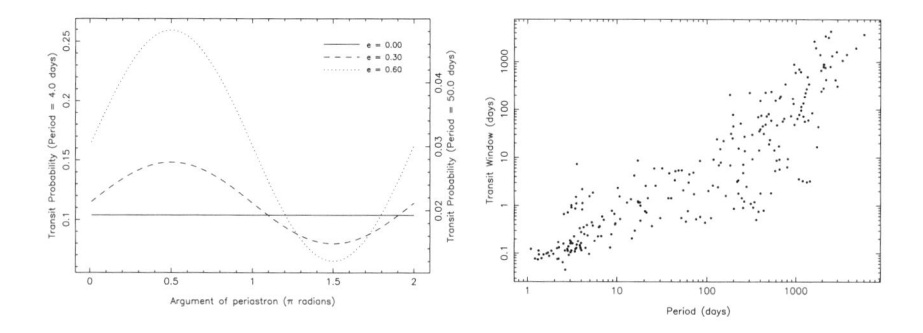

Figure 1. *Left:* Dependence of geometric transit probability on the argument of periastron for eccentricities of 0.0, 0.3, and 0.6, plotted for periods of 4.0 days (left axis) and 50.0 days (right axis). *Right:* Ephemeris calculations for a sample of 245 exoplanets, showing the dependence of transit window size on period.

of 245 exoplanets. The transit windows of the short-period planets tend to be significantly smaller than those of long-period planets since, at the time of discovery, many more orbits have been monitored to provide a robust estimate of the orbital period. Targets chosen for TERMS tend to be those which have small transit windows, medium-long periods, and a relatively high probability of transiting the host star.

3. TERMS Strategy

A considerable number of high transit probability targets are difficult to adequately monitor during their transit windows because the uncertainties in the predicted transit mid-points are too high (months or even years). The acquisition of just a handful of new radial velocity measurements at carefully optimised times can reduce the size of a transit window by an order of magnitude. This is described in more detail by Kane et al. (2009).

Due to the wide range of stars monitored, both in sky location and brightness, TERMS collaborates with a variety of existing groups to take advantage of transit window opportunities. Note that the observations from this survey will lead to improved exoplanet orbital parameters and ephemerides even without an eventual transit detection for a particular planet. The results from this survey will provide a complementary dataset to the fainter magnitude range of the *Kepler* mission, which is expected to discover many transiting planets including those of intermediate to long-period.

References

Kane, S. R., 2007, MNRAS, 380, 1488
Kane, S. R. & von Braun, K., 2008, ApJ, 689, 492
Kane, S. R. & von Braun, K., 2009, PASP, 121, 1096
Kane, S. R., et al. 2009, PASP, 121, 1386

Pathways Towards Habitable Planets
ASP Conference Series, Vol. 430, 2010
Vincent Coudé du Foresto, Dawn M. Gelino, and Ignasi Ribas, eds.

Flux and Polarization Signals of Water Clouds on Earth–Like Exoplanets

T. Karalidi,[1,2] D. M. Stam,[1] and J. W. Hovenier[3]

[1] SRON - Netherlands Institute for Space Research, Sorbonnelaan 2, Utrecht, the Netherlands

[2] Astronomical Institute, Utrecht University, Princetonplein 5, Utrecht, the Netherlands

[3] Astronomical Institute "Anton Pannekoek", University of Amsterdam, Science Park 904, Amsterdam, the Netherlands

Abstract. We present numerically simulated flux and polarisation signals of starlight that is reflected by Earth–like exoplanets covered by liquid water clouds at different altitudes, at wavelengths from 0.3 to 1.0 μm. Our results show that the degree of polarisation of this reflected light is more sensitive to the cloud top pressure than the flux. The results clearly illustrate the advantages of polarimetry for the characterisation of clouds in exoplanet atmospheres.

1. Introduction

The detection of rocky exoplanets has fueled the quest for signs of habitability. A crucial factor for a planet's habitability is its climate, in which clouds play an important role. The effect of clouds depends on parameters such as the cloud particles microphysical properties, the cloud's optical thickness and spatial dimensions. Thus the detection, and especially the characterisation of clouds on exoplanets will provide crucial information on the local climate and surface conditions. In particular water clouds may indicate liquid water on the surface, which is considered to be a crucial factor for habitability and life.

From remote-sensing of solar system planets, it is well-known that cloud characterisation is best done by measuring both the flux F and the degree of (linear) polarisation P of starlight that is reflected by a planet. Especially P is very sensitive to the cloud particles microphysical properties and the cloud top pressure (see e.g. Hansen 1971; Hansen & Travis 1974; Hansen & Hovenier 1974). Here, we present results of simulations of F and P of starlight that is reflected by Earth–like exoplanets, following calculations as described by Stam (2008).

Our model planet's atmosphere consist of homogeneous, plane–parallel layers, one of which is a liquid water cloud layer with an optical thickness of 10 at $\lambda = 0.55$ μm, and we use Earth–like pressure and temperature profiles (see Stam 2008). The planet's surface is black. In our simulations, we use cloud top pressures p_{top} ranging from 0.802 to 0.628 bar, and calculate F and P for wavelengths λ from 0.3 to 1.0 μm, and for planetary phase angles α from 0° to 180°. The cloud particles are described in size by a standard distribution (Hansen

Flux and Polarization Signals of Water Clouds   467

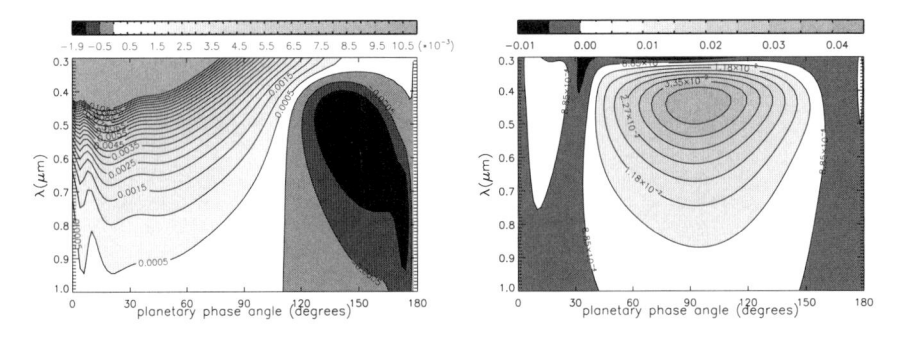

Figure 1. The absolute changes in F (left) and P (right) as p_{top} decreases from 0.802 to 0.628 bar (thus, $F_{0.802} - F_{0.628}$, and $P_{0.802} - P_{0.628}$) as functions of λ and α. The physical properties of the cloud layer are described in Section 1.

& Travis 1974) with an effective radius of 2.0 μm and an effective variance of 0.1, assuming a wavelength independent refractive index, their single scattering properties are calculated using Mie-theory.

2. The Results of our Numerical Simulations

Figure 1 shows the absolute changes in flux F and polarization P when the cloud top pressure p_{top} decreases from 0.802 to 0.628 bar (the cloud top thus increases in altitude). Flux F is normalized as described in Stam (2008). The flux decreases with decreasing p_{top}, except at the largest phase angles and for $\lambda > 0.4$ μm. The change of F with p_{top} is very small, especially when $\lambda > 0.4$ μm or $\alpha > 60°$. We do not expect such small differences to be measurable anytime soon, because of the extremely small number of photons that will be received from the exoplanet. The polarisation P appears to be quite sensitive to p_{top}; when p_{top} decreases to 0.628 bar, P decreases about 0.045 (4.5%) when $\lambda \approx 0.45$ μm and $\alpha \approx 90°$ (where light from the exoplanet will be easiest to detect). This decrease of P is due to the decrease of the amount of molecules above the cloud. Because light that is scattered by molecules is generally highly polarized (except at the smallest and largest scattering angles and when there is a lot of multiple scattering, such as at the shortest wavelengths), the decreasing amount of molecules above the cloud results in a decrease of P. Because P is a relative measure, we expect differences of a few percent will be measurable with the next generation of polarimeters.

3. Summary and Future Work

It appears that polarimetry will allow the retrieval of cloud top pressures on exoplanets, especially because P is most sensitive to p_{top} at phase angles where light of the planet will be easiest to detect. The sensitivity of F and P to other physical parameters of clouds (particle sizes and composition, optical thicknesses), the cloud and surface coverage will be subject of further study.

References

Hansen, J. E. 1971, J. Atmos. Sci., 28, 1400
Hansen, J. E. & Hovenier, J. W. 1974, J. Atmos. Sci., 31, 1137
Hansen, J. E. & Travis, L. D. 1974, Space Sci.Rev., 16, 527
Stam, D. M. 2008, A&A, 482, 989

Pathways Towards Habitable Planets
ASP Conference Series, Vol. 430, 2010
Vincent Coudé du Foresto, Dawn M. Gelino, and Ignasi Ribas, eds.

Using Polarization to Detect and Characterize Exoplanets

L. Kedziora-Chudczer and J. Bailey

School of Physics, UNSW, Sydney, NSW, Australia

Abstract. Polarization changes at different stages of a planet's orbit are complementary and enhance the information obtainable using spectroscopy. We describe how broadband polarimetry can be used in the future search for liquid water on extrasolar planets by detecting rainbow scattering from cloud droplets and a "glint" from the surface of the oceans.

1. Introduction

Light scattered from planetary atmospheres and surfaces is linearly polarized. In contrast, the light of normal stars is unpolarized with typical polarization levels of a few x 10^{-6}. Therefore polarization observations provide a means of distinguishing the planet from the residual light of the star.

In addition, polarization and its variation with phase angle of a planet can provide evidence for the existence of liquid water either in clouds, through rainbow scattering, or from the "glint" off oceans at the surface, and thus aid the search for habitable terrestrial planets. The three processes discussed here show strong polarization peaks at different phase angles of the planet.

2. Rayleigh Scattering from Molecules

Rayleigh scattering from molecules in a clear atmosphere produces a strongly polarized, wavelength-dependent signal that peaks at phase angles around 90° and is strongest at blue and UV bands. At such phase angles the planet is half-illuminated and at its maximum distance from the star. Rayleigh scattering can provide a good estimate of the atmospheric pressure at the reflecting surface, which is either the actual surface for a clear atmosphere or the cloud top for a cloud covered planet.

In contrast, it is more difficult to acquire such information from spectroscopy at the low resolution and S/N likely to be feasible in observations of exoplanets. The surface pressure depends on the potentially important gases, such as N_2 and H_2 abundantly present in giant planets of our solar system, that have no spectral features in the typically used spectral range of visual and infrared bands. A proper estimate of pressure is essential to recognize the amount of pressure broadening in observed spectral features of planetary atmospheres.

The signature of Rayleigh scattering was clearly seen in the polarization phase curve of Venus and it was used by Hansen & Hovenier (1974) to determine a cloud top pressure of 50mb for the planet's atmosphere.

3. Glint off the Surface Oceans

Oceans have very low albedo under most conditions, but reflect efficiently under specular reflection conditions producing a "glint" spot. This "glint" light is highly polarized and provides a way of detecting oceans on a planetary surface. Glint polarization peaks at crescent phases. This means that the planet appears fainter in its orbital cycle making signal detection more challenging. However the degree of polarization can reach up to 70% for the cloud free and entirely covered by water planet (Williams & Gaidos 2008). For the planet with a similar geography as the Earth one can expect 30% polarization in the cloud free conditions.

4. Rainbows - Liquid Water Clouds

Polarized light scattered at the primary rainbow angle (about 40deg for water) is a clear indicator of spherical (and hence probably liquid) cloud droplets. The phase angle of the rainbow peak is determined by the refractive index and thus provides an indication of the composition of the liquid. The width of the peak is determined by particle size. Importantly the rainbow scattering is seen at phase angles around $40°$ (gibbous phases) when the planet is bright.

Bailey (2007) estimated the global rainbow signal from Earth to be $\sim$ 12-16% for two different cloud cover scenarios – and that it should be detectable with a TPF-C style mission. Hansen & Hovenier (1974) used the polarization phase curve of Venus to determine that the Venus clouds were made of spherical liquid droplets with particle size about $1\mu m$, composed of 75% H_2SO_4 and 25% H_2O. This result was confirmed by in-situ spacecraft data. Polarized reflectance from clouds was also observed by the POLDER Instrument on the ADEOS satellite showing a rainbow signal that distinguishes water and ice clouds.

5. Observations Techniques and Modeling

In view of the benefits offered by polarimetric observations for planet characterization, the importance of polarimetric capability for new instruments designed for exo-planet searches is clear. A new astronomical polarimeter (PLANETPOL), built at the University of Hertfordshire, is capable of measuring polarization with a fractional sensitivity of one part per million (Hough et al. 2006). It was designed for differential polarimetry where the polarized, scattered light from hot Jupiter-type planets in the combined light of the star and the planet can be detected at the levels of a few x 10^{-6} (Seager et al. 2000). The first observations with the 4.2m William Herschel Telescope demonstrated its ability to reach this target, despite the effect of dust in the Earths atmosphere that can produce spurious polarization at low levels (Bailey et al. 2008; Ulanowski et al. 2007).

The polarization effects described above require modeling with a planetary atmosphere radiative transfer code that includes a full solution of the vector radiative transfer equation for all four Stokes parameters. Such a model is being developed by incorporating polarized radiative transfer into VSTAR package (Bailey 2006). Possible methods of polarized radiative transfer, such as adding-doubling (Stam et al. 2004), discrete ordinate (Rozanov & Kokhanovsky 2006) and Monte-Carlo analysis are being evaluated.

References

Bailey, J. 2006, Second Workshop on Mars Atmopshere Modeling and Observations, ed. F. Forget, M. .A. et al., 148
Bailey, J. 2007, Astrobiology, 7, 320
Bailey, J. et al. 2008, MNRAS, 386, 1016
Hansen, J. E. & Hovenier, J. W. 1974, J. Atmos. Sci. 31, 1137
Hough, J.H. et al. 2006, PASP, 118, 1302
Rozanov, V. V. & Kokhanovsky, A. A. 2006, Atm. Res., 79, 241
Seager, S., Whitney, B. A., & Sasselov, D. D. 2000, ApJ, 540, 504
Stam, D. et al. 2004, A&A, 428, 663
Ulanowski, Z. et al. 2007, Atmos. Chem. Phys. 7, 6161
Williams, D. & Gaidos, E. 2008, Icarus, 195, 927

Pathways Towards Habitable Planets
ASP Conference Series, Vol. 430, 2010
Vincent Coudé du Foresto, Dawn M. Gelino, and Ignasi Ribas, eds.

Signatures of Resonant Terrestrial Planets in Long-Period Systems

G. F. Kennedy[1] and R. A. Mardling[2]

[1] Institut de Ciències del Cosmos, Universitat de Barcelona, Barcelona, Spain

[2] School of Mathematical Sciences, Monash University, Clayton, Australia

Abstract. The majority of extrasolar planets discovered to date have significantly eccentric orbits, some if not all of which may have been produced through planetary migration. During this process, any planets interior to such an orbit would therefore have been susceptible to resonance capture, and hence may exhibit measurable orbital period variations. Here we summarize the results of our investigation into the possibility of detecting low-mass planets which have been captured into the strong 2:1 resonance. Using analytical expressions together with simulated data we showed that it is possible to identify the existence of a low-mass companion in the internal 2:1 resonance by estimating the time-dependant orbital period for piecewise sections of radial velocity data. This works as long as the amplitude of modulation of the orbital period is greater than its uncertainty, which in practice means that the system should not be too close to exact resonance. Here we provide simple expressions for the libration period and the change in the observed orbital period, these being valid for arbitrary eccentricities and planet masses. They in turn allow one to constrain the mass and eccentricity of a companion planet if the orbital period is sufficiently modulated.

1. Resonance Theory

The following summarizes the results of Kennedy & Mardling (in preparation). A resonant system is characterized by the libration of one or more resonance angles. For a coplanar system the quadrupole 2:1 resonance angle is given by

$$\phi = \lambda_i - 2\lambda_o + \varpi_i \qquad (1)$$

where λ_i and λ_o are the inner and outer mean longitudes, and ϖ_i and ϖ_o are the corresponding longitudes of periastron. The maximum possible variation in $\sigma = \nu_i/\nu_o$, the ratio of inner to outer orbital frequencies, is given by (Mardling 2008)

$$\Delta\sigma = 3\left[f(e_i)\,g(e_o)\left((m_o/m_*) + 2^{2/3}(m_i/m_*)\right)\right]^{1/2} \qquad (2)$$

where e_i and e_o are the inner and outer orbital eccentricities, m_i and m_o are the corresponding planet masses,

$$f(e_i) = 3e_i - \frac{13}{8}e_i^3 - \frac{5}{192}e_i^5 + \mathcal{O}(e_i^7) \quad \text{and} \quad g(e_o) = 1 - \frac{5}{2}e_o^2 + \frac{13}{16}e_o^4 + \mathcal{O}(e_o^6) \qquad (3)$$

For stable systems inside the 2:1 resonance the libration period is approximately

$$P_{lib} = 2P_o/\Delta\sigma = \alpha[f(e_i)]^{-1/2}P_o, \qquad (4)$$

where α depends on observed parameters when $m_i \ll m_o$. The amplitude of variation of the outer (observed) orbital period depends on the distance from exact resonance, $\delta\sigma$, and is given by

$$\delta P_o/P_o = -2\left[1 + (m_o/m_i)\sigma^{-2/3}\right]^{-1}\delta\sigma/\sigma \simeq -2^{2/3}\left(m_i/m_o\right)\delta\sigma, \qquad (5)$$

where $\delta\sigma = \sigma - 2$. If the period and amplitude of variation of the orbital period of a known system can be measured, (4) and (5) can be used to estimate e_i, m_i and $\delta\sigma$.

2. Data Analysis

To illustrate our procedure we used the orbital parameters of HD 216770b for which $m_o \sin i = 0.65\,M_J$, $M_* = 0.9 M_\odot$, $e_o = 0.37$, $P_o = 118.45$ days and $P_{lib} \simeq 25 P_o$. Using synthetic data for a system with and without a 10 Earth-mass planet in the interior 2:1 resonance, we were able to infer the presence of the super-Earth as follows. First we determined the range of values of e_i for which stable orbits exist using direct three-body integrations. Sampling at the rate of around 16 synthetic radial velocity data points per orbit and five orbits per segment, we found single-orbit solutions to each segment using a Levenberg-Marquardt algorithm. The resulting successive orbital periods are then fit by a sinusoid and the period and amplitude compared with (4) and (5). Using this procedure we were able to recover the true libration period and amplitude to within 30%. This is illustrated in Figure 1.

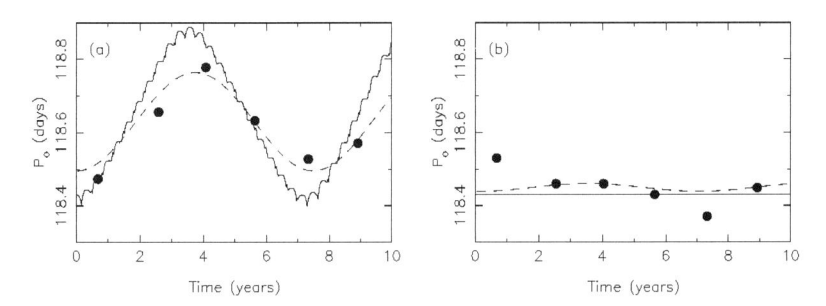

Figure 1. Piecewise estimates (filled circles) of the orbital period for a simulated HD 216770-type system (a) with and (b) without a $10 M_\oplus$ companion in the interior 2:1 resonance. The solid curves are from three-body integrations while the dashed curves are sine curve fits to the points, constrained by knowledge of possible values for the libration period. Simulated radial velocity data includes instrument and stellar jitter noise.

References

Mardling, R. A. 2008, in Lecture Notes in Physics, 760, The Cambridge N-body Lectures, eds. S. J. Aarseth, C. A. Tout, & R. A. Mardling, (Berlin, Heidelberg: Springer-Verlag), 59

Influence of Clouds on the Emission Spectra of Earth-like Extrasolar Planets

D. Kitzmann,[1] A. B. C. Patzer,[1] P. von Paris,[2] M. Godolt,[1] J. L. Grenfell,[1] and H. Rauer[1,2]

[1] Zentrum für Astronomie und Astrophysik, Technische Universität Berlin, Germany

[2] Institut für Planetenforschung, Deutsches Zentrum für Luft- und Raumfahrt (DLR), Germany

Abstract. The climate of Earth-like planets results from the energy balance between absorbed starlight and radiative losses of heat from the surface and atmosphere to space. Clouds reflect sunlight back towards space, reducing the stellar energy available for heating the atmosphere (albedo effect), but also reduce radiative losses to space (greenhouse effect). Clouds also have a large effect on the emission spectra of planetary atmospheres, by either concealing the thermal emission from the surface or dampening the spectral features of molecules, which is, of course, also true for biomarkers (e.g., N_2O and O_3). We present first results on the impact of multi-layered clouds in the atmospheres of Earth-like extrasolar planets orbiting different types of central stars on the planetary IR emission spectra.

A one-dimensional steady state radiative-convective atmospheric climate model is used here to study the effects of clouds on the climate and emission spectra of Earth-like planets orbiting different types of central stars (see Pavlov et al. 2000; Segura et al. 2003; Grenfell et al. 2007 and Kitzmann et al. 2010 for details of the model). In particular, the influence of two different cloud layers (low-level water and high-level ice clouds) are included in the model. The model uses the measured value of the Earth surface albedo. Details about the developed cloud description and the calculation of the particle size and wavelength dependent optical properties of both cloud types are given in Kitzmann et al. (2010). The chemical composition of the atmospheres is chosen to represent the modern Earth atmosphere and was calculated using a photochemical model (see Grenfell et al. 2007). Four different types of central stars are considered: F2V, G2V, K2V, and M4.5V-type stars. The incident stellar fluxes are scaled by varying the orbital distances, such that the energy integrated over each stellar spectrum equals the solar constant at top of the atmosphere of the corresponding Earth-like planet.

Due to enhanced cloud absorption and scattering of the outgoing radiative IR flux, the presence of clouds generally results in an overall decrease of IR emission compared to the respective clear-sky conditions. These results have been confirmed by measurements of the emission spectra of Earth published by Hearty et al. (2009). However, assuming clear sky conditions, the calculated planetary surface temperature differs for all four central stars from the measured mean Earth surface temperature of 288.4 K. Consequently, the mean Earth surface temperature can only be obtained by different combinations of low and

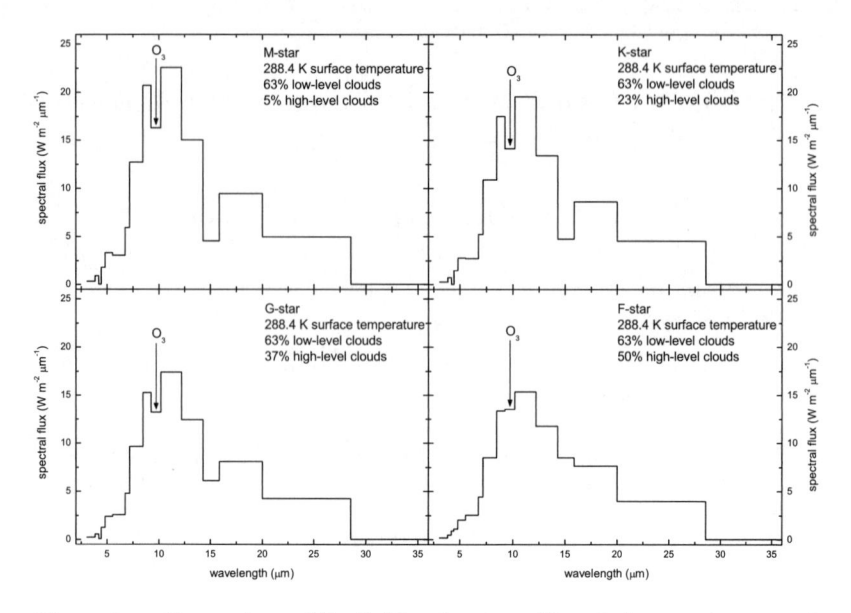

Figure 1. Comparison of Earth-like planetary IR emission spectra around different stars.

high-level cloud coverages for each stellar type. Figure 1 shows four different computed low-resolution IR emission spectra for 63% low-level clouds with the required high-level cloud cover, for which the planets reach a surface temperature of 288.4 K. Besides the overall decrease of IR emission, the individual spectral features of molecules are also dampened by the presence of clouds. The spectral absorption band feature of the biomarker ozone (at $\approx$ 9.6 μm) is especially affected by the high-level cloud. With increasing high-level cloud cover the depth of the band feature decreases. Even though the atmospheric profile of ozone is considered to be the same for all atmospheres, the absorption feature of ozone can disappear due to the effects of clouds (e.g., for the F-type star case). To sum up: the detectability of molecular absorption signatures is strongly influenced by the presence of clouds and may become impossible for high cloud coverages. This is especially apparent also for biomarkers in case of the F-type star where no absorption features are visible at all for high ice cloud coverages in the calculated low-resolution spectrum.

Acknowledgments. This work has been partly supported by the Forschungsallianz *Planetary Evolution and Life* of the Helmholtz Gemeinschaft (HGF).

References

Grenfell, J. L., Stracke, B., von Paris, P., et al. 2007, Planet. Space Sci., 55, 661
Hearty, T., Song, I., Kim, S., & Tinetti, G. 2009, ApJ, 693, 1763
Kitzmann, D. et al. 2010, A&A, (in press), ArXiv e-prints, 1002.2927
Pavlov, A. A., Kasting, J. F., & Brown, L. L. 2000, JGR, 105, 11 981
Segura, A., Krelove, K., Kasting, J. F., et al. 2003, Astrobiology, 3, 689

Pathways Towards Habitable Planets
ASP Conference Series, Vol. 430, 2010
Vincent Coudé du Foresto, Dawn M. Gelino, and Ignasi Ribas, eds.

A Wavefront Correction System for the SPICA Coronagraph Instrument

T. Kotani,[1] K. Enya,[1] T. Nakagawa,[1] L. Abe,[2] T. Miyata,[3] S. Sako,[3] T. Nakamura,[3] K. Haze,[1,4] S. Higuchi,[1,5] and Y. Tange[6]

Abstract. We describe the present status of the development of a wavefront correction system for the SPICA (Space Infrared telescope for Cosmology and Astrophysics) project. SPICA is a next-generation space infrared observatory with a 3-meter class telescope, One of the goals of SPICA is the detection and characterization of Jupiter-like extra-solar planets around nearby stars at mid-infrared wavelengths. We have started to develop a wavefront correction system for SPICA which is a critical component in order to achieve a very high dynamic range (over 10^6) to detect young Jupiter-like planets. It includes a 1020-element MEMS deformable mirror and a tip-tilt compensation system. We present a laboratory demonstration of a contrast enhancement experiment by using a deformable mirror and a checker-board type coronagraph mask. This system will largely relax the surface quality requirements for the telescope primary mirror.

1. Introduction

SPace Infrared telescope for Cosmology and Astrophysics (SPICA), a Japanese-led, joint JAXA-ESA mission aims at realizing a 3-meter class cryogenic space telescope (Nakagawa 2008) to be launched around 2018. All SPICA's optics, including the telescope itself, will be cooled down to 5 K, which will allow us to achieve an excellent sensitivity from mid- to far-infrared wavelengths. SPICA will have a coronagraph instrument (SCI) that is dedicated to the detection and characterization of Jupiter-like planets from 5 to 27 μm. The unique features of SCI include high-contrast imaging up to 10^6 contrast and spectroscopy (R=200). The SPICA coronagraph will enable us to characterize detailed structures and chemical compositions of planetary atmospheres. We plan to employ a checker-board type pupil mask (Enya et al. 2008) to suppress diffracted halo from a star.

---

[1]ISAS/JAXA, 3-1-1 Yoshinodai, Sagamihara, Kanagawa, Japan

[2]Laboratoire Fizeau, Université de Nice Sophia-Antipolis, Parc Valrose, Nice Cedex 02, France

[3]Institute of Astronomy, University of Tokyo, 2-21-1 Osawa, Mitaka, Tokyo, Japan

[4]Department of Space and Astronautical Science, The Graduate University for Advanced Studies, 3-1-1 Yoshinodai, Sagamihara, Kanagawa, Japan

[5]Department of Physics, Graduate School of Science, University of Tokyo 7-3-1 Hongo, Bunkyo-ku, Tokyo, Japan

[6]EORC/JAXA, 3-1-1 Yoshinodai, Sagamihara, Kanagawa, Japan

2. Wavefront Correction System

Although SPICA will achieve diffraction limited imaging from mid- to far-infrared wavelengths, a telescope mirror will not provide a good enough surface quality for very high-contrast imaging. To achieve 10^6 contrast, which is the goal of SPICA's coronagraph, a wavefront quality of the order of $\lambda/100$ is required. For that, SCI employs active wavefront correction systems, including a MEMS deformable mirror (DM). Thanks to active optics and a monolithic mirror of SPICA, SCI will have a potential to perform higher contrast than coronagraphs of JWST. A challenge is to realize the active optics working at cryogenic temperatures (5K). The design and test of cryogenic active optics is under way and a cryogenic coronagraph test-bed is being built at JAXA.

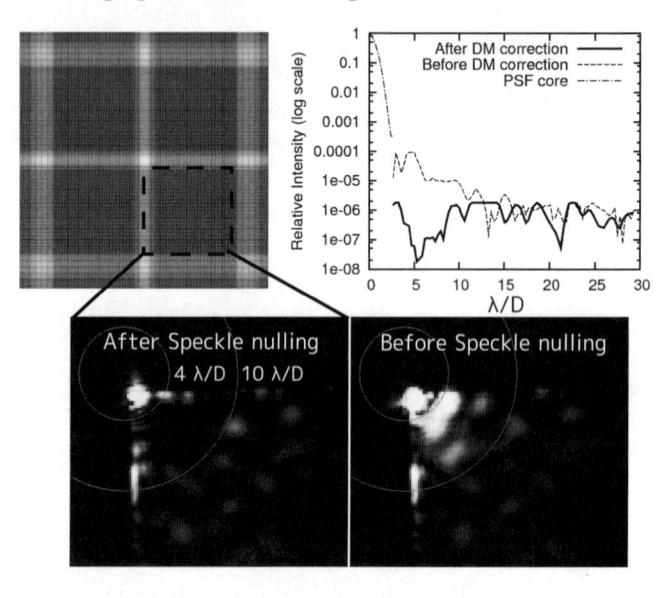

Figure 1. Results of the speckle nulling experiment. *Top left:* Simulated PSF. *Top right:* Measured diagonal cut of the PSF. *Bottom:* Images before and after speckle nulling obtained in the laboratory.

3. Laboratory Demonstration of Speckle Nulling

We have constructed a laboratory testbed (at room temperature) in order to demonstrate a wavefront correction capability by using a DM. For this, we used a MEMS-based 1020 element DM (Boston MicroMachines) and a stand-alone type checker-board pupil mask. A He-Ne laser light source was collimated thorough a pin-hole and a collimator lens and the light was reflected by the DM. At the position of the pupil mask, the DM was re-imaged then the light was focused on the detector. The iterative speckle nulling method (Malbet et al. 1995) was used to eliminate bright speckle patterns on the final image. The advantage of this method is that it does not require any additional wavefront sensor which may introduce non-common path errors and its implementation is relatively straightforward. Figure 1 shows the images before and after speckle nulling. One can clearly see that the bright speckles were nulled out especially near the central core of the PSF. A contrast of over 10^6 was achieved from 3.5 to 10 λ/D regions. This demonstrated performance almost meets the requirements of the SPICA coronagraph. The next step will be to demonstrate these performances at cryogenic temperatures (5K).

References

Enya, K., et al. 2008, A&A, 480, 899
Malbet, F., Yu, J., & Shao, M. 1995, PASP, 107, 386
Nakagawa, T. 2008, SPIE, 7010, 70100H

Pathways Towards Habitable Planets
ASP Conference Series, Vol. 430, 2010
Vincent Coudé du Foresto, Dawn M. Gelino, and Ignasi Ribas, eds.

A Laboratory Demonstration of High Dynamic Range Imaging using the Single-mode Fiber Pupil Remapping System (FIRST)

T. Kotani,[1,2] E. Choquet,[1] S. Lacour,[1] G. Perrin,[1] P. Fedou,[1] F. Marchis,[3,4] and G. Duchêne[3,5]

Abstract. We present the laboratory demonstration of a very high dynamic range, diffraction limited imaging instrument FIRST (Fibered Imager foR Single Telescope). FIRST will enable us to achieve very high dynamic range imaging very near the central object (down to a fraction of λ/D) at visible to near-infrared wavelengths without using adaptive optics, which makes this instrument very complementary to classical coronagraph systems. With FIRST, a dynamic range improves with the number of photons and it is no longer limited by static aberrations which can be totally filtered out thanks to spatial filtering and pupil remapping with single-mode fibers. A raw dynamic range up to 10^6 will be possible, but differential techniques which are under study would further enhance the dynamic range. We successfully demonstrated that the original image can be reconstructed with the expected high dynamic range through a pupil remapping system despite the existence of piston noises between fibers. We also describe the current status of the development of a prototype system. A first technical test will be performed at the Paris Observatory 1-meter telescope. The system will be set-up at the Lick Observatory 3-meter Shane telescope for operational tests in mid-2010.

1. Introduction

The fundamental idea of FIRST (Fiber ImageR for Single Telescope) is to apply techniques developed for long baseline interferometry to the case of a single aperture telescope (Perrin et al. 2006; Lacour et al. 2007). A telescope pupil is divided into sub-apertures each feeding a single-mode fiber in order to filter out atmospheric turbulence, then output sub-apertures are rearranged non-redundantly. A diffraction-limited image is reconstructed from the fringe pattern analysis with aperture synthesis techniques. The removal of atmospheric turbulences and redundant noise allows the calibration of degraded wavefront almost perfectly. Therefore a "truly" diffraction limited image with very high-dynamic range can be obtained with this technique.

---

[1]ISAS/JAXA, 3-1-1 Yoshinodai, Sagamihara, Kanagawa, Japan

[2]LESIA, Observatoire de Paris, 5 Place Jules Janssen, Meudon, France

[3]Department of Astronomy, 601 Campbell Hall, University of California, Berkeley, CA, USA

[4]SETI Institute, 515 Whisman Rd, Mountain View, CA, USA

[5]Université Joseph Fourier - Grenoble 1/CNRS, LAOG UMR 5571, Grenoble, France

2. Laboratory Demonstration

We have constructed a laboratory testbed in order to demonstrate our concept. For this demonstration, we used an artificial binary star (the contrast ratio was ~ 12) at the He-Ne laser wavelength. The input telescope pupil was divided into 9 sub-apertures by a microlens array. The beam from each sub-apertures were very precisely injected into the single-mode fibers by using the segmented deformable mirror (IRIS AO). The two dimensional fiber array (Fiberguide Industries) was used to arrange the fibers in the hexagonal configuration. The output pupils from the fibers were remapped non-redundantly by using a silicon v-groove chip. The beams were combined on the image plane to measure the interferometric fringes. The original image was reconstructed by using the image reconstruction algorithm, MIRA from Thiebaut (2008). The two point-like sources are clearly visible in both the reconstructed and the original images (see Figure 1a and 1b). The position of the companion and the contrast ratio measured with the CCD closely matched the values from the reconstructed image (Kotani et al. 2009).

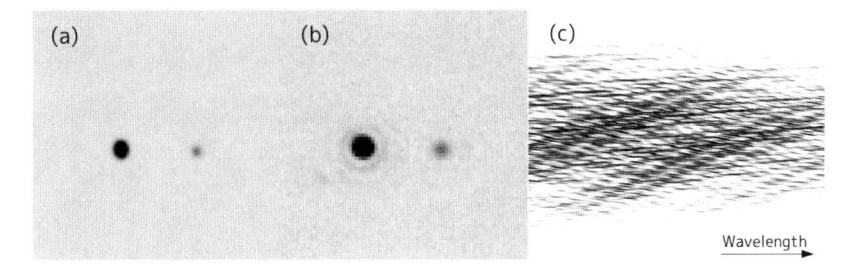

Figure 1. (a) Reconstructed image from complex visibility measurements from the testbed. (b) Direct CCD image of the simulated binary star source. (c) Measured spectrally dispersed interferograms. The abscissa is the wavelength (600-800nm).

3. Fiber Length Equalization

To use this system over a broad wavelength range, the effect of optical path difference (OPD) between fibers needs to be minimized. For this purpose, we equalized the fiber lengths with a precision of better than 100 μm by using the technique developed at LAOG in France. In addition, spectral dispersion of the output beams reduces the effect of OPD. Figure 1c shows the spectro-interferogram obtained in our laboratory using a white light source (600-800nm) and the 9-fiber system. The spectral resolution was about 100. The fringe patterns are clearly seen over a wide wavelength range.

4. Conclusion

We demonstrated the principle of a pupil remapping imaging system using a single-mode fiber by the laboratory experiments. We also showed that FIRST is capable of operating over a wide wavelength range. These are important steps in realizing very high dynamic range imaging using a pupil remapping system. We will carry out a first technical test at the Paris observatory 1-meter telescope in 2010. The system will be set-up a the Lick Observatory 3-meter Shane telescope for operational tests in mid-2010.

References

Kotani, T., et al. 2009, Optics Express, 17, 1925
Lacour, S., et al. 2007, MNRAS, 374, 832
Perrin, G. T., et al. 2006, MNRAS, 373, 747
Thiébaut, E. 2008, SPIE, 7013, 70131I

Pathways Towards Habitable Planets
ASP Conference Series, Vol. 430, 2010
Vincent Coudé du Foresto, Dawn M. Gelino, and Ignasi Ribas, eds.

Stellar Activity Characteristics at FUV and Radio Wavelengths

M. Leitzinger,[1] P. Odert,[1] A. Hanslmeier,[1] I. Ribas,[2]
A. A. Konovalenko,[3] M. Vanko,[4] M. L. Khodachenko,[5] H. Lammer,[5]
and H. O. Rucker[5]

[1] Institute of Physics, Department for Geophysics, Astrophysics and Meteorology, Karl-Franzens University, Universitätsplatz 5, Graz, Austria

[2] Institut d'Estudis Espacials de Catalunya/CSIC, Campus UAB, Facultat de Ciences, Torre C-5 -parell-2a planta, Bellaterra, Spain

[3] Institute of Radio Astronomy, Ukrainan Academy of Sciences, Chervonopraporna 4, Kharkov, Ukraine

[4] Astrophysikalisches Institut und Universitäts-Sternwarte, Schillergäßchen 2-3, Jena, Germany

[5] Space Research Institute, Austrian Academy of Sciences, Schmiedlstrasse 6, Graz, Austria

Abstract. Since stellar activity can affect atmospheres of close-in habitable exoplanets, knowledge of a star's activity level is crucial. Different wavelength ranges yield different possibilities on investigating stellar activity phenomena such as flares and coronal mass ejections (CMEs). In this context we present two approaches to this topic using observations from the far-ultraviolet (FUV) and radio domains. The FUV provides density sensitive line ratios, which show enhancements during stellar flaring. The question if these could be correlated to mass expulsions is investigated by analyzing time series of solar UV full-disk measurements using data from the SORCE and TIMED missions. The second approach is dedicated to the decameter wavelength domain, where we use the known correlation between radio decameter type II bursts and CMEs on the Sun. We present the detection of promising events on the active M-dwarf AD Leo which have a high probability of being of stellar origin. These bursts have parameters similar to solar decameter type III bursts which are fast drifting bursts usually correlated with flares on the Sun. Both approaches are discussed and results are presented.

1. Introduction

We are searching for signatures of mass expulsions on the active M star AD Leonis. This dwarf star is known to be a promising target for stellar activity investigations. One of the rare detections of mass motions on stars was found on AD Leo (Houdebine, Foing & Rodono 1990). Since these phenomena are not detectable directly, we use signatures known from the Sun. In the radio data (UTR-2, Kharkov/Ukraine) we search for decameter type II bursts which are known to be correlated on the Sun to shock waves driven by CMEs. In the FUV (NASA/FUSE) data we search for Doppler-shifted spectral enhancements and enhanced line ratios.

2. Results and Conclusion

In ten nights of decametric observations, where we had only one night of coordinated photometry (Tatranska Lomnica/Slovakia), we found about ten events which show a high propability of being of stellar origin, according to our selection criteria. All of them show a high drift rate of about 0.5-2 MHz/s. These are drift rates known from solar decametric observations of type III bursts. The optical photometry showed a distinct flare on AD Leo but no correlated variation in the radio data.

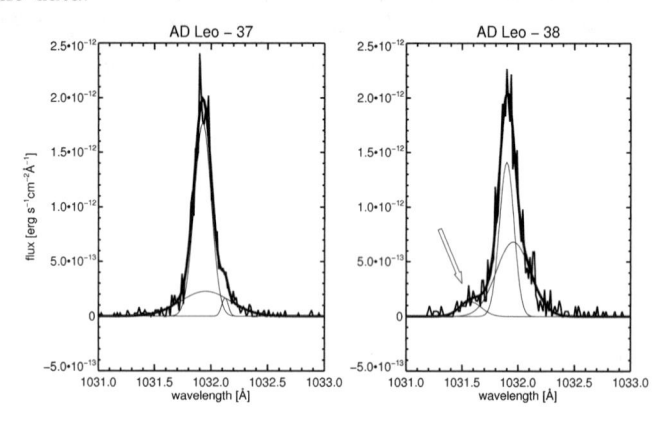

Figure 1. Shown are two FUSE spectra of AD Leo. The corresponding light curve of the left spectrum showed a flare, whereas the light curve of the right spectrum didn't. The enhancement in the blue wing of the OVI line is clearly visible in exposure nr. 38. The spectra were fitted using a multigaussian function.

The analysis of the FUSE data showed beside two flares, which were already presented in Christian et al. (2006), a blue shifted enhancement in the wing of the transition region line OVI (indicated by an arrow in the right panel of Figure 1). This enhancement (nr. 38) ocurred one exposure after a flare (37). Although the velocity of this event is about 90 km/s, one can speculate about a projected velocity of a mass expulsion event. Both approaches show that there are more or less unresolved structures on AD Leo which can be related to mass motions.

Acknowledgments. P.O., M.L., and A.H. gratefully acknowledge the Austrian *Fonds zur Förderung der wissenschaftlichen Forschung* (FWF grant P19446 -N16) for supporting this project.

References

Christian, D.J., et al., 2006, A&A, 454, 889
Houdebine, E.R., Foing, B.H. & Rodono, M. 1990, A&A, 238, 249

Pathways Towards Habitable Planets
ASP Conference Series, Vol. 430, 2010
Vincent Coudé du Foresto, Dawn M. Gelino, and Ignasi Ribas, eds.

Visible Nulling Coronagraph Progress Report

R. G. Lyon,[1] M. Clampin,[1] R. A. Woodruff,[2] G. Vasudevan,[2]
P. Thompson,[1] P. Petrone,[3] T. Madison,[1] M. Rizzo,[4] G. Melnick,[5] and
V. Tolls[5]

[1] NASA Goddard Space Flight Center, Greenbelt, MD, USA

[2] Lockheed-Martin Corp., USA

[3] Sigma Space Corp., 4801 Forbes Blvd., Lanham, MD, USA

[4] Institut Supérieur de l'Aéronautique et de l'Espace, Toulouse, France

[5] Smithsonian Astrophysical Observatory/Center for Astrophysics, 60 Garden St., Cambridge, MA, USA

Abstract. We report on recent laboratory results with the NASA Goddard Space Flight Center Visible Nulling Coronagraph (VNC) testbed. We have achieved focal plane contrasts of 10^8 and approaching 10^9 at inner working angles of 2 λ/D and 4 λ/D, respectively. Results were obtained with a broadband source and 40 nm filter centered on 630 nm. A null control breadboard (NCB) was also developed to assess and quantify MEMS based deformable mirror technology (DM), and to develop and assess closed-loop null control algorithms. We have demonstrated closed-loop performance at 27 Hz.

1. Principle of Visible Nulling Coronagraph

The VNC relies on destructive interference to null on-axis starlight but such that off-axis planet light sees an additional phase shift, introduced by the beam shear, allowing it to pass. This separates starlight into one output channel and planet light into the other – increasing the contrast of the planet to the star. Internally, the VNC consists of two Mach-Zehnder interferometers. Light from the telescope enters it and is relayed to the X-nuller (Figure 1) and then to the Y-nuller. A VNC acts like a Bracewell interferometer in that two (or more) apertures are mixed such that combinations of phase shifts cause destructive interference on-axis. In a VNC, the two Bracewell apertures are emulated by a single aperture, but by beam-splitting it into two beams with a π-phase shift between, and introducing a beam shear (i.e., lateral translation) and recombining them. The inner working angle (IWA) is the minimal separation at which a planet can be detected and given at the first transmission maximum by $IWA = \lambda/2sD$ where s is the fractional shear, λ is the wavelength and D the telescope diameter. For $s = 0.25$ the $IWA = 2\lambda/D$.

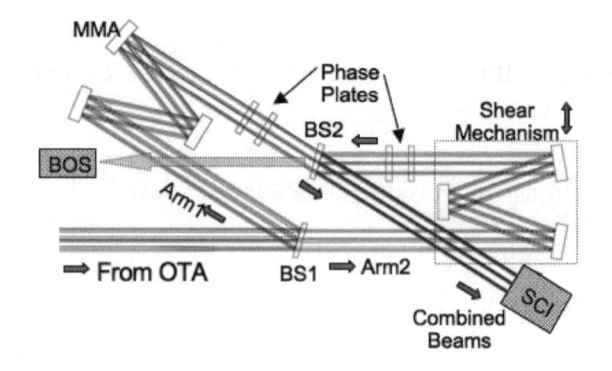

Figure 1. Schematic of Single VNC Nuller

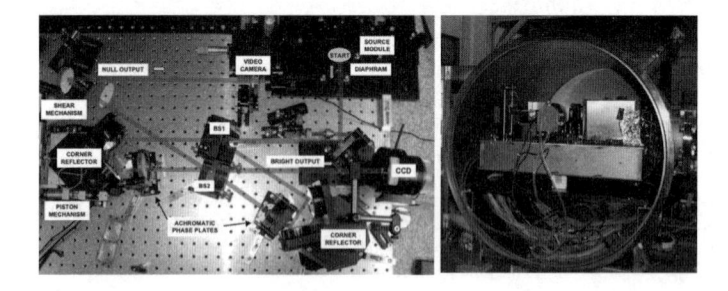

Figure 2. GSFC Vacuum VNC Testbed

2. Experimental Results

NASA GSFC with Lockheed-Martin and Sigma Space developed both the vacuum VNC testbed (Figure 2) and Null Control Breadboard (NCB) and is working on a compact nuller. The vacuum VNC is to assess performance in a stable environment and the NCB is to develop and assess control algorithms for deformable mirrors (DM) and quantify their performance. The compact nuller is textbook-sized and combines lessons learned from the first two with a detailed structural/thermal/optical model, along with dimensionally and thermally stable materials and mounts and is designed in a pseudo-flight like fashion.

Results to date include data collection runs over 14 hours of closed-loop where we have achieved long term stability of 2 nm rms with short term stabilities of 0.7 nm rms (few 10s of seconds) with segmented deformable mirrors. Figure 3 shows results from the NCB at 15 Hz, the left image is the pupil plane with all actuators driven such that the interferogram is as bright as possible; in the right image the actuators are driven such it is as dark as possible. Results were obtained with a 40 nm spectral filter centered on 630 nm. These segments are flat to 6–8 nm rms and no print through is visible, but residual leakage is evident towards the edges of the ∼500 μm diameter segments.

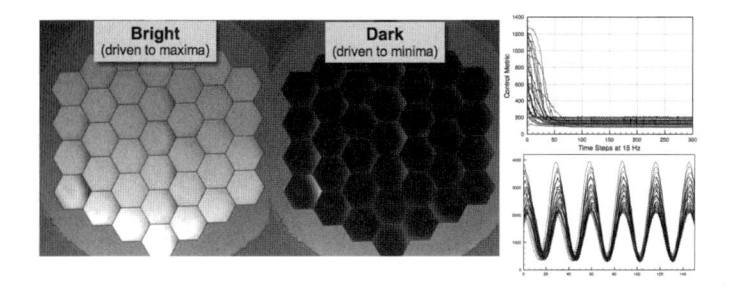

Figure 3. Pupil plane nulling with IRIS-AO 37 DM

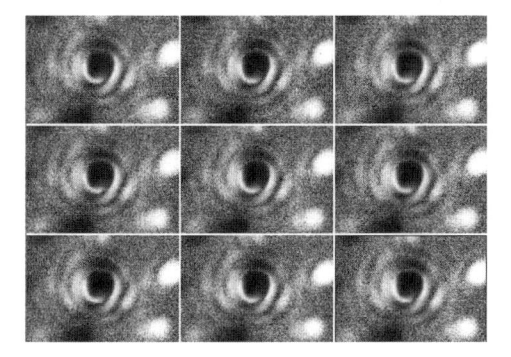

Figure 4. Time sequence of focal plane point spread function nulling

The vacuum VNC was brought to focus after nulling to assess contrast without control as the null control from the focal plane is still in progress. Figure 4 shows a time sequence separated by 0.5 seconds with a 40 nm filter – a linear polarizer was also used. This figure shows the average core suppression per Airy spot within a 1 λ/D circle (diameter). The suppression of the core shows that without DM control the PSF suppression at 1 cycle per aperture is at best average $\sim 10^{-4}$ and if the best minimal point is chosen it is $\sim 10^{-6}$ where best minimal is the minimal point within the core. In nulling, the goal is to uniformly suppress the point spread function and if best minimum is $\sim 10^{-6}$ at 2 λ/D, it would be inferred to be reduced to $\sim 10^{-8}$ and at 4 λ/D reduced to $1.3 \cdot 10^{-9}$ yielding focal plane contrasts of $\sim 10^8$ and $8 \cdot 10^8$ at 2 and 4 λ/D respectively. This implies that with DM control, we should be able to achieve our required contrasts of 10^9 at 2 λ/D, i.e. there is nothing physically limiting the result needed to achieve contrasts required for Jovian planet detections. We expect to achieve this result after integrating our latest DM into the system.

Pathways Towards Habitable Planets
ASP Conference Series, Vol. 430, 2010
Vincent Coudé du Foresto, Dawn M. Gelino, and Ignasi Ribas, eds.

Checking Stability of Planet Orbits in Multiple-planet Systems

F. Malbet,[1] J. Catanzarite, M. Shao, and C. Zhai

Jet Propulsion Laboratory/California Institute of Technology, 4800 Oak Grove Drive, Pasadena, CA, USA

Abstract. The SIM Lite mission will undertake several planet surveys. One of them, the Deep Planet Survey, is designed to detect Earth-mass exoplanets in the habitable zones of nearby main sequence stars. A double blind study has been conducted to assess the capability of SIM to detect such small planets in a multi-planet system where several giant planets might be present. One of the tools which helped in deciding if the detected planets were real was an orbit integrator using the publicly available HNBody code so that the orbit solutions could be analyzed in terms of temporal stability over many orbits. In this contribution, we describe the implementation of this integrator and analyze the different blind test solutions. We also discuss the usefulness of this method given that some planets might be not detected, but still affect the overall stability of the system.

A double blind study has been conducted in the framework of the SIM Lite mission (Traub et al. 2009, see also Traub et al. in this volume) to assess the capability of SIM to detect such small planets in a multi-planet system where several giant planets might be present. Five independent teams of dynamics experts generated ensembles of hundreds of plausible planetary systems that might form around solar-type stars. A data simulation team generated realistic simulated astrometric and radial velocity measurement data sets for 60 stars chosen from the SIM target list, by perturbing each with a planetary system randomly drawn from the ensembles generated by the dynamics teams, and then adding instrument noise. Next, four independent analysis teams processed the simulated data sets, detecting and fitting the orbits and masses of the planets orbiting each of the 60 stars. Finally, the analysis teams submitted their consensus solution for each star, chosen after cross-comparison and discussion of the solutions, and before being allowed to see the true solutions.

We used HNBody, a symplectic integration package for hierarchical N-Body systems (version 1.0.3) developed by Rauch & Hamilton (2002). It integrates the motion of particles in self-gravitating systems where the total mass is dominated by a single object; it is based on symplectic integration techniques in which two-body Keplerian motion is integrated exactly. HNBody is primarily designed for systems with one massive central object and has been used previously for extrasolar planet simulations (Veras & Armitage 2005, 2006). We used the parameters given by the different teams, namely the mass of the planets, their period translated in semi-major axis using the Keplerian formula, the orbit

---

[1]Now at the Université de Grenoble, J. Fourier/CNRS, Laboratoire d'Astrophysique de Grenoble, BP 53, Grenoble, France

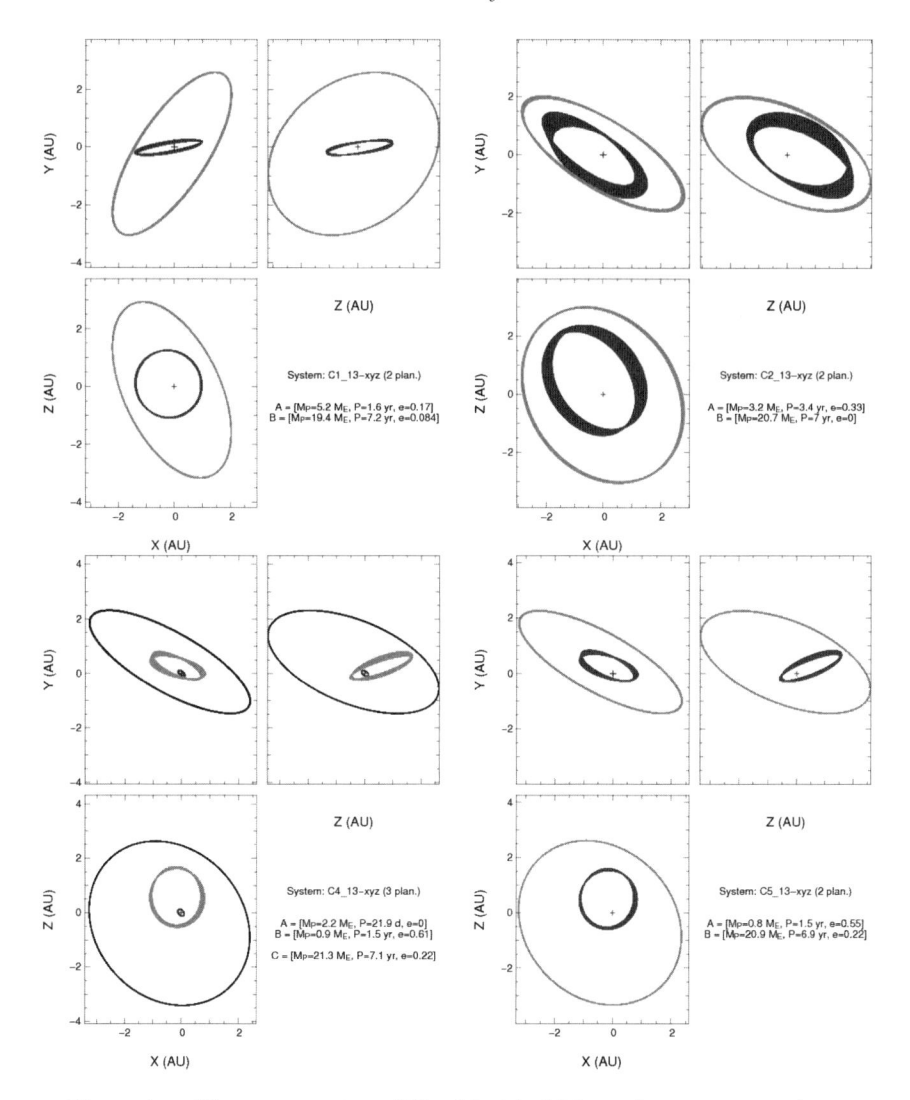

Figure 1. Planetary system #13 of batch 2/phase 2: xy-, xz- and yz-projections of the orbits for the solutions of the 4 teams.

eccentricity, the orbit inclination, the longitude of the ascending node, the argument of periapse in years, and for the periapse passage time, since HNbody only accepts the epoch and time, we rounded the time of periapse to get the Epoch and the time was then just the remaining decimals. In order to ensure that we correctly sample the orbit evolution, the total integration time has been set to 1 million years while the elementary step has been set up to 1/20 of the shortest period of the system. HNbody then computes the relative errors in energy and angular momentum exhibited by the integration. If the error mean and rms are less than 10^{-4} over the total integration time, the system is considered stable.

The plots presented in Figure 1 are representative of the temporal evolution of the orbit parameters found by the different teams and their 3-D orbits. Figure 1 display the system for which there was intense discussion about the different solutions, and for which the stability would be decisive in choosing the solution. This system was found stable (energy and momentum errors less than 10^{-8} in 10^6 yrs) even though there were secular changes eccentricity for the closest planet.

The question arises whether such a stability test is appropriate for validating a solution. Obviously, if a system misses large and massive planets orbiting further away, then a stable system might be found unstable because we have not identified and properly characterized all the planets. This is the case of the system #3 in batch #3 of phase 2. The unraveled solution is in fact a 9-planet system which is stable when HNBody integrates the orbit over 1 million years. The reverse is true especially if only one planet is discovered, because integration of only one planet is very likely to be stable. Nevertheless, the orbit integration tool appeared to be useful to choose the final solutions.

Acknowledgments. FM thanks Caltech/JPL and CNRS for funding for his stay during the year 2008-09 when this work was performed.

References

Rauch, K. P. & Hamilton, D. P. 2002, BAAS, 34, 938
Traub, W. A., Beichman, C., Boden, A. F., et al. 2009, ArXiv e-prints 0904.0822
Veras, D. & Armitage, P. J. 2005, ApJ, 620, L111
Veras, D. & Armitage, P. J. 2006, ApJ, 645, 1509

Spectroscopic Observations of Nearby Cool Stars: The DUNES Sample

J. Maldonado,[1] C. Eiroa,[1] R.M. Martínez-Arnáiz,[2] and D. Montes[2]

[1] Universidad Autónoma de Madrid, Dpto. Física Teórica, Facultad de Ciencias, Campus de Cantoblanco, C/ Francisco Tomás y Valiente 7, Spain

[2] Universidad Complutense de Madrid, Dpto. Astrofísica y Ciencias de la Atmósfera, Facultad de Físicas, Madrid, Spain

Abstract. The detection of faint dusty exo-zodies and exo-EKBs around mature stars is a direct proof of planetesimal systems. Relating the properties of such structures with the hosting stars is fundamental to get clear clues concerning how common planetary systems are, and how the form and evolve. DUNES (DUst around NEarby Stars[1]) is a Herschel Open Time Key Project with the aim of detecting cool faint exo-solar analogues to the Edgeworth-Kuiper Belt (EKB). Since the success of DUNES depends on very accurate determination of the stellar properties and age, we have started a high resolution observing program of the DUNES targets, with the first results are presented here.

1. The DUNES Sample

The DUNES sample consists of 133 main-sequence (luminosity classes V/IV-V) FGK stars. The sample is volume limited (d < 20 pc), only constrained by background confusion; stars with known planets and Spitzer-discovered faint debris disks up to 25 pc are also included. No further biases limit the DUNES sample, which includes binary and single stars, variable stars, and a broad range of stellar ages, from $\sim$ 0.1 to $\sim$ 10 Gyrs. In addition another 106 stars (including some A and M spectral type stars) will be observed in collaboration with Debris (another Herschel OTKP).

At the Universities of Madrid, we are carrying out a systematic analysis of the spectroscopic properties of the nearby (d < 25 pc) late-type stellar population (Maldonado et al. 2009; Mártinez-Arnáiz et al. 2009) We are using this high resolution spectra to study the properties of the DUNES targets. The spectra were obtained between 2005 and 2009 by using the FOCES spectrograph at the 2.2 m telescope of the Calar Alto Observatory, the SARG spectrograph at the Telescopio Nazionale Galileo (3.56m), and the FIES instrument at the Nord Optical Telescope (3.4 m) in La Palma Observatory. Additional spectra stars have been obtained in public archives and libraries like S4N (Allende Prieto et al. 2004).

---

[1] http://www.mpia-hd.mpg.de/DUNES/

2. Spectroscopic Analysis

The spectra have been analyzed following the next steps:

- *Kinematics (ascription to young stellar kinematic groups):* Heliocentric radial velocities have been determined by using the cross-correlation technique. We have used these radial velocities together with precise measurements of proper motions and parallaxes taken from Hipparcos and Tycho-2 catalogues to calculate the galactic space-velocity components (U,V,W) The (U,V) and (W,V) planes are used to analyse the membership of stars to different stellar kinematic groups (Montes et al. 2001).

- *Stellar ages (comparison with stars in clusters with known ages):* Lithium abundance, chromospheric activity measured by the R'_{HK} index or coronal emission quantified by X-ray activity levels are commonly used as age diagnostics in late-type stars. An age estimation can been obtained by comparing these quantities with those of stars in well known young open clusters of different ages. Additional age estimates, including rotational periods or the fractional dust luminosity, have also been applied.

- *Activity-rotation-age relations:* Rotational velocities have been obtained by using the cross-correlation technique, whereas the activity has been analyzed by using the spectral subtraction technique (Montes et al. 1995) in the main optical tracers: Hα, Hβ, Hγ, Hδ, Hϵ, Ca II H &K, Ca II IRT, Na I D_1, D_2. Using the computed fluxes and the obtained rotational velocities, we have analysed the dependence of the activity levels with the rotational rates.

Acknowledgments. This work has been supported by the Spanish Ministerio de Ciencia e Innovación, Plan Nacional de Astronomía y Astrofísica under grant AYA2005-00954.

References

Allende Prieto C., Barklem, P. S., Lambert, D. L. & Cunha, K. 2004, A&A, 420, 183
Maldonado, J., Martínez-Arnáiz, R. M., Eiroa, C. & Montes, D. 2009, in 15th Cambridge Workshop on Cool Stars, Stellar Systems and the Sun, AIP Conf. Proc. 1094, 413.
Martínez-Arnáiz, R. M., Maldonado, J., Montes, D., Eiroa, C., Montesinos, B., Ribas, I., & Solano, E. 2009, in AIP Conf. Proc. 15th Cambridge Workshop on Cool Stars, Stellar Systems and the Sun, ed. E. Stempels, 1094, 465
Montes, D., López-Santiago, J., Fernández-Figueroa, M. J. & Gálvez, M. C., 2001, A&A, 379, 976
Montes, D., Fernández-Figueroa, M. J., de Castro, E. & Cornide, M. 1995, A&A, 294, 165

A Mid-infrared Space Observatory for Characterizing Exoplanets

S. R. Martin, D. P. Scharf, R. Wirz, G. Purcell, and J. Rodriguez

Jet Propulsion Laboratory, California Institute of Technology, Pasadena, California, USA

Abstract. The TPF-Darwin planet-finding concept is a space-based mid-IR nulling interferometer consisting of four formation flying reflecting mirrors focusing light into a separate beam combiner spacecraft. As a flagship mission it is a highly capable instrument with the principal science goal of detecting Earth-like planets in the habitable zone of nearby stars (up to 15 parsec) from their thermal emission, and characterizing their physical properties. Measurements can be made of the size, temperature, orbit, and of the presence of an atmosphere with moderate resolution (R = 50) spectra to reveal bio-markers such as ozone, carbon dioxide, methane, and water vapor. The sensitivity is such that a whole planetary system down to half Earth-size planets can be detected in a single day and in a five-year mission life, revisits and repeat detections would reveal the system dynamics. For the most interesting objects, longer measurement series would reveal the detailed thermal spectrum; these measurements would take place in the later part of the mission. The system also has a general astrophysics capability for investigations of distant compact objects and the mission timeline provides ample opportunities for such observations. The mission would use a single heavy launch vehicle to place the system at L2 Sun-Earth halo orbit. This poster shows the main mission elements, the spacecraft design and formation flying, and xenon-ion thruster technologies. Many of the ideas have been developed with European colleagues and the mission has potential to be a collaborative effort between the US and European space agencies.

1. Introduction

There are now over 400 known exoplanets and these are mainly gas giants found by radial velocity methods. Optical techniques have been successful too, notably CoRoT and now Kepler are at work observing exoplanet transits from space. Larger space missions such as TPF-Coronagraph (Traub et al. 2006), TPF-Interferometer (Henry et al. 2004) and Darwin (Cockell et al. 2008) are proposed to find Earth-analogues in the habitable zone. For example, TPF-Emma (Martin et al. 2007) can detect and characterize Earth-size planets in the habitable zone using infrared nulling. TPF-Emma has five formation flying spacecraft, four of which are reflecting mirrors directing starlight towards a fifth beam-combiner spacecraft. The objective is spectral characterization of Earth-size exoplanets in the habitable zone with a science goal of ∼150 planets observed and ∼15 fully characterized. The mission duration would be 5 years with a 10 year goal. Launched on a Delta IV Heavy vehicle, the spacecraft system would orbit at the Sun-Earth L2 Halo. The four reflector spacecraft would be positioned 1200 m from the beam combiner spacecraft and would be arrayed in a rectangular formation with a 6:1 aspect ratio (Lay et al. 2007). The short baseline would vary between approximately 20 and 65 m, depending on the target star. Over one year, 98.8% of sky is observable.

2. Observatory Overview

The four reflector spacecraft would each carry a lightweight 3 m diameter SiC mirror formed from a double arch backing structure covered by a CVD facesheet polished to 5 Å roughness and 3 μm rms wavefront error. A similar mirror was recently made for Herschel (Pilbratt 2003) by braising twelve petals together to form a single circular mirror. The mirror is passively cooled to below 50K. On the beam combiner spacecraft, actively cooled to 40K and below, incoming beams are combined in pairs to null the starlight; the resulting two beams are cross-combined and the planet light is modulated, to enable detection, using path length (phase) chopping. The science detector is an arsenic doped silicon array cooled to 7K and the radiation is dispersed over some 100 pixels to provide coverage of the 6 to 20 μm band. Stellar fringe tracking is performed using 2 to 3 μm starlight and pointing and alignment are maintained using 0.6 to 1.0 μm starlight. Laser metrology systems, both internal to the beam combiner and running out to the reflectors, are used for fine control of the optical path and pointing (Martin 2005).

TPF-Emma uses a single propulsion system: all formation maneuvers and pointing are done with miniature Xe-ion thrusters. During observations the reflector spacecraft formation rotates to describe a circle under the combiner spacecraft and mirrors on the combiner spacecraft track the reflector spacecraft. Each reflector directs light from the target star towards the combiner using a combination of information from the star tracker and from the combiner spacecraft. The FACS (formation attitude and control) system uses radar and the laser systems for overall formation control. Figure 1 illustrates an artist's impression of the spacecraft array (considerably foreshortened for clarity) and the beam combiner and reflector spacecraft for TPF-Emma. Science observations would take typically 10 hours or more for a single rotation with the science data rate being approximately 1 to 10 planet photon/second. For investigational planet hunting observations, low spectral resolution spectroscopy (R=10) would be performed, while for deeper spectral characterization observations with higher spectral resolution at R=100, many rotations (10-30) would be needed.

2.1. Technology Development

Broadband nulling has been demonstrated to the null depth required in flight, over a 34% bandwidth centered on 10 μm. Single-mode mid-IR fibers have been demonstrated at 10 μm using both chalcogenide glass (Ksendzov et al. 2007) and silver halide materials (Ksendzov et al. 2008), providing the required single spatial mode filtering performance. A near flight-specification cryogenic delay line has been demonstrated by the European Space Agency (Ergenzinger et al. 2006). Formation flying algorithms have been demonstrated in the laboratory to have the performance required for flight (Scharf et al. 2007). The architecture trades studies in the US and Europe converged in 2007, leading to the Emma X-Array. Target catalogs and observatory performance models developed inde-

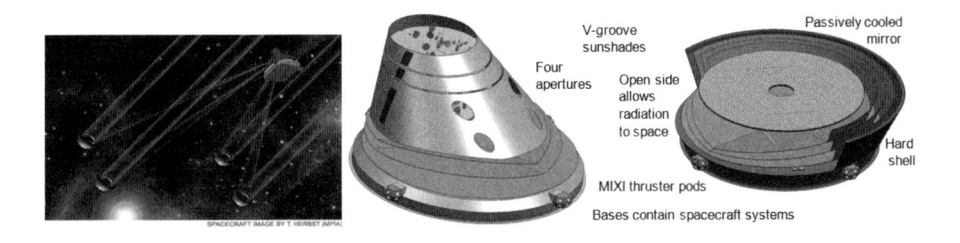

Figure 1. *Left*: artist's impression of spacecraft array. *Center*: beamcombiner spacecraft. *Right*: reflector spacecraft.

pendently in Europe and the US were shown to be in close agreement. Precursor formation flying missions are now in formulation (ESA's Proba-3, eg. Vives et al, 2008) and in preparation for launch (Swedish Space Corporations Prisma, eg. Bodin et al. 2008) to test RF metrology, optical metrology, and cold-gas and electrical micro-propulsion.

3. Summary

There can be few more compelling objectives in astronomy than the quest to find and characterize exoplanets. For the vast majority of the 400 or so (and constantly growing) list of such objects, only the most basic information such as orbital radius and period is known, while the more interesting data such as the type and constitution of the exoplanetary atmosphere is unknown. A large scale nulling infrared interferometer observatory such as illustrated here would yield exoplanet data on unprecedented scale. In discovery mode, the capability of imaging an entire exoplanet system in one day could produce hundreds of new exoplanets in a few weeks of operation. These exoplanets would also be coarsely characterized for their spectra at R = 10. In high resolution spectroscopy mode, the observatory can yield simultaneously the spectra (at R=100) of all the major and minor (down to one half Earth diameter) planets of an exoplanet system to reveal bio-markers such as ozone, carbon dioxide, methane, and water vapor.

The optical technology is well-advanced; the required starlight nulling performance (Peters et al. 2008, Gappinger et al. 2009) has been achieved in the lab and the technologies for formation flying, thrusters (Wirz and Katz 2005) and optical and optomechanical devices (Booth et al. 2008) are at TRL 4 and above. The design has evolved from a more complex set of collector telescopes to the current simple reflector set. The layout, equipment, power, and thermal design requirements for the beam combiner and reflector spacecraft have been studied and outline engineering designs for the sunshades, spacecraft buses, thruster systems, and control systems are in place. The basic scenario for launch and mission operations over a five year mission life is known. Many of these ideas have been developed with European colleagues and the mission has potential to be a collaborative effort between the U.S. and European space agencies.

Acknowledgments. This work was conducted at the Jet Propulsion Laboratory, California Institute of Technology and funded under the NASA Exoplanet Exploration Program.

References

Bodin, P. et al., 2008, AIAA 2008-6662
Booth, A. J., Martin, S. R., & Loya, F., 2008, SPIE 7013, 701320
Cockell, C. S. et al., 2008, Astrobiology, 9, 1
Ergenzinger, K., et al., 2007, SPIE 6268, 62682M
Gappinger, R.O. et al., 2009, Appl. Opt. 48, 868
Henry, C., et al., 2004, SPIE 5491, 265
Ksendzov, A., Lay, O., Martin, S. et al., 2007, Appl. Opt. 46, 7957
Ksendzov, A., Lewi, T., Lay O. P. et al., 2008, Appl. Opt. 47, 5278
Lay, O. P., Martin, S. R., & Hunyadi, S. L., 2007, SPIE 6693, 66930
Martin, S. R. et al., 2007, SPIE 6693, 669309
Martin, S. R., 2005, SPIE 5905, 590503
Peters, R. D., Lay, O. P., & Jeganathan, M., Appl. Opt. 47, 3920
Pilbratt, G. L., 2003, SPIE 4850, 586
Scharf, D. P., et al., 2007, SPIE 6693, 669307
Traub, W. A., et al., 2006, SPIE 6268, 62680T
Vivès, S., Lamy, P., Venet, M., Lavacher, P., & Boit, J.L., 2007 SPIE 6689, 66890F
Wirz, R. & Katz, I., 2005, AIAA 2007-690

Pathways Towards Habitable Planets
ASP Conference Series, Vol. 430, 2010
Vincent Coudé du Foresto, Dawn M. Gelino, and Ignasi Ribas, eds.

The Subaru Coronagraphic Extreme AO Project

F. Martinache and O. Guyon

Subaru Telescope, 650 N. A'ohoku Place, Hilo, HI, USA

Abstract. The Subaru Coronagraphic Extreme AO (SCExAO) Project is an upgrade to the newly commissioned coronagraphic imager HiCIAO for the Subaru Telescope, in the context of a massive survey for exoplanets and disks called SEEDS. SCExAO combines a high-performance coronagraph PIAA coronagraph and non-redundant aperture masking interferometry to a MEMS-based wavefront control system to be used in addition to the 188-actuator Subaru Adaptive Optics (AO188) system. The upgrade is designed as a flexible platform with easy access to both pupil and image planes to allow quick implementation of new high-angular resolution techniques, using a combination of interferometry and coronagraphy. The SCExAO system will enhance SEEDS by offering access to smaller separations and improved PSF calibration, and will therefore allow high quality follow-up observations of challenging SEEDS candidates. SCExAO will also enable new science investigations requiring high contrast imaging of the innermost (< 0.2 arcsecond) surrounding of stars.

1. Introduction

The Subaru Telescope, with its 8-meter primary mirror and efficient 188-actuator AO system, is a formidable tool for high angular resolution astronomy. The Subaru Coronagraphic Extreme AO (SCExAO) project is an open platform harboring a complete suite of devices designed to take full advantage of this angular resolution. Downstream a MEMS-based 1024-actuator deformable mirror (DM), SCExAO combines: a low noise fast readout detector for diffraction-limited imaging in the visible, as well as a high-efficiency near-infrared Phase-Induced Amplitude Apodization (PIAA) coronagraph to be used with the newly commissioned HiCIAO camera.

2. Infrared PIAA-based Coronagraph

The SCExAO Project capitalizes on the research and technological developments made at the Subaru Telescope for high contrast imaging, starting with the Phase Induced Amplitude Apodization (PIAA). Figure 1 provides an overview of the modules used in SCExAO. Since its first description by Guyon 2003, the PIAA has imposed itself as an attractive apodization technique, since it preserves both angular resolution and throughput (Guyon et al. 2005; Martinache et al. 2006; Pluznik et al. 2006). Laboratory experiments (Guyon et al. 2009a) have achieved raw contrast of 1.6×10^{-7} at $1.65\lambda/D$. SCExAO also addresses the problem of the 30% central obscuration and thick spider vanes of Subaru with a

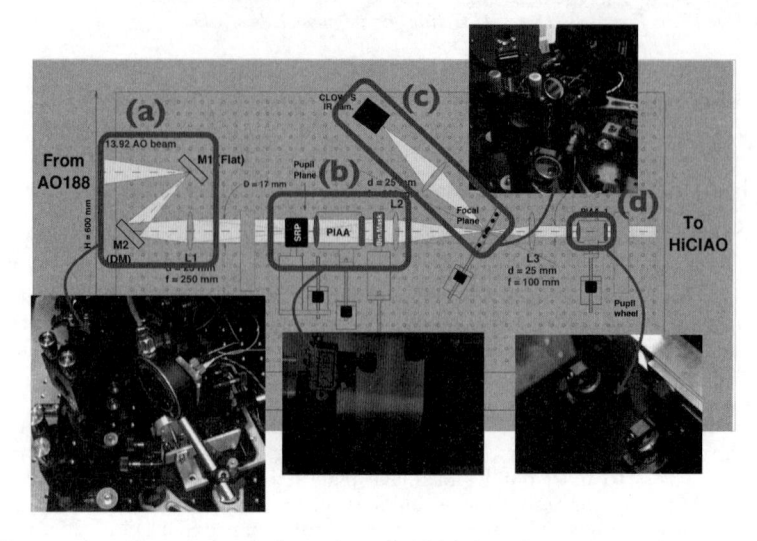

Figure 1. Overview of the infrared PIAA-based coronagraphic arm of SCExAO. The four labeled boxes highlight the key modules used in the project. (a):with two tip-tilt and a 1024-actuator DM, the AO beam adapter module receives the f/14 converging beam from AO188, and provides a well-corrected collimated beam, ready for apodization. (b):the beam shaping module suppresses the spider vanes as well as the central obscuration and apodizes the beam for coronagraphy, with a nearly 100% throughput. (c):the light of the bright on-axis source stopped by the focal plane mask is redirected toward the low-order wavefront sensor for high-precision tip-tilt and focus measurements. (d):scaled down copy of the first PIAA lens-set, plugged backwards restores the pupil after coronagraphy for good imaging of off-axis sources.

specially designed PIAA to be used in conjunction with a Spider Removal Plate (Lozi et al. 2009; Martinache et al. 2009).

Because SCExAO attempts at detecting companions located at very small angular separations ($\sim 1-5\lambda/D$), much attention must be paid to pointing errors as well as other low-order aberrations that generate strong coronagraphic leaks. The Coronagraphic Low Order Wavefront Sensor was designed in that scope. It uses the light reflected by the partially reflecting occulting mask reimaged on a camera and has demonstrated pointing sensitivity as good as $10^{-3}\lambda/D$ (Guyon et al. 2009b).

References

Guyon, O., 2003, A&A, 404, 379
Guyon, O., et al., 2005, ApJ, 622, 744
Guyon, O., et al., 2009a, PASP, accepted
Guyon, O., et al., 2009b, ApJ, 693, 75
Lozi, J., et al., 2009, PASP, 121, 1232
Martinache, F., et al., 2006, ApJ, 639, 1129
Martinache, F., et al., 2009, SPIE, 7440, 20
Pluzhnik, E., et al., 2006, ApJ, 644, 1246

Spectral Imaging with Nulling Interferometer

T. Matsuo,[1,2] W. A. Traub,[2] M. Tamura,[1] and H. Makoto[3]

[1] National Astronomical Observatory of Japan, Osawa, Mitaka, Tokyo, Japan

[2] Jet Propulsion Laboratory, 4800 Oak Grove Dr., Pasadena, CA, USA

[3] Graduate School of Science, Tohoku University, Aoba, Aramaki, Aoba-ku, Sendai, Japan

Abstract. Various methods of synthesis imaging with nulling interferometers have been proposed. These methods are based on the principle that brightness distribution in the sky is reconstructed through Fourier transformation of complex visibility. Here, we present a novel imaging interferometry technique for direct detection of exoplanets. According to this new method, only two measurements of the complex visibility are required for reconstruction of the sky. Spectra of the planets can be also estimated from information on their positions and the complex visibilities, its number corresponding to the number of the planets. This paper shows the theory of this new imaging method.

1. Theory

We briefly explain the theory of the new spectral imaging method. The dual Bracewell configuration, which is composed of two single nulling interferometers, overcomes the disadvantages of the original single Bracewell configuration. This type of interferometer can observe both amplitude and phase of the visibility by phase chopping techniques (e.g., Mennesson and Mariotti 1997) and clearly identify the location of the planet through the maximum correlation method (e.g., Woolf and Angel 1997). The maximum correlation method is analogous to Fourier transformation used for standard radio interferometers. Fourier transformation of the imaginary part of complex visibility on u-v plane gives the estimated brightness distribution, called dirty image:

$$\sum_n^N \hat{B}_n(\vec{\theta}_n, \nu) = FT[\hat{CV}(\vec{B}, \vec{b}, \nu)] ** FT[U(\vec{B}, \nu)] \quad (1)$$

where $\vec{\theta}$ is position vector, ν is observational frequency, and $\vec{B}$ and $\vec{b}$ are baseline vectors for imaging and nulling, respectively. In addition, ** is two dimensional convolution, $\hat{CV}(\vec{B}, \vec{b}, \nu)$ is the estimated complex visibility, and $U(\vec{B}, \nu)$ is the synthesized beam.

On the other hand, according to the new theory, only two measurements of the complex visibility are enough for estimation of the position of each planet. Given that the intensity of each point source is independent of the frequency for simplicity, Fourier transformation of the complex visibility along the frequency gives

$$FT[CV(\vec{B},\vec{b},\nu)] = \int d\nu CV(\vec{B},\vec{b},\nu)exp(-2\pi i\nu\tau)$$
$$= 2\sum_{n}^{N} B_n(\vec{\theta_n})[2\delta(c\tau - \vec{B}\cdot\vec{\theta_n}) - \delta(c\tau - \vec{B}\cdot\vec{\theta_n} - \vec{b}\cdot\vec{\theta_n})$$
$$- \delta(c\tau - \vec{B}\cdot\vec{\theta_n} + \vec{b}\cdot\vec{\theta_n})], \qquad (2)$$

where $B_n(\vec{\theta_n})$ is brightness distribution of the n-th planet, c is velocity of the light, and τ is delay time due to the optical path difference between the two beams. Thus, we can estimate the number of the planets, N, and the positions of planets without changing the imaging and null baselines. Here, we perform N measurements of the complex visibility, rotating the baselines. For separation of the planets from symmetrical components such as the central star and debris disk, only the imaginary part of the complex visbility should be used. The N measured imaginary parts can be written as follows:

$$\begin{pmatrix} CV_1(\vec{B},\vec{b},\nu) \\ \vdots \\ CV_N(\vec{B},\vec{b},\nu) \end{pmatrix} = \begin{pmatrix} t_{11} \cdots t_{1n} \\ \vdots \ddots \vdots \\ t_{n1} \cdots t_{nn} \end{pmatrix} \begin{pmatrix} I_1(\theta_1,\nu) \\ \vdots \\ I_N(\theta_N,\nu) \end{pmatrix}, \qquad (3)$$

where the term for the k-th measurement and the l-th planet is

$$t_{kl} = sin^2(\frac{\pi\nu}{c}\vec{b_n}\cdot\vec{\theta_m})sin(\frac{2\pi\nu}{c}\vec{b_n}\cdot\vec{\theta_m}) \qquad (4)$$

From Equation (3), the spectra of planets are

$$\begin{pmatrix} I_1(\theta_1,\nu) \\ \vdots \\ I_N(\theta_N,\nu) \end{pmatrix} = \begin{pmatrix} t_{11} \cdots t_{1n} \\ \vdots \ddots \vdots \\ t_{n1} \cdots t_{nn} \end{pmatrix}^{-1} \begin{pmatrix} CV_1(\vec{B},\vec{b},\nu) \\ \vdots \\ CV_N(\vec{B},\vec{b},\nu) \end{pmatrix}. \qquad (5)$$

Thus, the spectra of planets can be estimated with the N measurements of the complex visibility.

2. Future work

We perform a one-dimensional simulation for the validation of this method and estimate the signal to noise ratio. This new imaging interferometry will be compared with the current imaging interferometry such as Maximum correlation method and double Fourier interferometry. This work will revisit the TPF-I and Darwin concept.

References

Mennesson, B. & Mariotti, J.-M., 1997, Icarus 128, 202
Woolf, N. J. & Angel, J. R. P., 1997, ApJ, 475, 373

Pathways Towards Habitable Planets
ASP Conference Series, Vol. 430, 2010
Vincent Coudé du Foresto, Dawn M. Gelino, and Ignasi Ribas, eds.

Multiple Aperture Imaging and its Application to Exo-Earth Imager

T. Matsuo,[1,2] W. A. Traub,[2] M. Tamura,[1] and H. Makoto[3]

[1] *National Astronomical Observatory of Japan, Osawa, Mitaka, Tokyo, Japan*

[2] *Jet Propulsion Laboratory, 4800 Oak Grove Dr., Pasadena, CA, USA*

[3] *Graduate School of Science, Tohoku University, Aoba, Aramaki, Aoba-ku, Sendai, Japan*

Abstract. We present a new concept for the Exo-Earth Imager. According to this concept, only two apertures (nullers) allow us to take a spatially resolved image of an exo-earth. This concept can be also applied to classical interferometers.

1. Theory

The concept of the Exo-Earth Imager is presented by Labeyrie (1996). This new type of the interferometer, called "hyper-telescope", has a sparse array of many mirrors for acquiring a 10 by 10 pixel snapshot of an exo-Earth discovered by the TPF-C and TPF-I/Darwin mission. The Exo-Earth Imager will be launched long after the TPF-C and TPF-I/Darwin missions characterize exoplanets. In this paper, we present a new imaging interferometry for the Exo-Earth Imager.

Given that spectrum of the exo-Earth and its position are revealed by the TPF-C and TPF-I/Darwin mission, we introduce the theory of this interferom-

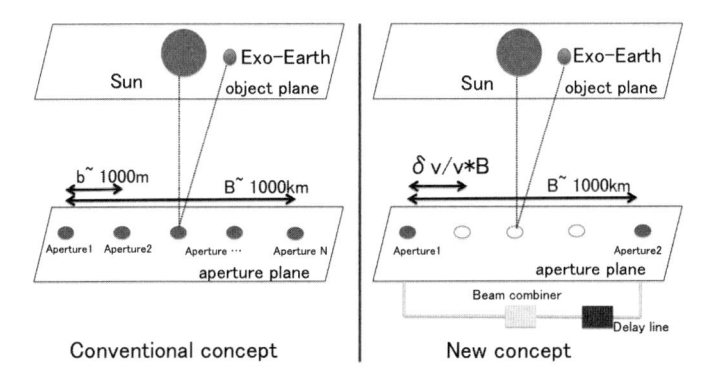

Figure 1. *Left:* Conventional concept of the Exo-Earth Imager. *Right:* New concept of the Exo-Earth Imager.

etry. Figure 1 shows the comparison of this new method with the previous one. First, we adjust the time delay between the two beams from the position of the Exo-Earth Imager before measurements of the complex visibilities in order to move the center of the coherent filed of view to the center of the exo-Earth. This is because the coherent field of view is very limited due to the long baseline length. Next, we measure the complex visibility, rotating the baseline. Finally, using velocity of light, c, observational frequency, ν, imaging and null baseline vectors, $\vec{B}$ and $\vec{b}$, complex visibility, CV, and position on the sky, $\vec{\theta}$, we perform Fourier transformation of νCV, along the frequency gives the averaged intensity of the exo-Earth over the observational band from ν_{min} to ν_{max}:

$$\begin{aligned} FT[CV(\vec{B},\vec{b},\nu)] &= \int_0^{2\pi} d\phi \int_{\nu_{min}}^{\nu_{max}} \nu d\nu CV(\vec{B},\vec{b},\nu) \exp(\frac{2\pi i \nu}{c}\vec{B}\cdot\vec{\theta'}) \\ &= \int_{\nu_{min}}^{\nu_{max}} d\nu \frac{c^2}{b^2} \int_\Omega B(\vec{\theta})O(\vec{B},\vec{b},\nu) \exp[\frac{2\pi i \nu}{c}\vec{B}\cdot(\vec{\theta}-\vec{\theta'})] \\ &= \int_{\nu_{min}}^{\nu_{max}} d\nu \frac{c^2}{b^2} \int_\Omega B(\vec{\theta})O(\vec{B},\vec{b},\nu) \exp[\frac{2\pi i \nu}{c}\vec{B}\cdot(\vec{\theta}-\vec{\theta'})] \\ &= \frac{c^2}{b^2}[B(\vec{\theta'})]_\nu[O(\vec{B},\vec{b})]_\nu * R\left(2\pi\frac{\delta u \theta}{2}\right) * \overline{f}(\nu), \end{aligned} \quad (1)$$

where $[B(\vec{\theta})]_\nu$ is averaged brightness distribution over the frequency, $[O(\vec{B},\vec{b})]_\nu$ is averaged null responce over the frequency, $R\left(2\pi\frac{\delta u \theta}{2}\right)$ is finite spatial resolution due to the finite observational band, and $\overline{f}(\nu)$ is Fourier transformation of the spectral structure of the Exo-Earth.

The spectrum of the exo-Earth actually has complicated structures such as absorption lines. From Equation (1), the brightness distribution of the exo-Earth is a convolution with Fourier transformation of the complicated spectrum structures. In other words, the absorption and emission lines make artificial peaks on the reconstructed image. On the other hand, only the specific frequency structures affect the reconstructed brightness distribution of the exo-Earth because the exo-Earth is a point source. In other words, the reconstructed image of the exo-Earth is little affected by the structures of spectrum. If the specific features affects the brightness distribution, Fourier transformation of the complex visibility should be performed after removing the specific structures. In this case, the brightness distribution of the exo-Earth contains the artificial components.

This study is not clearly beneficial for the Exo-Earth Imager at this time. In the future, we will perform numerical simulations of this concept for the validation and examine its usability.

References

Labeyrie, A. 1996, A&AS, 118, 517

Pathways Towards Habitable Planets
ASP Conference Series, Vol. 430, 2010
Vincent Coudé du Foresto, Dawn M. Gelino, and Ignasi Ribas, eds.

The Systemic Console Package

S. Meschiari,[1] A. S. Wolf,[2] E. Rivera,[1] G. Laughlin,[1] S. Vogt,[1] and P. Butler[3]

[1] *UCO/Lick Observatory, Department of Astronomy and Astrophysics, University of California, Santa Cruz, CA, USA*

[2] *Department of Geophysical and Planetary Sciences, California Institute of Technology, Pasadena, CA, USA*

[3] *Department of Terrestrial Magnetism, Carnegie Institute of Washington, Washington, DC, USA*

Abstract. The *Systemic Console* is a software package for the fitting of Doppler radial velocity (RV) and transit timing observations arising from arbitrarily complex planetary systems. To illustrate its capabilities, we analyze a new RV dataset and synthetic datasets for the HD128311 planetary system and show that integrated fits that combine radial velocities and a small number of transit timing observations in a self-consistent fashion can greatly constrain the orbital parameters of a perturbing body.

1. Introduction

The radial velocity method (e.g., Udry et al. 2007) remains the dominant technique for detecting extrasolar planets, both in terms of productivity and the ability to reach into the vicinity of the terrestrial mass region (e.g., Mayor et al. 2009). However, follow-up characterization using transit photometry can constrain planetary orbital parameters to very high precision (e.g., Bean et al. 2008; Borucki et al. 2009), given the extremely small central timings errors achieved by both ground and space programs.

Furthermore, non-transiting planets can exert perturbations that will result in transit timing variations (TTV) (Holman & Murray 2005; Agol et al. 2005). In presence of interacting dynamical configurations such as 2:1 resonances, TTVs can be substantial and easily exceed the timing error. This may represent an effective alternative route to probing into the terrestrial mass range (Meschiari et al., in preparation), in addition to greatly constraining the orbital parameters of the non-transiting planet. Deriving fits that model TTVs require (1) allowing for mutual gravitational interactions, and (2) efficient, global minimization algorithms.

The *Systemic Console* (Meschiari et al. 2009) is the only publicly available software package for deriving self-consistent Newtonian fits using the joint χ^2 statistic that includes both RV and transit timing datasets. It provides a comprehensive library of tools, including best-fit (Lomb-Scargle periodogram, Levenberg-Marquardt and Simulated Annealing) and error (bootstrap refitting and Markov Chain Monte Carlo) estimation routines.

2. Sample Application: Resonance Characterization through Transit Timing

The planetary system HD128311 (Vogt et al. 2005) is locked in a deep 2:1 mean motion resonance, which ensures the long-term stability of the system. Knowledge of the resonant configuration of such systems (e.g., whether they are participating in *apsidal corotation*) provides fundamental clues to their formation history (Sándor & Kley 2006). However, previous analyses using Monte-Carlo generated orbital fits resulted in a roughly even proportion of systems in apsidal corotation (R2) to systems with only one resonant argument librating (R1).

We analyzed this system using the Console software package. Further Doppler measurements, taken using the HIRES spectrometer at Keck and spanning three additional years, do not significantly improve the characterization of the planets' dynamical state. To characterize the system as either R1 or R2 with > 90% confidence required a 30-year stretch of RV observations. Nevertheless, synthetic RV datasets including a small number of transit timing measurements (4) showed that if a system like HD128311 was found to be transiting, then its dynamical configuration would be completely characterized. This was confirmed by running several Monte Carlo chains comprising 5×10^5 systems or more (Meschiari et al. 2009).

3. Conclusions

The Console software package provides a set of tools for analyzing data originating from a variety of sources (RV surveys and transits) and can recover best-fitting stable planetary configurations even in the presence of strong mutual perturbations. In the future, we plan to improve the software by allowing non-coplanar fits, full photometry fitting, and modeling of astrometric data.

Acknowledgments. S.M. acknowledges the support of the AAS and the NSF ITG program for funding his travel to the *Pathways* conference. The Console software package may be downloaded for free at http://www.oklo.org.

References

Agol, E., Steffen, J., Sari, R., & Clarkson, W. 2005, MNRAS, 359, 567
Bean, J. L., Benedict, G. F., Charbonneau, D., Homeier, D., Taylor, D. C., McArthur, B., Seifahrt, A., Dreizler, S., & Reiners, A. 2008, A&A, 486, 1039
Borucki, W. J., et al. 2009, Sci, 325, 709
Holman, M. J., & Murray, N. W. 2005, Science, 307, 1288
Mayor, M., Udry, S., Lovis, C., Pepe, F., Queloz, D., Benz, W., Bertaux, J.-L., Bouchy, F., Mordasini, C., & Segransan, D. 2009, A&A, 493, 639
Meschiari, S., Wolf, A. S., Rivera, E., Laughlin, G., Vogt, S., & Butler, P. 2009, PASP, 121, 1016
Sándor, Z., & Kley, W. 2006, A&A, 451, L31
Udry, S., Fischer, D., & Queloz, D. 2007, in Protostars and Planets V, ed. B. Reipurth, D. Jewitt, & K. Keil, (Tuscon: Univ. of Arizona Press), 685
Vogt, S. S., Butler, R. P., Marcy, G. W., Fischer, D. A., Henry, G. W., Laughlin, G., Wright, J. T., & Johnson, J. A. 2005, ApJ, 632, 638

Pathways Towards Habitable Planets
ASP Conference Series, Vol. 430, 2010
Vincent Coudé du Foresto, Dawn M. Gelino, and Ignasi Ribas, eds.

The Earthshine Project: Applications to the Search of Exoearths

P. Montañés-Rodríguez[1,2] and E. Pallé Bagó[1,2]

[1] *Instituto de Astrofísica de Canarias, Vía Láctea, La Laguna, Tenerife, Spain*

[2] *Departamento de Astrofísica, Universidad de La Laguna, Avenida Astrofísico Francisco Sánchez, La Laguna, Spain*

Abstract. To be able to detect a biosphere in an extrasolar planet, life in that planet should have been able to alter the original composition of the planetary atmosphere. In this way, an external observer could detect the chemical disequilibrium introduced by living organisms in the planet. The earthshine technique has allowed us to determine the best disk-integrated planetary features that we could use to find life in an exoplanet similar to Earth. Different observing methods have been investigated. In this poster, we summarize the scientific goals that could be reached using a variety of observational methods.

1. Introduction

Earthshine, or ashen light, is sunlight that has been reflected by the dayside region of the Earth towards the Moon and, from the dark region of the Moon, reflected back towards the Earth. The brightness ratio between the dark and bright lunar regions can be used to derive the Earth large-scale reflectance (Danjon 1928). This technique allows for the observation of Earth as a distant planet, which is essential for characterizing Earth-like planets.

2. Photometry

Daily photometric earthshine observations help to validate models of the Earth albedo (Qiu et al. 2003; Pallé et al. 2004). Models use real daily satellite data of the total cloud amount at each surface location from the International Satellite Cloud Climatology Project (ISCCP)

Photometry of a hypothetical distant Earth (including super-Earths) allows, despite its dynamic weather patterns, the inference of the following information (Pallé et al. 2008): (i) Measurements of the rotation period with an accuracy of minutes with just a few days of photometric integration; (ii) Light curve and regional properties of the planetary surface or atmosphere; (iii) The presence of short-lived clouds (from deviations from a periodic light curve; (iv) The presence of active climatology possibly related to the presence of liquid water in its surface.

3. The Reflection Spectrum

Ground-based observations of the Earth's spectrum and its simulation, including daily satellite clouds data and an atmospheric radiative transfer model (Montañés-Rodríguez et al. 2004) allows the measurement of terrestrial spectral reflectivity (Montañés-Rodríguez et al. 2005, 2006) in the visible and near-infrared, from 0.4 to 2.5 μm (Pallé et al. 2009). Data of the Earth's spectrum reveal the coexistence of atmospheric O_2, O_3, H_2O vapour, CO_2, and CH_4, which is only maintained by life. Features of the planetary surface, such as chlorophyll, may also be detectable. The vegetation signature, or red edge, is normally too weak on Earth, however models show that it becomes prominent under certain orbital geometries (Montañés-Rodríguez et al. 2006).

4. The Transmission Spectrum

A significant fraction of the terrestrial planets that will be discovered in the coming years will be detected by transit methodology (CoRoT, Kepler).

We have analyzed the transmission spectra of planet Earth as reflected from the Moon during a lunar eclipse (Pallé et al. 2009). A detailed atlas of its transmission spectrum with identifications of the main atomic and molecular absorption features has been provided. From here, it is possible to assume a state of chemical and thermodynamic disequilibrium attributable to life.

Atmospheric features that are weak in the reflection spectrum such as O_3, O_2, CO_2, and CH_4, are much stronger in the transmission spectrum. The spectroscopic observation of a planetary transit will be more effective than the direct observation of the planet in determining the main atmospheric components. Besides this, the observation of the transit allows the detection of dimers (oxygen collision complexes), the terrestrial ionosphere and N_2.

References

Danjon, A. 1928, Recherches sur la photométrie de la lumière cendrée et l'albedo de la terre, Ann. Obs. Strasbourg, 2, 165
Montañés-Rodríguez, P., Pallé, E., Goode, P.R., Hickey, J., Qiu, J., Yurchyshyn, V., Chu, M-C, Kolbe, E., Brown, C.T., Koonin, S.E. 2004, Adv. Sp. Res., 34, 293
Montañés-Rodríguez, P., Pallé, E., Goode, P. R., Hickey, J. & Koonin, S. E. 2005, ApJ, 629, 1175
Montañés-Rodríguez, P., Pallé E. & Goode, P. R. 2006, ApJ, 651, 544
Montañés-Rodríguez, P., Pallé E. & Goode, P. R. 2007, AJ, 134, 1145
Pallé, E., Goode, P. R., Montañés-Rodríguez, P. & Koonin, S. E. 2004, Sci, 304, 1299
Pallé E., Ford, E. B., Seager, S., Montañés-Rodríguez, P. & Váquez, M. 2008, ApJ, 676, 1319
Pallé, E, Zapatero Osorio, M. R., Barrena R., Montañés-Rodríguez, P. & Martín E. L. 2009, Nat, 459, 814
Qiu, J., Goode, P.R., Pallé, E., Yurchyshyn, V., Hickey, J., Montañés Rodríguez, P., Chu, M.-C., Kolbe, E., Brown, C.T., & Koonin, S.E. 2003, JGR, 108, D22, 4709

Pathways Towards Habitable Planets
ASP Conference Series, Vol. 430, 2010
Vincent Coudé du Foresto, Dawn M. Gelino, and Ignasi Ribas, eds.

Estimating the Age of Exoplanet's Host Stars by their Membership in Moving Groups and Young Associations

D. Montes,[1] J. A. Caballero,[1] C. J. Fernández-Rodríguez,[2] L. J. Alloza,[2]
S. Bertrán de Lis,[3] A. Garrido-Rubio,[3] R. Greciano,[2]
J. E. Herranz-Luque,[2] I. Juárez-Martínez,[2] E. Manjavacas,[3] F. Ocaña,[3]
B. Pila-Díez,[2,3] A. Sánchez de Miguel,[1,3] and C. E. Tapia-Ayuga[3]

[1] Dept. Astrofísica, Fac. Físicas, Univ. Complutense de Madrid, Madrid, Spain

[2] Grupo de Ciencias Planetarias de Madrid, Madrid, Spain

[3] Asociación de Astrónomos Aficionados de la Universidad Complutense de Madrid, Madrid, Spain

Abstract. We present a detailed study of the kinematics of known exoplanets host stars with known parallactic distance, and precise proper motion and radial velocity measurements, from where the Galactic space motions (U, V, W) were computed. For the stars with U and V velocity components inside or near the boundaries that determine the young disc population, we have analyzed the possible membership in the classical moving groups and nearby loose associations with ages between 10 and 600 Ma. For the candidate members, we have compiled the information available in the literature in order to constrain their membership by applying age-dating methods for late-type stars. We identify several dozen young exoplanet host star candidates, many of which were considered to have solar-like ages.

Context, Aims, and Methods

Young exoplanetary systems with ages $\tau < 600$ Ma (i.e., Hyades-like or younger) can provide constraints on the time scale and mechanism of planet formation, and on planet evolution. Apart from the very young planet candidates found by direct imaging (around e.g. HR 8799, 2M1207-39 or AB Pic), some young planet candidates have been found with the radial velocity method, such as HD 70573b (Setiawan et al. 2007) in the Hercules-Lyra association or the controversial TW Hya b (Setiawan et al. 2008).

We search for bright *Hipparcos* stars with radial-velocity planets that are member candidates in young ($\tau = 100 - 600$ Ma) moving groups (Montes et al. 2001), such as the Hyades, IC 2391, Ursa Majoris, Castor and the Local Association, and very young ($\tau < 100$ Ma) nearby loose associations (Torres et al. 2008). Generally, these stars are discarded from accurate radial-velocity searches based on activity indicators, but there might be young stars that passed the rejection filter (e.g. HD 81040, $\tau \sim 700$ Ma; Sozzetti et al. 2006).

On 2009 Sep 1, the Extrasolar Planets Encyclopaedia (exoplanet.eu) tabulated 346 planet candidates in 295 planetary systems detected by radial velocity (35 multiple planet systems). Of them, 228 have Hipparcos stars as host stars.

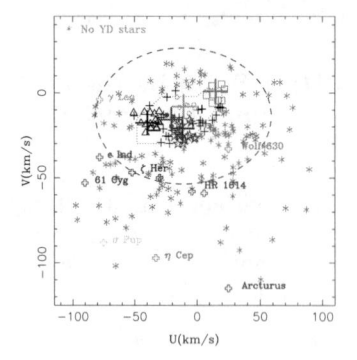

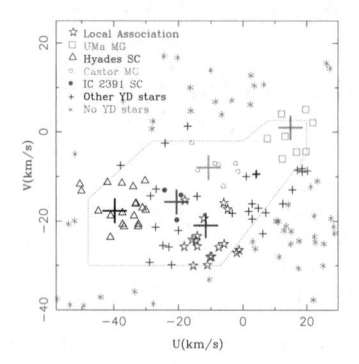

Figure 1. *Left*: position in the UV-plane of all the exoplanets host stars. *Right*: zoom of UV-plane in the region of the young disk stars.

We have computed Galactocentric space velocities UVW derived from star coordinates, proper motions, and parallactic distances (from van Leeuwen 2007), and systemic radial velocities, V_r (γ), from a number of works, including Nordström et al. 2004, Famaey et al. 2005 and planet discovery papers. To date, we have collected UVW velocities for 215 planetary systems (94%).

Results and Future Work

In Figure 1 (Böttlinger diagram) we plot the computed UV velocities. A total of 69 planet host stars satisfy the Eggen criterion for the young disc population (Figure 1 right) i.e., they are young star candidates. Five of them have values of vertical Galactocentric space velocity, W, that are too large with respect to young stars in the thin disk. The remaining 64 stars are the subject of a dedicated data compilation, including published values of effective temperature, T_{eff}, lithium abundance, $\log \epsilon(\text{Li})$, rotational velocities, $v \sin i$, activity indicators ($\log R'_{\text{HK}}$) and membership in a moving group. Interestingly, a relatively large number of stars have been tabulated as probable nearby young stars. Most of them are candidate and confirmed members in the Hyades Supercluster, such as ι Hor, HD 50554, HD 108147 or τ Boo, but there also candidate stars in the IC 2391 (94 Cet, HD 168746) or Castor (HD 217107) moving groups and the Local Association (HD 130322, V376 Peg - the transiting star HD 209458).

The data compilation will finish soon, and we will check if stellar kinematics are consistent with the other spectroscopic age indicators.

Acknowledgments. This work was supported by the projects AYA2008-00695 and S-0505/ESP-0237 (ASTROCAM).

References

Famaey, B., Jorissen, A., Luri, X., et al. 2005, A&A, 430, 165
Montes, D., López-Santiago, J., Gálvez, M. C., et al. 2001, MNRAS, 328, 45
Nordström, B., Mayor, M., Andersen, J., et al. 2004, A&A, 418, 989
Setiawan, J., Weise, P., Henning, et al. 2007, ApJ, 660, L145
Setiawan, J., Henning, T., Launhardt, R., et al. 2008, Nat, 451, 38

Sozzetti, A., Udry, S., Zucker, S., et al. 2006, A&A, 449, 417
Torres, C. A. O., Quast, G. R., Melo, C. H. F., & Sterzik, M. F. 2008, in Handbook of
    Star Forming Regions, Volume II, eds B. Reipurth, (San Francisco: ASP), 757
van Leeuwen, F. 2007, A&A, 474, 653

Eight-octant Phase-mask Coronagraph for Detecting Earth-like Exoplanets around Partially Resolved Stars

N. Murakami,[1] J. Nishikawa,[2] K. Yokochi,[2,3] M. Tamura,[2] N. Baba,[1] N. Hashimoto,[4] and L. Abe[5]

Abstract. We present recent progress in an eight-octant phase-mask (8OPM) coronagraph. We manufactured a phase-mask utilizing a nematic liquid crystal (NLC), which can be switched between the 8OPM and a four-quadrant phase-mask (FQPM) modes. We carried out laboratory experiments on the coronagraphs by using the NLC mask. The experiments demonstrate that the 8OPM coronagraph has a better tolerance of the tip-tilt error than the FQPM one. We manufactured also a fully achromatic 8OPM made of a photonic crystal (PC) device. The PC-8OPM shows good coronagraphic performance (less than 10^{-5} at peak) for two laser sources (wavelengths of 532 nm and 633 nm).

1. Eight-octant Phase-mask (8OPM) Coronagraph

An eight-octant phase-mask (8OPM) is a family of a well-known four-quadrant phase-mask (FQPM) (Rouan et al. 2000). Mathematically, the 8OPM can be expressed as a weighted sum of evenly charged optical vortex masks, each of which causes zero intensity within the pupil in a Lyot-stop plane (Murakami et al. 2008). Thus, the 8OPM coronagraphs can perfectly eliminate light from point-like unresolved stars. In addition, it is expected that the 8OPM coronagraph has a 4th-order tolerance of tip-tilt errors, while that of the FQPM one is 2nd order. This higher-order tolerance of the 8OPM will be advantageous for detecting Earth-like exoplanets within habitable zones of partially resolved nearby stars.

2. Nematic Liquid-crystal Phase-mask

We manufactured a nematic liquid-crystal (NLC) phase-mask, which is composed of eight-segmented variable retarders (Figure 1, *left*). The phase retardations of the retarders can be adjusted by applied voltages via flexible printed

---

[1]Division of Applied Physics, Hokkaido University, Sapporo, Japan

[2]National Astronomical Observatory of Japan, 2-21-1 Osawa, Mitaka, Tokyo, Japan

[3]Graduate School of Engineering, Tokyo University of Agriculture and Technology, 2-24-16 Naka-cho, Koganei, Tokyo, Japan

[4]CITIZEN Holdings Co., 840 Shimotomi, Tokorozawa, Saitama, Japan

[5]Laboratoire Hippolyte Fizeau, UMR 6525, Université de Nice-Sophia Antipolis, 28 avenue Valrose, Nice, France

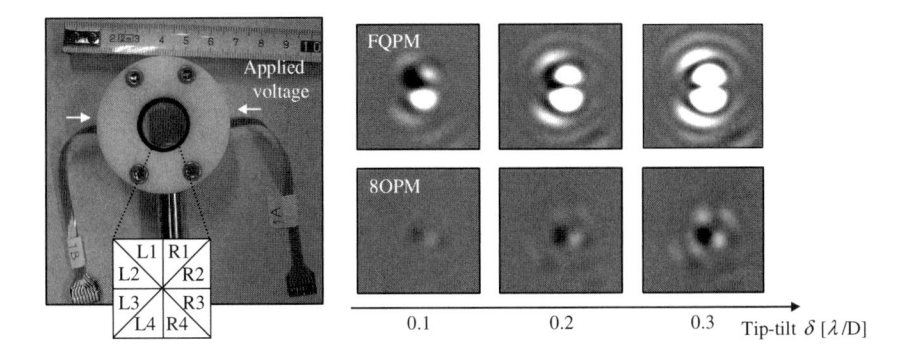

Figure 1. *Left:* Picture of a manufactured NLC mask. *Right:* Acquired FQPM- and 8OPM-coronagraphic images for tip-tilt errors δ =0.1, 0.2, and 0.3 λ/D.

circuits (two belts equipped on both sides of the NLC-part). The NLC mask can be switched between FQPM and 8OPM modes by applying voltages to appropriate segments. We carried out laboratory experiments of the both coronagraphs. As a light source, we used a He-Ne laser (a wavelength of $\lambda = 633$ nm) because the NLC mask has a chromatic characteristic in its retardation. Figure 1(*right*) shows the acquired FQPM- and 8OPM-coronagraphic images for tip-tilt errors δ =0.1, 0.2, and 0.3 λ/D. As can be seen, the 8OPM coronagraph suppresses the model star light even for the large tip-tilt errors. We also fit the experimental data with a function $I(\delta) = a\delta^b$, where $I(\delta)$ is a residual intensity integrated over a field of view. As a result, we obtained fitting parameters $b = 2.0$ for the FQPM mode and $b = 4.0$ for the 8OPM mode. The experimental results suggest that the 8OPM coronagraph has a 4th-order tolerance of the tip-tilt errors while that for the FQPM is 2nd order.

3. Photonic-crystal Phase-mask

We also manufactured a photonic-crystal (PC) phase-mask, which is composed of eight-segmented half-wave plates whose fast axes are $\pm 45°$. When $0°$ linearly polarized lights enter into the PC mask, the mask rotates the azimuth angles of the input polarized lights to opposite directions $\pm 90°$. Thus, $0/\pi$ phase shift can be realized achromatically due to polarization interferometry (Baba et al. 2002). A manufacturing technique of the photonic crystal realized a precise eight-segmented pattern of the phase-mask. We roughly estimated widths of the boundaries between the segments to be 100 nm. We carried out laboratory experiments of the 8OPM coronagraph using two laser light sources (a diode-pumped solid-state laser of $\lambda = 532$ nm and a He-Ne laser of $\lambda = 633$ nm). As a result, speckle-noise limited images with very high contrasts were obtained for the both laser light sources (Murakami et al. submitted to ApJ).

Acknowledgments. This research was partly supported by the National Astronomical Observatory of Japan (NAOJ), the Japan Aerospace Exploration Agency (JAXA) airborne and spacecraft instruments development program, and the Japan Society for the Promotion of Science (JSPS) through a Grant-in-Aid for JSPS Fellows (20-4783). This research was also partly supported by the Japanese Ministry of Education, Culture, Sports, Science and Technology, through a Grant-in-Aid for Scientific Research on Priority Areas, "Development of Extra-solar Planetary Science". We thank the Advanced Technology Center of the NAOJ for helpful support.

References

Baba, N., Murakami, N., Ishigaki, T., & Hashimoto, N. 2002, Opt. Lett., 27, 1373
Murakami, et al., 2008, PASP, 120, 1112
Rouan, D., Riaud, P., Boccaletti, A., Clénet, Y., & Labeyrie, A. 2000, PASP, 112, 1479

Pathways Towards Habitable Planets
ASP Conference Series, Vol. 430, 2010
Vincent Coudé du Foresto, Dawn M. Gelino, and Ignasi Ribas, eds.

A Coronagraph with an Unbalanced Nulling Interferometer and Adaptive Optics

J. Nishikawa,[1] K. Yokochi,[2,3] N. Murakami,[3] L. Abe,[4] T. Kotani,[5] M. Tamura,[1] T. Kurokawa,[2] A. V. Tavrov,[6] and M. Takeda[7]

Abstract. A coronagraph system with an unbalanced nulling interferometer (UNI) is described. It consists of an adaptive optics (AO), the UNI, a second AO, and a coronagraph. Wavefront corrections and star light rejections are made twice. The UNI provides a magnification of wavefront aberrations and makes it possible to compensate for the wavefront aberrations beyond the AO system capabilities. Experimental results proved the principle of the UNI by wavefront measurements in the pupil plane.

1. Configuration of the Coronagraph

In order to directly detect the light from exoplanets close to a star, we need to suppress the diffracted light and scattered light (speckle noise) which are produced by the aperture and the wavefront aberrations of the telescope optics, respectively. A novel coronagraph technique of light suppression was developed (Nishikawa et al. 2006), which uses first adaptive optics (AO), an unbalanced nulling interferometer (UNI), and another AO system with two deformable mirrors for phase and amplitude correction (PAC) as useful pre-optics of a last coronagraph. This technique was formally described by Nishikawa et al. (2008a). A brief description of the principle of the four-stage coronagraph is as follows.

We assume the first AO system compensates for the wavefront as good as $\lambda/1000$ rms. In the UNI, the star light is reduced to about two orders of magnitude and its electric field is a one-tenth by the unbalanced nulling of the two beams. As an effect, wavefront aberrations are magnified by the reduction of the unaberrated electric field. The PAC-AO stage compensates for the magnified wavefront aberrations in the amplitude as well as the phase by the two deformable mirrors. One of many kind of coronagraph can be placed at the last stage.

---

[1] National Astronomical Observatory of Japan, Mitaka, Tokyo, Japan

[2] Tokyo University of Agriculture and Technology, Koganei, Tokyo, Japan

[3] Graduate School of Engineering, Hokkaido University, Sapporo, Hokkaido, Japan

[4] Lab. Hippolyte Fizeau, Universite de Nice-Sophia Antipolis, Parc Valrose, Nice, France

[5] Institute of Space and Astronautical Science, JAXA, Sagamihara, Kanagawa, Japan

[6] Space Research Institute (IKI), 84/32 Profsoyuznaya Str, Moscow, Russia

[7] The University of Electro-Communications, Chofu, Tokyo, Japan

In this four-stage coronagraph system, the wavefront compensation (at the first AO system and the PAC stage, with at least three deformable mirrors in total) and the starlight suppression (at the UNI and the last coronagraph) are made twice in turn. The UNI provides the magnification of the wavefront aberrations, and the residual aberrations at the downstream aberration-magnified stages are equivalent to very precise wavefront at the initial wavefront by the magnification factor. At the final focal plane, we expect that the diffracted light of the star and speckle noise level would be lower than the off-axis planet.

2. Experiment

We made experiments of the coronagraph system without the first AO (Nishikawa et al. 2008b). By a modified Michelson interferometer, two beams were generated from a collimated beam of a laser light and destructively interfered in the unbalanced manner by polarization control. Two deformable mirrors and a Shack-Hartman wavefront sensor was placed as the PAC-AO.

We found that the behavior of the measured electric field at the UNI and the PAC stage by the wavefront sensor followed the principle of the configuration, i.e., the wavefront aberrations were magnified by the reduction of the unaberrated electric field and again re-compensated in the amplitude and the phase by the two deformable mirrors, for instance, the phase aberrations of about $\lambda/180$ was magnified to $\lambda/40$ and recorrected to $\lambda/100$ which was considered equivalent to $\lambda/450$ virtually by the magnification factor of 4.5.

We need to confirm the effects of the reduction of the diffracted and the scattered light at the focal plane, where a 3-D Sagnac interferometric coronagraph is placed as the last stage to reject the Airy pattern which can null out a point source achromatically (Yokochi et al. 2009).

Acknowledgments. The authors thank the Advanced Technology Center of the NAOJ for helpful support. A part of this work is supported by Grants-in-Aid (Nos. 21360037, 20656013) from the JSPS/MEXT and by airborne and spacecraft instruments development program of JAXA.

References

Nishikawa, J., Murakami, N., Abe, L., Kotani, T., Tamura, M., Yokochi, K., & Kurokawa, T. 2006, SPIE, 6265, 62653Q1
Nishikawa, J., Murakami, N., Abe, L., & Kotani, T. 2008a, A&A, 489, 1389
Nishikawa, J., Yokochi, K., Abe, L., Murakami, N., Kotani, T., Tamura, M., Kurokawa, T., Tavrov, A. V., & Takeda, M. 2008b, SPIE, 7010, 70102A1
Yokochi, K., Tavrov, A. V., Nishikawa, J., Murakami, N., Abe, L., Tamura, M., Takeda, M., & Kurokawa, T. 2009, Opt. Lett., 34, 1985

M-Type Stars as Hosts for Habitable Planets

P. Odert,[1] M. Leitzinger,[1] A. Hanslmeier,[1] H. Lammer,[2]
M. L. Khodachenko,[2] and I. Ribas[3]

[1]*Institute of Physics/IGAM, Universitätsplatz 5, Graz, Austria*

[2]*Space Research Institute, Austrian Academy of Sciences, Schmiedlstraße 6, Graz, Austria*

[3]*Institut d'Estudis Espacials de Catalunya/CSIC, Torre C5-parell, 2a planta, Bellaterra, Spain*

Abstract. Several planets orbiting M dwarfs have been discovered during the last years. However, it is still a matter of debate if these numerous, low-mass stars could be suitable hosts for habitable planets. Many M stars exhibit high levels of activity (high XUV fluxes, powerful flares etc.) during extended periods of time that could be harmful to the evolution of life. To address this topic we are compiling a catalog of nearby M dwarfs that could be suitable targets for habitable planet searches. It will include all data necessary to characterize the stars and to derive their fundamental properties. Special attention is turned towards data related to their activity (e.g., X-ray/EUV emission, and associated data like rotation periods). These data allow us to estimate important stellar characteristics (e.g., ages, flare rates, mass-loss rates) that could have a major impact on planets inside the habitable zone. Here we summarize the current status of this work.

1. Introduction

Since M dwarfs are promising targets for planet searches, there is increasing interest in a better determination of their properties. We compiled a sample of nearby M stars ($d \leq 15$ pc) and collected numerous data from literature to estimate stellar properties relevant to exoplanet studies. Here we focus on the determination of stellar ages.

The age of a star is difficult to determine, yet it is an important quantity, not only for stellar evolution studies but also for gaining insight into planetary formation and evolution. There are various methods used for constraining the age of a star, but not all of them are applicable to M dwarfs. Large uncertainties and often contradicting results make stellar age determination even more complex. Here, we use three methods to estimate ages for part of our sample, namely moving group membership, Ca II R'_{HK} index age, and gyrochronology. Moving group membership and corresponding ages were taken from literature (e.g., Montes et al. 2001). Ca II ages and gyrochronology ages were calculated with the calibrations of Mamajek & Hillenbrand (2008) using compiled data from literature.

2. Results and Conclusions

Our sample of nearby M dwarfs currently consists of more than 800 objects. For about 30% we have found sufficient data in literature to derive at least one of the above mentioned age estimates. For M dwarfs in systems with an early-type companion, the age of this companion has been derived and assigned to the M dwarf. Figure 1 shows the resulting age distributions.

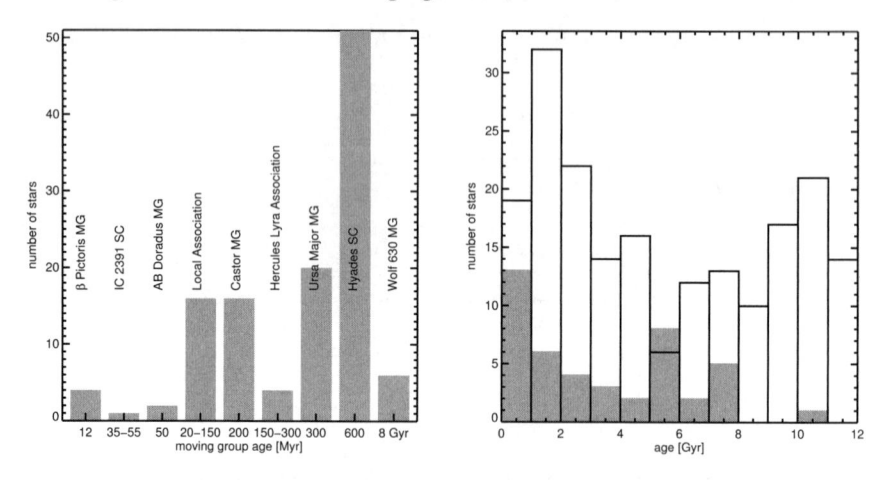

Figure 1. Histograms of nearby M stars with estimated ages derived by three different methods. The left panel shows M stars in moving groups with corresponding ages, the right panel shows the distribution of Ca II ages (outline) and gyro ages (shaded area).

In case more than one age estimate could be derived for a star, we compared the results of the different approaches. We find a rather good agreement between Ca II and gyro ages, whereas comparison with moving group ages shows that these frequently result in younger ages than the other methods. This could be an indication that several older stars share the kinematics of a young moving group by chance only. A combination of different age estimation methods is therefore important to gain reliable results.

Acknowledgments. P.O., M.L. and A.H. gratefully acknowledge the Austrian *Fonds zur Förderung der wissenschaftlichen Forschung* (FWF grant P 19446-N16) for supporting this project. This research has made use of the SIMBAD and VizieR databases, operated at CDS, Strasbourg, France.

References

Mamajek, E. E. & Hillenbrand, L. A. 2008, ApJ, 687, 1264
Montes, D., López-Santiago, J., Gálvez, M. C., Fernández-Figueroa, M. J., De Castro, E. & Cornide, M. 2001, MNRAS, 328, 45

Wide Angle Telescope Transit Search (WATTS): A Low-Elevation Component of the TrES Network

B. Oetiker,[1] M. Kowalczyk,[1] B. Nietfeld,[1] G. Mandushev,[2] and E. W. Dunham[2]

[1]Sam Houston State University, Huntsville, TX, USA

[2]Lowell Observatory, 1400 W Mars Hill Rd, Flagstaff, AZ, USA

Abstract. The Wide Angle Telescope Transit Search (WATTS) is a low-elevation, small-aperture, wide-field (5.5 deg ×5.5 deg) transit search instrument capable of achieving the photometric precision needed to detect giant extrasolar planets. In one year of operation, WATTS has identified over 20 candidates. As the second active component of the TrES network, WATTS significantly increases the efficiency of the survey when data from both sites are combined.

1. System Description

The WATTS (Oetiker et al. 2010) telescope is designed to be part of the TrES network (Alonso et al. 2004). To facilitate data sharing within TrES, the WATTS system has a similar field of view, bandpass, and magnitude range to the PSST (Dunham et al. 2004), Sleuth (O'Donovan et al. 2004), and STARE (Brown & Charbonneau 2000) systems.

WATTS is assembled primarily from off-the-shelf components. The WATTS detector is a PI Acton PIXIS2048B thinned, back-illuminated CCD with 13.5μ pixels, readnoise of $18e^-$, full well of $110000e^-$ and non-linearity of 1%. The chip is cooled to $-60 \deg$ C using built-in thermoelectric cooling with water assist, resulting in dark current of $\sim 1e^-$ pixel^{-1}min^{-1}. The telescope mount is a modified 12″ Meade LX200 GPS telescope.

A Canon EF300mm F2.8L camera lens is used for imaging through a Bessel R filter. The image scale is 9.7 arcsec/pixel with a FWHM of 1.8 pixels.

Unlike most systems, WATTS does not use an autoguider to keep the field centered. Instead, a periodic error correction (PEC) is used to compensate for periodic tracking errors while the system is imaging. Short-period PEC calibration and corrections are done using features included in the telescope mount system. The long-term errors are monitored and corrected using the shifts between a reference image and the data images.

2. Data Processing and Analysis

Data analysis for WATTS is an implementation of the difference image analysis (DIA) described in Dunham et al. (2004). DIA is a robust data analysis and photometry pipeline that compensates for PSF variations resulting from focus drift and sub-pixel image drift.

3. Low-Elevation Observing: Scintillation and Sky Brightness

Two major concerns when making precise photometric measurements at a low-elevation site near a population center are increased errors associated with scintillation and sky brightness. For the 0.1 m aperture and 90 s exposures of WATTS (elevation 100 m), we estimate that scintillation noise is increased by 35% compared to a traditional astronomical site (elevation 2500 m), adding 0.001 mag RMS to the error budget. Despite the increased noise, a significant fraction of stars in the WATTS sample have RMS error less than 1% (Figure 1).

Increased sky brightness is a more significant contributor to photometric noise, especially for the fainter stars in the sample. For a moderately dark-sky site like WATTS (20.6 mag/arcsec^2 in R), an RMS error of 1% corresponds to a star with $R \approx 12.0$, compared to $R \approx 12.3$ for a dark-sky site (21.5 mag/arcsec^2 in R). This reduces the number of stars with RMS error less than 1% by 23%. Therefore, it is critical to choose a site with the darkest sky possible when conducting a wide field photometric survey for extrasolar planets in order to maximize the number of low-noise light curves in the sample.

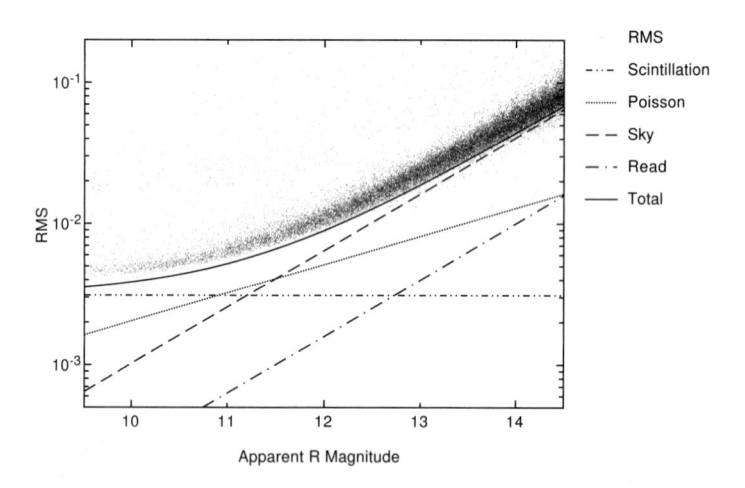

Figure 1. RMS error for unbinned decorrelated light curves for the Auriga field. Lines correspond to theoretical noise sources for the WATTS system.

References

Alonso, R., et al. 2004, ApJ, 613, L153
Brown, T. M., & Charbonneau, D. 2000, in ASP Conf. Ser. 219, Disks, Planetesimals, and Planets, ed. F. Garzón, C. Eiroa, D. de Winter & T. J. Mahoney, (San Francisco: ASP), 584
Dravins, D., Lindegren, L., Mezey, E., & Young, A. T. 1998, PASP, 110, 610
Dunham, E. W., Mandushev, G. I., Taylor, B. W., & Oetiker, B. 2004, PASP, 116, 1072
O'Donovan, et al. 2004, in AIP Conf. Proc. 713, The Search for Other Worlds, ed. S. S. Holt & D. Deming, 169
Oetiker, et al. 2010, PASP, 122, 41
Young, A. T. 1967, AJ, 72, 747

Where Can We Find Super-Earths?

E. Podlewska-Gaca and E. Szuszkiewicz

CASA and Institute of Physics, University of Szczecin, Szczecin, Poland*

Abstract. In recent years we have been witnessing the discovery of one extrasolar gas giant after another. Now the time has come to detect more low-mass planets like super-Earths and Earth-like objects. An interesting question to ask is: where should we look for them? We explore here the possibility of finding super-Earths in the close vicinity of gas giants, as a result of the early evolution of planetary systems. For this purpose, we have considered a young planetary system containing a super-Earth and a gas giant, both embedded in a protoplanetary disc. We have shown that, if the super-Earth is on the internal orbit relative to the gas giant, the planets can easily become locked in a mean motion resonance. This is no longer true, however, if the super-Earth is on the external orbit. In this case we have obtained that the low-mass planet is captured in a trap at the outer edge of the gap opened by the giant planet and no first order mean motion commensurabilities are expected. Our investigations might be particularly useful for the observational transit timing variation (TTV) technique.

The population of super-Earths (planets with the mass in the range of 2-10 $M_\oplus$) should be numerous, according to theories of planet formation (Ida & Lin 2005). To date we know of 18 such objects around main sequence stars. In comparison, there are more than 300 gas giant planets in the solar neighbourhood. So, where are the super-Earths? Most of the known low-mass planets have been discovered within the past 2 years, thanks to the sensitive instruments and sophisticated observational techniques used from the ground (e.g. HARPS at La Silla Observatory, microlensing programmes OGLE, MOA) and from the space (CoRoT). Another space mission, Kepler, will soon help to improve the statistics. Following the successful hunting for super-Earths, we have undertaken the task of answering the question whether there are any preferred planetary configurations in which super-Earths are involved. We have started our investigation from a particular configuration, namely from a super-Earth orbiting a Sun-like star close to a gas giant. We have given most emphasis to the possible resonant structures in such a system.

In our studies we have performed a series of 2-D hydrodynamic simulations of a pair of interacting planets: a super-Earth and a Jupiter, both embedded in a gaseous protoplanetary disc. The migration process induced by the disc-planet interactions will determine the final architecture of this system. Particularly interesting for us here, is the convergent migration, which is one of the most promising mechanisms for bringing planets into mean motion resonances.

The outcome of our simulations is illustrated in Figure 1. The Jupiter is located at the distance 1 (in dimensionless units) from the star, which corresponds to 5.2 AU in the Solar System. The results suggest that we can find super-Earths either close to the locations of the inner first order mean motion

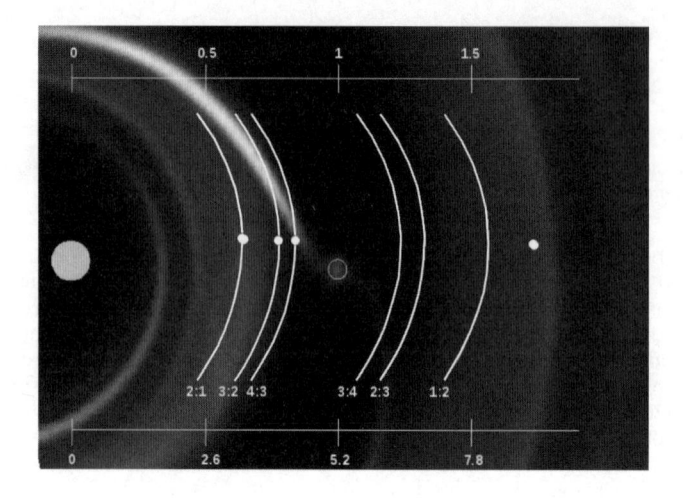

Figure 1. The possible locations of the super-Earth (denoted as white dots) in the system containing the Jupiter mass gas giant.

resonances with a gas giant (Podlewska & Szuszkiewicz 2008) or at the position of the external edge of the gap opened by the Jupiter-like planet (Podlewska & Szuszkiewicz 2009). The reason for the latter case is that the super-Earth is caught in a trap at the edge of the gap. Similar planet behaviour has been found by Masset et al. (2006) and Pierens & Nelson (2008) and more recently has been explored in detail by Paardekooper & Papaloizou (2009). Our investigations might be particularly useful for the observational TTV technique (Agol et al. 2005) which is sufficiently powerful to detect a planet as small as our Earth, especially if it is locked in the mean motion resonance with the gas giant.

Acknowledgments. This work has been partially supported by grants: MNiSW (N203 026 32/3831) and ASTROSIM-PL. The simulations reported here were performed using the HPC cluster HAL9000 of the Computing Centre of the Faculty of Mathematics and Physics at the University of Szczecin.

References

Agol, E., et al., 2005, MNRAS, 359, 567
Ida, S. & Lin, D. N. C., 2005, ApJ, 626, 1045
Masset, F.S, Morbidelli, A., Crida, A., & Ferreira, J., 2006, ApJ, 642, 478
Paardekooper, S-J., & Papaloizou, J. C. B., 2009, MNRAS, 394, 2283
Pierens, A. & Nelson, R.P., 2008, A&A, 482, 333
Podlewska, E. & Szuszkiewicz, E., 2008, MNRAS, 386, 1347
Podlewska, E. & Szuszkiewicz, E., 2009, MNRAS, 397, 1995

CARMENES: Calar Alto High-Resolution Search for M Dwarfs with Exo-earths with a Near-infrared Echelle Spectrograph

A. Quirrenbach,[1] P. J. Amado,[2] H. Mandel,[1] J. A. Caballero,[3] I. Ribas,[4] A. Reiners,[5] R. Mundt,[6] and the CARMENES Consortium[1,2,3,4,5,6,7,8,9]

Abstract. CARMENES, *Calar Alto high-Resolution search for M dwarfs with Exo-earths with a Near-infrared Echelle Spectrograph*, is a study for a next-generation instrument for the 3.5 m Calar Alto Telescope to be designed, built, integrated, and operated by a consortium of nine German and Spanish institutions. Our main objective is finding habitable exoplanets around M dwarfs, which will be achieved by radial velocity measurements on the m s^{-1} level in the near-infrared, where low-mass stars emit the bulk of their radiation.

1. Introduction

So far, radial velocity exoplanet searches have mainly focused on Solar-like main sequence stars. However, searches for exoplanets around M dwarfs have also been successful, and knowing the frequency of these objects places important constraints on planet formation scenarios. Some of the least massive exoplanets known orbit low-mass M dwarfs (e.g. GJ 581 and GJ 876 – Rivera et al. 2005; Udry et al. 2007; Mayor et al. 2009).

M dwarfs are the most common stars in the solar neighborhood, and their habitable zones lie close to them. This means that "habitable" planets around M dwarfs have large radial-velocity signatures and a large probability of showing transits. In spite of their interest, M dwarfs have not been searched for planets as extensively as late-F, G, and K stars, because of their faintness in the optical, where most radial velocity searches are being performed. A near-infrared (NIR) spectrograph with a radial velocity accuracy on the m s^{-1} level would be more efficient to detect Earth-like planets around stars with spectral types later than

---

[1] Landessternwarte Königstuhl, Heidelberg, Germany

[2] Instituto de Astrofísica de Andalucía, Granada, Spain

[3] Universidad Complutense de Madrid, Madrid, Spain

[4] Institut de Ciències de l'Espai, Barcelona, Spain

[5] Insitut für Astrophysik, Göttingen, Germany

[6] Max-Planck-Institut für Astronomie, Heidelberg, Germany

[7] Instituto de Astrofísica de Canarias, Tenerife, Spain

[8] Thüringer Landessternwarte, Tautenburg, Germany

[9] Hamburger Sternwarte, Hamburg, Germany

about M3. Radial velocities measured in the near infrared would also be less susceptible to stellar radial-velocity noise. As a high-resolution near-infrared spectrograph dedicated to a planet survey does not exist yet, we have performed a study of such an instrument for the Calar Alto Astronomical Observatory.

2. CARMENES Design Overview

The CARMENES study was initiated as a joint Spanish-German response to a call for ideas for new instruments for Calar Alto. In this contribution, we describe a summary of the CARMENES configuration as it was presented in the Conceptual Design Review in early October 2009. (The configuration may have changed since then; visit our webpage[1] for later developments). CARMENES is expected to become operational in 2013.

CARMENES will be fiber-fed from a front-end at the prime focus of the 3.5 m Calar Alto Telescope. The three cross-dispersed echelle spectrograph channels (NIR, VIS, MOS; see below) will be located in the coudé room of the telescope, where they can be thermally and mechanically stabilized. The instrument will cover from 500 to 1800 nm in one shot with a near-IR radial velocity precision requirement of $3\,\mathrm{m\,s^{-1}}$.

Thanks to an optical design with a mosaic of two $2\,\mathrm{k} \times 2\,\mathrm{k}$ detectors and an R2.9 echelle grating, the NIR channel will cover from 950 nm (Y band) to about 1800 nm (H band) with a spectral resolution R = 85 000 in 31 echelle orders. The cross disperser will consist of two S-NPH2 prisms. Most of the opto-mechanical components of the NIR channel will be located inside a vacuum tank at $T = -30°C$. The NIR channel needs an image slicer and an image scrambler.

The design of the visible (VIS) channel, which will cover from 500 to 900 nm with R = 60 000 in 42 echelle orders, will be based on the successful FEROS instrument. Simultaneous observations with the NIR and VIS channels will allow us to monitor the main activity indicators (Hα and the calcium triplet) with the same temporal sampling as the radial-velocity curve, which will help us discriminate between activity-induced and planet-induced radial-velocity variations.

The multiobject (MOS) visible channel takes advantage of the $\sim 0.8\,\mathrm{deg}^2$ field of the 3.5 m telescope to acquire the spectra of ~ 12 bright stars during the M dwarf observations; in this way a survey of G and K giants can be conducted without any additional telescope time.

CARMENES may share the telescope prime focus with another instrument. The design therefore, foresees a common front end mounted at the primary focus behind the K3 corrector. This front end will contain the fiber positioners for the CARMENES NIR, VIS, and MOS channels. Our primary channels (NIR and VIS) require two fibers each: one for the object and one for a ThAr lamp or sky.

Acknowledgments. The CARMENES study was funded by the Centro Astronómico Hispano-Alemán, which is operated jointly by the Max-Planck-Institut für Astronomie (Max-Planck-Gesellschaft) and the Instituto de Astrofísica de Andalucía (Consejo Superior de Investigaciones Científicas).

---

[1]http://www.ucm.es/info/Astrof/carmenes.

References

Mayor, M., Udry, S., Lovis, C. et al. 2009, A&A, 493, 639
Rivera, E. J., Lissauer, J. J., Butler, R. P. et al. 2005, ApJ, 634, 625
Udry, S., Bonfils, X., Delfosse, X. et al. 2007, A&A, 469, L43

Pathways Towards Habitable Planets
ASP Conference Series, Vol. 430, 2010
Vincent Coudé du Foresto, Dawn M. Gelino, and Ignasi Ribas, eds.

The Balloon Experimental Twin Telescope for Infrared Interferometry

S. A. Rinehart and the BETTII Team

NASA Goddard Space Flight Center, Greenbelt, MD, USA

Abstract. Astronomical studies at infrared wavelengths have dramatically improved our understanding of the universe, and Herschel and SOFIA will continue to provide exciting new discoveries. However, the relatively low angular resolution of these missions limits our ability to answer key science questions. Interferometry enables high angular resolution at long wavelengths - a powerful tool for scientific discovery. We will build the Balloon Experimental Twin Telescope for Infrared Interferometry (BETTII), and eight-meter baseline Michelson stellar interferometer to fly on a high-altitude balloon. BETTII, using a double-Fourier technique, will simultaneously obtain spatial and spectral information. Further, BETTII will serve as a technological pathfinder for future space-based interferometers such as FKSI, TPF-I, and Darwin.

1. What is BETTII?

The Balloon Experimental Twin Telescope for Infrared Interferometry (BETTII) is an eight-meter boom interferometer to operate on a high-altitude balloon. BETTII's scientific instrument is a Michelson interferometer operating in two wavelength bands, 30-50μm and 60-90μm, and using a double-Fourier technique that allows us to simultaneously achieve high angular resolution (0.5 arcsec at 40μm) and moderate spectral resolution (R 200). The angular resolution achieved by BETTII is 20x better than that achieved by *Spitzer* and 6x better than that of *Herschel*.

2. Scientific Objectives

BETTII enables new spatially-resolved spectroscopy. This capability will be of value for a wide range of important scientific questions, but have focused on three key science cases as drivers for the design and fabrication of BETTII. These cases include:

1. Star Formation: Does star formation in clusters differ from that in isolated regions? What far-infrared (FIR) emission arises from disks of individual stars? From inner envelopes?

2. Evolved Stars: How does dust form near evolved stars? How does it diffuse into the interstellar medium (ISM)?

3. Active Galactic Nuclei: What are the starburst energetics near the galactic nucleus? How do they differ from star formation in the outer starburst ring?

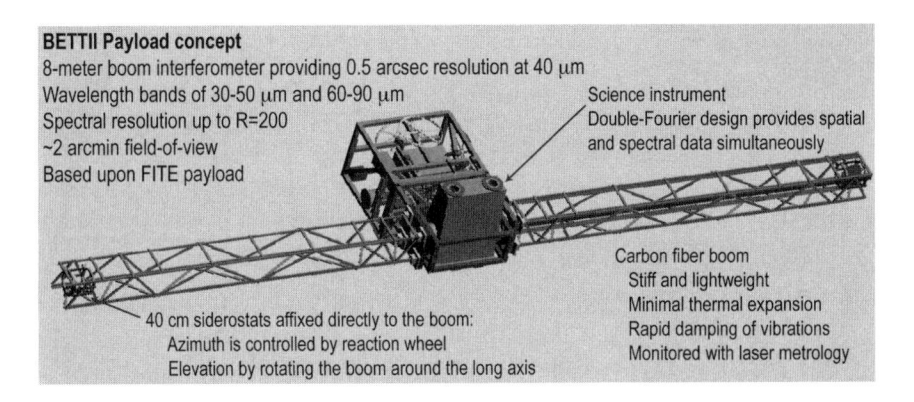

BETTII Payload concept
8-meter boom interferometer providing 0.5 arcsec resolution at 40 μm
Wavelength bands of 30-50 μm and 60-90 μm
Spectral resolution up to R=200
~2 arcmin field-of-view
Based upon FITE payload

Science instrument
Double-Fourier design provides spatial and spectral data simultaneously

40 cm siderostats affixed directly to the boom:
Azimuth is controlled by reaction wheel
Elevation by rotating the boom around the long axis

Carbon fiber boom
Stiff and lightweight
Minimal thermal expansion
Rapid damping of vibrations
Monitored with laser metrology

Figure 1. The combination of a stiff carbon fiber structure with external and internal metrology systems will allow BETTII to obtain high quality scientific data. Through collaboration with the Japanese FITE team, we greatly reduce both cost and risk while increasing scientific capability.

3. Technical Elements

BETTII's design is, in part, based upon the design of the FITE payload (see the FITE paper by Shibai in this volume). An overview of the BETTII design can be seen in Figure 1. Over the course of a flight lasting only a single night, BETTII can observe a source at multiple times, obtaining different baseline orientations by taking advantage of the Earth's rotation. This allows accurate derivation of the relative positions of sources that are unresolved by the BETTII primary beam. Simultaneously, the shape of the fringe patterns for each of the sources can be separated and transformed to provide spectral information.

4. Future Development

The far-infrared operation of BETTII is ideally suited to studies of cold and dusty objects. Further, the long wavelengths greatly simplify technical challenges for interferometry. In the future, we will build on the success of BETTII for studies of a wider range of astronomical objects. In particular, we will investigate using an interferometric nuller in the mid-infrared on a balloon-borne boom interferometer for studies of exoplanets and protoplanetary disks. Such an experiment, in addition to obtaining valuable scientific data, would provide a testing ground for nulling interferometers such as FKSI, TPF-I, or Darwin.

Acknowledgments. The author would like to thank the entire BETTII team, including: C. Allen, R. Barry, D. Benford, D. Fixsen, A. Kogut, D. Leisawitz, J. Mather, L. Mundy, R. Silverberg, and J. Staguhn. The BETTII team is also indebted to the FITE team in Japan, led by H. Shibai, for their contributions to the BETTII concept.

Detecting Planets around Very Cool Stars at Near Infrared Wavelengths with the Radial Velocity Technique

F. Rodler,[1] C. del Burgo,[2,3] E. L. Martín,[4] and C. Álvarez[5]

[1]*Instituto de Astrofísica de Canarias, La Laguna, Spain*

[2]*School of Cosmic Physics, Dublin Institute for Advanced Studies, Dublin, Ireland*

[3]*UNINOVA-CA3, Campus da Caparica, Caparica, Portugal*

[4]*Centro de Astrobiología (CAB-CSIC), Ctra. Ajalvir km. 4, Torrejón de Ardoz, Madrid, Spain*

[5]*Instituto de Astrofísica de Canarias, La Laguna, Spain*

Abstract. The radial velocity (RV) technique in the near-infrared is a powerful tool to search for low-mass planets around cool stars and sub-stellar objects. We investigate what precision can be achieved in RV measurements of an M9-dwarf in the Y-, J-, H-, and K-bands. Although late M-dwarfs emit the most flux in the J-band, we find that the highest RV-precision is attained in the Y-band.

Rationale The radial velocity (RV) technique in the near-infrared (NIR) is a powerful tool to search for low-mass planets around low-mass stars (Martin et al. 2006; Reiners et al. 2009).

In order to determine in which wavelength region we achieve the highest RV-precision, we run simulations for an M9-dwarf. To this end, we create our observation by adopting a theoretical M9-dwarf model spectrum with the parameters T=2,200K and $\log g = 4.5$, calculated with the PHOENIX code (del Burgo et al. 2009), then RV-shift this spectrum by a certain velocity v, and include then the telluric contamination by adopting a telluric spectrum, calculated with the HITRAN code (Rothman et al. 2005). Finally, we degrade the spectral resolution of this spectrum by a convolution with a suitable Gaussian, interpolate the resulting spectrum onto a chosen detector grid and add Poisson noise according to a certain signal-to-noise ratio (SNR).

To determine the RV, we first subtract the telluric lines from the observation, normalize the spectrum to one, and subdivide the spectrum into small, equidistant wavelength chunks. For each chunk, we determine the RV by cross-correlation and finally calculate the mean RV. The error of the mean of the RV is nothing but the standard deviation of the RV-measurements in the chunks.

Results Figure 1 shows that the highest RV-precision for an M9-dwarf is achieved in the Y-band with $v = 1.1$ m/s with a spectral resolution of R=80,000 and a SNR=274. It is interesting to note that although the peak of the spectral energy distribution of such a cool dwarf lies in the J-band, the richness of spectral features in the Y-band outweighs the flux difference. This plot demonstrates that it is important to take NIR spectra of high resolution to achieve high RV

Detecting Planets 527

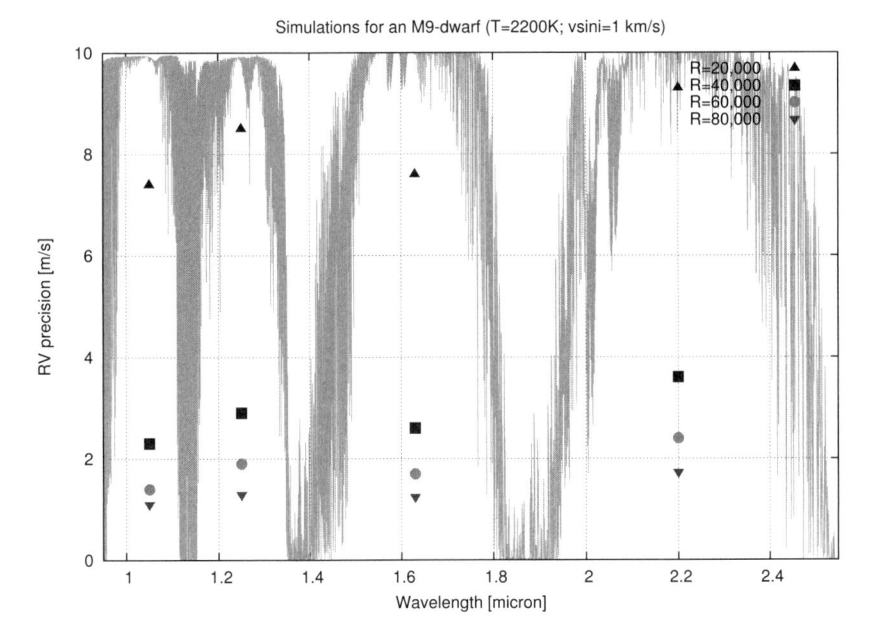

Figure 1. The dots depict the RV-precisions for an M9-dwarf in the different wavelength bands (Y, J, H, and K) and for different spectral resolutions. The grey spectrum represents the transmission of the Earth's atmosphere (the value 10 in the y-axis corresponds to 100% transmission). We simulated the case that the integration times for all the bands were the same, and the sensitivity of the detector was homogeneous throughout the NIR. Due to the spectral energy distribution of an M9-dwarf amd the transparency of the Earth's atmosphere, the SNRs obtained in the bands were different (Y: SNR=274; J: SNR=300; H: SNR=261; K: SNR=225).

precisions, which brings out the need for high-resolution cross-dispersed spectrographs (e.g. NAHUAL for the GTC (Martin et al. 2005), CARMENES for the CAHA (Quirrenbach et al. 2009)).

Acknowledgments. This work has been supported by the Spanish Ministerio de Eduación y Ciencia through grant AYA2007-67458.

References

del Burgo C., & Deshpande R., et al. 2009, in Proceedings of "New Technologies for Probing the Diversity of Brown Dwarfs and Exoplanets", China
Martín E. L., Guenther E., et al., 2005, AN, 326, 1015
Martín E. L., Guenther E., et al., 2006, ApJ, 644, L75
Reiners A., Bean J.L., et al., 2009, ApJ, 710, 432
Rothman L. S., Jacquemart D., et al., 2005, JQSRT, 96, 139
Quirrenbach A., Amado P. J., et al., 2009, arXiv:0912.0561 & in these proceedings

Pathways Towards Habitable Planets
ASP Conference Series, Vol. 430, 2010
Vincent Coudé du Foresto, Dawn M. Gelino, and Ignasi Ribas, eds.

Metallicity of M-dwarfs from NIR Spectra

B. Rojas-Ayala[1] and J. P. Lloyd[2]

[1] *208 Space Sciences Building, Cornell University, Ithaca, NY, USA*

[2] *230 Space Sciences Building, Cornell University, Ithaca, NY, USA*

Abstract. Above solar metallicity, the number of stars with planets and presence of multiple planets increases for solar-type stars (Santos et al. 2001; Fischer & Valenti 2005). M-dwarfs planet hosts known to date have shown abundances below solar metallicity when measured with optical spectra and photometry, suggesting that M-dwarfs have a planetary formation process that differs from F-, G- and K-dwarfs. This could also be due to statistics, with only few M-dwarfs with planetary candidates known, and/or the lack of reliable abundances estimates for this type of star. We present preliminary results from near-infrared spectra of M-dwarf secondaries (with a wide solar type star companion with known metallicity) obtained with the TripleSpec spectrograph operating on the Palomar 200-inch Hale Telescope.

1. A Near-Infrared Metallicity Indicator For M-Dwarfs

In order to estimate the metal content of M-dwarfs, we have followed a similar approach to that of Bonfils et al. (2005) and Johnson & Apps (2009), that consists of assuming that in binary systems, both components formed from the same original molecular cloud, and therefore they should share the same metallicity. But instead of doing a photometric calibration, using absolute magnitudes and colors of M-dwarfs, we have been working on a near-infrared (NIR) spectroscopic calibration using the TripleSpec instrument at Palomar 200-inch (Herter et al. 2008).

A spectroscopic method to estimate M-dwarf metallicities does not have a limitation on accurate parallax determinations, like the current photometric methods that rely on absolute magnitudes. Given that the spectral energy distributions (SEDs) of M-dwarfs peak in the NIR, studying them at these wavelengths is advantageous, not only because we have access to a larger sample in an efficient way with NIR instruments like TripleSpec than with optical instruments, but also because their J, H and K bands are rich with absorption features. Our preliminary results show that it is possible to distinguish between metal-rich and metal-poor M-dwarfs using some of the NIR features present in the spectra of these stars, such as Na I and Ca I in the K band, as seen in Figure 1.. Despite the fact that most of the features present in these bands are extremely temperature-dependent (Ali et al. 1995; Jones et al. 1996), and prone to non-local thermodynamic equilibrium (LTE) effects, the dispersion in equivalent width between objects with roughly the same temperature and gravity can provide insights on their metal abundance, allowing us to identify the most advantageous/metal-rich M-dwarfs for planetary search surveys around cool stars, such as TEDI (Triplespec Exoplanet Discovery Instrument, Edelstein et al. 2007), and MEarth (Irwin et al. 2009).

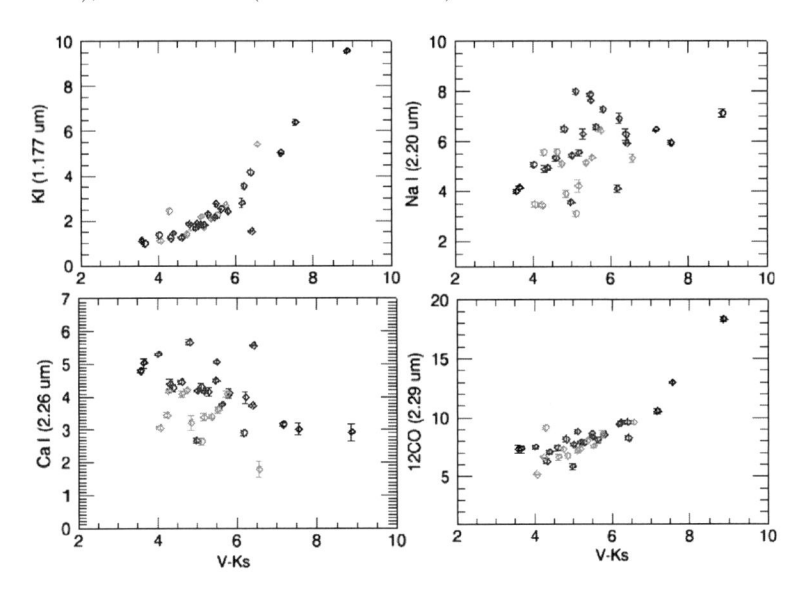

Figure 1. Equivalent widths of K I (1.177 μm), Na I (2.206 & 2.209 μm), Ca I (2.261, 2.263 & 2.265 μm) and ^{12}CO(2,0) (2.293 μm) features versus (V-K$_s$). The dark grey diamonds represent M-dwarfs with [Fe/H] > -0.05, and the light grey diamonds, M-dwarfs with [Fe/H] < -0.05, according to the photometric calibration by Johnson & Apps (2009). These preliminary results show that it is possible to distinguish between metal-rich and metal-poor M-dwarfs using some of the NIR features present in the spectra of these stars, such as Na I and Ca I in the K band. Both the equivalent widths of K I and ^{12}CO(2,0) are strongly temperature sensitive, making them useful for a NIR spectral classification of these stars.

References

Ali, B. et al. 1995, AJ, 110, 2415
Bonfils, X. et al. 2005, A&A, 442, 635
Edelstein, J. et al. 2007, SPIE, 6693
Fischer, D. & Valenti, J. 2005, ApJ, 622, 1102
Herter, T. L. et al. 2008, SPIE, 7014E, 30H
Irwin, J. et al. 2009, in IAU Symposium, "Transiting Planets", eds. F. Pont, D. D. Sasselov & M. J. Holman (Cambridge: CUP), 253, 37
Johnson, J. A. & Apps, K. 2009, ApJ, 699, 933J
Jones, H. R. A. et al. 1996, MNRAS, 280, 77
Santos et al. 2001, A&A, 373, 1019

Pathways Towards Habitable Planets
ASP Conference Series, Vol. 430, 2010
Vincent Coudé du Foresto, Dawn M. Gelino, and Ignasi Ribas, eds.

X-Exoplanets: An X-ray and EUV Database for Exoplanets

J. Sanz-Forcada,[1] D. García-Álvarez,[2,3] A. Velasco,[1,4] E. Solano,[1,4] I. Ribas,[5] G. Micela,[6] and A. Pollock[7]

Abstract. Extreme ultraviolet (EUV) and X-ray emission is of great importance in several phenomena related to the formation of planetary systems and the atmospheres of planets. The atmospheric composition and the mass of an exoplanet are partly dependent on the X-ray and EUV radiation received during the first stages of formation and even during main sequence of the star. Biological life developing on exoplanets would depend severely on the high energy radiation arriving from its parent star. Here we present a database of the X-ray and EUV emission of all the stars currently known to host exoplanets. The archive is public and accessible through the Spanish Virtual Observatory (SVO). The database gives the user the option to download observed X-rays and EUV spectra. Synthetic spectra covering the spectral range 1–912 Å are also available (present day telescopes do not give access to the EUV range at $\lambda > 180$ Å). These spectra are created using coronal models after fitting observed spectra.

Since the discovery of the first transiting exoplanet, it has been proposed that high energy stellar radiation should have some influence on the atmosphere of the planets. The effects of the high energy radiation (UV, EUV, and X-rays) are important at different stages. First, during protoplanetary disk dissipation. Later, when the stellar rotation is fast and the planet keeps most of its original atmosphere, this radiation is still strong, affecting the atmosphere. Finally, the development of biological life will partly depend on the high energy (EUV and UV in special) radiation arriving at the surface.

In recent years there has been increasing interest in modeling the effects of high energy radiation on the planet, such as vaporization or atmospheric "erosion" (see Lammer et al. 2009; Cecchi-Pestellini et al. 2009; Sanz-Forcada et al. 2010a and references therein). These models are limited to the X-ray or EUV fluxes calculated in the whole bands. The spectral energy distribution with the highest achievable resolution would be preferred. Such spectra are only available for the brightest X-ray sources. In most cases we are limited to low-resolution

---

[1] Centro de Astrobiología, CSIC-INTA, European Space Astronomy Center. Apartado 78, E-28691 Villanueva de la Canada (Madrid), Spain

[2] Instituto de Astrofísica de Canarias, Tenerife, Spain

[3] Grantecan CALP, Tenerife, Spain

[4] Spanish Virtual Observatory, Spain

[5] Institut de Ciències de l'Espai (CSIC-IEEC), Barcelona, Spain

[6] INAF- Osservatorio Astronomico di Palermo, Italy

[7] XMM-Newton SOC, European Space Agency, ESAC, Villanueva de la Cañada, Spain

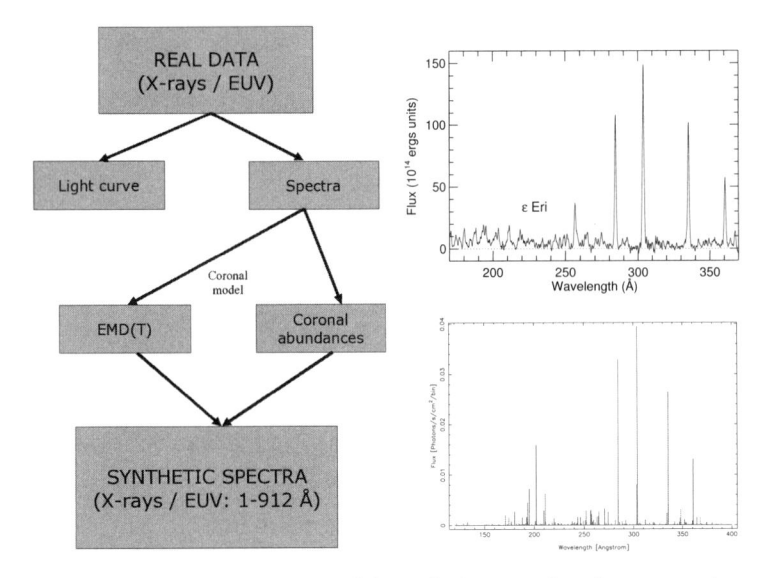

Figure 1. *Left:* schematic view of the technique employed to create the synthetic spectra from real data. *Right:* Real spectra (upper figure) of ϵ Eri observed with EUVE, in flux units. Synthetic spectra (lower figure) constructed with the coronal model successfully reproduces most spectral features.

spectra (range $\sim 1 - 125$ Å) at best. The situation is even worse in the EUV range ($\sim 100-900$ Å), where no current instrumentation can observe. In view of this situation we have constructed a database ("X-exoplanets") of real and synthetic spectra of stars with exoplanets. We create a set of high-resolution spectra constructed using coronal models as described in Figure 1 and Sanz-Forcada et al. (2010a). These models are calculated using the real spectra available, and we tested them with ϵ Eri, the only case with a good EUV spectrum available. Most emission in both bands (X-rays and EUV) is produced in the corona. The database covers all the planet-bearing stars that have been observed so far with XMM-Newton or Chandra, and will be updated regularly as new data are incorporated. The database is accessed through the Spanish Virtual Observatory at the address http://sdc.cab.inta-csic.es/xexoplanets/. The exploitation of the database has already given the first results, a proof of the long term effects of the planet erosion produced by coronal radiation (Sanz-Forcada et al. 2010b).

References

Cecchi-Pestellini, C., et al. 2009, A&A, 496, 863
Lammer, H., et al. 2009, A&A Rev., 17, 181
Sanz-Forcada, J. et al. 2010a, in IAU Symp. 264: Solar and Stellar Variability Impact on Earth and Planets; eds. A. Kosavichenko, A. Andrei, & J.-P. Rozelot, (Cambridge: Cambridge University Press), 478
Sanz-Forcada, J. et al. 2010b, A&A, accepted

Pathway toward an Infrared Interferometer

J. Schneider

LUTh - Paris Observatory, Meudon, France

Abstract. We propose a realistic pathway to satisfy two goals, thermal infrared studies of Earth-like exoplanets and interferometric architectures.

1. Interferometry: An Inevitable Need for the Long Term

After the first low resolution spectro-imaging of exoplanets, an obvious first step will be to increase the spectral resolution and the sensitivity of space missions. But, sooner or later, higher angular resolution will also be needed. Indeed, after the images of planets as single pixels, the imaging of planetary rings and moons will be the next step and, in a far but not unrealistic future, multi-pixel cartography of planetary surfaces will be necessary to understand the exact ocean/continents morphology of planets, importance of polar caps, geographic repartition of *exo-vegetation* etc.

2. A Bottleneck in the Interferometric Approach

Unfortunately, the first step of the interferometric approach in the mid-infrared, i.e. Bracewell-like nulling interferometry with a few apertures, presents some difficult aspects: (a) it requires high contrast/nulling performances (like any other approach); (b) the chosen nulling mechanism (Bracewell nulling) allows only a mono-pixel observation of the whole planetary system (with the subsequent problem of the background of exo-zodis); (c) it requires high performance free flyers metrology; and (d) by definition the sources (exoplanets) are extremely faint and contain only a few photons during the time scale of stability of the system. By themselves, each of these difficulties are not insurmountable. The difficulty comes from their occurring at the same time. One could imagine to start with a precursor, however there seems to be no simple precursor that is technologically "easy" and scientifically exciting.

3. Breaking the Deadlock: A Two Step Solution

Here we propose to decouple the needs for a thermal infrared approach and for future interferometers. First, a large IR coronagraph for nearby-by super-Earths and later on, an interferometric precursor that is scientifically valuable and technically feasible. Contrary to an interferometer with a few pupils, a large coronagraphic single aperture has an advantage: it provides a full 2-D image of the planetary systems instead of a single pixel signal, making the exozodi problem less severe. Other single aperture-like external occulters or Fresnel imagers are proposed, but they are better suited for the UV/visible approach. For the same angular resolution, the collecting area is at least 10 times larger than for the Darwin/TPF-I architecture.

The science requirements are the detection of H_2O, O_3 and CO_2 of the 70 nearest super-Earths (with a radius less than $2R_\oplus$) up to 5 pc. They translate into a 20 m telescope (a deployable set a few 3-4 m mirrors which would benefit from the JWST heritage) equipped with a coronagraph and with a spectral resolution of R = 20 to detect the central inversion peak of the CO_2 band at 15 microns. With these characteristics the SNR would reach a value of 5 per spectral channel. This type of architecture has already been studied by the former TRW Company (presently Northrop Grumman) and submitted to JPL (Lillie et al. 2001). The wavefront corrections and speckle calibration can be made by e.g., a "self-coherent camera" (Galicher, Baudoz & Rousset 2008). They should be adaptable to the thermal infrared. METIS, the Mid-infrared ELT Imager and Spectrograph, a proposed instrument for the European Extremely Large Telescope (E-ELT), is currently undergoing a phase-A study (Brandi et al. 2008), and could be a ground-based precursor.

4. A Scientifically Valuable Space Precursor

Instead of accumulating problems due to free-flying control, faintness of exoplanets, single pixel detection, and high contrast, we propose to have a simpler precursor instrument which would still produce first class science:

– No nulling/high contrast

– Bright targets

– No a priori wavelength constraints due to angular resolution (they are technically relaxed for free-flyers)

– Different science: general astrophysics (not mainly exoplanets). For instance: size and flattening of interesting stars, astrometric perturbation of stellar centroid by transiting planets, lensing (multiple quasars, stellar arcs, astrometry of planetary lensing), super hot super-Earths.

The idea would be to start with a two aperture interferometer, free-flying or not, with an evolution toward a multi-aperture interferometer similar to the stellar imager (Carpenter, Schrijver & Karovska 2009).

References

Brandl B., Lenzen R., Pantin E. et al., 2008. ETIS - the Mid-infrared E-ELT Imager and Spectrograph. SPIE 7014, 1N, astro-ph/0807.3271
Carpenter K., Schrijver C. & Karovska M., 2009, Astrophys. Space Sci. 320, 217
Galicher R., Baudoz P. & Rousset G., 2008, A&A, 488, L9
Lillie C. et al. 2001, TRW TPF Architecture. Phase 1 Study, Phase 2 Final Report.
http://planetquest.jpl.nasa.gov/TPF/TPFrevue/FinlReps/Trw/TRW12Fnl.pdf

Pathways Towards Habitable Planets
ASP Conference Series, Vol. 430, 2010
Vincent Coudé du Foresto, Dawn M. Gelino, and Ignasi Ribas, eds.

Characterization of Exoplanet Atmospheres in the Solar Neighbourhood with E-ELT/METIS

C. Schnupp,[1] W. Brandner,[1] C. Bergfors,[1] K. Geißler,[1] S. Daemgen,[1] S. Hippler,[1] F. Hormuth,[1] R. Lenzen,[1] T. Henning,[1] M. Janson,[2] and E. Pantin[3]

[1] Max-Planck-Institute for Astronomy, Königstuhl 17, Heidelberg, Germany

[2] University of Toronto, Department of Astronomy and Astrophysics, 50 St. George's Street, Toronto, Canada

[3] CEA/Saclay, DSM/DAPNIA/Service d'Astrophysique, Gif-sur-Yvette, France

Abstract. The Mid-Infrared ELT Imager and Spectrograph (METIS) will be the mid-infrared instrument for the European Extremely Large Telescope (E-ELT). It will provide diffraction limited imaging, high and low-resolution spectroscopy and phase mask coronography. Our science case for this instrument aims at direct detections of exoplanets and the study of their physical and chemical properties. We plan to observe a sample of M-dwarfs in the solar neighbourhood. In case of a direct detection of an exoplanet, we will conduct spectroscopic follow up observations with the low-resolution spectrograph provided by METIS.

1. METIS: The Mid-infrared Imager for the E-ELT

One of the instruments design studies for the E-ELT is METIS (Brandl 2008). The instrument will cover the wavelength range from 3 to 14 μm. It will provide diffraction limited imaging in L, M, and N band, a field of view of 20"×20", high and low-resolution spectroscopy, and phase mask coronography. All observing modes will be supported by Adaptive Optics to achieve these goals.

1.1. Science Case: Detections of Exoplanets

The detection of exoplanets is much more favorable in the mid-infrared than in the visual. Our aim of the exoplanets science case is to observe a sample of young, nearby, low-mass stars. An example for such targets are M dwarfs in the solar neighbourhood (Riaz 2006) which we observed and confirmed as single stars with Lucky Imaging by AstraLux (Hormuth 2008). For a M5V star at 4 pc we simulated the planet detection capability of METIS. Figure 1 shows the simulated observation of a planet-star system, with natural seeing of 0.7", subtracted by the exposure of a single star with a seeing of 0.75" to account for seeing variations during the observations. The images will be obtained with adaptive optics at 11 μm, i.e., they will have diffraction limited resolution. The

Characterization of Exoplanet Atmospheres 535

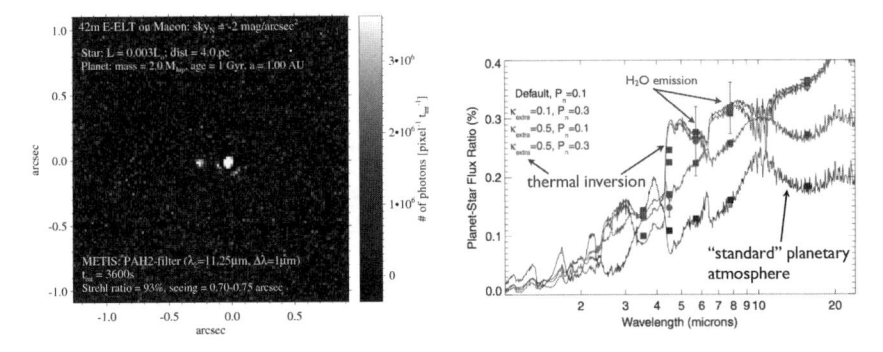

Figure 1. *Left:* Direct imaging simulation for a M5V star+planet with PSF subtraction in d=5 pc. *Right:* Planet-star flux ratios for TrES-4 received during Observation of the secondary eclipse with Spitzer/IRAC. Red dots are Spitzer/IRAC measurements.

$2\,M_{Jup}$ planet lies at a distance of $0.25''$ from the star, which corresponds to 1 AU in 4 pc.

Direct and coronographic imaging in LMN-band will reveal astrometric information (separation and position angle of the exoplanet with respect to its host star as a function of time), as well as the integrated flux of the exoplanet. The astrometric information will enable us to improve the orbit determination of the planet around its host star, and also yield an improved mass estimate for the exoplanet.

1.2. Spectroscopic Follow-up Observations

After the detection of a extrasolar planet, we want to characterize its atmosphere. Therefore spectroscopic follow-up observation will be performed by low-resolution slit spectroscopy. This will reveal molecular bands, and hence enable us to identify individual chemical species in the exoplanet atmosphere. Comparison with atmospheric models will reveal not only the chemical composition, but also the pressure-temperature structure of the exoplanet atmospheres. This would also reveal the presence of temperature inversions, as has been observed in the atmosphere of the transiting giant exoplanet TrES-4b (Knutson et al. 2009), see Figure 1.

References

Brandl, B. et al. 2008, in Ground-based and Airborne Instrumentation for Astronomy II, eds. I. S. McLean, & M. M. Casali, SPIE, 7014, 70141N
Hormuth, F. et al. 2008, in Ground-based and Airborne Instrumentation for Astronomy II, eds. I. S. McLean, & M. M. Casali, SPIE, 7014, 701448
Knutson, H. et al. 2009, ApJ, 691, 866
Riaz, B., Gizis, J., & Harvin, J. 2006, AJ, 132, 866

Pathways Towards Habitable Planets
ASP Conference Series, Vol. 430, 2010
Vincent Coudé du Foresto, Dawn M. Gelino, and Ignasi Ribas, eds.

US and European Technology Roadmap for a Mid-infrared Space Interferometer

P. A. Schuller,[1] P. R. Lawson,[2] O. P. Lay,[2] A. Léger,[1] and S. R. Martin[2]

[1] *Institut d'Astrophysique Spatiale (IAS), Université Paris-Sud, Bât. 121, Orsay, France*

[2] *Jet Propulsion Laboratory (JPL), MS 301-451, 4800 Oak Grove Drive, Pasadena, CA, USA*

Abstract. Studies of mid-infrared space interferometer concepts in the USA and in Europe have converged on a single architecture. We address the question of how the US and European communities could collaborate to advance technology efforts leading to a future space mission. We present the current state of the art in nulling interferometry, as demonstrated at ambient temperature and pressure in the lab, and outline required steps to demonstrate its performance under space conditions. Design studies of a cryogenic optical test bench under vacuum have already been carried out. We highlight pre-conditions and constraints of a collaborative effort, foreseeable practical and administrative challenges, and possible strategies to meet those challenges.

1. State of Technology Studies

JPL's Terrrestrial Planet Finder Interferometer (TPF-I) technology studies set the milestone #3 criteria for nulling performance to achieve repeatably an interferometric null of $\leq 10^{-5}$ in a spectral band width $\geq 25\%$ over 6 hours in an ambient environment. ***It was reached in early 2009 on the Adaptive Nuller Testbed (AdN)*** (Peters et al. 2010).

Milestone #4 set out to demonstrate the feasibility of four-beam nulling, the achievement of the required null stability, and the consequent detection of faint planets using approaches similar to the ones contemplated for a flight-mission. Experimental work was to implement this goal by the detection of a planet $\approx 10^6$ fainter than the star at a signal-to-noise ratio of ≥ 10 in an ambient environment. Phase chopping, averaging and rotation were to yield a factor 100 in residual starlight suppression. Rejection levels were therefore expected to reach $\leq 10^{-7}$ repeatably during 3 hours of operation.

The Planet Detection Testbed (PDT) was developed to work toward this goal.

All the data to meet Milestone #4 were acquired by summer 2009. ***Milestone #4 was formally reviewed and accepted in October 2009*** (Martin et al. 2009).

Milestone #5, planned for the year 2010, will demonstrate the effectiveness of spectral filtering to suppress instability noise. The expected gain in nulling depth is about one order of magnitude.

Once this milestone is reached, further development will have to demonstrate equivalent performance under space flight conditions, i.e., at *cryogenic temperatures under vacuum*.

IAS has been involved in *Darwin* technology studies on Achromatic Phase Shifters (APSs). As part of those studies, a *model for a cryogenic optical test bench under vacuum has been developed* (Labeque et al. 2004).

2. Joint Efforts

The next desired collaborative effort would be a joint study and implementation of interferometer technology and signal extraction procedures in a cryogenic system. It would address aspects such as the following:

In the study phase
1. Review and agree on overall performance goals and conditions; 2. Develop experimental layout: light sources, detectors, other independent analytic tools, phase shifting device, dispersive elements, cryogenic cooling, evacuation etc.; 3. Identify sub-systems and interfaces; 4. Define interface parameters, create error budget; 5. Identify technical solutions for sub-systems.

In the implementation phase
6. Acquire, test and fully characterize sub-systems; 7. Integrate sub-systems, test and fully characterize integrated experiment; 8. Demonstrate overall performance in compliance with goals.

3. Challenges and Questions

Given the required financial and human resources, forseeable challenges are: 1. site selection for system integration: will depend on surrounding infrastructure, availability of technical support, already available equipment etc.; 2. export/import of equipment: certain items may be rated as space and/or defense-relevant and therefore be restricted and subject to possible customs duties; 3. institutional property sovereignty: loan agreements vs. transfer of ownership; 4. visa regulations for visiting personnel and/or long-term exchange staff.

Further questions include: A. What can be learned from earlier agreements between space agencies regarding collaboration on space missions? Which elements can be applied to collaborations on an institutional level at the stage of technology studies and testing? How and to what extent can formal agreements be made between the institutions involved? B. Are there (international) funding bodies which may support an overarching project as a whole and can split their funds according to local requirements of the project partners? C. Are there other (international) bodies which can facilitate the exchange and transfer of equipment and/or personnel by being a third partner?

In order to present a strong case to funding agencies on both sides of the Atlantic, it is imperative to agree on common goals and build a scientific community in their support.

References

Labeque, A. et al., 2004, in SPIE Conf. Series, 5491, New Frontiers in Stellar Interferometry, ed. W. Traub, 999
Martin, S. R. et al., 2009, Exoplanet Interferometer Technology Milestone #4 Report: Planet Detection Demonstration, JPL D-63261, Nov. 2009.
Peters, R. D. et al., 2010, PASP, 122, 85

Pathways Towards Habitable Planets
ASP Conference Series, Vol. 430, 2010
Vincent Coudé du Foresto, Dawn M. Gelino, and Ignasi Ribas, eds.

Habitable Planets in Compact Close-in Planetary Systems

R. Schwarz,[1] E. Pilat-Lohinger,[1] B. Funk,[2] and G. Wuchterl[3]

[1] *Institute for Astronomy, A-1180 Türkenschanzstr. 17, Vienna, Austria*

[2] *Department of Astronomy, H-1117 Pázmány Péter sétány 1/A, Budapest, Hungary*

[3] *Institute for Astronomy, D- Thüringer Landessternwarte, Tautenburg, Germany*

Abstract. We investigate the dynamical stability of planetary systems in the CoRoT discovery space. The aim was to check whether they are stable within the habitable zone around M main-sequence stars. We place the first fictitious planet at a distance of 0.01 AU from its host-star and fill-up the CoRoT discovery space with planets so that they are tightly packed according to the Hill criterion supplying the initial minimum relative distances between the fictitious planets. The habitable zone of M-stars covers the region between 0.02 - 0.5 AU which is partly inside the CoRoT discovery space. Additionally we include a gas giant which perturbs the close-in planets, and compare the different results.

1. Introduction and Results

Today ground-based detection of Earth-like planets in extrasolar systems orbiting a star near 1 AU is beyond our opportunities. However CoRoT is able to find such small planets closer to the stars. Therefore, we focus our investigation on the dynamical stability of planetary systems within the CoRoT discovery space.

We investigated the dynamical stability of Earth-like planets, (i.e., planets with 1 Earth mass) moving around M main-sequence stars, where the habitable zone (HZ) is very close to the star and therefore partly in the discovery space of CoRoT. These stars are also of special interest since we know that most of the stars in the solar neighbourhood are M-stars. The planets are lined up according to the Hill-criterion (for further details see Funk et al. 2009). This criterion has been used very often as an approximation for orbital stability to identify the onset of orbital instability of planets in extra-solar planetary systems e.g., Chambers et al. (1996), Smith & Lissauer (2009) and Cuntz & Yeager (2009).

In this study the terrestrial planets were started on initially circular, coplanar orbits and perturbed by an outer Jupiter-like planet at 0.26 AU, which corresponds to a period of 50 days for a sun-like star. We studied 3 different models (A, B, and C shown in Figure 1 left graph): model A was used to determine the maximum number of planets inside the giant planets orbit, i.e., a planet with 1 Jupiter mass M_{Jup}, around a central star of 0.3 Solar mass M_{Sun}. In models B and C we investigated how many terrestrial planets may stay in the HZ. For a star of 0.3 M_{Sun}, (e.g., GJ 674) the HZ extends from 0.12 to 0.25 AU (model B) and for 0.5 M_{Sun}, (e.g., HD 285968) the HZ extends from 0.14 to 0.28 AU.

We can conclude that all 3 models are stable for an integration time of 1 Myrs. The results show that in model B, 5 terrestrial planets and in model C, 3 terrestrial planets stay close to or in the HZ. Additionally, we studied the stability of model B by varying the initial conditions of the perturbing giant planet, its mass (from 0.5 to 3.5 M_{Jup}) and eccentricity given as e_{Jup} (0-0.05), integrated for 10^5 years. The results are presented in Figure 1 right graph, where we can see a limit at 3 M_{Jup} and e_{Jup}=0.04. Hence it follows that the giant planet should have an almost circular orbit and its mass should not exceed a mass three times that of Jupiter. The stability limits of the perturbing planets (mass and eccentricity) are mainly influenced by the outermost terrestrial planet which moves close to the giant planet.

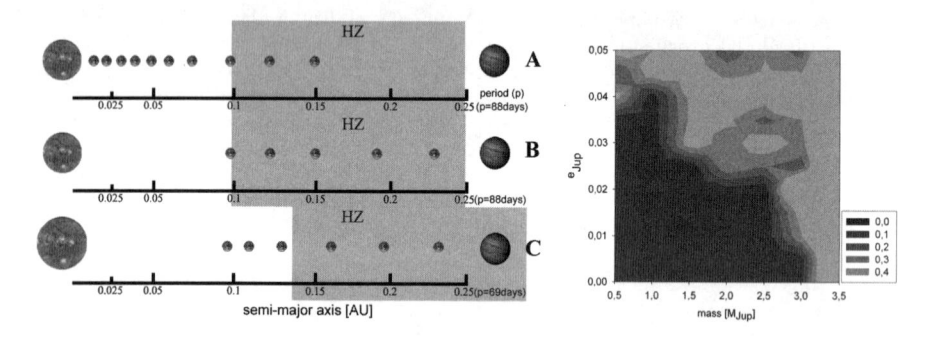

Figure 1. The scheme for the three different models: (left upper graph) stability for a compact close-in systems with 11 planets around a star of 0.3 M_{sun}, (left middle graph) stability of planets in the habitable zone for host stars with 0.3 M_{sun} and 0.5 M_{sun} (left lower graph) The right graph shows a stability map of model B. The grey-scale depicts the maximum eccentricity, where we checked the mean maximum eccentricity of all calculated orbits. The dark region depicts stable motion the light regions chaotic ones.

Acknowledgments. R. Schwarz wants to acknowledge the support by the Europlanet NA1 and the ÖFG project 06/11206. B. Funk wants to acknowledge the support by the FWF - project J2892. E. Pilat-Lohinger wants to acknowledge the support by the Austrian FWF - project P19569.

References

Chambers, J. E., Wetherill, G. W., & Boss, A. P. 1996, Icarus, 119, 261
Cuntz, M, & Yeager, K. E. 2009, ApJ, 697, L86
Funk, B., Wuchterl, G., Schwarz, R., & Pilat-Lohinger, E 2009, A&A, submitted
Smith, A. W. & Lissauer, J. J. 2009, Icarus, 201, 381

Pathways Towards Habitable Planets
ASP Conference Series, Vol. 430, 2010
Vincent Coudé du Foresto, Dawn M. Gelino, and Ignasi Ribas, eds.

Far-Infrared Interferometric Experiment (FITE): Toward the First Flight

H. Shibai,[1,3] M. Fukagawa,[1] E. Kato,[1] T. Kanoh,[1] T. Kohyama,[1]
Y. Itoh,[1] K. Yamamoto,[1] M. Kawada,[2] T. Watabe,[2] A. Nakashima,[2]
M. Tanabe,[2] R. Kanoh,[2] and M. Narita[3]

[1] *Graduate School of Science, Osaka Univ., Machikane-Yama, Toyonaka, Osaka, Japan*

[2] *Graduate School of Science, Nagoya Univ., Furo-Cho, Chikusa-ku, Nagoya, Japan*

[3] *ISAS, JAXA, Yoshino-Dai, Sagamihara, Kanagawa, Japan*

Abstract. FITE is a far-infrared interferometer on board balloon. In the far-infrared wave region, the spatial resolution is generally limited by the diffraction limit resolution of the primary aperture of a telescope. FITE is a Fizeau type interferometer with the baseline of 8 m corresponding to 4 arcsecond at 150 microns. The foci of the two parabola mirrors coincide at the primary focus that must be located at the input diaphragm of the cold optics in the cryostat. The two beams make an interference fringe at the focal plane. In December, 2008, the first launch of FITE was planned in Brazil as a collaboration between Japan and Brazil. FITE was assembled and checked out about its performance. However, the launch was postponed to 2010 due to unsuitable weather condition.

As the first attempt of a far-infrared interferometer in the world, we have developed the Far-Infrared Interferometric Telescope Experiment (FITE) to be launched by a balloon for the purpose of spatially resolved FIR observations of protoplanetary disks/envelopes. This is a Fizeau-type interferometer, and its baseline is 8 m corresponding 4 arcsecond resolution at 150 μm at the first flight. The baseline will be extended to 20 m as the next step so as to achieve one arcsecond resolution at 100 μm. The beam size of each aperture is 40 cm.

The interferometer is composed of two sets of collecting optics each of that consists two plane mirrors and an off-axis parabola mirror. The foci of the two parabola mirrors are coincided at the same position, and also are located at the input diaphragm of the cold optics in the cryostat. The two beams make an interference fringe at the focal plane. The optical system is described in Kato et al. (2009) in detail.

Kohyama et al. (2008) describes the design of the cold optics as well as of the cryostat. In the cryostat, the incident beam diverges again, and is collimated. The cold optics are equipped with a beam shutter, a Lyot stop, two dichroic beam splitters. The two beam splitters divide the beam into the far-infrared, the mid-infrared, and the visual light. The visual light goes out of the cryostat, and focused on the outer focus where a CCD camera is attached. The far-infrared and the mid-infrared beams are imaged on each array sensors in the cold part. In order to monitor the direction of the telescope, two CCD cameras are mounted on the telescope.

541

The newly developed far-infrared sensor has been optimized for FITE. This is a linear-array of 15 pixels. The sensor is installed in a liquid-helium-cooled cryostat together with a mid-infrared array and a cold optical system. The capacity of the LHe tank is 30 liters, and the holding time is 30 hours or more.

One of the most important techniques of FITE is the on-board optical adjustment system. During flight the ambient temperature decreases to - 50 degree C or lower. Therefore, even after precise adjustment on the ground before launch, during flight the optical adjustment of the interferometer becomes worse due to thermal contraction and adjustments must be made in flight.

Another important technique is the attitude control. FITE adopted a three-axis control like a satellite in free space in order to minimize the external turbulence from the pendulum motion of the balloon-gondola system. A detailed description is presented in a separate paper Nakashima et al. (2009).

A mid-infrared array sensor is employed for monitoring the optical path difference. The array itself is installed in the cold optics, and uses the same beam as FIR but the mid-infrared light is divided by a dichroic beam splitter. The array is made by Raytheon and is 240 x 320 pixels.

In December 2008, the first launch of FITE was planned in Brazil as a collaboration between Japan and Brazil. The launch base, operated by INPE, is located at Cachoeira Paulista, SP, Brazil. FITE was assembled and its performance was checked out. The astronomical targets are Neptune and one of the brightest mid-infrared sources, IRC+10214. These objects were selected because they have bright point-like sources in optical, mid-infrared, and far-infrared.

Almost all functions were acceptable for launch. However, the parabola mirrors were misaligned due to insufficient insulation for vibration during the transportation from Japan to the launch base in Brazil. Therefore, the first flight was regretfully postponed to 2010 or later. By adjusting the optics again in Japan, we are hoping to make its first flight as soon as possible.

Acknowledgments. The authors thank contributions by Ms. Yuka Matsumoto and Ms. Hirono Morishita, and support by Profs. Tetsuya Yoshida, Yoshitaka Saito, and the staff of ISAS balloon group. The observation is a collaboration with Drs. Antonio Mario Magalhaes and Jose William Villas-Boas. The development of FITE was supported by the grants-in-aid for scientific research (Specially Promoted Research, No. 15002006) of Japan Society for the Promotion of Science (JSPS).

References

Kato, E., Shibai, H., Kawada, M., Narita, M., Matsuo, T., Ohkubo, A., Suzuki, M., Kanoh, T., Yamamoto, K., & FITE team 2009, Trans. JSASS, Aerospace Tech. Jp., 7,47
Kohyama, T., Shibai, H., Kawada, M., Watabe, T., Matsuo, T., Ohkubo, A., Katoh, E., Kanoh, T., Suzuki, M., Mochizuki, et al. 2008, SPIE, 7013, 70133O
Nakashima, A., Shibai, H., Kawada, M., Matsuo, T., Narita, M., Kato, E., Kanho, T., Kohyama, T., Matsumoto, Y., & Morishita, H. 2009, Trans. JSASS, Aerospace Tech. Jp., submitted.

Pathways Towards Habitable Planets
ASP Conference Series, Vol. 430, 2010
Vincent Coudé du Foresto, Dawn M. Gelino, and Ignasi Ribas, eds.

Remote Detection of Biological Activity using Circular Polarization of Light

W. B. Sparks

Space Telescope Science Institute, Baltimore, MD, USA

Abstract. The identification of a universal biosignature that could be sensed remotely is critical to the prospects for success in the search for life elsewhere in the universe. A candidate universal biosignature is homochirality, which is likely to be a generic property of all biochemical life. Due to the optical activity of chiral biological molecules, this unique characteristic may provide a suitable remote sensing probe using circular polarization spectroscopy. Photosynthetic microbial organisms are of major importance to astrobiology, are plausibly commonplace, and are amenable to remote sensing. We show that their circular polarization spectra exhibit distinctive polarization features. Circular polarization spectroscopy could provide a powerful remote sensing technique for generic life searches.

1. The Advantages of Photosynthesis

Photosynthesis is extremely important in astrobiology, arising early in the Earth's history and conferring enormous evolutionary advantages to organisms that utilize it. The mechanism has been adopted as a successful life strategy ever since its emergence some 3 billion years ago, and oxygenic photosynthesis has had a major influence on the Earth's atmosphere and biosphere. The rapidity with which the phenomenon appeared after the formation of the Earth and the extraordinary success which it has enjoyed, imply a plausible and even likely scenario for randomly chosen habitable extrasolar planets: microbial dominated Worlds, and specifically photosynthetic microbes. The nature of photosynthesis requires that it be present at the surface since it depends on the harvesting of light from the host star. Further, windows of atmospheric transparency are favored, and regions of the spectra where the star is brightest offer the most promising source of photons. These qualities favor observability when viewed from a distance, and indeed the visible biosphere of the Earth is dominated by photosynthesis (vegetation, oceanic phytoplankton).

The biophysics of photosynthesis represent strong (electronic) interactions between light and the organism. Such interactions are polarization sensitive and hence the possibility of a measurable signal is enhanced.

2. Circular Polarization in Light Scattered from Biological Material

A unique characteristic of living material is its homochirality (handedness). The vast majority of organisms use only left-handed L-amino acids in proteins and, right-handed D-sugars in nucleic acids. Homochirality may be necessary for self-replication and may be a property of all biochemical life. Due to their chirality, the electronic absorption transitions of photosynthetic molecules have different strength for left and right circular polarization ("circular dichroism"). We designed a series of experiments to test whether this difference can introduce circular polarization in light scattered by photosynthetic material (Pospergelis 1969, Wolstencroft 1974, Wolstencroft et al. 2002).

We purchased a dedicated precision polarimeter, located at the National Institute of Standards & Technology (Wang et al. 2009). The polarimeter is optimized for the detection of circular polarization down to a polarization degree of 10^{-4} in the presence of significant linear polarization and operates across the optical spectral window 400 — 800 nm. Cultures of marine cyanobacteria were grown at the Center of Marine Biotechnology at the University of Maryland Biotechnology Institute (UMBI), and their scattered and transmitted polarization spectra were measured.

We detected significant circular polarization in marine cyanobacteria, and leaves (Sparks et al 2009a, b), with polarization features clearly related to spectral absorption features from chlorophyll and a variety of antenna pigments. A polarization sign change at the absorption maximum of chlorophyll in the scattered spectra is strong evidence that the polarization is due to the circular dichroism of the photosynthetic molecules.

3. Future Testing for a Potential Biosignature

Photosynthetic microbes and vegetation produce circular polarization in scattered light from one to three orders of magnitude higher than mineral controls. Oceans, with optics dominated by phytoplankton, and vegetation may hence produce a macroscopic circular polarization signature which could lend itself to remote sensing measurement for distant terrestrial planets, Earths, and super-Earths. We need to know if real-world terrestrial scenes in fact yield measurable polarization (oceans; forests; grassland; microbial pools and evaporites). Can a polarization signal survive in the real world, with dilution from extraneous scattered light and a mixture of sources? If it can, then circular polarimetry offers the intriguing prospect of being able to remotely sense life's chiral signature.

References

Pospergelis, M.M., 1969, Soviet Astronomy, 12, 973
Sparks, W.B., Hough, J.H., Germer, T.A., et al., 2009a, PNAS, 106, 7816
Sparks, W.B., Hough, J.H., Kolokolova, L., et al., 2009b, JQSRT, 110, 1771
Wang, B., Sparks, W.B., Germer, T.A., & Leadbetter, A., 2009, in SPIE Proc., Instruments and Methods for Astrobiology and Planetary Missions XII, ed. R. B. Hoover, G. V. Levin, A. Y. Rozanov, & K. D. Retherford (College Park: AIP), 7441, 744108
Wolstencroft, R.D., 1974, in Planets, Stars, and Nebulae Studied with Photopolarimetry, ed. Gehrels, T., University of Arizona Press, 495
Wolstencroft, R.D., Tranter, G., & Le Pevelen, D.D., 2002, in IAU Symp. 213, Bioastronomy 2002: Life among the stars, eds. R. Norris & F. Stootman, 149

The Inner Boundary of the Habitable Zone for Earth-like Planets

B. Stracke,[1] J. L. Grenfell,[2] P. von Paris,[1] A. B. C. Patzer,[2] and H. Rauer[1,2]

[1] *Institut für Planetenforschung, Deutsches Zentrum für Luft- und Raumfahrt, Rutherfordstr. 2, Berlin, Germany*

[2] *Zentrum für Astronomie & Astrophysik, Technische Universität Berlin, Hardenbergstr. 36, Berlin, Germany*

Abstract. The orbital region around a star where liquid water could exist on the surface of a terrestrial planet is usually defined as the Habitable Zone (HZ). We investigate the inner boundary of the HZ for different planetary scenarios with a one-dimensional radiative-convective model of the atmosphere. Our modeling approach involves the step-by-step increase of the incoming stellar flux and the subsequent iterative calculation of resulting changes in the atmospheric water vapour content and the radiative properties. Modelling results are presented for the influence of various planetary and atmospheric conditions on the inner boundary (water loss limit) of the HZs around a Sun-type star.

Liquid water is a commonly accepted fundamental requirement for the development of life. Based on this condition the habitable zone (HZ) is usually defined as the region of orbits around a star where liquid water could exist on the surface of a terrestrial planet. The inner boundary of the HZ can be defined in two different ways (Kasting et al. 1993). Firstly, the water loss limit occurs when an Earth-like planet loses its entire water content within the planet's lifetime due to the following mechanism. With increasing solar flux water can enter the stratosphere, where it is dissociated by solar ultraviolet radiation and the hydrogen atoms can be lost to space. Secondly, the runaway greenhouse limit marks the point where the greenhouse effect becomes unstable. This limit is reached when the surface temperature, T_{surf}, exceeds the critical temperature of water ($T_{surf} \geq 647$ K).

We present here a study of the water loss limit of the HZ for Earth-like planets around a Sun-type star using a one-dimensional radiative-convective model of the atmosphere from the surface to the mid-mesosphere based originally on Kasting et al. (1984). We performed several runs with increasing solar flux, which corresponds in principle to the decrease of the planetary distance to the central star. In contrast to Kasting et al. (1993), the model is run until radiative equilibrium under the increased flux conditions is reached. The infrared radiative transfer scheme used is based on von Paris et al. (2008). The water continuum absorption is only taken into account in the 8-12 μm region as in Kasting et al. (1993). The background atmosphere of 1 bar consists of N_2 and CO_2. The troposphere is assumed to be saturated. This overestimates the amount of water in the atmosphere but, it should be more realistic for a water dominated atmosphere than an Earth-like relative humidity profile.

The water loss limit is determined via the stratospheric water mixing ratio $c_{strat}(H_2O)$. When the stratospheric water vapour mixing ratio $c_{strat}(H_2O)$ exceeds the critical value $c_{strat,crit}(H_2O)$ this water loss limit is reached. We use the same approach to determine the critical stratospheric mixing ratio water vapour as Kasting et al. (1993), where the critical value is proportional to the diffusion-limited rate Φ_{esc} of hydrogen escape (Hunten 1973). With this critical value of stratospheric water vapour $c_{strat,crit}(H_2O)=1.5 \cdot 10^{-3}$ an Earth's ocean could be lost to space in less than the lifetime of the Earth.

We calculate temperature and H$_2$O profiles iteratively for increased solar fluxes. In our runs the critical mixing ratio of stratospheric water vapour is reached at a critical incoming flux $S_{crit}=1.29 S_0$ (S_0 solar flux at the top of the present Earth atmosphere). This flux corresponds to a water loss distance (d) of 0.88 AU. Our water loss limit lies between the water loss limit (d=0.95 AU) and the runaway greenhouse limit (d=0.84 AU) of Kasting et al. 1993.

In the context of the detection of Super-Earth planets, we investigate the effect of planetary mass on the inner limit of the HZ. To identify this effect we use two different cases to determine the background pressure. For case (i) the background pressure p_g is fixed at 1 bar for all planetary scenarios and for case (ii) all planets have the same atmospheric column mass (column mass $\propto p/g$) as present Earth. For the second case the necessary mass-radius relation to determine the gravity ($g \propto m/r^2$) is taken from Valencia et al. (2007) to calculate the appropriate background pressure.

Increasing the planetary mass to 10 Earth masses results in a decrease of the orbital distance where the water loss occurs of 6% for case (i) and about 1% for case (ii). The column mass of water decreases with increasing mass because the water vapour pressure is temperature dependent. Hence the greenhouse effect of water decreases with increasing planetary mass and the water loss limit moves closer to the star. The difference in case (i) and (ii) comes from the treatment of the background atmosphere. For case (i) the increasing planetary mass leads to a weaker greenhouse effect due to decreasing column mass (because greenhouse effect depends on column amount). For case (ii) the column mass of the background atmosphere is constant, hence the greenhouse effect of the background atmosphere is approximately constant. Therefore, the water loss limit occurs closer to the Sun for planets.

Acknowledgments. This research has been supported by the Helmholtz Gemeinschaft through the research alliance "Planetary Evolution and Life".

References

Hunten, D. M., 1973, J. Atm. Sci., 30, 1481
Kasting, J.F., Pollack, J.B., Ackerman, T.P., 1984, Icarus, 57, 335
Kasting, J.F., Whitmire, D.P., Reynolds, R.T., 1993, Icarus, 101, 108
Valencia, D., Sasselov, D.D., O'Connell, R.J., 2007, ApJ, 656, 545
von Paris, P., Rauer, H., Grenfell, J.L. et al., 2008, Plan. Space Sci., 56, 1244

The Role of the Solar XUV Flux and Impacts in the Earliest Atmosphere of the Earth

J. M. Trigo-Rodríguez,[1] I. Ribas,[1] and J. Llorca[2]

[1] *Institute of Space Sciences (CSIC-IEEC), Barcelona, Spain*

[2] *Institut Tècniques Energètiques, UPC, Barcelona, Spain*

Abstract. During the formation stages of the terrestrial planets, the Sun's high energy emissions were about 100 times stronger than today. At that early times, the Earth also underwent a violent bombardment phase by planetesimals. We have explored different compositional scenarios to know the properties of the atmosphere in that period. We discuss the implications that these processes could have in the atmospheric evolution and ultimate appearance of life on Earth.

Strong evidence gathered from the study of solar proxies indicates that the Sun was much more active in the past. The results from the Sun in Time program (Ribas et al. 2005) show that the high energy emissions (XUV; 0.1-120 nm) of the ZAMS Sun, were about 100 times stronger than today. That active phase of the Sun is supported by lunar and meteorite fossil records (Newkirk 1980). At the same time, the Earth underwent a violent bombardment phase by large planetesimals (Chyba & Sagan 1996). We explore the influence that these processes had on atmospheric evolution. Typical accretionary rate uncertainties are several orders of magnitude (Chyba & Sagan 1996), also exacerbated by the unknown number of large impacts with energy capable of causing erosion of the primeval atmosphere and hydrosphere (Genda & Abe 2003). The population of impactors was presumably formed by ice-rich bodies that enriched the volatile-depleted content of terrestrial planets (Oró et al. 1990). These bodies were scattered from the outer regions of the solar system as a consequence of the giant planets migration invoked by the Nice model (Gomes et al. 2005).

The extreme XUV and bombardment environment resulted in significant heating of the upper atmosphere of the Earth, probably promoting escape of atmospheric constituents. For simplicity in this work we have not studied the processes that occurred in the upper atmosphere, but only applied equilibrium calculations on the main constituents present in the lower atmosphere following the abundances given in the scientific literature. We performed equilibrium calculations using the thermodynamic values given in JANAF (1998). Several scenarios of reducing and oxidizing atmospheres are considered to study the stability of CH_4 and NH_2 and the formation of organic compounds. By considering the reaction: $H_2(g) + CO_2(g) \rightleftharpoons H_2O(g) + CO(g)$ with the partial pressure of molecular hydrogen typically assumed in literature $10^{-7} \leq pH_2 \leq 10^{-3}$ (Holland 1984) CO_2 seems to be the main oxidizing agent. To assess the stability of CH_4 and CO_2 we analyzed the reaction: $4H_2(g) + CO_2(g) \rightleftharpoons 2 H_2O(g) + CH_4(g)$. Assuming that the partial pressure of water can be similar to that in the present atmosphere (i.e., $pH_2O \sim 0.1$), we obtained Figure 1. The derived pCH_4/pCO_2 ratio suggests a key role of CH_4 as greenhouse gas.

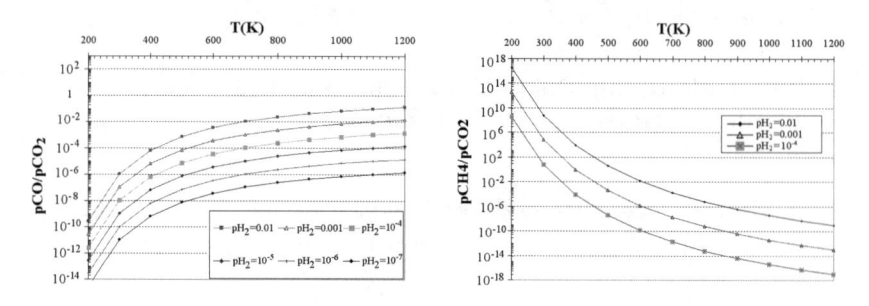

Figure 1. The pCO/pCO$_2$ (left) and pCH$_4$/pCO$_2$ (right) ratios plotted for presumable conditions in the Hadean.

Kasting 1997 proposed that the rapid weathering of the ejecta by large impacts could be a sink for the CO$_2$. In his model, methane and ammonia participated in a feedback cycle, working together in two main ways: (i) presence of methane produced an organic haze due its capacity to polymerize and (ii) this haze can protect ammonia from UV photodissociation. This cycle could yield longer lifetimes for these gases favoring the greenhouse effect on the primeval atmosphere during longer time scales than when each is considered separately. This is supported by a CH$_4$ mixing ratio of 10^{-4} in 2.8 Ga old paleosol data (see, e.g., Pavlov et al. 2000).

In conclusion, CH$_4$ probably played a key role in the Hadean atmosphere as was pointed out previously (Sagan & Chyba 1997). Not only for an outgassed reducing atmosphere, but for an atmosphere dominated by CO$_2$. The overall role of gaseous species was modeled by the intensity of UV radiation, but under a strong meteoric flux the lower atmosphere probably retained significant abundances of organics. The CH$_4$ estimated abundance: $10^{-4} \leq$ pCH$_4$/pCO$_2\leq 1$ has UV shielding implications for atmospheric stability and production of organics (Pavlov et al. 2000). In our scenario, meteoric metals could act as catalysts for the formation of complex organic compounds during ablation processes from precursors such as CO$_x$ and CH$_4$, thus promoting increasing complexity.

References

Chyba, C. F. & Sagan, C. 1996, in Comets and the Origin of Life, eds. P. J. Thomas et al., (Berlin: Springer), 147
Genda, H. & Abe, Y. 2003, Icarus, 164, 149
Gomes, R. et al. 2005, Nat, 435, 466
Holland, H. D. 1984, The Chemical Evolution of the Atmosphere and Oceans, (Princeton, N.J.: Princeton Univ. Press)
JANAF, 1998, Thermochemical Tables, Chase, M.W. (Ed.), (NY: AIP)
Kasting, J.F. 1997, Sci, 276, 1213.
Newkirk Jr., G. 1980, Geochi. Cosmochi. Acta Suppl., 13, 293
Oró, J. et al. 1990, Ann. Rev. Earth Planet. Sci., 18, 317
Pavlov A. et al. 2000, JGR, 105, 11981
Ribas, I., Guinan, E. F., Güdel, M., & Audard, M. 2005, ApJ, 622, 680
Sagan, C. & Chyba, C. 1997, Sci, 276, 1217

Pathways Towards Habitable Planets
ASP Conference Series, Vol. 430, 2010
Vincent Coudé du Foresto, Dawn M. Gelino, and Ignasi Ribas, eds.

Venus Near-Infrared Spectra: SCIAMACHY-Observations and Modeling

M. Vasquez, F. Schreier, M. Gottwald, S. Slijkhuis, S. Gimeno-García, E. Krieg, and G. Lichtenberg

DLR — Remote Sensing Technology Institute, Oberpfaffenhofen, Wessling, Germany

Abstract. SCIAMACHY (Scanning Imaging Spectrometer for Atmospheric Chartography) onboard ENVISAT successfully captured visible and near-infrared spectra of sunlight scattered and reflected from the Venus atmosphere. Only when the planet was rising above the Earth's limb, the line-of-sight geometry allowed the observation of Venus. Addtionally, Venus spectra were simulated using a line-by-line radiative transfer model. These modeled spectra can be used for a better understanding of the Venus atmospheric composition and therefore for helping in the explanation of the Venusian spectra measured by SCIAMACHY. In this work we present results for the Venus atmosphere, both experimental and simulated.

1. Introduction

The atmosphere of Venus consists mainly of CO_2 (96.5%) and N_2 (3.5%), with thick sulfur dioxide and sulfuric acid clouds ($\sim 20\,\text{km}$) ranging from 50 to 70 km (cf., e.g., Venus Express 2008). These clouds reflect about 76% of the sunlight that falls on them, letting only a small amount of sunlight to penetrate into the surface, which is what obscures it from regular imaging.

2. Radiative Transfer Model (GARLIC)

GARLIC (Generic Atmospheric Radiation Line-by-line Infrared Code) is a versatile model that computes radiative transfer in a spherical atmosphere for up-, down-, and limb-viewing geometry. Instrument effects important for remote sensing are modeled by convolution with spectral response and field of view functions.

3. Observations

In March 2009, Venus was close to passing right between the Earth and the Sun, i.e., inferior conjunction. During this phase SCIAMACHY (Gottwald et al. 2006) mainly saw the backside of Venus and only a very small part of the planet's atmosphere illuminated by sunlight. Thus, observing conditions mimicked a limb geometry. In June 2009, Venus was observed once again showing more than half of its disk illuminated by sunlight, corresponding to a nadir geometry.

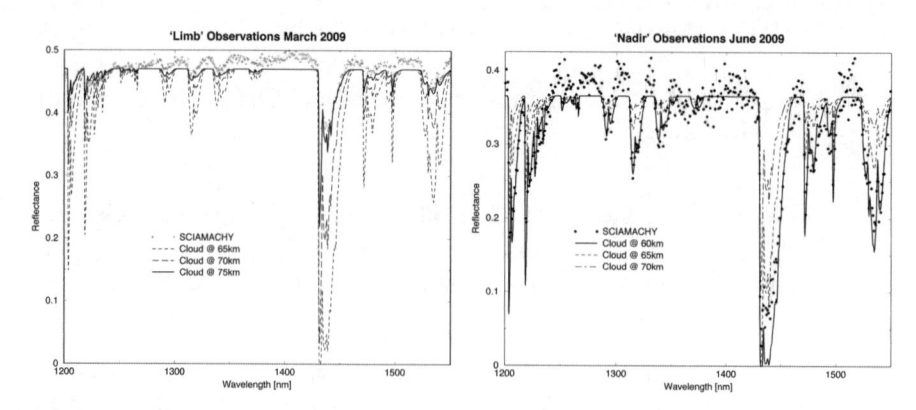

Figure 1. Comparison of SCIAMACHY measurements of Venus with model spectra assuming different cloud altitudes: March and June 2009. The solid line indicates the best match.

4. Results

The Venusian spectra were modeled with GARLIC using carbon dioxide (CO_2) concentrations for a 100 km atmospheric profile, cf. Figure 1. Solar and observer zenith angles were calculated to meet the measuring geometries by considering the distances to Venus and the Sun at the time of observation. In order to simulate the illuminated area of the disk, Venus phase angle was first determined in order to obtain the fraction contributing to the amount of radiance received by the satellite. Reflecting clouds at different altitude levels were tested to match the simulated spectrum to the measured one.

Thick clouds can hardly be penetrated by photons, so they can be considered as reflecting surfaces. Under these conditions, photon path lengths are directly related to the cloud top height: the lower the cloud, the longer the photon path lengths, which raises the probability of absorption events. Therefore, the depth of the absorption lines can be used as a proxy for the estimation of Venusian cloud top heights.

The line depth of the modeled spectrum better simulated the observations from March when placing the cloud at an altitude of approximately 70 km. It is possible to see the differences in the absorption line depths of the spectra when using lower clouds below 65 km as it was expected. For the case of the data from June, the cloud altitude seems to be lower when comparing it to the previous case. The modeled spectrum with a reflecting cloud at an altitude between 60 and 65 km seems to simulate better the observation from June.

Acknowledgments. This study was performed within the frame of the alliance *Planetary Evolution and Life* funded by the HGF.

References

Gottwald, M. (ed.) 2006, SCIAMACHY — Monitoring the Changing Earth's Atmosphere, DLR

Venus Express Searching for Life — on Earth, ESA News, 10 October 2008

Pathways Towards Habitable Planets
ASP Conference Series, Vol. 430, 2010
Vincent Coudé du Foresto, Dawn M. Gelino, and Ignasi Ribas, eds.

Constraints on Secondary Eclipse Probabilities of Long-Period Exoplanets from Orbital Elements

K. von Braun and S. R. Kane

NASA Exoplanet Science Institute, California Institute of Technology, Pasadena, California, USA

Abstract. Long-period transiting exoplanets provide an opportunity to study the mass-radius relation and internal structure of extrasolar planets. Their studies grant insights into planetary evolution akin to the solar system planets, which, in contrast to hot Jupiters, are not constantly exposed to the intense radiation of their parent stars. Observations of secondary eclipses allow investigations of exoplanet temperatures and large-scale exo-atmospheric properties. In this short paper, we elaborate on, and calculate, probabilities of secondary eclipses for given orbital parameters, both in the presence and absence of detected primary transits, and tabulate these values for the forty planets with the highest primary transit probabilities.

1. Introduction

Secondary eclipses of exoplanets provide unique insight into their astrophysical properties such as surface temperatures, atmospheric properties, and efficiency of energy redistribution. In Kane & von Braun (2008, 2009), we demonstrate that the probability of detecting transits or eclipses among known radial velocity (RV) planets is sensitively dependent on the values of orbital eccentricity e and argument of periastron ω, with some combinations of e and ω making transit/eclipse searches among long-period planets viable. Though it is feasible to detect planetary eclipses from space (e.g., Laughlin et al. 2009) and even from the ground (e.g., Sing & López-Morales 2009), the difference in signal-to-noise ratio between transits and eclipses makes detections of the former much more straightforward. In this paper, we calculate the probability of planetary eclipses with or without the knowledge of the existence of a primary transit.

2. Planetary Eclipse Probabilities

The *a priori* geometric eclipse probability of an extrasolar planet, P_e, is

$$P_e = \frac{(R_{planet} + R_\star)(1 + e\cos(3\pi/2 - \omega))}{a(1 - e^2)}, \quad (1)$$

where R_{planet} and $R_\star$ are planetary and stellar radii, respectively (Kane & von Braun 2009). P_e is highest for $\omega = 3\pi/2$.

The presence of an observed transit imposes a lower limit to the orbital inclination angle i, which, using equations 8–11 in Kane & von Braun (2009), provides the following (conditional) lower limit of P_e:

$$P'_e \geq \frac{(R_\star + R_{planet})}{(R_\star - R_{planet})} \frac{(1 - e\cos\omega)}{(1 + e\cos\omega)}. \tag{2}$$

3. Discussion

Figure 1 plots the *a priori* values (open circles) and conditional lower limits (crosses) of P_e for 203 planets as functions of e and ω values tabulated in Butler (2006). The *left* panel clearly shows that, for low values of e, the existence of a transit practically guarantees an observable eclipse, whereas for higher eccentricities, this is not the case due to the correspondingly weaker constraint on inclination angle imposed by an existing transit (cf. Equation 8 in Kane & von Braun 2009). The *right* panel demonstrates that, for $\omega \sim 3\pi/2$, a detected transit ensures the existence of an observable eclipse, whereas for $\omega \sim \pi/2$, this constraint is much weaker (cf. Figure 1 in Kane & von Braun 2009). Thus, the presence of a planetary transit greatly affects the likelihood of existence of a secondary eclipse. Both the *a priori* value and the conditional lower limit of P_e remain sensitively dependent on the combination of e and ω.

Table 1 shows the forty extrasolar planets with the (currently) highest transit probabilities, their orbital elements P, e, and ω, and the explicit values for eclipse probabilities (P_e: *a priori* value; P'_e: conditional lower limit). For instance, HD 118203 b has $P_e = 7.05\%$, but *if a primary transit is observed* ($P_t = 9.11\%$), then the existence of a secondary eclipse is almost certain ($P'_e \geq 94.65\%$). See Table 1 in Kane & von Braun (2009) for the equivalent inverse case of $P_t = f(P_e)$.

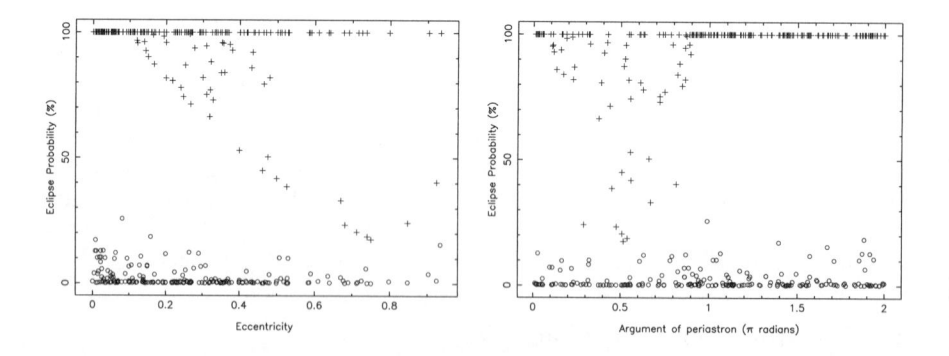

Figure 1. The *a priori* values of P_e (Equation 1; open circles) and conditional lower limits of P_e (Equation 2; crosses) for 203 planets from the Butler (2006) catalog, plotted as a function of e and ω. For purpose of comparison, we assume solar and Jupiter masses and radii for all systems.

Table 1. P_t, a priori P_e, and conditional (revised) P'_e for the 40 exoplanets from Butler (2006) with highest P_t values. See §3.

Planet	P (days)	e	ω(deg)	P_t (%)	P_e (%)	P'_e (%)
HD 41004 B b	1.33	0.08	178.50	25.81	25.70	100.00
HD 86081 b	2.14	0.01	251.00	16.93	17.19	100.00
GJ 436 b	2.64	0.16	339.00	16.50	18.50	100.00
55 Cnc e	2.80	0.26	157.00	15.17	12.33	99.36
GJ 674 b	4.69	0.20	143.00	14.93	11.72	95.96
HD 46375 b	3.02	0.06	114.00	13.58	12.11	100.00
HD 187123 b	3.10	0.01	5.03	12.81	12.78	100.00
HD 83443 b	2.99	0.01	345.00	12.77	12.82	100.00
HD 179949 b	3.09	0.02	192.00	12.74	12.86	100.00
HD 73256 b	2.55	0.03	337.00	12.66	12.95	100.00
HD 102195 b	4.12	0.06	109.90	10.85	9.69	100.00
HD 76700 b	3.97	0.09	29.90	10.82	9.84	100.00
HD 75289 b	3.51	0.03	141.00	10.47	10.03	100.00
51 Peg b	4.23	0.01	58.00	10.35	10.12	100.00
τ Boo b	3.31	0.02	188.00	10.21	10.27	100.00
HD 88133 b	3.42	0.13	349.00	10.16	10.68	100.00
BD -10 3166 b	3.49	0.02	334.00	10.15	10.32	100.00
HAT-P-2 b	5.63	0.51	184.60	9.44	10.24	100.00
HD 17156 b	21.20	0.67	121.00	9.14	2.47	33.05
HD 118203 b	6.13	0.31	155.70	9.11	7.05	94.65
υ And d	4.62	0.02	57.60	8.69	8.38	100.00
HD 68988 b	6.28	0.15	40.00	8.20	6.76	100.00
HIP 14810 b	6.67	0.15	160.00	7.86	7.10	100.00
HD 162020 b	8.43	0.28	28.40	7.84	6.02	93.77
HD 217107 b	7.13	0.13	20.00	7.76	7.11	100.00
HD 168746 b	6.40	0.11	17.40	7.63	7.16	100.00
HD 185269 b	6.84	0.30	172.00	7.30	6.72	100.00
HD 49674 b	4.94	0.29	283.00	6.68	11.94	100.00
HD 69830 b	8.67	0.10	340.00	6.24	6.68	100.00
HD 130322 b	10.71	0.03	149.00	5.76	5.62	100.00
HD 33283 b	18.18	0.48	155.80	5.31	3.56	82.03
HD 38529 b	14.31	0.25	100.00	5.21	3.17	74.40
HD 55 Cnc b	14.65	0.02	164.00	4.67	4.63	100.00
HD 13445 b	15.76	0.04	269.00	4.47	4.85	100.00
HD 27894 b	17.99	0.05	132.90	4.43	4.12	100.00
HD 108147 b	10.90	0.53	308.00	4.14	10.09	100.00
HD 6434 b	22.00	0.17	156.00	4.02	3.50	100.00
HD 190360 c	17.11	0.00	168.00	3.94	3.93	100.00
HD 20782 b	585.86	0.93	147.00	3.92	1.29	40.33
GJ 876 c	30.34	0.22	198.30	3.85	4.44	100.00
HD 99492 b	17.04	0.25	219.00	3.83	5.29	100.00

References

Butler, R. P., et al. 2006, ApJ, 646, 505
Kane, S. R., & von Braun, K. 2008, ApJ, 689, 492
Kane, S. R., & von Braun, K. 2009, PASP, 121, 1096
Laughlin, G., et al. 2009, Nat, 457, 562
Sing, D. K., & López-Morales, M. 2009, A&A, 493, L31

Pathways Towards Habitable Planets
ASP Conference Series, Vol. 430, 2010
Vincent Coudé du Foresto, Dawn M. Gelino, and Ignasi Ribas, eds.

Extrasolar Planets in the Gliese 581 System - Model Atmospheres and Implications for Habitability

P. von Paris,[1] B. Stracke,[1] A. B. C. Patzer,[2] H. Rauer,[1,2]
J. L. Grenfell,[2] P. Hedelt,[1] M. Godolt,[2] S. Gebauer,[2] and D. Kitzmann[2]

[1] *Institut für Planetenforschung, Deutsches Zentrum für Luft- und Raumfahrt, Rutherfordstr. 2, Berlin, Germany*

[2] *Zentrum für Astronomie & Astrophysik, Technische Universität Berlin, Hardenbergstr. 36, Berlin, Germany*

Abstract. The planets Gliese (GL) 581 c and d were the first objects which merited a detailed study of their potential habitability. The orbital distances of the two planets have recently been revised. We investigated the habitability of GL 581 c and d under these new conditions with a 1-D radiative-convective atmospheric model, varying the surface pressure and atmospheric composition. GL 581 c was found to be uninhabitable for all scenarios considered. GL 581 d was also found to be uninhabitable for low and medium CO_2 atmospheres ($3.55 \cdot 10^{-4}$ and 5% CO_2 vmr, $1 \leq p_{surf} \leq 20$ bar).

The first published studies regarding the potentially habitable planets around GL 581 (Selsis et al. 2007, von Bloh et al. 2007) concluded that GL 581 c was too hot for habitable conditions, whereas GL 581 d was located just beyond the outer edge of the habitable zone. However, since then, their orbital distances have been revised, putting them about 10% closer to the star (Mayor et al. 2009). We apply a 1-D radiative-convective cloud-free model (von Paris et al. 2008) to potential atmospheric scenarios, taking the new orbits into account.

The model calculates the temperature, pressure and water profiles. Model atmospheres contained water and CO_2 as greenhouse gases and N_2 as a background gas. There is no complete stellar spectrum of GL 581 available, hence we used a spectrum of the M4.5V star AD Leo (appropriately scaled). The planetary parameters used in the runs are taken from Mayor et al. (2009). We varied CO_2, N_2 and H_2O concentrations as well as background/initial surface pressures. For GL 581 c, the surface reservoir of liquid water was varied. Due to their high eccentricity, the two planets are not assumed to be tidally locked.

We use the following definition of habitability:

(i) The surface temperature T_{surf} remains between 273 K and the critical temperature of water (273 K$<T_{surf}<$647 K);

(ii) The available water reservoir (i.e., ocean with p_{ocean}) can sustain high vapor pressures p_{vap} at high temperatures ($p_{vap}(T_{surf}) < p_{ocean}$).

Condition (i) for surface habitability is not fulfilled for GL 581 c, except for small ocean reservoirs on the order of less than 1 bar (Earth: 270 bar). However, in these runs the entire water reservoir resided in the atmosphere, so condition (ii)

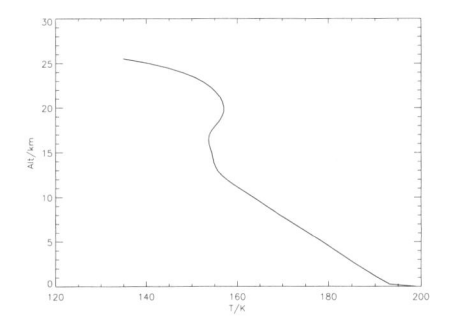

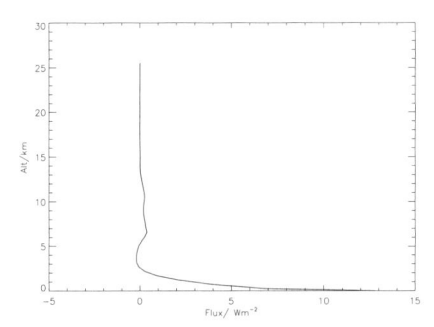

Figure 1. Temperature (left) and total radiative flux (right) profile for GL 581 d for the low CO_2 20 bar run.

is not fulfilled. Hence, GL 581 c cannot be classified as habitable, in agreement with previous studies (Selsis et al. 2007, von Bloh et al. 2007).

Figure 1 shows the temperature (left) and the total radiative flux (right) profile for the GL 581 d 20 bar run with low ($3.55 \cdot 10^{-4}$ vmr) CO_2 concentration. The surface temperature is 200 K, i.e., well below the required 273 K for liquid water. This surface temperature is only a little higher than the equilibrium temperature of 191 K, hence the greenhouse effect provided by the atmosphere is rather small (9 K, compared to about 30 K on Earth). This is partly due to the low water concentrations (less than 10^{-7} vmr at the surface).

Due to the high gravity of GL 581 d (23.76 m s^{-2}), the dry adiabatic lapse rate is very high (24 K km^{-1} compared to 10 K km^{-1} on Earth). Hence, the model atmospheres are stable against convection, unlike the atmosphere of the Earth where a well-developed convective troposphere exists. The shown temperature profile is mostly determined by radiative equilibrium, as demonstrated in the right-hand part of Figure 1.

Even for runs with a higher CO_2 concentration of 5% vmr, surface temperatures did not reach 273 K (increasing from 203 to 248 K when increasing the surface pressure from 1 to 20 bar). However, still higher CO_2 concentrations (95% vmr) could lead to a much more enhanced greenhouse effect and more stellar radiation absorbed, and hence could possibly result in habitable surface conditions on GL 581 d. This is the subject of further investigation.

Acknowledgments. This research has been supported by the Helmholtz Gemeinschaft through the research alliance "Planetary Evolution and Life".

References

Mayor, M., Bonfils, X., Forveille, T. et al., 2009, A&A, 507, 487
Selsis, F., Kasting, J.F., Levrard, B., Paillet, J., Ribas, I., & Delfosse, X. 2007, A&A, 476, 1373
von Bloh, W., Bounama, C., Cuntz, M. & Franck, S. 2007, A&A, 476, 1365
von Paris, P., Rauer, H., Grenfell, J.L., Patzer, B., Hedelt, P., Stracke, B., Trautmann, T., & Schreier, F. 2008, P&SS, 56, 1244

Pathways Towards Habitable Planets
ASP Conference Series, Vol. 430, 2010
Vincent Coudé du Foresto, Dawn M. Gelino, and Ignasi Ribas, eds.

Detection of Transiting Super-Earths around Active Stars

J. Weingrill,[1] H. Lammer,[1] M. L. Khodachenko,[1] and A. Hanslmeier[2]

[1] *Institut für Weltraumforschung, Austrian Academy of Sciences, Graz, Austria*

[2] *Institut für Physik, Karl-Franzens-Universität, Graz, Austria*

Abstract. We studied the influence of stellar activity by G-, K- and M-type stars on the detection of transiting planets in the size range of Neptunes down to super-Earths. The main goal is to improve transit detection algorithms by analyzing the stellar activity like stellar spots or flares. We analyzed measurements of space-based missions like CoRoT as well as ground-based observations of solar-like stars with extrasolar planet candidates. We realized that ground based observations have limited capability to detect short-term stellar variations due to atmospheric effects. Otherwise space-based observations tend to measure higher activity of solar like stars than usual. We present some filtering methods to increase the signal to noise ratio for the detection of Neptune- to super-Earth-class planets.

1. Introduction

The search for extrasolar planets by looking for a transiting object supersedes the radial velocity method by simplicity and parallelism. A measured lightcurve consists of different sources: the transit signal, stellar variations, instrumental effects, and photon noise. The transit signal itself represents only a small fraction in case of Neptunes and super-Earths. Variations of the host star, such as the signature of the rotation, stellar spots, and flares, adulterate the lightcurve. When searching for exoplanets around G-, K- and M-type stars, we encounter variability chiefly caused by eclipsing binaries and stellar spots in combination with the stellar rotation.

2. Methods

A normalization of the lightcurve is done in order to minimize numerical problems. Stellar variability caused by oscillation as seen in Figure 1 is usually in a different frequency domain as the planet. Especially when looking for habitable planets with periods around 100 days or more, this effect is no longer relevant. Cooler M-type stars usually have a closer semi-major axis of the habitable zone, where the orbital periods can interfere with the periods of the activity of the star.

Prewhitening of the lightcurve by removing the most dominant frequencies removes stellar oscillation as demonstrated by Alonso et al. (2009) and increases the signal-to-noise ratio of the transit signal. Long term stellar variability e.g., a decay in brightness, can be removed by a floating median filter with a length

Detection of Transiting Super-Earths around Active Stars 557

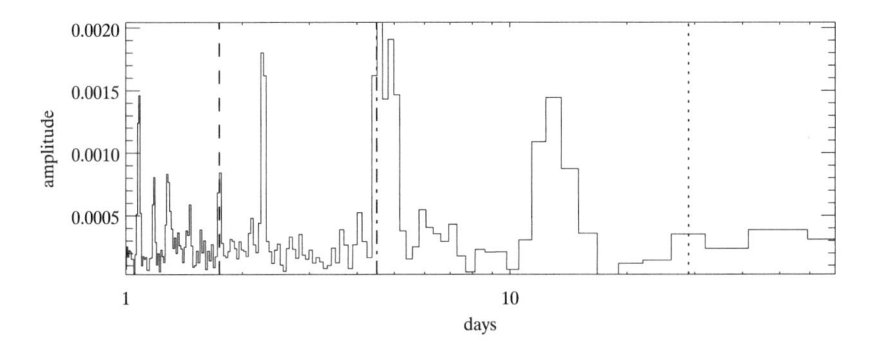

Figure 1. Periodogram of 1 to 70 days. The orbital period (1.7 d) is indicated by the dashed line, stellar rotation is marked with a dashed-dotted line and stellar activity is represented by a dotted line (28.9 d). Data taken from CoRoT-2b, see Alonso et al. (2008).

of 12 hours. The length of the filter is critical and is chosen to be at least twice as long as an expected transit signal and shorter than one day.

3. Results

The detection of habitable planets can be affected by the periodic stellar activity of cool M-type and K-type stars. Filtering methods like Fourier-based filters in general, and prewhitening in particular, help to reduce intrinsic stellar variability. Especially the stellar rotation can be derived from the Fourier autocorrelation.

4. Conclusions and Outlook

Stellar activity and the detection of transit signals from habitable planets are well separated in the frequency domain. The activity of M-type stars with their close-in habitable zone can degrade the detection. Denoising of lightcurves can be done with wavelet filtering (Carter & Winn 2009). Ground-based observations are limited by exposure times and observational windows – a difficult pathway to habitable planets.

Acknowledgments. The authors express their gratitude for funding by Europlanet NA1.

References

Alonso, R., Auvergne, A., and 41 coauthors, 2008, A&A, in press, arXiv:0803.3207
Alonso, R. et al., 2009, in IAU Symp. 253, Transiting Planets, ed. F. Pont, D. Sasselov, & M. Holman (Cambridge: Cambridge Univ. Press), 91
Carter, J. A. & Winn, J. N., 2009, ApJ, 704, 51

Pathways Towards Habitable Planets
ASP Conference Series, Vol. 430, 2010
Vincent Coudé du Foresto, Dawn M. Gelino, and Ignasi Ribas, eds.

Habitability on the Outer Edge: Three-dimensional Climate Modeling of Early Mars and Gliese 581d

R. D. Wordsworth,[1] F. Forget,[1] E. Millour,[1] J.-B. Madeleine,[1]
V. Eymet,[2] and F. Selsis[3]

[1] Laboratoire de Météorologie Dynamique, Institut Pierre Simon Laplace, Paris, France

[2] Laboratoire Plasma et Conversion de l'Energie, Université Paul Sabatier, Toulouse, France

[3] Laboratoire d'Astrophysique de Bordeaux, Bordeaux, France

Abstract. We have developed a new three-dimensional climate model for the study of primitive terrestrial atmospheres and habitability. Here we discuss preliminary modeling results for the early Martian atmosphere and the recently discovered exoplanet Gl581d. For the latter, our results indicate that a sufficently dense pure CO_2 atmosphere could allow for local surface temperatures above 273 K in the warmest regions. However, without other warming mechanisms, such an atmosphere would be extremely vulnerable to collapse on the poles and/or dark side of the planet. Investigation of the effects of other gases and unknowns such as topography is currently underway.

The recent discovery of exoplanets close to the predicted habitable zones of their stars represents an unprecedented challenge for theorists and modelers (Selsis et al. 2007). As direct measurement of the spectra of terrestrial exoplanets is not yet possible, assessments of potential habitability must for now proceed through a combination of accurate modeling of the basic physics, analogies with the solar system, and constraints provided by planetary formation theories. Here we focus on the outer edge of the habitable zone, taking the Martian climate (present and past) as a starting point.

A new 3-D climate model. To investigate the climates of planets with varying orbits, stellar insolation, and atmospheric composition, we have developed a new general circulation model (GCM) with "universal" radiative transfer, microphysics and dynamics. Our radiative scheme is based on the correlated-k method, with the absorption data calculated directly from high resolution spectra. Because of its importance for early Mars, we have paid particular attention to the radiative properties of dense CO_2, using new experimental data to produce a more accurate parameterisation of CO_2-CO_2 collision-induced absorption (CIA). We include the effects of clouds, aerosols, and Rayleigh scattering in the radiative transfer through the scheme of Toon et al. (1989). For CO_2 clouds, the optical properties are calculated using simplified microphysics, including sedimentation and particle sizes that vary with location.

Early Mars. While Mars today is dry and cold, its early climate remains enigmatic, as geological evidence strongly suggests that the planet underwent a period warm enough to allow for the long-term, possibly episodic presence of running water on the surface. Preliminary results from simulations with a

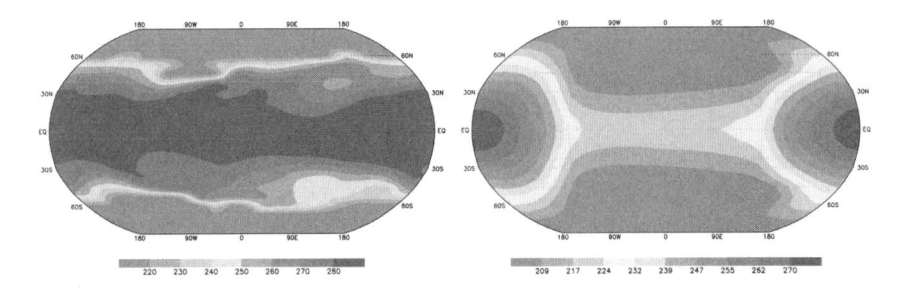

Figure 1. (left) Snapshot surface temperature for a hypothetical Gl581d with a 5-bar CO_2 atmosphere, diurnally averaged heating, and a circular orbit ($a = 0.22$, $e = 0.0$, $L = 0.0135 L_{sol}$). (right) Yearly average surface temperature for a planet with the same insolation but 2:1 tidal resonance, $e = 0.38$, and a 2-bar atmosphere. In both cases, CO_2 is beginning to condense on the surface in the cold regions of the planet.

2-bar CO_2 atmosphere and reduced solar constant $F = 0.75 F_0$ show that with the more accurate CIA data surface warming is significantly reduced, making the so-called "faint young Sun" problem more challenging than before. As the impact of water vapour and clouds on the radiative budget may also be extremely important, we are currently extending the model to include a water cycle.

Gliese 581d. To assess the potential habitability of Gl581d, we have performed a range of simulations, starting from the simplest possible hypothesis of a planet with negligible atmosphere (Figure 1). Two effects in particular cause significant differences from habitability calculations in the solar system: the spectrum of Gliese 581, which is an M-class star, and tidal resonance of the planet's rotation. The former reduces Rayleigh scattering and hence increases surface warming, but also appears to reduce the warming efficiency of CO_2 clouds. The latter effect leads to dramatic contrasts in insolation between dark and light sides of the planet, which in turn causes large differences in surface temperatures even for atmospheres of several bar.

Although Gl581d is potentially habitable locally if its atmosphere contains sufficient CO2, condensation in cold regions severely constrains the maximum achievable greenhouse warming. Next, we plan to investigate the atmospheric stability at higher CO_2 pressures, the effects of H_2O on the radiative transfer, and the effects of topography on local habitability.

References

Toon, O., McKay, C., Acherman, T., & Santhanam, K. 1989, JGR, 94, 16287
Selsis, F., Kasting. J., Paillet, J., Levrard, B., & Delfosse, X. 2007, A&A, 476, 1373

Pathways Towards Habitable Planets
ASP Conference Series, Vol. 430, 2010
Vincent Coudé du Foresto, Dawn M. Gelino, and Ignasi Ribas, eds.

The SEDs of Circumsubstellar Protoplanetary Disks

O. V. Zakhozhay[1] and V. A. Zakhozhay[2]

[1] Main Astronomical Observatory NAS of Ukraine, Kiev, Ukraine

[2] V.N. Karazin Kharkiv National University, Kharkiv, Ukraine

Abstract. We present a model of the spectral energy distribution for substars with disks. This model is based on the observational data and theoretical configurations of physical properties of the protoplanetary disks around the solar-like stars and the stars whose masses are smaller. We consider the case of a plane disk with different sizes of inner hole and without an inner hole. The analysis of the correct selection of the initial conditions is made.

1. Introduction

At the beginning of this century a physical model for evolution of substellar objects (brown dwarfs) was created by a research group from the Astronomical Institution of the V.N. Karazin Kharkiv National University (Pisarenko et al. 2007). This model shows the evolution of substellar objects with masses from 0.01 to 0.08 of solar mass and with ages in interval from 1Myr to 10Gyr.

Based on these investigations, we simulate the physical models for disks that will surround such substellar objects and then calculate a spectral energy distribution (SED). Here we present the results for the case where: substellar masses are in the limits $(0.01 - 0.08)M_{sun}$; protoplanetary disks are in the plane of the sky; substars and protoplanetary discs radiate like a black body; distance from Sun to substar equals to 10 parsec; disk's inner radius varies from 1 to 10 central object radii.

2. SEDs Calculation

The disks sizes are calculated by the relation between the mass of central object M (in solar masses M_{sun}) and disk outer radius R_{out} (Zakhozhay 2005):

$$R_{out}(a.u.) = 150 \left(\frac{M}{M_{sun}}\right)^{0.75}$$

This formula was obtained based on the analysis of 107 protoplanetary disks around single main sequence and T Tauri stars.

Based on the heat balance equation we have received the formula for disks temperature T_d on the definite distance r from central object:

$$T_d(r) = T_{ef} \left[\frac{1-A}{4}\frac{R_{ss}^2 R_{in}}{r^3}\right]^{-0.25}$$

The SEDs of Circumsubstellar Protoplanetary Disks. 561

where T_{ef} - substellar effective temperature, R_{ss} - substellar radius, R_{in} - inner disk radius, A - spherical albedo ($A = 0.1$ for the cases when $R_{in} > R_{ss}$ and $A = 0$ when $R_{in} = R_{ss}$).
The disks fluxes F_λ were calculated with the formula:

$$F_\lambda = \Omega \frac{2hc^2}{\lambda^5} \int_{R_{in}}^{R_{out}} rB_\lambda(T_d(r))dr$$

where h - Plank constant, k - Boltzmann constant, c - speed of light, λ - wavelength, $B_\lambda(T_d(r))$ - Planck function and Ω - is a parameter responsible for disk geometry:

$$\Omega = \frac{2\pi}{d^2}cos\alpha$$

where d - is a distance to the object (in our calculations it equals to 10 parsec) and α - angle between disk normal and the direction to the observer (here $\alpha = 0$).
Consequently a number of curves were generated. Results for the substellar mass equals 0.08 are shown on Figures 1 and 2. These figures represent SEDs for the systems with different disk inner radii. Each figure consists of four curves for the different ages from 1 to 30 Myr.

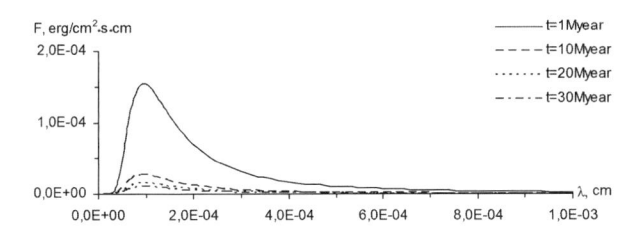

Figure 1. SEDs for systems "substar-disk" with different ages and for a case when $R_{in} = R_{ss}$.

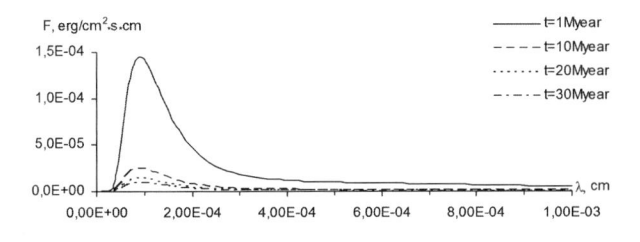

Figure 2. SEDs for systems "substar-disk" with different ages and for a case when $R_{in} = 10R_{ss}$.

Every presented SED is a sum of fluxes from substar and protoplanetary disk. On Figure 3 the disk contribution in the total flux "substar-disk" is shown.

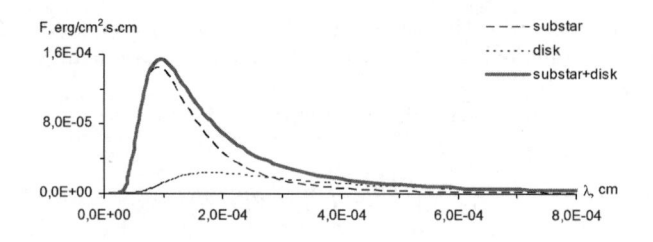

Figure 3. Disk's and substellar contribution in the total flux for the substellar mass equals 0.08 M_{sun}, in the age 1 Myr and for the case when $R_{in} = R_{ss}$.

3. Analysis of the Calculation Results

The initial data that are taken from substellar models (Pisarenko et al. 2007) could have different values in comparison with observed data. With this aim, calculations for small variations of substellar radii and temperatures have been performed. Figures 4 and 5 represent the results for such calculations for substars with 0.08 solar masses with the age of 10 Myr and for the case when $R_{in} = R_{ss}$. Figure 4 describes the behavior of total flux (from the substar and disk) when the radius of central object changes up to twice. Curves on Figure 5 correspond to the total flux variation when the temperature of the substar changes up to 20%. On both figures the thick black line is the total flux that was calculated based on the initial data. The variations of total fluxes maxima with the age and substellar mass (for the case when $R_{in} = R_{ss}$) is shown on Figures 6 and 7. SEDs maxima are changing with the time and with the substellar masses from $1.42 \cdot 10^{-9}$ to $2.31 \cdot 10^{-5}$ $erg/cm^2 \cdot s \cdot cm$ for disks and from $9.48 \cdot 10^{-9}$ to $1.55 \cdot 10^{-4}$ $erg/cm^2 \cdot s \cdot cm$ for the systems "substar-disk".

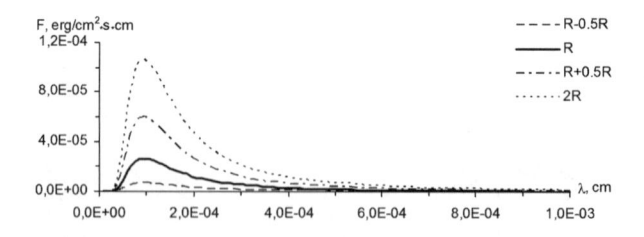

Figure 4. SEDs variation with different substellar radii.

4. Conclusions

The total fluxes from substars and its disks were calculated. The SED calculations were performed for substellar mass intervals from 0.01 to 0.08 solar masses, for age intervals from 1 Myr to 30 Myr, and for systems that are located in the plane of the sky.

If the radius of the substar is twice as small or twice as big than in our calculations, the peak value of total flux for substars with minimal and maxi-

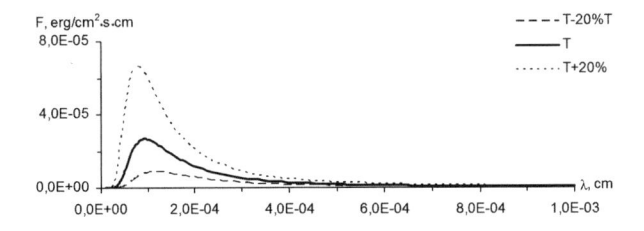

The SEDs of Circumsubstellar Protoplanetary Disks. 563

Figure 5. SEDs variation with different substellar temperature.

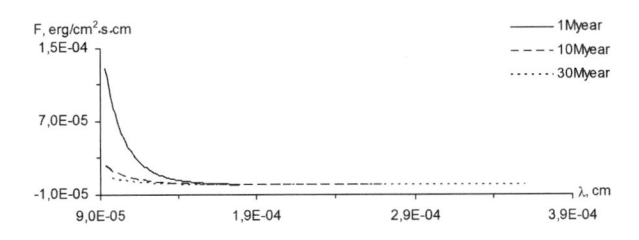

Figure 6. Dependence of disk flux maxima from the substellar masses for systems "substar-disk" at different ages.

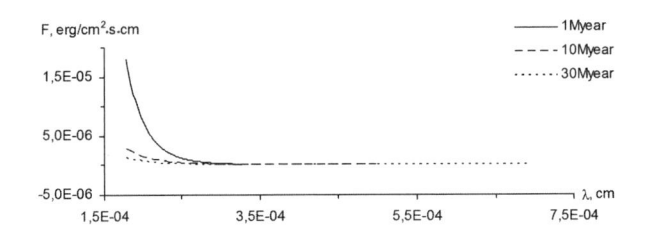

Figure 7. Variations of total flux maxima from the substellar masses for disks at different ages.

mal masses in the age of 10 Myr would be 4 times smaller and 4 times larger, respectively.

If the temperature of the substar is 20% smaller or 20% bigger than in our calculations, then the peak value of total flux for substars with minimal and maximal masses when the age equals 10 Myr would be 3 times smaller and 2.5 times larger, respectively.

References

Pisarenko, A. I., Yatsenko, A. A., & Zakhozhay, V. A. 2007, Astron. Rep., 51, 8, 605
Zakhozhay, V. A. 2005, Vysnyk astronomichnoi shkoly, 4, 2, 55

Conference dinner at the Palau Nacional, home of the
Museu Nacional d'Art de Catalunya (Catalan Art Museum).

Author Index

Abe, L., 231, 284, 457, 477, 510, 513
Absil, O., **29**, 188, **293**, 403, 422
Afonso, C., 45
Alexander, J., 375
Alloza, L. J., 507
Alonso, R., 45
Álvarez, C., 181, 526
Amado, P. J., 521
Amans, J. P., 460
Amaro-Seoane, P., **399**
Anadarao, B. G., 413
Angerhausen, D., 201
Anglada-Escudé, G., **401**
Antichi, J., 231
Arentoft, T., 260
Augereau, J.-C., 188, 293, 403

Baba, N., 284, 510
Baena, R., 428
Baffa, C., 181
Bailey, J., 469
Balan, S. T., **122**
Barge, P., 158
Barillo, M., 460
Barnes, R., **133**
Barry, R. K., 188, **403**
Baruffolo, A., 231
Batista, V., 266
Baudoz, P., **207**, 231
Baumgardt, H., 399
Bean, J., 127
Beaulieu, J. P., **266**
Becklin, E. E., 201
Beichman, C. A., 293
Béjar, V. J. S., 181
Belikov, R., 375, 453
Belu, A., 443
Bennett, D. P., 266
Bergfors, C., **405**, 534
Bernagozzi, A., 420

Bertán de Lis, S., 507
Bertolini, E., 420
Beust, H., 188, 403
Beuzit, J.-L., **231**, 416
Bierden, P., 284
Blanco, V., 428
Blank, D. L., 45
Boccaletti, A., 37, 207, 231
Bonavita, M., 37, 416
Bonfils, X., 188, 403
Booth, M., **407**
Bordé, P., 188, 403
Bortolozzo, U., 239
Boss, A. P., 401
Bouchy, F., 158
Brandner, W., 405, 534
Briot, D., **409**
Bronowick, A., 375
Brown, R., 361, 375
Brugarolas, P., 375
Burningham, B., 434
Butler, P., 503

Caballero, J. A., 181, 507, 521
Cagigal, M. P., **411**
Cahoy, K., 390
Calcidese, P., 420
Campanella, G., 139
Canales, V. F., 411
Carbillet, M., 231
Cash, W., **353**
Cassaing, F., 460
Cassan, A., 266
Catalán, S., 437
Catala, C., 45, **260**
Catanzarite, J., 488
Chakraborty, A., **413**
Charton, J., 231
Choquet, E., 480
Ciardi, D. R., 464

565

Clampin, M., **167**, 361, **383**, 485
Clarke, J. R. A., 434
Claudi, R., 231, **416**
Connelley, M., 453
Conturie, Y., 375
Coudé du Foresto, V., **15**, **418**, 460

Daemgen, S., 405, 534
Damasso, M., **420**
Danchi, W. C., 21, **188**, 293, 403
Dawson, O., 375
Day-Jones, A. C., 434
Deba, P., 278
Deeg, H., 45
Defrère, D., 188, 293, 403, **422**
del Burgo, C., 181, 526
Delplancke, F., 84
Demangeon, O., **424**
Deming, D., 21
den Hartog, R., 422
Deshpande, R., 181
Desidera, S., 37, 231, 416
Dohlen, K., 231
Domagal-Goldman, S. D., **152**
Dong, S., 266
Downing, M., 231
Dreizler, S., **127**
Dressing, C. D., 109
Dressler, A., 21
Duchêne, G., 480
Dunham, E. W., 517

Ealey, M., 375
Egerman, R., 375
Eggl, S., 430
Eiroa, C., 293, 491
Elias, J., 375
Elias, N., 84
Encrenaz, T., **65**
Endl, M., 272
Enya, K., **284**, 457, 477
Esparza, P., 181
Eymet, V., 558

Fabron, C., 231
Feautrier, P., 231
Fedou, P., 480
Fedrigo, E., 231
Feldt, M., 231

Fernández-Rodríguez, C. J., 507
Fernandez, J. M., **426**
Folkes, S. L., 434
Forget, F., **55**, 558
Fors, O., **428**
Fossey, S. J., 139
Fouqué, P., 266
Fridlund, M., 158, 260, 293
Fukagawa, M., 284, 541
Funk, B., **430**, 539
Fusco, T., 231

Gabor, P., **432**
Gach, J.-L., 231
Galicher, R., 207
Gálvez-Ortiz, M. C., **434**
Gappinger, R., 375
Garcés, A., **437**
García Muñoz, A., **441**
García Pérez, A. E., 434
García-Álvarez, D., 530
García-Melendo, E., **439**
Garcia-Vargas, M. L., 411
Garrido-Rubio, A., 507
Gaudi, B. S., 21, 266
Gebauer, S., **443**, 554
Gehrz, R. D., **201**
Geißler, K., 534
Giacobbe, P., 420
Gili, R., 278
Giménez, A., **317**
Gimeno-García, S., 549
Giro, E., 231
Godolt, M., 443, **445**, 475, 554
Goicoechea, R. R., **448**
Gottwald, M., 549
Gould, A., 266
Goullioud, R., **450**
Gratton, R., **37**, 231, 416
Greciano, R., 507
Greenberg, R., 133
Greene, T., 390, **453**
Greenhouse, M. A., 201
Grenfell, J. L., 45, 443, 445, 475,
    545, 554
Griffith, C. A., 115
Güdel, M., **76**
Guenther, E., 181
Guyon, O., 284, 375, **390**, 453, 497

Author Index

Hamann-Reinus, A., 445
Hanot, C., 422
Hanslmeier, A., 483, 515, 556
Hara, T., **455**
Hartman, H., 127
Hashimoto, N., 510
Hatzes, A., 158
Hayano, H., 284
Haze, K., 284, **457**, 477
Hedelt, P., 443, 554
Hejal, R., 375
Heller, R., 133
Hellier, C., 45
Hénault, F., 460
Henning, T., 231, 405, 534
Henry, C., 375
Henry, T. J., 127
Herranz-Luque, J. E., 507
Higuchi, S., 284, 457, 477
Hinz, P., 21
Hippler, S., 405, 534
Honda, M., 284
Hormuth, F., 405, 534
Horner, S., 201
Houairi, K., **460**
Hovenier, J. W., 466
Howard, A., 464
Hoyer, S., **462**
Hubin, N., 231
Huignard, J.-P., 239

Ida, S., 284
Itoh, Y., 284, 541

Jackson, B., 133
Jacquinod, S., 460
Jacquinot, J., **336**
Janson, M., 405, 534
Jenkins, J. S., 434
Jones, E. G., **145**
Jones, H. R. A., 434
Joos, F., 231
Juárez-Martínez, I., 507

Kajiura, D., 455
Kane, S. R., **464**, 551
Kanoh, R., 541
Kanoh, T., 541
Karalidi, T., **466**

Kasdin, J., 375
Kasper, M., 37, 231, 416
Kasting, J. F., **3**
Kataza, H., 284
Kato, E., 541
Kawada, M., 541
Kedziora-Chudczer, L., **469**
Kennedy, G. F., **472**
Kerber, F., 37
Kern, P., 188, 403
Kerrins, E., 266
Khodachenko, M. L., 483, 515, 556
Kipping, D. M., **139**
Kitzmann, D., **475**, 554
Koechlin, L., **278**
Kohyama, T., 541
Kokubo, E., 284
Komatsu, K., 284
Konovalenko, A. A., 483
Kotani, T., 284, 457, **477**, **480**, 513
Kowalcyzk, M., 517
Krabbe, A., 201
Krieg, E., 549
Krist, J., 361, 375
Kubas, D., 266
Kudryavtseva, N., 405
Kunze, M., 445
Kurokawa, T., 513

Labeyrie, A., **239**
Lacour, S., 480
Lagrange, A.-M., 231
Lahav, O., 122
Lammer, H., 483, 515, 556
Langematz, U., 445
Langlois, M., 231
Latham, D. W., 45
Lattanzi, M. G., **253**, 420
Laughlin, G., 464, 503
Launhardt, R., 84
Lawson, P. R., 21, 188, 536
Lay, O. P., 536
Lazio, J., 21
Lazorenko, P., 84
Le Coroller, H., 239
Le Duigou, J. M., 460
Lee, D., 375
Léger, A., 158, 188, 403, 536
Leitzinger, M., **483**, 515

Léna, P. J., **345**
Lenzen, R., 231, 534
Lever, G., 122
Levine, B. M., 368
Levison, H. F., 407
Lichtenberg, G., 549
Lillie, C., 375
Lindberg, R., 260
Lineweaver, C. H., 145
Liseau, R., **219**
Llorca, J., 547
Lloyd, J. P., 528
Lo, A., 361
Löckmann, L., 399
Lopez, B., 188, 403
López-Morales, M., 181, 462
Lozi, J., 460
Lynch, D., 453
Lyon, R. G, 383, **485**

Madeleine, J.-B., 558
Madison, T., 485
Mahadevan, S., **272**, 413, 464
Makoto, H., 499, 501
Malbet, F., **84**, 302, **488**
Maldonado, J., **491**
Maldonado, M., 411
Mandel, H., 521
Mandushev, G., 517
Manjavacas, E., 507
Mao, S., 266
Marchis, F., 480
Marcy, G., 375
Mardling, R. A., 472
Martín, E. L., **181**, 441, 526
Martínez-Arnáiz, R. M., 491
Martin, S. R., **493**, 536
Martinache, F., **497**
Mas-Hesse, J. M., 260
Matsuo, T., 284, **499**, **501**
Matthews, G., 375
Maurel, M.-C., **331**
Mawet, D., 375
McKelvey, M., 453
Meadows, V. S., 152
Melnick, G., 485
Menou, K., 109
Merino, M., 428
Meschiari, S., **503**

Meyer, M. R., **93**
Micela, G., 260, 530
Millour, E., 558
Minniti, D., 45
Miralda-Escudé, J., 266
Mireles, V., 375
Mitchell, J. L., 109
Miyata, T., 284, 477
Moitinho, A., 181
Monin, J.-L., 188, 403
Montañés-Rodríguez, P., **505**
Montes, D., 181, 491, **507**
Montgomery, M. M., 181
Montojo, F. J., 428
Montri, J., 460
Moody, D., 375
Morbidelli, A., 407
Morcillo, R., 428
Moro-Martín, A., 407
Mouillet, D., 231
Mountain, M., 361
Mourard, D., 188, 403
Moutou, C., 158, 231
Muiños, J. L., 428
Mundt, R., 521
Murakami, N., 284, **510**, 513
Muterspaugh, M. W., 21, 84

Nakagawa, T., **195**, 284, 457, 477
Nakamura, T., 284, 477
Nakashima, A., 541
Narita, M., 541
Narita, N., 284
Nietfeld, B., 517
Nilsson, H., 127
Nishikawa, J., 284, 510, **513**
Núñez, J., 428

Ocaña, F., 507
Odert, P., 483, **515**
Oegerle, W., 361
Oetiker, B., **517**
Ollivier, M., 188, 403, 460
Oya, S., 284

Pallé, E., 181, **311**, 441, 505
Pantin, E., 534
Park, P., 375
Patzer, A. B. C., 475, 545, 554

Pavlov, A., 231
Perrin, G., 480
Petit, C., 231
Petrone, P., 485
Petrov, R., 188, 403
Pila-Díez, B., 507
Pilat-Lohinger, E., 430, 539
Pinfield, D. J., 434
Pluzhnik, E., 453
Podlewska-Gaca, E., **519**
Polidan, R., 375
Pollacco, D., 260
Pollock, A., 530
Pont, F., 45
Poretti, E., 260
Postman, M., **361**
Pragt, J., 231
Price, T., 375
Pueyo, L., 375
Puget, P., 231
Purcell, G., 493

Queloz, D., 158
Quirrenbach, A., 84, **521**

Rabou, P., 231
Raksasataya, T., 278
Ramsey, L., 272
Rauer, H., 45, 260, 443, 445, 475, 545, 554
Raymond, S. N., 109, 133
Redman, S., 272
Reess, J. M., 460
Reffert, S., 84
Reiners, A., 127, 521
Residori, S., 239
Riaud, P., 239
Ribas, I., **302**, 399, 437, 439, 483, 515, 521, 530, 547
Rigal, F., 231
Rinehart, S. A., 188, 403, **524**
Rivera, E., 503
Rivet, J.-P., 278
Rizzo, M., 485
Roberge, A., 293
Rochat, S., 231
Rodler, F., 181, **526**
Rodriguez, J., 493
Roelfsema, R., 231

Rojas-Ayala, B., **528**
Rojo, P., 462
Rouan, D., **158**
Rousset, G., 231
Roxburgh, I., 260
Rucker, H. O., 483

Sánchez de Miguel, A., 507
Ségransan, D., 84
Saisse, M., 231
Sako, S., 284, 477
Saldivar, N., 375
Samuel, B., 158
Sanchez-Blanco, E., 411
Sanz-Forcada, J., **530**
Sato, T., 457
Scharf, C. A., 109
Scharf, D. P., 493
Schmid, H.-M., 231
Schneider, J., 139, 158, 207, **324**, 375, 409, **532**
Schnupp, C., **534**
Schreier, F., 549
Schuller, P. A., **536**
Schwarz, R., 430, **539**
Seifahrt, A., 127
Selsis, F., 443, 558
Serabyn, E., 375
Serre, D., 278
Shaklan, S., 375, 390
Shao, M., **213**, **368**, 488
Shibai, H., 284, **541**
Slijkhuis, S., 549
Smart, R., 420
Solano, E., 530
Sorrente, B., 460
Soummer, R., 21, 361
Sozzetti, A., **45**, 84, 253, 420
Sparks, W. B., **543**
Spergel, D., 375
Spiegel, D. S., **109**
Spittler, C., 375
Stadler, E., 231
Stam, D. T., 466
Stankov, A., 260
Stapelfeldt, K., 361, 375
Stark, C., 422
Stracke, B., **545**, 554
Süli, Á., 430

Sumi, T., **225**
Swain, M. R., 201
Swinyard, B., 448
Szuszkiewicz, E., 519

Takagi, T., 455
Takami, M., 284
Takeda, M., 513
Tamura, M., 284, 499, 501, 510, 513
Tanabe, M., 541
Tange, Y., 477
Tapia-Ayuga, C. E., 507
Tata, R., 181
Tavrov, A. V., 513
Tenerelli, D., 390
Thalmann, C., 231
Thompson, P., 485
Tinetti, G., **115**, 139, 207, 416
Tolls, V., 485
Torres, G., 21
Traub, W. A., **21**, 188, **249**, 361, 375, 499, 501
Trauger, J.., **375**
Trigo-Rodríguez, J. M., **547**
Turatto, M., 231

Uchida, H., 284
Udry, S., 231, 260
Unwin, S. C., 21, 188

Vakili, F., 188, 231, 403
Valdvielso, L., 181
Valle, P. J., 411
Vallone, P., 375
van Belle, G., 84
Vanderbei, B., 375
Vanko, M., 483
Vasquez, M., **549**
Vasudevan, G., 485
Velasco, A., 530
Venet, M., 284
Vigan, A., 231
Vogt, S., 503
von Braun, K., 464, **551**
von Paris, P., 443, 475, 545, **554**
Voyer, P., 375

Wakayama, T., 457
Wambsganss, J., 266
Watabe, T., 541

Waters, R., 231
Weinberger, A. J., 401
Weingrill, J., **556**
Wiedemann, G., 127
Wildi, F., 231
Wirz, R., 493
Wiseman, J., 361
Witteborn, F., 453
Wolf, A. S., 503
Wolszczan, A., 272
Woodgate, B., 375
Woodruff, R. A., 485
Wordsworth, R. D., 55, **558**
Wright, J., 272
Wuchterl, G., 539
Wyatt, M. C., 407

Yamamoto, K., 541
Yamamuro, T., 284, 457
Yamashita, T., 284
Yokochi, K., 510, 513
Young, E. T., 201

Zakhozhay, O. V., **560**
Zakhozhay, V. A., 560
Zapatero Osorio, M. R., 181
Zarka, P., **175**
Zhai, C., 488
Zhao, B., 272
Zinnecker, H., **300**
Zonak, S., 272

ASTRONOMICAL SOCIETY OF THE PACIFIC

THE ASTRONOMICAL SOCIETY OF THE PACIFIC is an international, nonprofit, scientific, and educational organization. Some 120 years ago, on a chilly February evening in San Francisco, astronomers from Lick Observatory and members of the Pacific Coast Amateur Photographic Association—fresh from viewing the New Year's Day total solar eclipse of 1889 a little to the north of the city—met to share pictures and experiences. Edward Holden, Lick's first director, complimented the amateurs on their service to science and proposed to continue the good fellowship through the founding of a Society "to advance the Science of Astronomy, and to diffuse information concerning it." The Astronomical Society of the Pacific (ASP) was born.

The ASP's purpose is to increase the understanding and appreciation of astronomy by engaging scientists, educators, enthusiasts, and the public to advance science and science literacy. The ASP has become the largest general astronomy society in the world, with members from over 70 nations.

The ASP's professional astronomer members are a key component of the Society. Their desire to share with the public the rich rewards of their work permits the ASP to act as a bridge, explaining the mysteries of the universe. For these members, the ASP publishes the Publications of the Astronomical Society of the Pacific (PASP), a well-respected monthly scientific journal. In 1988, Dr. Harold McNamara, the PASP editor at the time, founded the ASP Conference Series at Brigham Young University. The ASP Conference Series shares recent developments in astronomy and astrophysics with the professional astronomy community.

To learn how to join the ASP or to make a donation, please visit http://www.astrosociety.org.

ASTRONOMICAL SOCIETY OF THE PACIFIC
MONOGRAPH SERIES

Published by the Astronomical Society of the Pacific

The ASP Monograph series was established in 1995 to publish select reference titles.
For electronic versions of ASP Monographs, please see
http://www.aspmonographs.org.

INFRARED ATLAS OF THE ARCTURUS SPECTRUM, 0.9-5.3μm
eds. Kenneth Hinkle, Lloyd Wallace, and William Livingston (1995)
ISBN: 1-886733-04-X, e-book ISBN: 978-1-58381-687-5

**VISIBLE AND NEAR INFRARED ATLAS
OF THE ARCTURUS SPECTRUM 3727-9300Å**
eds. Kenneth Hinkle, Lloyd Wallace, Jeff Valenti, and Dianne Harmer (2000)
ISBN: 1-58381-037-4, e-book ISBN: 978-1-58381-688-2

ULTRAVIOLET ATLAS OF THE ARCTURUS SPECTRUM 1150-3800Å
eds. Kenneth Hinkle, Lloyd Wallace, Jeff Valenti, and Thomas Ayres (2005)
ISBN: 1-58381-204-0, e-book ISBN: 978-1-58381-689-9

**HANDBOOK OF STAR FORMING REGIONS: VOLUME I
THE NORTHERN SKY**
ed. Bo Reipurth (2008)
ISBN: 978-1-58381-670-7, e-book ISBN: 978-1-58381-677-6

**HANDBOOK OF STAR FORMING REGIONS: VOLUME II
THE SOUTHERN SKY**
ed. Bo Reipurth (2008)
ISBN: 978-1-58381-671-4, e-book ISBN: 978-1-58381-678-3

A complete list and electronic versions of ASPCS volumes may be found at
http://www.aspbooks.org.

All book orders or inquiries concerning the ASP Conference Series, ASP Monographs, or International Astronomical Union Volumes published by the ASP should be directed to:

Astronomical Society of the Pacific
390 Ashton Avenue
San Francisco, CA 94112-1722 USA
Phone: 800-335-2624 (within the USA)
Phone: 415-337-2126
Fax: 415-337-5205
Email: service@astrosociety.org

For a complete list of ASP publications, please visit
http://www.astrosociety.org.